gph	gallons per hour
GPS	global positioning system
GR	grade resistance
GVW	gross vehicle weight
hr	hours
in. Hg	inches of mercury
ISO	International Organization for Standardization
lcy	loose cubic yard
lf	linear foot
msec	millisecond
NSPE	National Society of Professional Engineers
O & O	ownership and operation cost
OSHA	Occupational Safety and Health Act (Administration)
pcf	pounds per cubic foot
PCSA	Power Crane and Shovel Association
PI	plasticity index
psf	pounds per square foot of pressure
psi	pounds per share inch of pressure
PWCAF	present worth compound factor
RR	rolling resistance
SAE	Society of Automotive Engineers
sec	second
sf	square foot
sta.-yd	station-yards
sy	square yards
TR	total resistance
USCRF	uniform series capital recovery factor
USSFF	uniform series sinking fund factor
yr	year

Construction Management Fundamentals

McGraw-Hill Series in Civil Engineering

CONSULTING EDITORS

George Tchobanoglous, University of California, Davis

Raymond E. Levitt, Stanford University

Construction Management Fundamentals

Kraig Knutson, Ph.D.

Senior Lecturer
Del E. Webb School of Construction
Arizona State University
Tempe, Arizona

Clifford J. Schexnayder, P.E., Ph.D.

Eminent Scholar Emeritus
Del E. Webb School of Construction
Arizona State University
Tempe, Arizona

Christine Fiori, Ph.D.

Assistant Director of Undergraduate Programs
Myers-Lawson School of Construction
Virginia Polytechnic Institute and State University
Blacksburg, Virginia

Richard E. Mayo, P.E., Ph.D.

Late Associate Professor
Arizona State University
Tempe, Arizona

McGraw-Hill Higher Education

Boston   Burr Ridge, IL   Dubuque, IA   New York   San Francisco   St. Louis
Bangkok   Bogotá   Caracas   Kuala Lumpur   Lisbon   London   Madrid   Mexico City
Milan   Montreal   New Delhi   Santiago   Seoul   Singapore   Sydney   Taipei   Toronto

McGraw-Hill
Higher Education

CONSTRUCTION MANAGEMENT FUNDAMENTALS, SECOND EDITION

Published by McGraw-Hill, a business unit of The McGraw-Hill Companies, Inc., 1221 Avenue of the Americas, New York, NY 10020. Copyright © 2009 by The McGraw-Hill Companies, Inc. All rights reserved. Previous edition © 2004. No part of this publication may be reproduced or distributed in any form or by any means, or stored in a database or retrieval system, without the prior written consent of The McGraw-Hill Companies, Inc., including, but not limited to, in any network or other electronic storage or transmission, or broadcast for distance learning.

Some ancillaries, including electronic and print components, may not be available to customers outside the United States.

This book is printed on acid-free paper.

5 6 7 8 9 0 DOC/DOC 10 9 8 7 6 5 4 3

ISBN 978-0-07-340104-1
MHID 0-07-340104-8

Global Publisher: *Raghothaman Srinivasan*
Sponsoring Editor: *Debra B. Hash*
Director of Development: *Kristine Tibbetts*
Developmental Editor: *Lorraine K. Buczek*
Senior Marketing Manager: *Curt Reynolds*
Senior Project Manager: *Kay J. Brimeyer*
Senior Production Supervisor: *Sherry L. Kane*
Associate Design Coordinator: *Brenda A. Rolwes*
Cover Designer: *Studio Montage, St. Louis, Missouri*
(USE) Cover Image: *BioDesign Building on the Arizona State University campus: photo by Kraig Knutson; Insert of crane on Pennsylvania Avenue, Washington DC during construction of the Newseum: photo by Clifford J. Schexnayder*
Lead Photo Research Coordinator: *Carrie K. Burger*
Compositor: *Lachina Publishing Services*
Typeface: *10.5/12 Times Roman*
Printer: *R. R. Donnelley Crawfordsville, IN*

Library of Congress Cataloging-in-Publication Data

Construction management fundamentals / Kraig Knutson … [et al.]. – 2nd ed.
    p. cm
  Rev. ed. of: Construction management fundamentals / Clifford Schexnayder, Richard E. Mayo. c2004.
  Includes index.
  ISBN 978-0-07-340104-1 — ISBN 0-07-340104-8 (hard copy : alk. paper) 1. Construction industry—Management. I. Knutson, Kraig. II. Schexnayder, Clifford J. Construction management fundamentals.
  HD9715.A2S363 2009
  624.068—dc22
                                                                            2008030000

www.mhhe.com

Richard E. Mayo was a good friend, the kind of friend who was always there to help you. We shared many joys, triumphs, and times of stress together, and through it all, his positive attitude supported us. Along with this book, his legacy lives on through his wife and seven children.

CLIFF SCHEXNAYDER

CONTENTS

CHAPTER **3**

Construction Management Functions 58

CHAPTER **4**

Scheduling Techniques for Construction Projects 94

CHAPTER **8**

Construction Contract Administration **272**

CHAPTER **9**

Construction Accounting **316**

CHAPTER **17**

Trends **546**

PREFACE

Construction is one of America's most important industries and a critical asset in helping U.S. industry succeed in a globally competitive market. Today's construction engineers are faced with unprecedented challenges and opportunities in planning, designing, building, and managing public and private facilities to meet the needs of society. The construction industry is everywhere society looks. Modern society relies on the construction industry for producing commercial and industrial facilities for business, civil infrastructure for public and private needs, and housing for residents.
Jeffrey S. Russell et al., "Education in Construction Engineering and Management Built on Tradition: Blueprint for Tomorrow," ASCE 2007.

Construction management is about controlling time, money, and quality, and producing work in a safe manner. This text is intended as an introduction to both technical and business sides of construction. It is intended primarily for use in undergraduate civil engineering curriculum or for construction management graduate students who need a text that covers the fundamentals of construction in a logical, simple, and concise format. Photographs from actual projects are included in the text to illustrate ideas and construction methods. Additionally, the use of examples to reinforce the concepts through application has been incorporated in the text. Based on professional practice, standard formats for analyzing common problems are presented. Many companies use such formats to avoid errors when estimating production during the fast-paced efforts required for bid preparation and closing.

There are questions that require research in construction publications or Web-based sources to locate basic information in support of construction management decisions. These questions expose one to the processes of independently locating information. At the close of each chapter, there are names, addresses, and in many cases the Internet address, for sources of additional information.

Engineering News-Record (ENR), the primary weekly publication serving the construction industry, is a recommended information source to support a construction management course using this book. *ENR* provides the latest information concerning developments in the industry. Use of the magazine provides the student with an exciting view of construction and an up-to-date source of critical information. Students should be particularly encouraged to read the

"Legal" column that regularly appears in *ENR* as it provides practical management information. Visit *www.construction.com* for links to *ENR* and other McGraw-Hill construction resources.

This book benefited from the advice provided by many people actively engaged in the construction industry. Additionally, many professors who are teaching construction in universities around the world gave freely of their counsel. Without the many hours of guidance by my good friend William A. (Wink) Ames of the Minard-Ames Insurance Group, Chapter 9, "Construction Accounting," would not be a reality. Likewise Chapter 4, "Scheduling Techniques for Construction Projects," was much improved by the efforts of Sandra Weber of the Del E. Webb School of Construction. She was one of my lifelong friends and collaborators. Her critical review and insightful advice added much to this text. Aviad Shapira of the Technion-Israel Institute of Technology, Haifa, Israel, helped draft the section on tower cranes in Chapter 11, "Equipment Selection and Utilization." Many other individuals and firms supplied information and illustrations for this text; however, we take full responsibility for all material.

We would like to express our thanks for many useful comments and suggestions provided by the following reviewers:

Second Edition Development Reviewers:

Jim Burati, Clemson University

Daniel Castro-Lacouture, Ohio University

Gary Bruce Gehrig, University of North Carolina at Charlotte

Sanjiv Gokhale, Vanderbilt University

Daphene Cyr Koch, Indiana University–Purdue University Indianapolis

Rose Marie Kuehni, University of Minnesota

Gunnar Lucko, Catholic University of America

John J. Robinson Jr., Penn State Harrisburg

Doug Schlagel, University of Notre Dame

Ralph Sinno, Mississippi State University

Wesley C. Zech, Auburn University

First Edition Development Reviewers:

Hovel Babikian, California State University, Pomona

Bruce Bassler, Iowa State University

Lansford Bell, Clemson University

Hyman Brown, University of Colorado

Paul Chinowsky, University of Colorado

Jesus M. de la Garza, Virginia Polytechnic Institute and State University

Paul Goodrum, University of Kentucky

Bud Griffis, Polytechnic University

Carl Haas, University of Waterloo, Ontario, Canada

Paul Harmon, University of Nebraska, Lincoln

Eric LaChance, U.S. Military Academy

Stephen Lee, Carnegie Mellon University

Charles McIntyre, North Dakota State University

John Messner, Pennsylvania State University

Douglas Mills, Arizona State University

Jerry Parish, Texas A&M University–Commerce

James Pocock, U.S. Air Force Academy

Jeffery Russell, University of Wisconsin–Madison

Stephen Thomas, University of Texas, Austin

Jorge Vanegas, Texas A&M University

Significant changes have been made to this edition based on the comments and suggestions from students and because of industry changes in the last six years. The Construction Specifications Institute's new MasterFormat 2004 edition, expanded to 50 Divisions, is incorporated in this text. To stress the importance of safety, the authors have completely rewritten the Safety chapter and added more information on OSHA-required safe practices and reporting requirements. All chapters have undergone revision, ranging from simple clarification to major modifications, depending on the need to improve organization and presentation of concepts. The pictures in all of the chapters have been updated and drawings added so that the important ideas are clearly communicated. We have also found that since the last edition the Web has become a critical information resource. When you see the website icon in the margin, visit our website at *www.mhhe .com/schexnayder* for additional resources available on the World Wide Web.

Most importantly, we express our sincere appreciation to our wives, Judy and Susan, who typed chapters, proofread too many manuscripts, and who otherwise got pushed further into the exciting world of construction than they probably wanted. Without their support, this text would not be a reality.

Comments on this edition are solicited.

Cliff Schexnayder and Kraig Knutson
Del E. Webb School of Construction
Tempe, Arizona

Electronic Textbook Options

This text is offered through CourseSmart for both instructors and students. CourseSmart is an online browser where students can purchase access to this and other McGraw-Hill textbooks in a digital format. Through their browser, students can access the complete text online for one year at almost half the cost of a traditional text. Purchasing the eTextbook also allows students to take advantage of CourseSmart's Web tools for learning, which include full text search, notes and highlighting, and email tools for sharing notes between classmates. To learn more about CourseSmart options, contact your sales representative or visit *www .CourseSmart.com*

ABOUT THE AUTHORS

Kraig Knutson is a senior lecturer in Del E. Webb School of Construction at Arizona State University. He received his Ph.D. in industrial engineering, Master of Science degree in construction, and Bachelor of Science degree in construction at Arizona State University. He has held various positions including estimator, safety director, project manager, and owner on construction projects ranging from small residential and commercial to large industrial projects.

He has taught construction courses at Arizona State University, ITESM Hermosillo, Sonora, Mexico Campus, Mesa Community College, and the Superintendent Career Training Program for the United Brotherhood of Carpenters and Joiners. He is a member of American Society of Safety Engineers (ASSE) and American Association of Cost Engineers (AACE).

Clifford J. Schexnayder held the Eminent Scholar position at the Del E. Webb School of Construction, Arizona State University from 1994 to 2003. He received his Ph.D. in civil engineering (construction engineering and management) from Purdue University, and master's and bachelor's degrees in civil engineering from Georgia Institute of Technology. As a construction engineer, he worked with major construction contractors as field engineer, estimator, and corporate chief engineer. Additionally, he served with the U.S. Army Corps of Engineers on active duty and in the Reserves, retiring as a colonel. His last assignment was as executive director, Directorate of Military Programs, Office of the Chief of Engineers, Washington, DC.

He has served as a consultant to the Autoridad del Canal de Panamá for the Third Lane expansion of the Panama Canal; the Secretary, Business, Transportation & Housing Agency, State of California, to review the cost and risks of constructing the main East span of the San Francisco-Oakland Bay Bridge; and the California, Georgia, and Minnesota Departments of Transportation on project cost estimating.

He is a registered professional engineer in five states. He served as chairman of the American Society of Civil Engineers' Construction Division and on the task committee, which formed the ASCE Construction Institute. From 1997 to 2003, he served as chairman of the Transportation Research Board's Construction Section.

Christine M. Fiori is the assistant director of undergraduate programs, industry relations and outreach in the Myers-Lawson School of Construction at Virginia Polytechnic Institute and State University. She received her Ph.D. and master's and bachelor's degrees in civil engineering from Drexel University. A construction engineer with over 18 years of practical experience, Dr. Fiori has worked with major construction contractors and served with the U.S. Air Force in the civil engineering career field on active duty, managing military housing projects and major base facility renovations.

She is a certified occupational safety and health administration outreach trainer, authorized to conduct 10- and 30-hour construction outreach training in accordance with guidelines provided by the OSHA Training Institute. She has taught construction and safety courses at Virginia Polytechnic Institute and State University, Arizona State University, Universidad de Piura in Peru, the U.S. Air Force Academy, and various companies.

Dr. Fiori is a registered professional engineer as well as a member of the American Society of Civil Engineers. She serves an assistant specialty editor for the American Society of Civil Engineers' *Journal of Construction Engineering and Management* and writes a regular blog for *Engineering News-Record.*

Richard E. Mayo (1940–2002) was a structural engineer for Communications Services, Inc. (CSI), Mesa, Arizona, and taught in the Del E. Webb School of Construction at Arizona State University. He received his Ph.D. in civil engineering (construction management) from Stevens Institute of Technology. He also held a Master of Science degree in management from Rensselaer Polytechnic Institute and a Master of Science degree in civil engineering from Purdue University. He was a 1962 graduate of the U.S. Military Academy at West Point, New York.

Dr. Mayo was a construction engineer with nearly 40 years of experience. He served in the Corps of Engineers for 22 years, completing his service as the deputy district engineer in New York City. Following his retirement from the army, he was the director of construction services for the RBA Group in Morristown, New Jersey, for seven years. At RBA, he supervised the widening of the New Jersey Turnpike and bridge repairs on the Garden State Parkway.

In 1975, *Engineering News-Record* recognized Dr. Mayo at the Construction Man of the Year Dinner for his service in the construction of a fibrous concrete army tank motor pool pavement at Fort Hood, Texas. At that time, the project was the largest fibrous concrete facility in the country.

Dr. Mayo taught construction management at Stevens Institute, Roger Williams University, and in the Del E. Webb School of Construction at Arizona State University. He was a registered professional engineer in several states.

1

Historical Perspective

The engineering efforts of humans are recorded by constructed systems such as works to bring water to the population of a city. The purpose of those built systems and structures is to support our civilization. Today engineers marvel at the skill of the Roman and Inca builders who constructed great aqueducts and roads across mountains and barren deserts. Such systems and structures, built in the past, provide a historical record of how engineering and construction have developed. To understand the profession you are entering, it is helpful to understand the history of construction. It is useful to have an appreciation of how the disciplines of architecture, engineering design, and construction come together to raise great works.

CONSTRUCTION

In 1828, the Royal Charter for the Institute of Civil Engineers in Great Britain defined civil engineering as "the art of directing the Great Sources of Power in Nature for the use and convenience of man." In less formal terms, the definition might read: engineers, by providing technical foresight—design—and supervision—construction—address the needs of humans. But definitions fail to capture the excitement that those in the profession experience as buildings rise and civil works are created.

To capture the excitement and to gain an appreciation of how construction is accomplished, let us consider the works of the great builders that have preceded us. Such a study of construction can serve one well in meeting today's construction challenges. Between 1852 and 1856, Frank and Walter Shanly achieved fame in directing the construction of the western division of the Grand Trunk Railway of Canada [19]. The quality of their work is confirmed by the fact that the substructures of several bridges they constructed are still in use today. After this achievement, the brothers followed diverse career paths.

Frank went on to build many other railroads as a contractor. He acquired a reputation for building railways of good quality. An examination of his papers reveals that not once is there criticism of his workmanship. It was noted that he was methodical and thorough in preparing his estimates of construction costs. He was a good manager, employing technically competent people and placing them where their skills were needed. He had a very good accounting system for his contracts that recorded every expense and that balanced credits and debits monthly. But Frank quit the construction contracting business $150,000 in debt. He would have been bankrupt if his brother Walter had not cosigned his bank loans. Frank never learned to schedule his projects and to keep them on schedule; every job he undertook was late in finishing. He missed obtaining bonuses because of his operational tardiness. Those bonuses would have gone a long way toward making Frank Shanly a successful builder. Additionally, schedule delays caused his construction costs to increase. By not finishing one project before cold weather set in, he was forced to use **black powder** to blast the frozen earth in order to finish the excavation. His brother Walter had warned him early in his career, when they were still working together, "that the only way to finish on time was to plan for the unexpected." You should use the lessons of the past as guides to your success.

Frank Shanly worked in a competitive contracting environment very similar to the one contractors experience today. Owners using fixed price contracts for all of the required work passed responsibility for managing the work to general contractors just as they do today. Such contracts place the risk of project cost and schedule upon the contractor. Therefore, to be successful, builders must have more than knowledge about the mechanics of building. Management of trades or subcontractors, estimating, scheduling, and financial management are all important parts of a constructor's job.

black powder

An explosive mixture composed of potassium nitrate, sulfur, and charcoal.

Early History

Civilizations are built by construction efforts (Fig. 1.1). Each and every civilization of the past had a construction industry that fostered its growth. About 2 million years ago, humans began to produce identifiable tools and use fire [5]. Construction had its rudimentary beginning at that point in the nonliterate ages of human antiquity when humans built their first shelters. The Mesopotamians used the term *batu,* which means builder, and Hammurabi (Hamarabi), the ruler of Babylon, (1795–1750 B.C.), in his celebrated code [8], clearly set forth the responsibilities of a builder.

> 228. If a builder build a house for some one and complete it, he shall give him a fee of two shekels in money for each sar of surface.
> 229. If a builder build a house for some one, and does not construct it properly, and the house, which he built, fall in and kill its owner, then that builder shall be put to death [8].

architect

A person who designs structures; must have the ability to conceptualize and communicate ideas effectively to clients, engineers, government officials, and construction crews.

Up to and through the Renaissance, the engineer or **architect** was the builder. It is only in the modern era that a division of duties developed between the designer and the builder. But in the last 20 years, that trend has been reversed with the increased use of design-build construction contracts.

Fortifications

All early builders were called upon to construct fortifications. Marcus Vitruvius Pollio (70–25 B.C.) was a Roman architect (builder). His book *de Architectura, librum* was the chief reference on architectural matters until the Italian Renaissance. The book dealt with building materials including bricks, sand, lime, pozzolan concrete, stone, and timber. He counseled that architecture is highly technical, and proved his point by critiquing construction of the Forum, the basilica, the baths, and harbors. In the final chapter, he discussed military construction matters. "I shall now proceed to an explanation of those instruments, which have been invented, for defense from danger, and for the purposes of self-preservation; I mean the construction of scorpions, catapultae, and ballistae."*

Leonardo da Vinci served as a military engineer in 1502 for Cesare Borgia. Sebastien la Prestre de Vauban (1633–1707), the great French military engineer, suggested the formation of a Corps of Engineers within the Army for the purpose of building roads and fortifications. It was during his lifetime that the French term *ingénieur,* meaning engineer, was first used. In French, the new military machines were referred to as *engins,* whence the word *ingénieur* being applied to those who looked after their construction and use. The first

---

*These are siege artillery machines; catapulta means "shield piercer," and ballista generally means "projectile shooter" in which the projectile is typically a stone shot. The smallest iron bolt and arrow throwing engines were called scorpions.

FIGURE 1.1 The monumental Roman **aqueduct** in Segovia, Spain, 20,400 stone blocks that are not bonded by mortar or concrete, still delivers water to the city today.

aqueduct
A bridge or channel for conveying water, usually over long distances.

governmental organization for the advancement of bridge building, the Corps de Ponts et Cahaussées (Corps [of engineers] for Bridges and Roads), was organized in 1716, and in 1747, L'Ecole des Ponts et Cahaussées was founded. L'Ecole des Ponts was the first engineering school in the world. Napoleon authorized the founding of L'Ecole Polytechnique in 1795. These two engineering schools served as the models for engineering schools across Europe and later for those in the United States. Benoit Claudius Crozet, who graduated from the Polytechnique in 1807, brought its methods to West Point when he migrated to the United States in 1816.

Engineering Education in the United States

Just as in England, early American engineers received their training through apprenticeships and work experience. Very few had any formal education in engineering subjects. Their experience was similar to that of their English brothers. Most were artisans or practical craftsmen from humble homes and lacking formal education. They served apprenticeships as **millwrights**, mechanics, or stonemasons. Colonel Rufus Putnam, chief engineer for the Continental Army, in 1776 was a millwright by trade. James Brindley, the great canal builder in England, was first a millwright.

millwright

A carpenter having knowledge of gear ratios so that he could construct the machinery of a mill.

In 1724, the Carpenter's Company of the City and County of Philadelphia was established [2]. The master carpenters who constituted the membership were comparable to the master builders of a European guild. In 1734, the company began to assemble a professional library of English language treatises on building. The company's library probably constituted the first organized study of construction in the colonies. The company members became trained professionals who knew arithmetic, geometry, surveying, and drafting. In their professional roles, they were responsible for designing, estimating, and supervising the construction of buildings.

At Harvard College, Professor John Winthrop offered lectures on experimental philosophy (physics) prior to the revolution. His lectures covered dioptics, optics, lenses, refractions, hydrostatics, and hydraulics and pneumatics. In the colonies, these lectures were the closest thing to a formal engineering education.

During the American Revolution, a long list of European soldiers aided the Continental cause, including Tadeusz Kosciuszko. Kosciuszko had been a cadet at the Military Academy in Warsaw, and he later obtained an engineering education in France. By the time Kosciuszko arrived in America from Poland in 1776, he was a skilled engineer. He planned the defense for Saratoga and engineered the construction of fortifications around West Point. It was Kosciuszko who suggested to General George Washington that West Point become a military academy.

When New York State was ready to begin its Erie Canal, it looked first to England for a chief engineer because there were so few Americans trained in constructing civil engineering works. Even the U.S. government had to bring an engineer from France to plan coastal defenses in 1816. Because of the bitterness and strained relationships caused by the War of 1812, no English engineer could be persuaded to accept leadership of the Erie Canal work. Therefore, New York turned to Benjamin Wright, a judge and one of its own. Wright had extensive land surveying experience but only limited familiarity with engineering work.

In the early days of the Republic, judges and lawyers were, as a rule, surveyors, for they found surveying useful in determining questions of deeds and leases, and it was from this class of men that many early engineers sprang. Wright and the others, who assisted him in building the Erie Canal, had to rely to a large extent on ingenuity to solve many of the technical difficulties of building a canal across central New York. Benjamin Wright, however, did have the advantage of experience gained by working with the English engineer William Weston on the Western Inland Lock Navigation Company surveys of the Mohawk River in the mid 1790s.

Wright and those who toiled with him in building the Erie Canal learned their engineering by doing, and 363 miles of canal served as their diplomas. Thus, the project is often referred to as the first American school of engineering. Though West Point had been organized in 1802, it was not until 1817, the year construction of the Erie Canal began, that Colonel Sylvanus Thayer introduced a formal civil engineering curriculum. This was the first comprehensive engineering curriculum offered in the United States.

During the first 10 years of West Point's existence, there was no consistent system of instruction. In 1815, President James Madison sent Colonel William McRee and Major Sylvanus Thayer to study French educational establishments. They purchased roughly 1,000 books in Europe, and soon after his return, Thayer was made superintendent of the Military Academy. It was only at that point that a 4-year system of classes was instituted with regular terms of study.

During its early decades, West Point produced most of the academically trained engineers found in the United States. In 1820, Dennis Hart Mahan entered West Point. Born in New York City, the son of Irish immigrants, Mahan was raised in Norfolk, Virginia. He graduated at the top of his class in 1824, and the army sent him to Europe to study engineering in France. He returned home in 1832 to take charge of the Department of Civil and Military Engineering at West Point. Mahan taught engineering at the academy until 1871. But more important, as a member of the academic board, Mahan guided the development of engineering education at West Point.

The U.S. Army Corps of Engineers traces its origins to the construction of fortifications at Bunker Hill in 1775, through Mahan's curriculum, to an engineering mission for nation building in the 19th and 20th centuries. The corps helped a young nation map the frontier and expand westward by surveying roads and canals. But it was 1818 before the first West Point graduate entered civilian engineering practice. The first nonmilitary engineering educational programs (Table 1.1) imitating the West Point curriculum model came into being in 1834 when the University of the City of New York and Norwich University in Vermont began their programs. Later, with the establishment of land grant colleges in the 1860s and 1870s, formal engineering education expanded greatly.

TABLE 1.1 Early American civil engineering programs.

Institution	Year
University of the City of New York	1834
Norwich University, Vermont	1834
Rensselaer School at Troy, New York*	1835
The University of Alabama	1837
Virginia Military Institute	1838
Union College	1845
Lawrence Scientific School at Harvard	1846
Sheffield Scientific School at Yale	1846

*Rensselaer Polytechnic Institute (originally Rensselaer School) was founded in 1824, however, it did not initially emphasize engineering. The first degree in civil engineering was granted in 1835 and until 1850 the curriculum did not exceed 1 year.

It is clear that prior to the establishment of engineering schools, the only way to gain engineering expertise was by an apprenticeship. One of the most interesting individuals to rise by apprenticeship training to the top of the profession as an engineer-builder was Horace King, a slave belonging to Robert Jemison of Tuscaloosa, Alabama. He was a man with little or no education who became the best structural engineer in the southeast United States. From his first master, John Godwin, he learned bridge building. Godwin and King's initial project was the first bridge over the Chattahoochee River at Columbus, Georgia. King later built the self-supporting spiral wooden staircase for the Alabama State Capitol, which is still in use 175 years later. His master later made him a freeman, something very rare for a black man residing in the South prior to the Civil War.

Canals

Before steam locomotives and railroads, the amount of goods a person could move over land was limited to what he or she could carry or pack on a mule. Water transportation was the primary mode of commerce, and canals were of critical importance to a country's economic development. Waterways therefore served as important transportation links for commerce. Man-made canals were a major factor in the development of the early civilization that sprang up around the Nile. As early as 3200 B.C., Herodotus [7] recorded that Menes banked up the Nile at a bend below Memphis, laid the ancient channel dry, and dug a new course for the stream halfway between two lines of hills.

The ancient Greeks considered the possibility of digging a canal through the Isthmus of Corinth. Such a canal linking the Corinthian and Saronic gulfs would eliminate the 185-nautical-mile voyage circumnavigating the Peloponnese. The first to look into the matter was Periander, tyrant of Corinth, who drafted a canal plan in 602 B.C. Subsequent planners included Demetrius Poliorcetes, Julius Caesar, and Caligula. These plans were later adopted by

FIGURE 1.2   The Corinth Canal today.

Nero, who in A.D. 67 announced to the spectators at the Isthmian Games that he was going to join the two seas by digging a canal through the Isthmus. He went so far as to cut the first turf himself, with a golden pick, and to carry the first basket of earth on his back. But his plans came to nothing, as did those of Herodes Atticus, the Byzantines, and the Venetians in later times. Finally, in 1893, after an 11-year joint effort by Greek and French engineers, the Corinth Canal was completed (Fig. 1.2).

Barge Canals   Construction of the Santee Canal in South Carolina started in 1793. It was the first such project in the United States. The work was accomplished under the direction of John Christian Senf. Senf, a Swedish engineer, had served in the Continental Army after being captured with his British employers at Saratoga. He returned to Europe in 1774–85, inspecting hydraulic works in Holland before accepting the engineering position in South Carolina.

In 1793, a group of men were successful in petitioning the General Court of Massachusetts to grant it incorporation for the purpose of building a canal from the Merrimack River to Boston. Their first difficulty was in securing someone with civil engineering knowledge to direct the work. Eventually, Loammi Baldwin, a borer of logs for crafting pumps and sometime builder of cabinets, undertook the building of the Middlesex Canal and, as a result, became the most influential civil engineer in New England. All of the engineering and construction supervision of the Middlesex Canal rested on Baldwin's shoulders. He had to become his own researcher engineer, developing new construction methods and techniques adaptable to the resources he found available in the new republic. He realized the value of engineering texts and, over the years, he and his sons collected an outstanding personal library. The Baldwin library is now part of the Rare Book Collection of the MIT Libraries.

In 1794, the Carondelet Canal, a shallow ditch 2 miles long and 15 feet wide connecting New Orleans with Lake Pontchartrain, was completed. Carondelet, the Spanish governor of the Louisiana and West Florida provinces, used the city's convicts and 150 slaves provided by area plantations, to construct the canal that provided a shortcut by sea to Spanish settlements in Mobile (Alabama) and Pensacola (Florida). The Carondelet Canal was enlarged after the Louisiana Purchase when the United States took control of New Orleans.

President Thomas Jefferson realized that a national system for internal improvements was required for commerce and to tie the nation together. In 1808, Jefferson's secretary of the treasury, Albert Gallatin, submitted his famous report, *Subject of Public Roads and Canals,* recommending federal aid for a great system of roads and canals to link the Atlantic Ocean and the interior of the country. The following year, Congress appropriated $25,000 to lengthen the Carondelet Canal and to deepen the Mississippi channel at New Orleans. These were the first federal funds appropriated for such purposes. It should be noted that this first appropriation was strictly for waterway development.

Ship Canals   In the middle of the 19th century, a French company under Ferdinand de Lesseps undertook construction of the Suez Canal. De Lesseps' canal uniting the Mediterranean and Red Seas opened in 1869. Today the names Voisin Bey or Alexandre Lavalley are seldom remembered when speaking of the canal, but they were the men who brought the actual engineering talent to the work. Bey was the chief engineer and Lavalley was the man who built the steam dredges used for completing the excavations. The acclaim went to de Lesseps, a retired diplomat. He had connected two bodies of water that were at practically equal elevations and the deepest cut had pierced a ridge only 54 feet above sea level.

Ten years later, at the Congres International d'Études du Canal Interoceanique, the 74-year-old de Lesseps convinced the assembly that construction of a sea-level canal at Panama was a practical venture [12]. Less than a quarter of the delegates at the congress were engineers. Preliminary work in Panama began in 1880. The Compagnie Universelle du Canal Interoceanique was formally incorporated in 1881, and the work was given to one general contractor. This did not prove satisfactory, so in 1883 Canal Interoceanique began acting as its own general contractor. That organization was also unsatisfactory, and in 1886 the work was let to six large construction firms. In 1888, the company went into the hands of receivers. Possibly the effort was simply ahead of its time, considering the difficulties of the site, but it is also evident that de Lesseps failed to heed the advice of his engineers.

Baron Godin de Lépinay, chief engineer with the Corps des Ponts et Caussées, stated that the idea of digging down to sea level was thoroughly unrealistic if one understood the terrain and the rivers of tropical America. He was one of the few delegates at the congress possessing actual construction experience in tropical America, having built the railroad between Córdoba and Veracruz (Mexico) in 1862. Garcia Aniceto Meniocal, an American delegate and coauthor of American surveys in Nicaragua and Panama, accused the technical subcommittee of imaginary schemes traced on imperfect maps. When the congress voted on the sea-level canal plan, the Suez engineer Alexandre Lavalley was conveniently absent, and engineers Lépinay and Gustave Eiffel voted no. Eiffel is best known as the creator of the Eiffel tower in Paris.

The new Republic of Panama came into being in November 1903, and the United States took over building the canal. The first chief engineer, John F. Wallace, stayed at the task only 1 year. John F. Stevens [17] succeeded him in 1905 and organized the massive excavation work as a continuous railroad dump train operation. When Stevens first went to Panama, he assumed that he would be building a sea-level canal. It was the vision that had dominated thinking since 1879 and persisted in both the popular and official imagination. Stevens undertook a careful study of the situation and concluded that only a **lock** canal was feasible because of the necessity to control the torrential flows of the Chagres River. He testified to that effect before Senate and House committees studying the issue. The majority vote in the Senate was close, but because of Stevens, the American efforts in Panama were directed to building a practical lock-type canal.

Stevens submitted his resignation in 1907 and President Theodore Roosevelt turned to the Army engineers for a chief engineer. Major George Washington Goethals (later major general), the third chief engineer, remained until the canal was completed in 1914. The U.S. government, using day hires, did practically all the work, though contractors built a few items such as lock gates and emergency dams.

Loammi Baldwin's Middlesex Canal connecting Boston and Lowell, Massachusetts, which had been a commercial but not financial success since its opening in 1803, was superseded in 1853 by the Boston & Lowell Railroad. The barge canal transportation systems (Table 1.2), which had played such an important role in the early development of the country, were succeeded by the railroads.

lock
At a location where the water elevations up- and downstream in a waterway are different (at a dam), a structure closed off by gates in which, by controlling its inner water level, it is possible to raise and lower vessels.

Railroads

The railroads were the engine of growth and expansion to the west. Before the railroads, western cities and towns survived in virtual isolation except for primitive transportation such as the horse-drawn wagon, which carried mostly

TABLE 1.2 Early American canals.

Canal	Construction dates	Route	State
Santee	1793–1800	Connecting the Cooper and Santee Rivers	South Carolina
Carondelet	1794	Connecting New Orleans and Lake Pontchartrain	Louisiana
Middlesex	1794–1803	Connecting the Charles River (at Boston) with the Merrimac River (at Lowell)	Massachusetts
Erie	1817–1825	Connecting the Hudson River (at Troy) with Lake Erie (at Buffalo)	New York
Union Canal	1821–1827	Connecting the Schuylkill River (at Reading) with the Susquehanna River (at Middletown)	Pennsylvania
Wabash and Erie	1832–1851	Connecting the Maumee River, which emptied into Lake Erie at Toledo, Ohio, with the Wabash River, into the Ohio River (at Evansville)	Indiana
Morris	1841–1845	Connecting the Delaware River (opposite Easton, Pennsylvania) with the Hudson River (at Jersey City)	New Jersey

agricultural products over short distances. The railroads would spread their net of iron rails rapidly and influenced just about every aspect of American history and life from the mid-19th through the mid-20th centuries.

George Stephenson's steam engine, the *Rocket,* traveled at an average speed of 36 miles per hour from Liverpool to Manchester, England, in 1830. In South Carolina, the first American train, *The Best Friend of Charleston,* made its initial run on Christmas Day 1830. Horatio Allen was the project's chief construction engineer. He built a 136-mile line that ran from Charleston to Hamburg passing through Summerville, Branchville, and Aiken.

Early American railroads petitioned the government for engineering assistance in surveying and constructing their lines. President John Quincy Adams was empowered by the General Survey Act of 1824 to authorize surveys only of roads and canals but by 1826 he was permitting examination of railroad routes. He assigned the army engineers of West Point to make the surveys, the plans, and the designs for railroads. Eventually, 60 railroads were planned in that way by army engineers.

Therefore, West Point graduates had a great influence on early railroad construction. One of the most prominent was George Washington Whistler who was born on May 19, 1800, at Fort Wayne, Indiana, where his father was serving in the army. Whistler graduated from West Point in 1819 and was commissioned a second lieutenant in the corps of artillery, but he served on topographic duty between 1819 and 1821. He was an assistant professor of drawing at West Point from 1821 to 1822. In 1828, while on loan to the Baltimore and Ohio Railroad (B&O), he was sent to England with Major William G. McNeill and Jonathan Knight to examine railroad construction in that country. Upon his return in October 1829, he served as superintending engineer for the B&O.

In December 1833, he resigned from the Army and became chief engineer for the Boston and Lowell Railroad in Massachusetts. While at Lowell, he directed the manufacture of one of the first locomotives built by an American shop. He took an English locomotive imported from the Stephenson works at Newcastle and studied how it was constructed. He then fabricated patterns from which the railroad's shop manufactured its own locomotive.

From 1840 till 1842, he was chief engineer of the Boston and Albany (Western) Railroad, with his headquarters at Springfield, Massachusetts. English railroad engineers designed and built solid masonry bridges with stone foundations. Whistler, being a student of English practice, designed his railroads in a similar manner. The Western Railroad, as it pushed westward from Springfield, followed the Westfield River into the Berkshires of western Massachusetts. The course of the railroad made necessary ten keystone arch bridges over the winding Westfield River. A Scottish stonemason named Alexander Birnie was contracted for the work. Whistler's stone-arch railroad bridges built in the early 1840s are still in service on the CSX mainline in western Massachusetts (Fig. 1.3).

Whistler went to Russia in 1842 as a consultant for the building of the Moscow to St. Petersburg line. This opportunity resulted from his hospitality to the czar's colonels P. P. Mel'nikov and N. O. Kraft when they visited the United States in 1840 studying American railroad practices. The two colonels spent time with Whistler while he was constructing the Western Railroad and he impressed them as the engineer with the greatest ability and fullest knowledge. He helped with engineering the road and manufacturing of the track, the locomotives, cars, and everything appertaining to the road czar's new railroad. While working on this project, he contracted cholera and died in St. Petersburg on April 7, 1849, before the line was completed. Today, more people know of his son who left West Point at the end of his third year and became a renown artist, James McNeill Whistler.

Congress passed the Pacific Railroad Act in 1862. The act authorized the Central Pacific (CP) and Union Pacific (UP) Companies to build a transcontinental rail line along the 42nd parallel and provided public lands and subsidies for every mile of track laid. Chief engineer Grenville Dodge (a graduate of the civil engineering program at Norwich University) organized, pushed, and bossed the UP construction survey and construction. As a union general and civil engineer during the Civil War, Dodge had built or rebuilt railroads so fast that they used

FIGURE 1.3 One of George W. Whistler's stone-arch bridges on the Boston and Albany Railroad.

to say of him, "We don't know where he is, but we can tell where he has been." The Union Pacific's hard-driving construction boss, Jack Casement, pushed his crews to unparalleled effort during winters on the prairie and across the Wyoming desert in the summer. It was a great adventure, but Samuel B. Reed, the UP's chief of construction, wrote, "I have too much for any mortal man to do." It was all done by mortal men except a small bit of excavation in Echo Canyon, Utah, where the UP used a steam shovel. The very first steam shovel had been put in service by William Otis some 30 years earlier on the Boston & Providence Railroad in Massachusetts. Otis later worked his machine on a section of Whistler's Western Railroad east of Springfield (Fig. 1.4).

Building east from California, the CP's Charles Crocker managed the construction. He overcame shortages of manpower and money by hiring Chinese immigrants to do much of the backbreaking and dangerous labor. Samuel Skerry Montague was the CP's construction engineer and James Harvey Strobridge was employed as superintendent. Strobridge, who had started as a tracklayer on the Fitchburg Railroad in Massachusetts, showed his ability in pushing the very difficult CP construction over the High Sierras. The work included heavy cuts and fills, and tunnels in granite rock. A few minutes before noon on May 10, 1869, Strobridge of the CP and Reed of the UP laid the last tie in place and the golden spike was tapped into a predrilled hole. The United States had its transcontinental railroad.

During the next two decades, the railroads experienced their greatest growth, adding 110,000 miles to the system and eventually constructing seven

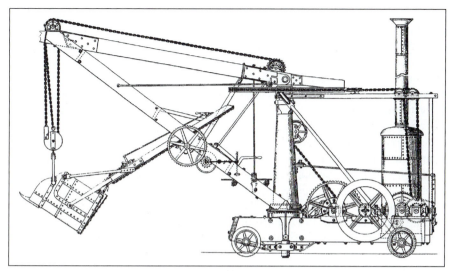

FIGURE 1.4 Drawing of an early Otis steam shovel.

transcontinental rail routes. The Southern Pacific out of San Francisco reached New Orleans in 1882. The Great Northern out of St. Paul, Minnesota, reached Everett, Washington, in 1893.

In western Massachusetts, a few brave and daring men began digging the Hoosac Tunnel in 1851. Twenty-four years later in 1875, after moving 2,000,000 tons of rock excavation, the first train moved through the 5-mile bore connecting Boston, Massachusetts, and Albany, New York. While not a major contribution to transportation like the transcontinental railroad, it was a proving ground that advanced construction techniques. The first American pneumatic **jackhammer** was used there in 1866. The black powder of the era could barely crack the gneiss and quartz veins the miners faced, so Professor Mowbray, an English chemist, concocted that era's most powerful explosive, **trinitroglycerin**, on the project. The project history is a very interesting study in problem solving.

Highways

Perhaps the first U.S. road was the National Road built in the early 19th century. It extended from Cumberland, Maryland, to St. Louis, Missouri, and was the great highway of western migration. Congress approved the route and appointed a committee to plan details in 1806, although construction did not begin until 1815. The first section from Cumberland, Maryland, to Wheeling, West Virginia, which is usually referred to as the Cumberland Road, opened in 1818. The work continued sporadically but finally the road reached St. Louis. Control of the road was turned over to the states through which it passed, and the old route became part of U.S. Highway 40.

jackhammer
An air- (originally) or electric- (today) driven reciprocating percussion tool used for drilling or breaking rock, concrete, brick, or asphalt.

tri-nitroglycerin
An extremely sensitive-to-shock explosive compound produced by mixing glycerine with sulphuric and nitric acid.

In 1895, there were four registered automobiles using the country's road system. In the next 10 years, the number grew to 78,000. The first use of asphalt concrete for street paving was in St. Louis. There, in 1873, asphalt blocks were used to support the increasing vehicle loads. Bellefontaine, Ohio, installed the first concrete roadway, a 200-ft-long, 10-ft-wide strip in front of the courthouse in 1892. But it was not until 1909, when Wayne County, Michigan, inaugurated a systematic program of concrete highway construction, that concrete roads began to appear across the country.

The Federal Aid Road Act of 1916 established the federal-aid highway program and transformed America's roads from dusty or muddy trails, depending on the weather, to a comprehensive road network. The act required the creation of state highway departments as a prerequisite to receiving federal funds. The program got off to a slow start. The initial problem was that America entered World War I in April 1917. Personnel shortages were compounded by shortages of road-building materials and railroad cars to ship materials to project sites. Furthermore, because the railroads were unable to keep up with military shipping, the fledgling trucking industry seized its opportunity to secure interstate shipping. As a result, the roads that the states did not have the resources to improve deteriorated even more under the unexpected and unplanned for weight of loaded trucks.

In the summer of 1919, Lieutenant Colonel Dwight D. Eisenhower participated in an army expedition that was to make a lasting impression on the young officer, and that was to have a great consequence for the United States when he later became the 34th president. Eighty-one motorized army vehicles crossed the United States from east to west. The convoy traveled 3,251 miles, from Washington, D.C., to San Francisco in 62 days at an average speed of 6 mph. The average progress per day was a little over 58 miles. Half the distance traveled by the convoy, some 1,800 miles, was over dirt roads, wheel paths, desert sands, and mountain trails (Fig. 1.5).

After World War I, there were advocates of long-distance roads and advocates of farm-to-market roads. The 1921 Federal Road Act rejected the view of long-distance road advocates who wanted the federal government to build a national highway network. The act limited federal support to a system of *federal-aid* highways, not to exceed 7% of all roads in the state, but three-sevenths of those had to consist of roads that were "interstate in character." It was a very limited beginning.

As automobiles began to give Americans mobility in the 1920s and 1930s, it became increasingly clear that greatly improved highway networks were needed. The first segment of the German autobahn opened to traffic in 1935. By the early 1940s, the success of superhighways, the autobahn, and especially the Pennsylvania Turnpike sparked considerable interest in developing a national toll road system. But World War II and the policy priorities of the immediate postwar years prevented substantial progress until much later. Congress first approved the interstate highway system in 1944, but it was 1956 before a comprehensive program was enacted to build the system. By this time, it had become apparent that post-

FIGURE 1.5 Photograph with Eisenhower's comments.
(Courtesy Dwight D. Eisenhower Library)

war affluence had produced a huge increase in highway demand and that rapidly expanding highway freight movements required a much improved system.

The Federal-Aid Highway Act of 1956 authorized the *interstate highway system,* which was later formally named the Dwight D. Eisenhower System of Interstate and Defense Highways. The act authorized 41,000 miles of interstate highways to tie the nation together. By 1960, more than 10,000 miles were opened. By 1980, 40,000 miles had been constructed nationwide.

The Dwight D. Eisenhower System of Interstate and Defense Highways is dated from June 29, 1956—the day President Eisenhower signed the Federal-Aid Highway Act. On August 2, 1956, Missouri became the first state to *award a contract* with the new interstate construction funding. The Missouri State Highway Commission worked on three contracts that day, but the *first signed contract* was for work on U.S. Route 66—now Interstate 44—in Laclede County. As soon as that contract was signed, S. W. O'Brien, district engineer for the Bureau of Public Roads, called his headquarters in Washington, D.C., and confirmed that the contract was the first in the nation.

In February 1994, the American Society of Civil Engineers (ASCE) designated the Dwight D. Eisenhower System of Interstate and Defense Highways as one of the "Seven Wonders of the United States." Other ASCE designated "wonders" include the Golden Gate Bridge, the Hoover Dam, and the Panama Canal. As ASCE noted, the interstate system has often been called "the greatest public works project in history." Every state possesses an extraordinary interstate highway section of engineering and construction ingenuity.

I-10 in Arizona   The Papago Freeway in Phoenix, which opened in August 1990, was built partly as a depressed freeway covered by 19 side-by-side bridges, which formed the foundation for a 30-acre urban park. The Margaret T. Hance Park was built on the bridge decks to establish a connection between the neighborhoods bisected by the freeway.

I-10 in Louisiana   In Louisiana, the 37 miles across the Atchafalaya Swamp received design awards even before it opened in March 1973. The elevated roadway was constructed from precast concrete pieces, which were cast at a plant in Mandeville on Lake Pontchartrain and then floated by barge through a network of streams and canals to the Atchafalaya River basin. There, barge-mounted cranes placed the roadway deck on top of previously driven piles.

I-70 in Colorado   "I-70 follows the Colorado River past manicured mountains, fertile valleys, and breathtaking canyons." That is the way the American Automobile Association describes I-70 between Grand Junction and Denver. The construction of this segment of I-70 involved a series of spectacular feats.

The 1.7-mile-long Eisenhower Memorial Tunnel under the Continental Divide is the longest tunnel built as part of the interstate program. The first bore opened in March 1973, and the second bore in December 1979.

Glenwood Canyon in western Colorado was one of the most challenging projects in the interstate highway system. The engineers created a world-class scenic byway and an outstanding environmental and engineering project. Forty-plus bridges and viaducts—including precast box girders; precast I-beams; cast-in-place, post-tensioned box girders; welded steel box girders; and a tunnel—were used to minimize damage to the setting.

Airports

The Wright Brothers achieved powered, controlled, and sustained manned flight at Kitty Hawk, North Carolina, on December 17, 1903. Fourteen years later when the United States entered World War I, in April 1917, the army had two aviation fields and 55 serviceable airplanes. A year and a half later when the armistice was signed, there were 27 fields in operation. Aviation training schools in the United States graduated 8,602 men from elementary courses and 4,028 from advanced courses. More than 5,000 pilots and observers were sent overseas. The pilots for the budding airmail and airline services had been trained by the necessity of war.

On May 15, 1918, the Post Office inaugurated the nation's first continuous scheduled airmail service between cities (Washington, Philadelphia, and New York). After initial help from the army, the postal authorities took full responsibility for airmail operations and began to expand the system. They undertook the establishment of lighted airways and pioneered the techniques of airway flying. The Kelly Act of 1925 encouraged commercial aviation by ordering Post Office delivery of airmail through competition by bids from private airlines. This led to the launch of commercial airline routes. The DC-3, the first aircraft to make money carrying passengers rather than mail, was introduced in 1935. By 1940, 90% of the commercial air traffic was flying on DC-3s.

In the beginning, an airfield was just that—any level field with no obstructions. Fullerton Municipal Airport in Orange County, California, is a good example. It traces its origins back to early 1913 when barnstormers and crop dusters used the then vacant site as a makeshift landing strip. The land, once used for raising pigs, had been the city's sewer farm for many years. So deeply rutted and unpromising was the parcel that a city councilman, despairing of any chance to sell it, declared it good only "for raising bullfrogs."

After World War I, commercial aviation began using larger aircraft, which required compacted landing areas. Grand Central Air Terminal at Glendale, California, was developed considering the new construction requirements. The approval to purchase the land for the field was given by the city council on December 9, 1922. From this facility, Charles A. Lindbergh piloted the first airline service from Southern California to New York.

The field had a concrete runway 72 feet wide and 3,000 feet in length. This runway even had 10-ft-wide shoulders of asphaltic paving 3 inches deep. The concrete runway was 6 inches thick at the edges and 5 inches in the center. The *landing strip* paralleled the concrete runway. In those days, the runway was strictly for takeoffs and there was a separate landing area. The landing area was 300 feet wide by 1,500 feet in length and consisted of 6-inch-thick, oil-treated sand and silt. The *Engineering News-Record* report of April 25, 1929, declared the oil to be a "steam-distilled product of 60 to 70 percent asphalt content." The oil was put on in five or six applications and the area was disked after each application. "Finally an 8-ton tandem roller went over the entire area once, thus giving a finished surface not too solidly compacted." Lindbergh "pronounced the landing strip the finest of its kind he had seen. It yields readily, thus decreasing rebound in rough landings, and its depth and structure insure permanent water tightness."

Between January 1942 and May 1943, the United States acquired the field for military training purposes. After World War II, the field proved too small for the larger jet planes, and it closed in 1955.

New York City started work in 1937 on a world-class airport. The North Beach Airport (now known as La Guardia) on Long Island was the largest and most modern airport in the United States when opened in 1940. Approximately half of the area for the field was created by hauling 12 million cubic yards (cy) of old rubbish from the Riker Island dump. When originally constructed, the four runways were 6,000, 5,000, 4,500, and 3,500 feet long. The two longest were 200 feet wide while the other two were 150 feet wide. Concrete paving was considered, but because of the soft foundation it was decided to use 12 inches of **asphalt macadam** instead. During the initial construction phase, six land-plane hangers were constructed. Later seaplane hangers were constructed.

It was soon realized that even this large facility would not handle the growing aircraft traffic, so in 1942, 2 years after La Guardia opened, the work of

asphalt macadam
Compressed layers of broken rock held together by spraying the layers with liquid asphalt.

placing 61 million cy of sand **hydraulic fill** over the marshy tidelands at the head of Jamaica Bay began. The new airport was officially dedicated in July 1948, as New York International Airport, but was generally referred to as Idlewild. It was renamed the John F. Kennedy Airport in 1963. The longest runway at La Guardia was 6,000 feet, the same as the shortest proposed at Idlewild. The longest at Idlewild went to 11,200 feet.

The first decade of Idlewild's existence coincided with the introduction of large piston-powered airliners, such as Boeing's 377 Stratocruiser. As traffic increased in the 1950s, nine passenger terminals were built in various distinctive architectural styles to handle the larger jet aircraft that entered passenger service beginning in 1958.

As in New York, a modern air-transport facility had been built for Washington, D.C. in the late 1930s on a hydraulic fill placed in the Potomac River. National Airport (later renamed Ronald Reagan) opened for service in 1940. Shortly after World War II, it became apparent that the fast-growing Washington, D.C., area needed a second airport. Construction began in September 1958, with architect Eero Saarinen providing a prize-winning design for the terminal complex. An airport is more than just runways, and at Washington's Dulles International, the terminal's 65-ft main facade, with its great windows, is a breathtaking architectural statement. Dulles International Airport stands as a vivid reminder that beautiful architecture can serve the community well.

When Denver International Airport opened in 1995, it was the first new major airport to open in the United States since the Dallas-Fort Worth Airport in 1974. It is another striking architectural statement with its stretched membrane roofs reaching 126 feet above the terminal floor. Thirty-four masts and 10 miles of steel cable support the roof system. The construction effort to build a modern airport is enormous. Denver's airport occupies 53 square miles of prairie. This is a space about twice the size of Manhattan Island. The builders had to move over 110 million cy of earth.

Industrial Buildings

When the power to drive the factory machinery was derived from falling water, manufacturing facilities were located near rivers. With the advent of steam boilers and electrical power, owners were freed from this restriction, and other factors could be taken into consideration when locating facilities.

Mill buildings of the mid-19th century tended to be narrow enclosures with load-bearing walls. The interior space had closely spaced wood columns supporting the floor joists above. In the period between about 1874 and 1904, many advances were made in construction methods and materials, and fire protection. One important advance was the use of steel trusses for framing. The George N. Pierce automobile plant in Buffalo, New York, was constructed with skylights in the roof to provide natural lighting for workers on the assembly line. This was in 1906 as the Pierce Arrow car company sought energy efficient structures.

---

hydraulic fill
Constructed by transporting in a water suspension the soil material for an embankment or dam. Placement of the material is accomplished by allowing it to settle out of suspension in a confined area.

By the 1920s, industrial buildings were being constructed with skeleton framing and curtain walls. Steel framing with wide-flange sections, long-span trusses, or open-web joists is now common. Concrete building systems were developed. Universally, cast-in-place, tilt-up, and precast facilities are now common. Today a typical factory building has a one-story manufacturing area with an office area that can be multistory.

Evolution of the Skyscraper

The skyscraper and the technology that made it possible was an American creation. Tall buildings have always captivated the public's imagination. Until the mid-19th century, few buildings rose more than 100 feet high and aesthetic treatments consisted of false fronts with a profusion of ornamentation and imitations of classic Gothic, Greek, and Roman styles. The structural materials were brick, stone, timber, and wrought or cast iron. The exterior walls of buildings were load bearing. Therefore, in the case of tall buildings, wall thickness was 5 feet or more.

Because of the **load-bearing walls**, penetrations for doors and windows were kept to a minimum. As a result, interiors were poorly illuminated. To create such a building, engineers had to overcome many challenges and four specific engineering innovations had to come together before the height of buildings would rise to a level where they would be called skyscrapers: (1) safe elevators, (2) steel framing, (3) fireproofing, and (4) caisson method foundation construction.

Safe Elevators   E. G. Otis installed the first passenger elevator with an automatic brake in the Haughwout Store Building in New York City in 1857 (Fig. 1.6). Early elevators were steam-powered, then hydraulic (1878), and later electric (1889). The nine-story (180-ft) Home Insurance Building in Chicago is considered by architects to be the first skyscraper because it had skeletal construction in all essential respects, bolted connections, cast iron columns and lintels, wrought iron beams up to the sixth floor, and steel above. The building permit was issued in 1884 and the structure completed the following year. This building incorporated hydraulic elevators. The first electric elevator was installed in the Demarest Building in New York City. The development of the elevator culminated with the high-speed electric gearless traction elevator of 1902, and in 1903 the first such elevator was installed in the 182-ft Beaver Building in New York City. It operated at much higher speeds (600 feet per minute) than earlier elevators and could be employed in buildings of any height.

Steel Framing   The knowledge and use of **iron** can be traced to the Greeks. They used iron in the temple of Zeus Olympios at Akragas (470 B.C.). Iron beams of considerable size were included in the masonry to help support the upper cornices of the building. The Romans acquired their knowledge of iron from the Greeks. During the Middle Ages, **wrought iron** was used, though **cast iron** is a medieval invention. St. Anne's Church in Liverpool (1770–1772) was an important step in iron framing as it was the first building in England to have cast iron columns.

---

load-bearing wall
Any wall that carries some of the weight of the structure above.

---

iron
A cheap, abundant, useful, and important metal but pure iron is only moderately hard. Iron is used to produce other alloys, including steel.

---

wrought iron
A tough and malleable metal. Wrought iron contains only a few tenths of a percent of carbon compared to steel, which contains about 1% carbon.

---

cast iron
A hard, brittle, nonmalleable iron-carbon alloy. It is cast into shapes.

FIGURE 1.6   Otis's patented hoisting machine.
(from Scientific American, *Dec. 13, 1862)*

steel

Carbon steel is an alloy of iron with small amounts of Mn, S, P, and Si. Carbon is the major variable that distinguishes between the properties of iron and steel, a very strong metal.

The 151-ft-high framing of the Statue of Liberty (1883–1886) was an impressive lesson to American builders. It posed a framing problem on a scale with the new skyscrapers. The frame was designed and fabricated by Gustave Eiffel's engineering firm and erected on the site by the Keystone Bridge Company. The system uses heavy wrought iron girders and **steel** posts.

The first all-steel skeleton (1889) was used for the Rand-McNally Building in Chicago. During the 1880s, architects in New York and Chicago were pioneering the use of iron and steel-framing techniques. The first structure in New York to embody complete interior framing without bearing walls was the Tower Building at 50 Broadway. It was completed in September 1889.

Fireproofing   Steel framing received a transitory setback after a series of buildings with wood flooring experienced severe structural damage from fires. Steel columns and beams loaded to normal design limits will collapse when the steel temperature reaches approximately 1000°F. Unprotected columns and beams can collapse after only 10–15 minutes exposure to fire. Because of this danger, fireproofing is very important.

John B. Cornell made an early attempt to overcome the fire problem. In 1860, he patented a column consisting of two cast iron tubes, one inside the other with the space between filled with fire-resistant clay. In 1871, Balthasar

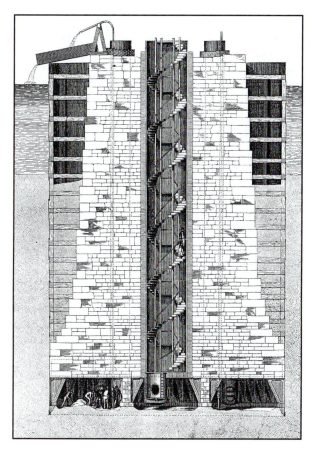

FIGURE 1.7 Pneumatic caisson for the east pier of the Ead's Mississippi River Bridge in St. Louis.
(from Scientific American, *April 15, 1871)*

Kreischer received a patent for fireproofing floor beams by enveloping them in refractory tile. In 1871, the Nixon Building was under construction when the great Chicago fire destroyed four square miles of the city. The builders of this structure had attempted to make it completely fireproof by covering the tops of the beams with 1 inch of concrete and including a 1-inch coat of plaster on the ceilings. Except for wood molding and trim, it survived undamaged and opened for use a week after the fire. These efforts demonstrated how to fireproof steel buildings, and encouraged further development of steel-framed structures.

Caisson Method Foundation Construction   In constructing a high-rise building, a solid foundation is always necessary. Beginning about 1901, the pneumatic caisson method, which was adapted from bridge foundation work (Fig. 1.7), was applied to building construction. A **caisson** (French for big box) is a cylinder or box with an airtight bulkhead and a cutting edge around its

caisson
A watertight, dry chamber in which people can work below the level of the water table.

bottom. As excavation proceeds and the caisson moves downward into the ground, air pressure in the chamber beneath the bulkhead keeps water from entering the interior work area while shafts through the bulkhead permit the passage of workers, equipment, and excavated material between the bottom and the surface. There can also be open-air caissons where the box is not sealed with a top.

Thanks to these developments—new building materials and systems, and new construction techniques—taller buildings began appearing around the country. By 1916, a record height of 60 stories was reached with the erection of the Woolworth Building in New York City. This building was built on a thick and highly unstable deposit of alluvial mud and sand. The main column footings of the main shaft were set on concrete caissons sunk 100 feet below grade.

The pinnacle in steel framing was the Empire State Building (1930) with 102 floors. This building was a true construction achievement as it was built in the very center of New York City in less than 18 months, under budget and ahead of schedule. By means of derricks on projecting frames, the builders erected 57,000 tons of steel in 6 months. The Empire State Building is still one of the tallest skyscrapers in the world (Table 1.3).

The Empire State Building was a riveted steel structure, but that method of tying members together was passing. In 1917, the Lincoln Electric Company introduced a gasoline-engine-driven welder on a truck-type undercarriage. The first use of welding for steel frame construction came in 1920 with the factory of the Electric Welding Company of America. The Pratt trusses supporting the roof were of welded construction. In 1926, the Westinghouse Company designed its five-story factory in Sharon, Pennsylvania, as a completely welded structure. The Austin Company realized the advantages of welded steel structures, and in 1928, using Lincoln welders, Austin completed the Upper Carnegie Building in Cleveland. Though this structure was only four stories high, it demonstrated welding technology and attracted attention because of the 15% reduction in steel as compared to a riveted design.

Zoning   Tall buildings, skyscrapers, were designed based on legislated zoning rules. To prevent one building from completely blocking the natural light to another building, cities propagated building setback rules. These are usually

TABLE 1.3 The world's tallest buildings.

Name	City	Height (ft)	Floors	Year Built
Taipei 101	Taipei	1,667	101	2004
Shanghai World Financial Centre	Shanghai	1,614	101	2007
Petronas Twin Towers	Kuala Lumpur	1,483	88	1997
Sears Tower	Chicago	1,450	110	1974
Jin Mao Building	Shanghai	1,379	88	1998
Empire State Building	New York	1,250	102	1931
Eiffel Tower	Paris	1,043	318	1887

based on the height of the building and the width of the surrounding streets. Many skyscrapers consequently have a vertical form, resembling a flight of stairs. Buildings in New York City provide some good examples of this design. The other skyscraper form is the full-height slender tower.

Form   Cesar Pelli, former dean of Yale University's School of Architecture, led the design of the Petronas Towers (Fig. 1.8). Pelli was preceded by a long line of notable architects who changed the face of buildings. Toward the end of the 19th century, Louis Sullivan demonstrated that ornamentation and false fronts, the earlier building tradition, were not essential to the beauty of tall buildings. He allowed the structure to be plainly visible. A good example of this is his Gage Building in Chicago (1899). The facade of the building emphasized the outer columns, and windows were continuous. Following Sullivan, Frank Lloyd Wright expressed form and function in his designs. His Guggenheim Museum in New York City is a curved building that follows the outline of an interior spiral ramp.

Wright also did residential structures. His *Robie House* (1909) in Chicago was one of the first homes to integrate electric lighting and mechanical heating

FIGURE 1.8   The Petronas Twin Towers in Kuala Lumpur, Malaysia.

as part of the interior design. One of his more famous residential efforts is Falling Water located in Bear Run, Pennsylvania.

Concrete Buildings

Steel is not the only material used for constructing tall buildings. Concrete, the oldest building material artificially prepared by humans, is often used in building construction. Concrete was a Roman invention. Marcus Vitruvius Pollio, an architect and engineer under Julius Caesar, had documented such a mixture, but the art of using concrete seems to have been lost until the latter part of the 12th century when medieval builders began to use it for the foundation footings of large cathedrals. In America, it appears that the first use was by the Spanish builders of the Castillo de San Marcos in St. Augustine, Florida (begun in 1672).

In the 1790s and early 1800s when Loammi Baldwin was building the locks for the Middlesex Canal, a consistent theory for cement had not evolved. English engineers of high repute, such as John Smeaton, held different opinions, but builders knew that a material manufactured by commingling volcanic pozzuolana and lime would harden under water. Baldwin, using trass from St. Eustatius, conducted his own experiments and formulated a set of mix proportions for achieving a good hydraulic cement. The experiments additionally proved important in developing a superior method for processing the trass. By this work, Baldwin demonstrated his understanding of mixing processes. He chose to first grind the dry trass stones into a fine powder before mixing it with the lime and water.

At Chittenango, New York, Canvass White had a mill during the 1820s and used natural hydraulic cement to make what he called "water lime." Extensive use was made of this material in concrete for facework and the abutment walls on the Erie Canal.

The use of concrete for the foundation of the Statue of Liberty (1883–1886) demonstrated the versatility of the material and led to interest in its further use. The pioneer work in reinforced concrete was by French builders. In 1861, Parisian gardener Joseph Monier developed a practical method of wire-mesh reinforcing that he used in the construction of concrete tanks. After the Paris Exposition of 1867, reinforced concrete construction appeared in the United States. The man most responsible for advancing the use of reinforced concrete construction was E. L. Ransome of San Francisco.

In 1892, Ransome designed two buildings for what was then known as LeLand Stanford Junior University in Palo Alto, California. One was a girls' dormitory (Roble Hall) and the other was the University Museum. The dormitory had columns, beams, joists, and floors of reinforced concrete. In the museum building, the stairways and roof, in addition to the structural elements, were of reinforced concrete. After that, reinforced concrete building became more common, but the achievement of greater height came slowly. The Ingalls Building in Cincinnati, completed in 1902, reached 16 stories, and the United Brethren Building in Dayton (1923–1924) reached 21 stories (274 ft). In 1958, the Executive House in Chicago was constructed with 39 stories but the total height was

FIGURE 1.9 Wooden stiffleg derricks placing stones, during construction of Roosevelt Dam. *(Courtesy Salt River Project Heritage, Walter Lubkin collection)*

only 371 feet because the ceiling height was kept at the standard 8 feet and the floor-to-floor distance was only 8 feet 10.5 inches.

Energy and Building

The construction of environmental systems with the specific purpose of meeting human needs traces to the first engineering efforts and is visually illustrated by the great Roman aqueducts that can still be found in Europe. The Romans built these structures in order to bring water to their cities. In a similar manner, the Incas of South America constructed very sophisticated water supply systems for their cities. The best known of these can be seen today in the abandoned structures at Machu Picchu, Peru [20].

In the United States, engineers have designed and constructed systems that while usually hidden, buried under the street, provide potable water and waste water systems to almost every dwelling. Systems for generating and delivering electricity have been constructed. To conserve these resources, engineers are constantly developing better designs.

Water and Electricity   The Roosevelt Dam in Arizona, was originally constructed between 1905 and 1911 as a masonry arch structure. It was designed to control the flooding of the Salt River and to store spring floodwater for irrigation. The Roosevelt Dam was the world's largest "cyclopean-masonry" dam, a Greco-Roman style of building that used huge, irregular stone blocks (see Fig. 1.9). The Salt River Project served as a model for other reclamation projects throughout the West. In 1971, the American Society of Civil Engineers recognized Roosevelt Dam with the National Historic Engineering Landmark Award. Hoover Dam, a major source of electric energy for the Southwest, was a monumental construction project. In fact, it was the most massive masonry structure

TABLE 1.4   The highest dams in the United States.

Name	River	State	Height (ft)	Year completed
Oroville	Feather	California	770	1968
Hoover	Colorado	Arizona/Nevada	727	1936
Dworshak	N. Fork Clearwater	Idaho	717	1973
Glen Canyon	Colorado	Arizona	710	1964
New Bullards Bar	North Yuba	California	645	1969
Seven Oaks	Santa Ana	California	632	1999
New Melones	Stanislaus	California	625	1979

since the Great Pyramid, requiring 4.4 million cy of concrete in the entire project and 3.25 million in the dam alone. At the time, it was the largest construction contract ever awarded in the United States, and it was the construction industry's first hard hat job. Many other major dams for flood control, water supply, and generation of electricity followed (Table 1.4).

Buildings   The Rockefeller Center in New York City was one of the last major buildings with windows that can be opened (really a group of 10 buildings). Later office skyscrapers were sealed and "climate controlled." The world's largest building, the vehicle assembly building at Cape Kennedy, could easily fit a 40-story skyscraper through its entry doors. To conserve energy, climate control is provided only in the zones where people work. In 1999, the Four Times Square skyscraper in New York City was completed. It is a model of energy-saving ingenuity. The building itself generates electricity with fuel cells and solar curtain wall panels. It was the first skyscraper to fully embrace standards for energy efficiency, indoor air quality, and the use of sustainable materials.

CONSTRUCTION INDUSTRY

The accomplishment of the great civil engineering works requires the massing and management of large labor forces.

Artisans and Day Laborers

Egyptians   The postulation that the Egyptians used slave labor as a workforce to construct their monuments is probably correct to a limited extent. There are surviving frescoes depicting masses of men straining under the threat of the whip while dragging massive stones in the blazing sun, and those pictures lead one to accept the slave labor proposition.

The Romans, and later during feudal times lords and/or the state, made a common practice of exacting a tax payable in hours or days of labor as a means of gathering workforces for major projects. The term for such required labor is *corvee,* which comes from the Latin *corrogare,* to summon together. It is known that the Egyptians used corvees to repair the dikes along their irrigation canals, which were so important to the bounty of the country. Based on that knowledge, it follows that corvees were used to construct pyramids and temples. However, by whichever means or combination of means, the Egyptians did organize and manage a large workforce to accomplish their construction projects.

Greeks   The use of slave labor to erect constructed works was still common in the time of the ancient Greeks as manual labor was considered beneath the dignity of a citizen. Mass manpower was needed to move the massive stones, but those who dressed the stones had to have special talents, and those who sculpted the capitals were skilled craftsmen. Some of these artisans were freemen, but many skilled craftsmen were probably slaves.

Romans   In the Roman period, the bottom line was still forced labor but there were subtle changes. There were slaves, mainly war prisoners; there were freemen working under a corvee system; and there were soldiers who were not campaigning. The skilled workmen who fashioned the decorative columns and other facade work were of a higher class, and they could be either freemen or slaves. This group of freemen and slaves formed themselves into collegia—free societies. The state imposed very strict rules on the collegia:

- Membership for skilled craftsmen was mandatory.
- Workers, and their descendants after them, had to follow the same trade for life.
- Wages were fixed.
- Craftsmen could be transferred to any location, as directed by the state.

Byzantine   Military labor and corvees were still employed during the Byzantine period. Emperor Justinian, however, gave more freedom to the workers. The oppressive restrictions tying the collegia members to the state were removed and they were given more privileges. During this period, work was done by a "master-worker" and his body of fellow workers. This was something new and these master-workers might be considered the first general contractors.

Master Builders

The origin of the word *architect* comes from a Greek word meaning a chief artifice, master-builder, or director of works. The Romans used the Latin form *architectus,* which prevailed through the Dark Ages. The scope of an architect's responsibilities was to design and superintend the construction of the work. The consensus of historians is that the architect of a medieval building was frequently styled "master," but the word cannot be taken as a specific title for an architect. The *master builder* answered to the owner and was responsible for both the design and construction of a structure. The first master builders were probably masons, because their trade represented so much of the building effort [4]. It was the master builder's responsibility to hire the other trades. This type of construction organization reached its apex during the Renaissance.

These builders produced some of the world's finest structures in terms of technical sophistication, engineering skill, grace in design, and sheer size. Pope

FIGURE 1.10   Notre Dame Cathedral in Paris, France.

Alexander III laid the first stone for Notre Dame Cathedral in 1163 and an army of master builders and medieval craftsmen toiled for 170 years to complete that magnificent structure in Paris (Fig. 1.10). The master builders used the principle of proportional geometry to develop the complicated design with a minimum of equipment.

Contractors

general contractor
Contractor who accepts cost estimates from subcontractors, and then signs the general contract with the owner. By doing so, the GC assumes responsibility for the entire project.

The term *contractor* comes from the legal agreement or contract that is negotiated and executed between the owner and the builder. As there can be many builders involved on a project, a hierarchy of responsibility and authority needs to be established to ensure proper coordination of the various efforts. Hence the term **general contractor** (GC), the individual or company that has a legal duty to the owner for satisfactory completion of the project.

Today, contractors bring together the hands-on building skills and professional management ability to execute the plans of design professionals (architects/engineers). The responsibility for building and designing has been divided. Contractors assume the responsibility for assembling, organizing, and managing the construction labor force; for placing the equipment appropriate to the work tasks on the project; and for procuring the necessary construction materials. They sell to a project owner their knowledge of construction techniques and their ability to manage large enterprises. The contractor is a problem-solver who, through interpretation of the designs produced by the architect/engineer, creates a physical structure.

In today's construction industry, there is radical change in the way many federal, state, and local governments and other organizations are awarding their construction contracts. In the past, most government agencies and other organizations used a competitive low-bid process. Today, owners are seeking and experimenting with other contracting methods such as design-build and best value.

Best value contracting has been gaining favor over traditional bid evaluation, where agencies go strictly with the bidder submitting the lowest price. The old low-bid method of contracting often produced shoddy work, late project completion, and costly change orders. With best value, officials look at a number of weighted factors before considering a contractor's bid price. Owners must develop a structure of well-defined selection criteria for any contract that will be let using best value.

Contractor Specialization

After World War II, many contractors began to concentrate on certain types of work and today a project can require many specialists who accomplish individual portions of the work. These specialty or trade contractors involve themselves with only a particular type of work, such as electrical, plumbing, or heating and air conditioning. A specialty contractor can work for a general contractor on a large project or may have a direct contract with an owner on a project where the owner handles the GC responsibilities of overall project coordination and control.

SUMMARY

Civilizations are built upon construction efforts. Each civilization had a construction industry that fostered its growth and quality of life. L'Ecole Polytechnique engineering school served as the model for engineering schools across Europe and later in the United States. Before steam locomotives and the railroads, what humans could carry over land was limited to what they could pack onto their backs or mules. Therefore, engineers devoted their attention to enhancing water transportation, as canals were the primary mode of commerce. Later, construction of the seven transcontinental railroads linked the country together. Automobiles gave Americans mobility in the 1920s and 1930s, and it became clear that America needed a greatly improved highway system. In 1937, New York City started work on a world-class airport, but only 2 years after it opened, a second was under construction.

Four engineering innovations had to come together before skyscrapers began to dot a city's skyline: (1) safe elevators, (2) steel framing, (3) fireproofing, and (4) the caisson method of foundation construction. Louis Sullivan demonstrated that ornamentation and false fronts, the earlier building tradition, were not essential to the beauty of tall buildings.

The accomplishment of the great civil engineering works requires the massing and management of large labor forces. Tomorrow you will be the leader of those forces as you build a magnetically levitated (Maglev) train or a new energy-efficient skyscraper for our civilization.

REVIEW QUESTIONS

1.1 What was the title of the famous text on construction authored by Marcus Vitruvius Pollio?

1.2 Name the famous engineering school authorized by Napoleon.

1.3 What American authored the first engineering texts for use in a U.S. school teaching engineering?

1.4 What U.S. canal was the first to receive a federal appropriation for its improvement?

1.5 The highest ridge pierced, while constructing the Suez Canal, was at what elevation above sea level?

1.6 Name the three chief engineers who worked on the Panama Canal during the American construction effort.

1.7 Name the chief engineer for the Union Pacific during construction of the transcontinental railroad.

1.8 Name the construction engineer for the Central Pacific during construction of the transcontinental railroad.

1.9 In what year did Lieutenant Colonel Eisenhower participate in the army convoy trip from Washington, D.C., to San Francisco?

1.10 What German construction endeavor influenced the development of the U.S. interstate highway system?

1.11 What is the name of the Interstate-70 tunnel under the Continental Divide?

1.12 From what airfield did Lindbergh pilot the first airline service from Southern California to New York?

1.13 In what year did New York start work on La Guardia Airport?

1.14 In what year did New York start work on Kennedy Airport?

1.15 Who designed the terminal complex at Dulles Airport in Washington, D.C.?

1.16 What four things were necessary before the first real skyscraper could be constructed?

1.17 Did Louis Sullivan believe that ornamentation was essential to the beauty of tall buildings?

1.18 Research these engineers on the Web and write a one-page paper about their accomplishments.

Guy F. Atkinson

Loammi Baldwin (1740–1807)

Loammi Baldwin (1780–1838)

Stephen D. Bechtel, Sr.

James Eads

A. Gustave Eiffel

George W. Goethals

General Leslie R. Groves

Benjamin Holt

Henry J. Kaiser

Peter Kiewit

R. G. LeTourneau

Robert Maillart

William Mulholland

Frei Otto (tensile structures)

Louis R. Perini

John F. Stevens

Benjamin Wright

1.19 Research these engineering accomplishments on the Web and write a one-page paper about their construction.

Chek Lap Kok Airport, Hong Kong

Chunnel Tunnel

Empire State Building

Golden Gate Bridge

Hoover Dam

Middlesex Canal

REFERENCES

1. *Civil Engineering History* (1996). Edited by Jerry R. Rogers, Donald Kennon, Robert T. Jaske, and Francis E. Griggs, Jr., Proceeding of the First National Symposium on Civil Engineering History, ASCE.

2. Condit, Carl W. (1968). *American Building,* The University of Chicago Press, Chicago.

3. Florman, Samuel (1987). *The Civilized Engineer,* St. Martin's Press, New York.

4. Follett, Ken (1989). *The Pillars of the Earth,* William Morrow & Co., New York, NY.

5. Garrison, Ervan (1998). *A History of Engineering and Technology Artful Methods,* 2nd ed., CRC Press LLC, Boca Raton, FL.

6. Hill, Forest G. (1957). *Roads, Rails, & Waterways, The Army Engineers and Early Transportation,* University of Oklahoma Press, Norman.

7. *The History of Herodotus* (1947). Translated by G. Rawlinson, Tudor Publishing Co., New York.

8. Horne, Charles F. (2002). "Introduction," The Code of Hammurabi, The Avalon Project at the Yale Law School, www.yale.edu/lawweb/avalon/medieval/hammint.htm.

9. *International Engineering History and Heritage* (2001). Proceeding of the Third National Congress on Civil Engineering History and Heritage, ASCE, Reston, VA.

10. Landberg, Lynn (2002). "Lincoln Electric and the History of Welding," *Construction Equipment,* March 2002.

11. McCullough, David (1972). *The Great Bridge,* Avon Books, New York.

12. McCullough, David (1977). *The Path Between the Seas,* Simon and Schuster, New York.

13. Middleton, William D. (2001). *The Bridge at Quebec,* Indiana University Press, Bloomington.

14. Morris, M. D. (2000). *Capstones of 20th Century Construction,* ASCE, Reston, VA.

15. Mulligan, Donald E., and Kraig Knutson (2000). *Construction and Culture: A Built Environment,* Stipes Publishing LLC, Champaign, IL.

16. Schexnayder, Cliff J. (2006). "Construction Forum, The Baldwins," *Practice Periodical on Structural Design and Construction,* ASCE, 11(4), pp. 179–195.

17. Schexnayder, Cliff (2000). "John F. Stevens—A Great Civil Engineer," *Journal of Construction Engineering and Management,* ASCE, 126(5), Sept/Oct, pp. 325–330.

18. Schexnayder, Cliff J., and Scott A. David (2002). "Past and Future of Construction Equipment," *Journal of Construction Engineering and Management,* ASCE, 128(4), pp. 279–286.

19. White, Richard (1999). *Gentlemen Engineers,* University of Toronto Press, Toronto.

20. Wright, Kenneth R., and Alfredo Valencia Zegarra (2000). *Machu Picchu: A Civil Engineering Marvel,* ASCE Press, Reston, VA.

21. Zarbin, Earl A. (1984). *Roosevelt Dam: A History to 1911,* Salt River Project, Phoenix, AZ.

22. "125 Years . . . 125 Innovations" (1999). *ENR* (*Engineering News-Record*), The McGraw-Hill Companies, New York, 18 October.

23. "125 Years . . . 125 Top People" (1999). *ENR* (*Engineering News-Record*), The McGraw-Hill Companies, New York, 30 August.

WEBSITE RESOURCES

1. bridgepros.com   The Bridgepros site is dedicated to the engineering, history, and construction of bridges.

2. www.agc.org   The Associated General Contractors (AGC), the nation's largest and oldest construction trade association.

3. www.eisenhower.utexas.edu   Eisenhower Presidential Library and Museum.

4. www.asce.org/history   The American Society of Civil Engineers history website.

5. whc.unesco.org   UNESCO World Heritage Sites, includes technological and industrially significant sites, including bridges, factories, and factory towns. The site also provides virtual tours.

6. http://www.ctbuh.org/   The Council on Tall Buildings and Urban Habitat studies and reports on all aspects of the planning, design, and construction of tall buildings.

7. www.rtri.or.jp/rd/maglev/html/english/maglev_frame_E.html   Overview of Maglev R&D, Railway Technical Research Institute of Japan (2002).

8. www.apti.org   Association for Preservation Technology International (APT), 4513 Lincoln Avenue, Suite 213, Lisle, IL 60532–1290. APT is a cross-disciplinary organization dedicated to promoting the best technology for conserving historic structures and their settings.

9. www.ex.ac.uk/~RBurt/MinHistNet/welcome.html#ToC   At the Department of Economic and Social History, University of Exeter, United Kingdom, a Mining History Network homepage is maintained.

Construction Management

2

Overview of the Construction Industry

The construction industry is the second largest goods-producing industry in the United States. It employs more than 6.4 million people in craft and management positions. Construction can be broken down by type of construction into residential, commercial/institutional building, industrial, and heavy/highway segments. Most contracts are awarded to a general contractor who awards subcontracts to specialty contractors. The most common project delivery system used in commercial construction, heavy/highway work, and nearly all government construction is design-bid-build, also known as competitive low bid; but that system is slowly being replaced by other project delivery systems such as design-build.

BUILDING YOUR FUTURE

Every person entering the construction industry needs to remember that we are in the business of building things, and the most important thing each of us will ever build is our own reputation. Reputation building is a one-person job that is inseparable from issues of personal **ethics**. It requires a little work every day. No one can build a reputation for someone else, so build yours carefully. Always ask yourself if you would be proud to read about the actions and decisions you make today in tomorrow's newspaper.

Philosophers have been discussing ethics for centuries, but people in business today need to have a practical working code of ethics. It is as Aristotle taught, moral judgments are not the product of reading moral treatises and applying them to case histories. He counsels that if you need moral guidance, seek out a person who has succeeded in living a moral life rather than someone who has succeeded in memorizing moral arguments.

Many students seem to believe that the ethical standards of the construction industry need to be improved. Along with that perception of less than perfect ethics in the industry, however, is the fact that most who hold such beliefs have very little direct exposure to the industry. The industry is undoubtedly more

ethics

All construction and design organizations publish a code of ethics to guide the behavior of their members. Ethics involves doing the right thing and protecting the public.

ethical than many tend to believe, but some improvements can be made. As in any industry, if changes in the overall reputation of the industry and the practices that form the foundation for that reputation are ever going to change, it is the *new* people coming into the industry, today's college students, who will ultimately be the influence that causes those changes. It can safely be assumed that today's members of the construction industry have already established their own ethical codes and are accustomed to the way the industry currently is. New entrants into the construction industry therefore carry a great responsibility for helping to raise the ethical standards of the industry during their careers.

The best working definition of ethics is "doing the right thing." Those who follow the rule of doing what they believe to be right will be ethical people. People who shop for bids, a technique called **bid shopping**, are aware that they are engaging in a practice that is considered to be unethical. Bid shopping means that a contractor (usually a general contractor) tells another contractor (usually a **subcontractor** or supplier) the amount of a third competing contractor's bid, and asks the second contractor to beat the other contractor's bid because, "I would rather give the work to you." The truth is, bids are submitted in confidence and deserve to be kept confidential. Bid shopping is not ethical.

Some owners have tried to prevent bid shopping by requiring contractors to submit their list of subcontractors along with their bid. Some cities have established bid depositories, where prospective subcontractors submit their bids, and general contractors may collect them. The subcontractor bid amounts are recorded with the bid depository and may not be changed. Some government agencies such as the city of New York and the state of Wisconsin use a system of multiple prime contractors to prevent bid shopping. Certain **specialty contractors** such as electrical and mechanical contractors submit their bids directly to the government to prevent their prices from being shopped.

Professional organizations such as the American Society of Civil Engineers (ASCE), the National Society of Professional Engineers (NSPE), the American Institute of Architects (AIA), and the American Institute of Constructors (AIC) publish a code of ethics, which their members promise to uphold. Students are encouraged to visit the websites of these organizations, study these codes, and begin an understanding of the importance of ethics in one's professional life. Contractor organizations, however, are reluctant to publish a code of ethics because the federal government may interpret policies of discouraging activities such as bid shopping as restraint of free trade, in violation of federal statutes. Even though ethics would require that we refrain from bid shopping, the law does not. Some actions may be lawful, but not ethical. There is a difference between ethics and law. Students need to know the difference.

There are still situations where making "under the table" payments to the right person of influence may help a company obtain work or get a proposed development approved. The AIA *Code of Ethics and Professional Practice,* Rule 2.201, states, "Members shall neither offer nor make any payment of a gift to a public official with the intent of influencing the official's judgment in connection with an existing or prospective project in which the Members are interested." All codes of ethics contain a similar statement or rule. Normally

bid shopping
Unethical procedure of requesting preferred subcontractors to lower bids to meet or beat bids submitted to the GC by another firm.

subcontractor
Specialty contractor performing under contract to the GC.

specialty contractors
See subcontractor.

there is a *quid pro quo* (Latin for "something for something" or "a favor for a favor") in the exchange of money and gifts. Involvement in an unethical transfer of funds leaves an indelible mark on one's reputation, and there is no way to remove it. The act cannot be undone. Every student is encouraged to make a personal commitment to seek out and work for an ethical company. You will know when something is simply a goodwill gift with no strings attached. You will know when your company is shopping bids and not negotiating a legitimate contract. You will know when your actions are wrong. *Do the right thing.*

The accused criminal is expected to plead not guilty. That is not an ethics issue. The politician may be expected to promise the voters anything during a campaign. However, bosses expect employees to say yes when they think yes and say no when they think no. Do not say yes when you think no. *Express opinions honestly.* Those who make a habit of saying what they think the boss wants to hear soon become redundant in their company. Failure to be candid in this situation may be an ethics issue. Bosses who do not tolerate employees with dissenting opinions lose the opportunity to compare ideas and weaken their own leadership. Be true to thine own self.

In some ways, business ethics is different from personal ethics. There are hundreds of ethical questions that confront a business. Is it ethical to build a project that adds to the air pollution during construction? Is it possible to avoid adding to the air pollution? Why is air pollution a matter of ethics? Is it ethical to construct a facility that causes people to lose their homes under eminent domain laws? How long can general contractors hold subcontractors' money before the abuse becomes an ethics issue? How does a company allow factors such as race and gender to impact on hiring and firing decisions? How harmless are those free baseball game tickets? What obligations does a company have regarding continuing education for employees? How safe are the working conditions on a project? Will the completed project be safe for the public? There are hundreds of questions that relate to ethics.

> Young construction engineers and construction managers who are faced with an unethical situation will normally have no problem recognizing the lack of ethics in the deal. Knowing how to handle the situation requires prior thought and a strong sense of ethical standards.

This text is not designed to cover the topic of ethics in detail. This introduction to ethics at the beginning of the text is meant to convey the importance of the topic. If ethical, competent people will work only for ethical companies, and refuse to work for unethical companies, then companies that have unethical practices will be forced to change or they will fail. To be an ethical person or an unethical person will always be an individual choice. Making ethical choices or accepting unethical ones is a habit, just like being early or being late is a habit. Everyone can develop either habit. The authors encourage you to make a personal study of business ethics, place it high on your list of life goals, and develop the habit of making the proper ethical choice all the time.

THE CONSTRUCTION INDUSTRY

Construction professionals sometimes argue about whether the construction industry is a service industry, just as consulting, barbering, taxi service, health care, and automobile repair are service industries. The **Bureau of Labor Statistics** classifies construction as a goods-producing industry to separate it from the service industries. There are certain segments of the construction industry that are service businesses; for example, designers who complete studies and designs for clients are providing a service. Contractors who build facilities are producing a product, just as an automobile maker or furniture manufacturer is providing a product. The most significant differences between the construction industry and most other product industries are the size and cost of the product, and its custom-designed, one-of-a-kind features. The automobile maker and the furniture maker develop their designs, produce their products, and make them available for customers who want to purchase them. Customers may specify colors and extra features in automobiles or the fabric to be used on a piece of furniture, but the basic product is not eligible for complete redesign or customization. The product required by any construction contract is also a manufactured product: a building, road, factory, church, or office building built in response to the needs of the customer. Contractors must service the needs of their customers, but the product of construction is not a service, it is a product, properly built in accordance with the **plans**, **specifications**, and expectations of the owner.

Construction in the United States can be described as a single industry, just as the manufacture and sale of automobiles is described as the automobile industry. But construction includes more than just the work of building structures on a construction site. It includes the engineers and architects who do the design work, the manufacturers and distributors of the materials and equipment that go into the structures, the construction managers who manage the site work, the unions and tradespeople who do the work, the municipal officials who review plans and enforce the building codes, the workers at concrete batch plants where concrete is proportioned for delivery to the construction site, the workforces in manufacturing facilities assembling prefabricated components that are used in many buildings, and many, many more participants. The construction industry is so large and so multifaceted that it may be better described as a sector of the economy instead of an industry.

The sheer size of the construction industry defies comprehension. The construction industry is about 6% of the national gross domestic product and employs 5% of the workers (see Fig. 2.1). Estimates of the size of the industry vary according to where the writer draws the boundary around the industry. In some states where the population is growing rapidly and there is a great deal of construction going on to support that growth, construction is actually more than 5% of the local economy and may be the largest single industry in that particular state or region. Worldwide, construction accounts for more than $4 trillion of in-place work annually.

Bureau of Labor Statistics
Part of the Department of Labor. Publishes statistics relating to labor in all industries, including construction.

plans and specifications
Contract drawings and written materials descriptions, which convey the detailed requirements to the contractor.

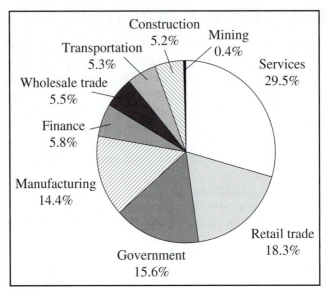

FIGURE 2.1   Industry size by percentage of workers.

Everyone is impacted by the construction industry. All citizens need the construction industry. We cannot manufacture products such as computer chips outdoors or in temporary facilities. We need offices and hospitals, manufacturing facilities and stores, roads and bridges, airports, water supply and communications facilities, and all the other facilities that make up the built environment. Construction has significantly changed the quality of every American's life.

Construction has been described as an easy in, easy out business. It is easy to get started in construction because the licensing requirements are not uniform from state to state and not difficult in most states. Nearly anyone with a pickup truck and a cell phone can get into the business. The number of people keeping their accounts receivable on one visor of the pickup and the accounts payable on the other is probably quite large.

For very small companies in any industry, getting out is not difficult either. All it takes is one bad year. Most construction companies are small, having only one or two permanent employees. Such companies are generally not able to withstand a down year, so the number of construction companies that fail every year is quite large when compared to other industries. About 10,000 construction companies, somewhere around 1–5% of the industry, fail every year with about $2 billion in liabilities. Profit margins in the construction industry are only about 1–2% according to many studies.

ORGANIZATION OF THE INDUSTRY

There are many ways to describe the organization of the construction industry, but it may be best to visualize it as organized around four types of construction

FIGURE 2.2 Residential construction.

FIGURE 2.3 Commercial construction.

projects. Owners who expect to earn a profit from the constructed facility usually build the first three types of projects. To fulfill a social need, the government usually builds the last category.

Residential Construction

This category includes contractors and developers who build individual homes (see Fig. 2.2), apartments, and assisted living facilities. Builders of individual homes generally fall into two categories: those who build custom homes individually suited to the buyer's needs and builders of "tract" homes, offering

FIGURE 2.4 Industrial construction.

standard designs and floor plans, and a limited number of options or upgrades. Residential construction is generally not bid work. Builders of custom homes and apartments negotiate a price with the owner. Developers of tract homes set the price based on market conditions.

Commercial Construction

Included in this category of construction are the office buildings (see Fig. 2.3), stores, schools, libraries, and other types of buildings except homes that make up the built environment. Projects in this category range in size from small convenience stores to multimillion-dollar office buildings. Both private owners and government owners contract for buildings in this category. Private owners generally negotiate the price of the buildings, while the government agencies normally use a low-bid process.

Industrial Construction

Examples of industrial facilities include manufacturing plants, refineries, pipelines, electricity-generating facilities (see Fig. 2.4), and high-tech facilities, such as hospitals and clean rooms. Contractors who do this kind of work are special-

FIGURE 2.5 Heavy/highway construction.

ists in the kind of projects they undertake, and they normally negotiate the price of the work.

Heavy/Highway Construction

In addition to highways, this category includes structures such as dams and levees, canals, water and waste water treatment projects, bridges (see Fig. 2.5), railroads, and tunnels. This kind of work is usually awarded by low bid. Usually the owner is a government agency. For some projects, owners prefer to prequalify bidders, meaning that contractors can bid only after they have submitted to a prequalification process to demonstrate their experience, qualifications, technical and managerial expertise, and financial capacity for the project.

PUBLIC AND PRIVATE WORKS

There are many ways to categorize the construction industry. One of the most convenient is to consider public work and private work. The type of owner, public or private, is key because of the procurement procedures and regulations that encumber the public process but have no application to private projects. Public work is defined in this book as all work done using the public's money. That includes government at all levels, including federal, state, and local. It also includes agencies such as water districts and school boards. The general rule is that public agencies are not spending their own money, but are spending money that belongs to the taxpayers. They have an obligation to spend the public's money properly. Such organizations have very bureaucratic rules and procedures. For example, projects must be awarded to the lowest responsible and responsive bidder. There is only one prescribed way to open bids, and exceptions are not allowed. There are strict rules that govern how projects are advertised. Anyone who can provide a bid bond is allowed to bid unless the project requires contractors to prequalify. There are hundreds (thousands?) of such rules and policies. They apply to all the construction of infrastructure and other facilities needed by the general population.

Private persons and corporations, on the other hand, are spending their own money. They are not as encumbered by bureaucracy and rules. They may open bids (or simply take proposals) in any way they select. They may decide who they will allow to submit bids. They may choose not to advertise the project at all, but to award it instead to their favorite contractor. It is their money. They make their own rules. The majority of construction is privately funded. **Public works** projects receive most of the publicity, but there is more private work than public work.

public works
Publicly funded construction.

HORIZONTAL AND VERTICAL CONSTRUCTION

Another way to categorize the construction industry is horizontal or vertical construction. With this categorization, the terms *horizontal* and *vertical* describe the direction of the construction. Examples of horizontal construction projects include roads, bridges, dams—those projects that are parallel to the horizon. Examples of vertical construction include schools, shopping centers, skyscrapers—those projects that tend to rise out of the ground. Typically, the type of construction, horizontal or vertical, indicates who the lead designer is, with engineers leading the design team on horizontal-type projects and architects leading the design team on vertical-type projects.

PROJECT PARTICIPANTS

No description of the industry would be complete without including the other professions, government organizations, and interested parties who are part of the total industry.

Owners

construction manager
A person who coordinates the construction process on behalf of the owner.

No construction would ever be accomplished without owners who must make the decision to build the facility, define the need, provide the financing, and manage the construction process. Owners are public (government) or private. Most public owners, such as the Corps of Engineers or state departments of transportation, are experienced **construction managers**. Most private owners, such as small manufacturing companies, have little or no construction management expertise and depend on consultants to help them through their projects. Besides providing the project funding, the primary responsibility of the owner is to define the scope of the work.

General Contractors

Most contracts are awarded to a general contractor (GC) who manages the project, and subcontracts portions of the work, such as the electrical and mechanical work, to subcontractors or specialty contractors. The primary job of the general contractor is to manage the job, keep it on schedule, control costs, ensure the work is well coordinated and performed in a safe manner, and coordinate with the owner on all matters since the GC is the only one of the contractors holding a contract with the owner.

Subcontractors or Specialty Contractors

Subcontractors or specialty contractors make up the largest portion of the construction industry. They do the work. All subcontractors have a specialty such as electrical or mechanical, steel erection, acoustical, drywall and painting, and carpeting. The construction of a typical building may require 20 or more subcontractors.

Designers, Architects, and Engineers

Project owners select an architect or engineer (A/E) to design their projects. The owner usually selects the A/E based on demonstrated ability to design the project in the time frame available and at a cost acceptable to the owner. The designer will prepare the construction documents (blueprints, specifications, contracts) for the project owner. Architects are usually the lead designers for buildings—vertical construction. They subcontract some of the work, such as the structural frame and mechanical/electrical/plumbing (MEP) systems, to engineers who specialize in such work. Engineers are normally the lead designers for heavy civil and highway projects—horizontal construction. They may subcontract part of the work such as train stations or office buildings to an architect.

The designer's involvement during construction will depend on the project delivery method used by the owner. In the case of a **design-bid-build** project, the owner may employ the designer to oversee the work of the construction contractor. In such a capacity, the designer will assess the quality of the contractor's work and approve progress payments. If it is necessary to issue change orders to the contract, the designer will assist the owner in negotiating with the contractor the magnitude of the resulting cost and time adjustments.

design-bid-build
Conventional procurement system in which the owner contracts separately with a designer and then the contractor.

Construction Managers

During the 1970s, the term *construction manager* became popular. Unfortunately today, construction manager has many different meanings. Some contractors now call themselves construction managers. Usually they subcontract 100% of the project work and prefer to be involved in both the design and construction processes. They bring construction expertise to the design process. Some engineering and architectural firms offer construction management services, indicating usually that they represent the owner during construction. To make the definition even more difficult, there are some government agencies that have employees who are construction managers. Not all people who call themselves construction managers work for a construction contractor.

Trades

open shop
Contracting business that does not require union membership.

The trades consist of both union and **open shop** workers. The rules of employment vary by state and locality. The unions are organized by craft, with specific unions representing workers within a certain trade. For example, it is not difficult to understand that electricians, those who belong to a union, are members of an electricians' union. The carpenters' union is somewhat more diverse. It represents workers who are carpenters, piledrivers, millwrights, tradeshow workers, shipwrights, exterior/interior specialists, scaffold erectors, insulators, and related craft workers. The specialty trades include over four million workers and nearly two-thirds of the total number of construction industry workers.

Labor Unions

Labor unions in the United States began about 150 years ago. Their initial purpose was to improve the lives of their members. The unions are generally formed around crafts and trades, but today many unions represent several trades, not only the one trade that carries their name. Unions are operated by the members, have constitutions, and are made up of locals. For instance, the carpenters' union, The United Brotherhood of Carpenters and Joiners of America, has over 2,500 locals. The locals serve as the contract-negotiating body for their members in their local area, negotiating labor agreements that are then accepted by the "signatory companies," meaning those that employ union workers and agree to the terms of the union's contract. Benefits to the union employers include training programs, pensions, and hiring halls.

Insurance Companies

Contractors are required to provide bid bonds as a condition of being allowed to bid, and then they must provide insurance, performance bonds, and payment bonds prior to award of the contract. Insurance companies provide bid bonds and performance and payment bonds, and they also service the liability and property insurance needs of contractors. The various types of insurance will be discussed in Chapter 3.

Banks

Banks provide working capital to contractors so the contractor can pay for materials, labor, equipment, and overhead expenses until the contractor is paid by the owner. Contractors earn their progress payments (usually monthly), and no up-front payments are made by the owner. The impact of this practice on a contractor's profit and loss will be discussed in Chapter 3. Banks also provide both short-term and long-term financing to the project owners.

Suppliers

Everything on a construction project, from concrete to paint, comes from suppliers. Many suppliers assist designers in material and equipment selection and then assist the contractors in preparing their bids, preparing shop drawings, and fabricating items specifically for individual projects. Designers rely heavily on standard specifications and standards such as those published by the American Society for Testing and Materials (ASTM). It is most important that designers understand the standard specifications and design standards they are using because there are design standards for nearly every level of product quality. When the owner wants a high-quality product, it is important to use a high-quality standard. It is equally important to ensure the quality of a construction project by using quality suppliers.

Permitting Agencies and Building Authorities

Permitting agencies and building authorities represent the interests of public safety. This group includes federal agencies such as the Federal Highway Administration and the General Services Agency, state departments of transportation, and local agencies such as municipal zoning boards and even home owners' associations. They administer publicly funded construction projects, and they ensure private construction projects comply with zoning laws and building codes.

PUBLIC

The public is impacted by every construction contract and those impacts can be both good and bad. Voters approve bond issues for new schools, roads, and other public facilities. The public is then inconvenienced while highways are widened or repaired. Individual members of the public wait while their homes are constructed. The public works in temporary offices while permanent facilities are under construction. The American public enjoys the best quality highways, residences, offices, infrastructure, and public facilities in the world. The country's economy is dependent on its infrastructure and built environment. For example, each additional $1 million spent on construction in the United States creates 46.8 new jobs in the construction, supplier, and service industries.

INDUSTRY ORGANIZATIONS

Most participants in the construction industry belong to one or more industry or professional associations. Membership in some of these associations is by the firm or corporation while in other cases membership is by the individual.

American Concrete Institute

The American Concrete Institute (ACI) was founded in 1904. It serves as a technical and educational society. The mission of ACI is to develop, share, and disseminate the knowledge and information needed to utilize concrete to its fullest potential. The institute has produced more than 400 technical documents, reports, guides, specifications, and codes for the best use of concrete and has 13 different certification programs for concrete practitioners [1].

American Institute of Constructors

The American Institute of Constructors (AIC), organized in 1971 as the professional society for the practicing constructor, is the sponsoring organization for the Constructor Certification Program. In 1994, the AIC Constructor Certification Commission was organized under the auspices of AIC to expand the constructor-qualifying process to include a written examination and to offer an internationally recognized certification process. The certification process is peer developed and is intended to set high standards for skills, knowledge, education, and conduct for the Certified Professional Constructor (CPC) [2].

American Institute of Architects

The American Institute of Architects (AIA) is a professional organization for architects in the United States. Organized in 1857, the institute conducts various activities and programs to support the profession and enhance its public image, including periodically awarding the AIA Gold Medal and the Architecture Firm Award.

American Institute of Steel Construction, Inc.

The American Institute of Steel Construction, Inc. (AISC) is a nonprofit trade association and technical institute established in 1921 to serve the structural steel industry in the United States. Its purpose is to promote the use of structural steel through research activities, market development, education, codes and specifications, technical assistance, quality certification, and standardization [3].

American Society of Civil Engineers

Founded in 1852, the American Society of Civil Engineers (ASCE) represents 125,000 members of the civil engineering profession worldwide, and is America's oldest national engineering society [4].

Associated Builders and Contractors

The Associated Builders and Contractors (ABC) is a national trade association representing about 23,000 contractors, subcontractors, material suppliers, and related firms from across the country and from all construction industry specialties. The ABC is the only national association devoted exclusively to the merit shop philosophy, the principle of providing the best management techniques, the finest craftsmanship, and the most competitive bidding and pricing strategies in the industry, regardless of labor affiliation [5].

Associated General Contractors of America

The Associated General Contractors of America (AGC) was formed in 1918 at a meeting in Chicago and it was the first major association of general contractors. Today the AGC has over 33,000 members, including general contractors, specialty contractors, suppliers, and service providers. Historically, the association has handled wage and contract negotiations with local building trades unions throughout the United States [6].

Construction Industry Institute

The Construction Industry Institute (CII) is a research organization with a singular mission: improving the competitiveness of the construction industry. The CII is a unique consortium of leading owners and contractors who have joined together to find better ways of planning and executing capital construction programs [8].

National Society of Professional Engineers

The National Society of Professional Engineers (NSPE) is the national society of engineering professionals from all disciplines that promotes the ethical and competent practice of engineering, promotes licensure, and enhances the image and well-being of its members. Founded in 1934, NSPE serves more than 54,000 members and the public through 53 state and territorial societies and more than 500 chapters. For more information visit www.nspe.org.

CONSTRUCTION LABOR FORCE

Twenty-two states are right-to-work states, meaning that it is not legal to require union membership as a condition of employment. On the average, 8.9% of the construction workers in the right-to-work states belong to a union. The other states are union states where the state laws permit union organization of employees, and union contractors agree to hire only through the union hiring halls. Nationwide, the percentage of workers in the construction industry who are union members fell from a high of nearly 80% in 1970 to about 20% today.

project labor agreement (PLA)
Agreement between the contractor and owner on a large project to hire workers through the union hiring hall in exchange for a no-strike agreement.

Project Labor Agreements

There is a growing trend among government agencies to require **project labor agreements (PLAs)** on their projects. PLAs come with many different sets of agreements, but the basic agreement is that contractors must hire workers only through the union hiring halls. In return, the unions agree not to **strike** for the duration of the job. By law, a PLA has no impact on a contractor's existing

strike
A work stoppage by a body of workers to enforce compliance with demands made on an employer.

employees, only on hiring new employees. According to recent court decisions, this kind of requirement does not exclude open shop contractors from bidding work. The number of union versus nonunion contractors who successfully bid projects that are covered by PLAs does not bear out that claim.

Construction Crafts

The trade unions all have very valuable training programs, which may be their most notable contribution to the industry. There are two unfortunate problems with trade union training programs that must be solved if the industry is to obtain maximum benefit from the union training programs. The first problem is that they are able to reach only a limited number of people. Craft training requires a great deal of "hands on" work. For some tools, the instructor needs to put his or her hands onto the hands of the student to demonstrate and teach the proper use of the tool. Because of the nature of the training, the size of classes must be limited. The second unfortunate aspect of union craft training is the general lack of knowledge about and interest in the craft union training programs at the high school graduate level. The result is that the industry needs more trained craftspeople than the unions can provide. Construction company owners believe the top challenge facing the construction industry in the next five years is shortage of trained labor. Labor and management need to work together to solve this problem.

Presented next are descriptions of the tasks each of the construction crafts perform.

Brickmasons, Blockmasons, and Stonemasons   The work of a mason varies in complexity, from laying a simple masonry walkway to installing an ornate exterior on a high-rise building. *Brickmasons* and *blockmasons,* who often are referred to simply as bricklayers, build and repair walls, floors, partitions, fireplaces, chimneys, and other structures with brick, precast masonry panels, concrete block, and other masonry materials. *Stonemasons* build stone walls as well as set stone exteriors and floors [12].

Carpenters   The tasks of cutting, fitting, and assembling wood and other materials for the construction of buildings, highways, bridges, docks, industrial plants, boats, ships, and many other structures are the responsibility of carpenters. Builders increasingly are using specialty trade contractors who, in turn, hire carpenters who specialize in just one or two activities. Some of these activities are setting forms for concrete construction; erecting scaffolding; or doing finishing work, such as installing interior and exterior trim. However, a carpenter directly employed by a general building contractor often must perform a variety of the tasks associated with new construction, such as framing walls and partitions, putting in doors and windows, building stairs, laying hardwood floors, and hanging kitchen cabinets [12].

Cement Masons and Concrete Finishers   One of the most common and durable materials used in construction is concrete. *Cement masons* and *concrete finishers* place and finish the concrete. They also may color concrete surfaces;

FIGURE 2.6 Concrete finisher leveling the freshly placed concrete.

expose aggregate (small stones) in walls and sidewalks; or fabricate concrete beams, columns, and panels. In preparing a site for placing concrete, cement masons first set the forms (formwork) for holding the concrete and properly align them. They then direct the placing of the concrete and supervise laborers who use shovels or special tools to spread it. Concrete finishers then guide a straightedge back and forth across the top of the forms to "screed," or level, the freshly placed concrete. Immediately after leveling the concrete, finishers carefully smooth the concrete surface (Fig. 2.6) [12].

Construction Equipment Operators   The machines used to move construction materials, earth, and rock, and to apply asphalt and concrete to roads and other structures are controlled by skilled operators. The operation of much of this equipment is becoming more complex as a result of computerized controls. Construction equipment operators may also set up and inspect equipment, make adjustments, and perform minor repairs and maintenance to their machines [12].

Construction Laborers   Although the term *laborer* implies work that requires relatively low levels of skill or training, many tasks that these workers perform require a fairly high level of training and experience. Construction laborers perform a wide range of physically demanding tasks involving building and highway construction, tunnel and shaft excavation, hazardous waste removal, and demolition. They dig trenches, mix and place concrete, and set braces and shoring to support the sides of excavations. Construction laborers may sometimes help other craft workers including carpenters, plasterers, and masons [12].

Electricians Electricity is essential for light, power, air-conditioning, and refrigeration. Electricians install, connect, test, and maintain electrical systems for a variety of purposes, including climate control, security, and communications. They also may install and maintain the electronic controls for equipment [12].

Glaziers Glass serves many uses in modern buildings. Insulated and specially treated glass keeps in warmed or cooled air and provides good condensation and sound-control qualities; tempered and laminated glass makes doors and windows more secure. In large commercial buildings, glass panels give office buildings a distinctive look while reducing the need for artificial lighting. *Glaziers* are responsible for selecting, cutting, installing, replacing, and removing all types of glass [12].

Painters Paint and wall coverings make surfaces attractive and bright. In addition, paints and other sealers protect outside walls from wear caused by exposure to the weather. Painters apply paint, stain, varnish, and other finishes to buildings and other structures [12].

Pipelayers, Plumbers, Pipefitters, and Steamfitters Although pipelaying, plumbing, pipefitting, and steamfitting sometimes are considered a single trade, workers generally specialize in one of the four areas. *Pipelayers* lay clay, concrete, plastic, or cast-iron pipe for drains, sewers, water mains, and oil or gas lines. *Plumbers* install and repair the water, waste disposal, drainage, and gas systems in homes and commercial and industrial buildings. Plumbers also install plumbing fixtures, bathtubs, showers, sinks, toilets, and appliances such as dishwashers and water heaters. *Pipefitters* install and repair both high- and low-pressure pipe systems used in manufacturing, in the generation of electricity, and in heating and cooling buildings. They also install automatic controls that are increasingly being used to regulate these systems. Some pipefitters specialize in only one type of system. *Steamfitters,* for example, install pipe systems that move liquids or gases under high pressure. *Sprinklerfitters* install automatic fire sprinkler systems in buildings [12].

Sheet Metal Workers The installation and maintenance of air-conditioning, heating, ventilation, and pollution control duct systems; siding; rain gutters; downspouts; skylights; restaurant equipment; outdoor signs; and many other products made from metal sheets is the responsibility of the sheet metal workers. They also may work with fiberglass and plastic materials. Although some workers specialize in fabrication, installation, or maintenance, most do all three jobs [12].

Structural and Reinforcing Iron- and Metalworkers Structures have frames made of steel columns, beams, and girders. In addition, reinforced concrete—concrete containing steel bars or wire fabric—is an important material. More-

FIGURE 2.7 Ironworkers erecting a steel frame building.

over, metal stairways, catwalks, floor gratings, ladders, window frames, and decorative ironwork increase the functionality and attractiveness of structures. Structural and reinforcing iron- and metalworkers fabricate, assemble, and install these products.

Even though the primary metal involved in this work is steel, workers often are known as *ironworkers*. Before construction can begin, ironworkers must erect steel frames and assemble the cranes and derricks that move structural steel, reinforcing bars, and other materials around the construction site. The structural metal arrives at the construction site in sections and it is then lifted into position by a crane. Ironworkers connect the sections and set the cables to do the hoisting (Fig. 2.7).

ORGANIZATION OF THE CONSTRUCTION BUSINESS

Architects and engineers accomplish design work. Normally for buildings, the architect will be the lead designer, which means it is the architect who has a contract with the owner. Structural engineers, mechanical engineers, or others may provide some design services to the architect; these engineering companies have a contract with the architect, not the owner. Usually the construction project itself is awarded to a general contractor. This is a contractor who manages the entire job, probably constructs part of the project, and subcontracts a large portion of the work out to specialty contractors such as mechanical, electrical, masonry, roofing, and landscaping subcontractors. The subcontractors have a contract with the general contractor, and have no contractual relationship with the owner or each other.

multiple prime contractors
Contracting system that eliminates the general or prime contractor. The owner contracts individually with the specialty contractors.

Some states and municipalities have laws permitting contracting using **multiple prime contractors**. On some projects such as bridges, it is beneficial to award separate contracts for different parts of the bridge. Contractors whose expertise qualifies them to build bridge pier foundations (the substructure) may not be qualified to build the superstructure (bridge deck and approaches). One of the effects of having multiple primes on a project is that the owner must perform the day-to-day management and coordination of the project because there is no general contractor to perform those functions. The owner must also perform the general conditions responsibilities such as providing project access, clean up, office space, and utilities. Owners prefer multiple primes for one of two reasons, either to save the general contractor's markup on subcontractor work or to finish the work sooner by starting specialty contractors sooner. The down side is that the owner must intensively manage the specialty contractors, a function normally done by the general contractor. Many of these owner-managed contracts finish late because of the owner's limited experience in controlling construction work. This is not a widespread contracting system.

JOB OPPORTUNITIES

requests for information (RFIs)
Formal documentation used by contractor to clarify contract requirements.

New college graduates who begin their careers by working for a construction contractor often start as a project engineer or assistant project manager. Project engineers are assigned a great variety of tasks such as tracking **requests for information (RFIs)**, managing shop drawings, maintaining the daily records of the project, calculating pay estimates, updating project schedules, and resolving errors in plans and specifications. As project engineers gain experience, they can expect to become estimators, schedulers, and project managers. Larger companies have a greater ability to meet bonding requirements, which leads to a greater likelihood of working on a large project. This provides the opportunity to work with a larger project team. In small companies, project engineers work with smaller projects as members of smaller project teams.

Construction is a team endeavor. The job of each individual on the project team is to make sure everyone else on the team always succeeds, because everyone is essential, and the project may fail if one person is allowed to fail. There are no jobs in construction where a person can work alone. Teams succeed or fail as a group. Unlike some athletic teams, there are no superstars on a failing construction project team. In construction, people skills are a major key to success.

General contractors tend to move their employees around more than specialty contractors because specialty contractors self-perform work and are regionally focused. General contractors tend to follow the work. There is always mechanical construction or electrical construction, for instance, in every city in the country, so specialty contractors have less need to move people.

A second factor new graduates need to consider is the ownership of a prospective company. A growing number of large companies are publicly held, with their stock traded in the stock market; some companies are privately held by a few individuals or a family; and some construction companies make ownership of company stock available as people are promoted up the management ladder. The impact of company ownership on an individual employee relates to areas such as potential for bonuses or profit sharing, potential ownership, or potential promotion limitations.

All reputable companies are incorporated and licensed. Small one- and two-person companies may not be either incorporated or licensed, but they have minimal potential for hiring new college graduates.

SUMMARY

Each person in the construction industry must build his or her own reputation. It is the most important thing any person will ever build. Understand the difference between lawful and ethical. Acquire a personal definition of ethics that is personally workable, such as, "ethics is doing the right thing."

It may be best to visualize the construction industry organized into four general categories of construction: residential, commercial/institutional building, industrial, and heavy civil and highway. The major participants in the industry, in addition to the owners of the projects, are the general and specialty contractors, designers, construction managers, tradespeople, insurance companies, and materials suppliers.

Twenty-two states are right-to-work states. But unions are strongly pushing project labor agreements with project owners. Most construction is done on private contracts, even though the public contracts gain the most attention and tend to be larger individual contracts. Most work is done by a general contractor and specialty contractors or subcontractors. Some public agencies require a system of multiple prime contracts.

Construction is a team endeavor. People skills are your key to success. Success in your career will depend on several factors such as college and technical education, continuing education, experience and hard work, ability to be a productive member of a project team, oral and written communication skills, leadership ability, and ethical conduct.

REVIEW QUESTIONS

2.1 Describe the essential differences between design-bid-build and design-build. Make a list of the major advantages and disadvantages of each.

2.2 What are the major types of construction contractors?

2.3 What are the major differences in contracting with the government and contracting with a private firm or individual?

2.4 Why is there a growing attitude that construction is a commodity? What does that mean?

2.5 What are the advantages and disadvantages of project labor agreements? What is causing their increased use?

2.6 Make a list of the major differences between construction and manufacturing. Which items do you think are the most significant?

2.7 How large is the American construction industry? Make a list of businesses that you consider part of the construction industry.

2.8 What are the major services to the industry that are performed by the trade unions?

2.9 Why do owners prefer the design-build form of contracts when many of them know they can obtain better-quality construction using other forms?

REFERENCES

1. American Concrete Institute, www.aci-int.org.

2. American Institute of Steel Construction, Inc., www.aisc.org.

3. American Society of Civil Engineers, www.asce.org/.

4. Associated Builders and Contractors, www.abc.org/.

5. Associated General Contractors of America (The), www.agc.org/.

6. Construction Financial Management Association, *Construction Industry Annual Financial Survey 2000,* 12th ed. Compiled by PricewaterhouseCoopers, Princeton, NJ.

7. Construction Industry Institute, www.construction-institute.org/.

8. Design/Build Survey 1999, Zweig White and Associates, Inc. Natick, MA.

9. "Industry Accounts Data," Bureau of Economic Analysis, U.S. Department of Commerce, www.bea.doc.gov/bea/dn2.htm.

10. Kern, Hans G. (2001). *Best Practice for Process Industry Project Delivery Systems: Design-Build,* Design-Build Institute of America, Washington, DC.

11. Nash, Laura L. (November–December 1981). "Ethics Without the Sermon," *Harvard Business Review,* Vol. 59.

12. *Occupational Outlook Handbook,* U.S. Department of Labor, www.bls .gov/oco/home.htm.

13. Tulacz, Gary J. "Business before Celebration," 5 June 2000, *ENR, Engineering News-Record,* McGraw-Hill, New York.

WEBSITE RESOURCES

1. www.icivilengineer.com   This site is designed for civil engineering professionals and students. Subject areas include construction materials, civil engineering news, software guide, and jobs.

2. www.aicnet.org   American Institute of Constructors website. This site contains information related to AIC membership and the Certified Professional Constructor (CPC) process, which provides the constructor with formal recognition of the education and experience that defines the constructor as a professional.

3. www.aiaonline.com   This home page of the American Institute of Architects contains information for the professional architect and for consumers. Professional architects who are members of AIA can purchase contract forms online. Consumers can seek assistance in identifying an architect in their local area.

4. www.aflcio.org   Home page of the AFL-CIO. This page contains political commentary, information about the AFL-CIO, and labor relations current events.

5. www.agc.org   This home page of the Associated General Contractors of America contains a description of the AGC, construction marketplace news, contract documents, and other services for members.

6. www.construction.st/indexreg.htm   This is a construction education website. It links to many sites with information relating to construction and education. This site has a wealth of information and would require hours to explore thoroughly. It may be used as a reference source for specific topics.

7. www.asaonline.com   American Subcontractors Association website. This site contains information for members and the public. Concentrations are on leadership, safety, and education.

8. www.enr.com   Contains general information from current events to pricing and trends in the construction industry.

9. www.dbia.org   Home page of the Design-Build Institute of America. Contains guidance for contracting using the design-build process.

3

Construction Management Functions

The purpose of operating a business is to earn a profit by providing a valuable service. In that respect, construction companies are no different from any other kind of company. They bid or negotiate for work to earn a profit. To be successful, they must know how to estimate the cost of construction projects accurately, predict the schedule of the work, control the progress and expenditures during construction, and complete projects safely and on time. They have a responsibility to construct the project in accordance with the plans and specifications, and to satisfy the customer's cost, quality, and time expectations. The construction project team is organized for the purpose of accomplishing those objectives.

PROJECT PLANNING AND DESIGN

The management of a construction project can be addressed on several levels. The owner of the project has construction-related functions such as defining the scope of the project, planning and financing the project, and ensuring the project team understands the project goals. There are construction company-level construction management functions such as selecting the right jobs to bid, preparing the cost estimate and submitting the bid, procuring the payment and performance bonds, scheduling the work, and securing project operating capital. At the project site, the construction management functions are setting the standards for quality and safety, planning the sequence of construction, controlling progress and expenditures, communicating effectively with the owner and architect/engineer, coordinating the work of the subcontractors, managing submittals, managing change orders, submitting **periodic pay estimates**, and closing out the project.

periodic pay estimates
Payments to the contractor during the work, normally monthly, and based on the progress to date.

MASTER PLANNING (CAPITAL BUDGETING)

All organizations, corporations, schools, shopping malls, and cities need to have a master plan. One easy way to understand the necessity for master planning is to consider the results of not having a master plan. When a corporation decides to expand its operations and it already has a large campus-type complex of buildings, there are many answers that are needed to support a decision of where to locate the new facility. Will placing the building within the existing campus strengthen or weaken its distribution capabilities? Can the locality provide the necessary new employees? Will the local zoning laws permit construction of another building? Are there any tax advantages? What are the engineering and architectural problems? Does the site analysis confirm that the utilities are capable of supporting the new facility? What kind of construction would be compatible with the surroundings? These questions, and hundreds more like them, would be addressed in a master plan. The purpose of master planning is to answer those questions ahead of time and to be ready to move into design and construction without delay if the decision is made to do so. Master planning is a preparedness function that small companies think they cannot afford, but large expanding companies consider a necessity. Many large companies such as UPS, Pet Smart, Wal-Mart, and Home Depot have buildings under construction all the time. They need master planning. They must continually identify new suitable locations, and determine the priority of facility construction.

The government spends a great deal of money on master planning but has a challenge that is very large. On a national scale, the government is usually not good at master planning. For instance, the country now needs to build 60 to 90 new major power plants per year for the next 10 years to meet the growing demands for electricity. Along with the new plants comes the need for the support structure to provide the fuel and power lines to transmit the generated power. At the same time, the need for new airport facilities to support the growing demand for air transportation is far ahead of our ability to construct the facilities. Some of these master-planning shortcomings may be caused by external impacts such as environmental regulation and privatization. These impacts should be considered in a good master plan. National master plans should have been started years ago to plan the needs of electricity generation and airport capacity. While there are many areas such as these where the national government has fallen short in the master-planning arena, the interstate highway system described in Chapter 1 is an example of successful master planning.

Scope Definition

The first major activity that must be accomplished in the planning stage is **scope definition**. In many ways, this is the most important responsibility of the owner in the entire process. The owner-defined scope of the project, which the owner defines with the help of the architect/engineer/construction manager, will ultimately be the controlling factor on the schedule and budget for the facility.

scope definition
Owner responsibility to develop and convey to the architect in sufficient detail to allow the design to progress.

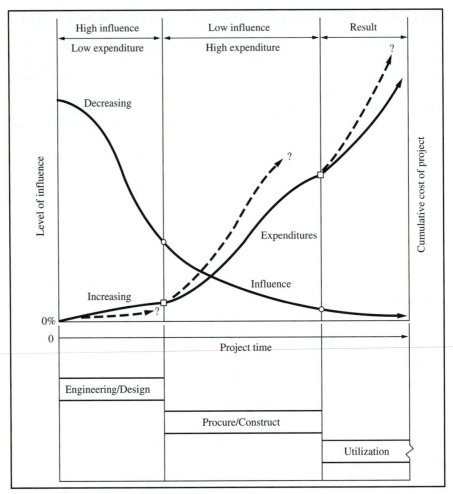

FIGURE 3.1 Level of influence on project cost.
(Paulson, Boyd C., Jr. (1976). "Designing to Reduce Construction Costs," *Journal of the Construction Division, ASCE*, Vol. 102, No. CO4. Reproduced by permission of ASCE.)

Scope definition will set the design objectives for the designer. In the scope definition phase, the owner determines what kind of a project will be built.

The purpose of the building must be clearly defined by the owner. If it is a manufacturing facility, there are numerous questions that must be answered. Examples would include the number of employees in the building, the size and quantity of the machinery, the storage requirements, the type and quantity of shipping facilities needed, the expected life of the building, the involvement with hazardous materials, and environmental requirements.

As shown in Fig. 3.1, which is reproduced from Boyd Paulson's [8] excellent paper "Designing to Reduce Construction Costs," it is much easier and much less costly to make changes early in the planning or design phase than it is later. As planning progresses, the expense of making changes increases. During the planning and design phases, changes require redesign and issuance of new drawings. Once construction starts, making changes requires change orders, rescheduling, and potentially the return of materials and procurement of new materials. After construction is complete, changes are virtually impossible because of the cost. It is in the owner's best interests to define the scope of the project early, clearly, and accurately. A properly scoped project will be large enough, will offer the proper access, will be designed for the useful life needed, and will fit the owner's budget and schedule. It is altogether too easy to scope projects that exceed the owner's budget, especially when the owner is not experienced with construction. The owner must depend on the design team for an accurate estimate of the construction cost of the facility as the scope is being developed. The designer must protect the owner from overscoping the project.

PLANNING PHASE

Project planning (separate from master planning) is also a function that is primarily the responsibility of the project owner, although most owners will contract with a designer to do the actual planning documents. The owner has a requirement and an idea for a construction project that will fulfill that requirement. The owner has several responsibilities that must be accomplished during the planning phase:

- Select the designer.
- Define the project goals.
- Ensure the availability of sufficient funds to complete the project.
- Obtain project site zoning changes if necessary.
- Select and purchase the project site.
- Perform environmental impact studies and obtain permits.
- Determine construction procurement system and the form of construction contract to be used.

It should be obvious that design cannot start until the site has at least been identified and access for the design team guaranteed. The designer needs to determine the locations of underground and overhead obstructions, and survey the site. The owner has the responsibility to provide legal access to the site so the designer can begin work.

The owner's goals are always essentially the same—to produce a quality facility that meets their needs, within budget, on time, safely. The owner must provide the most accurate possible guidance to the designer for project budget, type of facility, owner's priorities, and quality standards—both in workmanship and materials. Many owners lack the capability to provide the guidance that the project team needs, and must depend on the architect/engineer/construction manager for assistance. Everyone says they want high-quality construction, and it is always true, but high quality to one person means marble bathrooms and to another high quality means efficient layout of the floor plan to support the manufacturing process. Quality is not defined in terms of frills; it is defined in terms of satisfying the needs of the owner. The owner who can clearly communicate clear project goals has the best chance of realizing those goals.

The owner must make financing arrangements prior to the award of a construction contract. If the owner has sufficient funds internally available, there will be no financial institution involved in funding the project. Normally, the owner must seek a **construction mortgage** and later, a permanent mortgage or take out mortgage. A mortgage is a loan that uses the facility itself as collateral. If the owner fails to make the mortgage payments, the lender can seize the facility and sell it at auction to satisfy the terms of the mortgage loan. A construction mortgage will normally be in effect only during the period of actual construction. During that time, the project owner will make interest-only payments to the lender, with no payments on the principal of the loan. The amount of the construction contractor's progress payments will be drawn from the construction mortgage each month and paid to the contractor. When the facility is completed, the construction mortgage will be converted to a permanent mortgage (often called a take out mortgage), and thereafter the owner will be obligated to make sequential payments to cover both interest and principal. During construction, which is the period of greatest risk to the lender, the contractor will normally be required to provide performance and payment bonds for the lender and owner's protection. The **surety** offering the performance and payment bonds will prequalify the contractor for the bond and guarantee the performance of the contractor and the payment of the labor and materials bills. Bonds are discussed in greater detail in Chapter 8.

The owner must decide what form of construction delivery system is to be used. This is one of the owner's key **risk management** issues. For example, if the project is to be design-bid-build, the designer must be selected early so the design can be completed before the project is advertised for bids. If the project is to be a design-build contract, the owner needs to select the design-build contractor, one organization capable of arranging both the design and construction. Other forms of construction contracts, such as construction management or construction management at risk, require similar decisions. These decisions determine how the work must be accomplished during the design phase.

construction mortgage
Bank loan to finance construction, secured by the facility.

surety
A firm executing a bond, payable to the project owner, securing performance of the contract or securing payments for labor and materials.

risk management
Process of sharing risks, usually based on insurance for high-risk areas such as equipment loss or liability matters.

PROJECT DELIVERY METHODS

The major project delivery methods used in the United States are

■ Design-bid-build (commonly called low bid or traditional)
■ Design-build
■ Design-build-operate
■ Construction management at risk

Design-Bid-Build

The theory of this process is that any qualified contractor will produce the same product from the plans and specifications, provided the plans and specifications are complete and written properly. This process treats construction like a commodity. Design the product and assume the results of the process will be identical, no matter who constructs the product. Design-bid-build is the most common government-used process for construction services.

Some agencies complete the design phase using their own "in-house" designers, but most public agencies contract for design services from an outside architectural or engineering firm for at least a portion of their projects. The designer selection process starts with architects or engineers competing for the design assignment. Competition is based on qualifications. The government wants the best-qualified designer to design each project. It looks for prior experience with the same type of work. It evaluates the current workload of the potential designer. It looks for the availability of key personnel in the firm. It looks for an appropriately sized firm, located in the general proximity of the project. This process is called qualifications-based selection (QBS). The purpose is to find the best-qualified designer.

After the design has been completed and accepted by the owner, the owner will publicly advertise the project and award it to the lowest responsive and responsible bidder. These are legal terms that deserve a layman's definition. To be responsive, a bid must be submitted on time, on the required forms, with acknowledgments of all addenda, and with the required bid bond. The purpose of the bid bond is to ensure that the contractor will accept the contract if the firm is the low bidder and that the firm will provide the required payment and performance bonds. A bid bond also serves to ensure that the contractor does not simply submit a price without studying the project. A responsible bidder, for practical purposes, is one who is capable of providing the required payment and performance bonds. To be a responsible bidder, a contractor must be able to demonstrate the managerial and financial capability necessary to successfully complete the work.

In theory, design-bid-build should be the most economical form of construction procurement. The system is designed to identify the contractor who is willing to do the required work at a specified quality level for the lowest price. Most experts agree, however, that design-bid-build is one of the most expensive forms of construction procurement ever devised and that the products are the lowest quality the contract will permit. The system causes contractors to find the most economical way to do the work. They may even look for change orders so they can make a profit. It causes manufacturers to find even less expensive products to sell to the agencies that continue to use the system. It does not offer the builder the opportunity to contribute the company's construction expertise to the design process. As with any project delivery system, it is most successful when the low bidder is a quality contractor.

There are reasons that owners continue to use design-bid-build, including

- Low-bid regulations are firmly entrenched in most government systems for the purpose of promoting fairness.
- It is easy to justify the selection of the low bidder.
- Contractors understand the system.
- Voters understand the system.
- There is always resistance to change.

two-step procurement
Procurement process based on submission of a design, selection of the preferred design by the owner, and then submission of a price.

A variation of this process is **two-step procurement**, often used for more technically demanding jobs. In this process, contractors are prequalified to bid, and bids are only accepted from those who successfully prequalify. Contractors may be required to submit their financial records and evidence of their management capability. Those deemed acceptable will be allowed to compete for the project on a low-bid basis. Contractors may simply be required to demonstrate prior experience with similar work. There are many variations of two-step procurement, but they all have the same basic framework. In step 1, contractors must prequalify to be allowed to submit a bid. In step 2, the contract will be awarded to the low bidder.

In summary, the design-bid-build process is set up to find the best designer to design the product, and the lowest priced contractor to build it.

Design-Build

design-build
Procurement system in which the owner contracts with a contractor/designer as a single unity, under one contract.

The fastest growing construction delivery system in America is the **design-build** system, in which the designer and the builder are hired as a team under a single contract. One contractor is responsible for both the design and construction. This type of project delivery method is growing in popularity. A design-build contract provides distinct advantages to the owner since the owner no longer needs to referee disagreements between the designer and the contractor because they are working together under the same contract. Design-build contracts are often structured on a cost plus with a guaranteed maximum price (GMP) basis. Often the savings below the GMP is shared by the owner and contractor. There is extensive information on design-build contracts from the Design-Build Institute of America at www.dbia.org.

Design-build is an example of how improvements to the construction industry are often made; they come from outside the industry. Design-build was started as a project delivery method by owners who wanted to solve the problems associated with having to referee between the designer and the builder when each blamed the other for project problems. In the past, many large private corporations and essentially all government agencies had large engineering staffs. They represented a huge part of the company's overhead costs. Today many owners have downsized their internal engineering resources, and retained only enough staff to supervise the outsourcing of functions beyond their core competencies. Design-build provides the means for companies to manage their construction needs without a large internal staff, and still meet their needs for quality and timeliness. There is more than one form of design-build contract in common use, with reimbursable and lump sum being the most common. A reimbursable engineering contract pays the designer for the hours spent plus a fee. Lump sum design-build contracts require that the scope of the project be well defined early in the process. Other design-build contracts are negotiated on a **guaranteed maximum price (GMP)** basis. Design-build is well suited to today's business philosophy of reducing overhead by outsourcing what detracts from the company's primary business endeavor.

Design-build also enables construction expertise to play a role during design, and gives the designer an opportunity to be involved during construction. A recent study of construction owner desires showed that owners want better quality, less cost, and less hassle. Design-build, properly implemented, addresses all of these desires. The DBIA is an advocate of the design-build process and is not inclined to list any disadvantages of the system, such as perceived loss of checks and balances between the architect and the general contractor.

Usually, in a design-build contract, the contractor is the dominant partner. Designers normally do not have the financial strength to obtain payment and performance bonds, or liability insurance in the magnitude demanded by the owner. The contractors take over responsibility for those requirements, and thereby assume overall responsibility for the contract. One possible disadvantage of this relationship is that a dominant contractor will overrule the designer's recommendations of quality in the interest of cost.

guaranteed maximum price (GMP)
Type of contract generally used by construction managers, who are selected during or before the design process. The GMP is used because the incomplete design does not permit detailed estimating.

Design-Build-Operate

Design-build-operate projects are not as common as design-build projects. Examples of these projects are athletic arenas, water treatment plants, water purification facilities, and toll roads and bridges. The principle is that the contractor will retain some percentage of ownership in the facility, up to 100%, for a specified period of time, and operate the facility during that time to recoup the capital investment (total cost plus profit). The period of ownership by the contractor may vary from a few years to permanent. During the period of ownership, the contractor is responsible for all costs of ownership, and all profits resulting from ownership. In the case of athletic arenas, there is typically a revenue- or profit-sharing agreement with the team ownership or municipality.

Construction Management at Risk

Construction management at risk, or CM @ risk, is gaining in popularity, especially in the construction of schools and some large projects such as sports arenas, because it enables an owner to bring a construction manager (CM) with construction knowledge onto the project team early in the design process. The CM can then provide construction expertise to the project team in areas such as construction means and methods, material, and equipment selection. The CM works with the designer during the design phase of the project and acts as the general contractor during the construction phase. The purpose of CM @ risk is

construction management at risk
Construction manager who also acts as the general contractor, with a guaranteed maximum price.

to reduce the risk of cost overrun and schedule creep, and to expedite the construction process without compromising quality. The owner solicits proposals from a select group of CM firms and checks their work history, the qualifications of their staff, their system approaches, and experience with similar projects and proposed management structure. The owner selects the CM based on qualifications, references, and perceived "best value."

The CM becomes an integral part of the project team, providing advice and counsel to the owner. As each phase of the design is completed, the CM solicits preliminary and final bids from specialty contractors and performs these services:

- Reviews the design for buildability, including cost, time, quality, safety, and regional market trends impacting the work
- Prequalifies subcontractors
- Divides the scope of work for bidding purposes
- Recommends award of the subcontracts

The CM then prepares the final cost estimate for the total project including work that has not yet been bid. Based on the above information and the predefined scope of work, the CM submits a final guaranteed maximum price, including indirect costs, bonds, bid and unbid costs, and contingency. Unused contingency funds at the end of the job are either returned to the owner or split between the owner and the CM, depending on the terms of the contract. The CM manages the construction as a general contractor, assuming all the responsibilities for cost, time, safety, and quality. The process is subject to audit by the owner throughout. The GMP can be increased only if the architect or the owner changes the scope of work.

There are some advantages to the owner from using the CM @ risk approach:

- Risk is reduced for the architect and owner.
- Produces a more manageable and predictable project cost and schedule outcomes
- Centralizes responsibilities
- The owner benefits from the CM's experience during both design and construction.
- Allows for a possible early start to construction by phasing the work
- Results in better-quality construction because the selection of the CM is based on record of performance in the same type work

The disadvantages that have been experienced in the CM @ risk process include

- The CM is at risk of having to pay any costs that exceed the GMP but that do not result from owner scope changes.
- Disputes can arise regarding what was implied but was not in the contract documents at the time the CM submitted the GMP. As a result, CMs have learned to bid conservatively.

- The CM must rely on its own estimate because of the incomplete design documents.
- CM @ risk is not as suitable for small projects.
- CM @ risk may diminish the role of the architect and engineer (A/E) on the design team. This could be problematic if the A/E does not approve.

CM @ risk does create a collaborative and nonadversarial environment that uses the wisdom, experience, and creativity of the architect and the CM. The CM has the opportunity to review the design as it progresses and to offer suggestions based on his or her experience and expertise. The procedure is more interactive with all key project players than the design-bid-build method.

DESIGNER SELECTION

In the design-bid-build project delivery method, the project owner would select an architect or engineer to design the project, and then advertise the project for bids after the design documents have been completed. The owner should select the A/E based on demonstrated ability to design the project, in the time frame available, and at a cost acceptable to the owner. Designers are usually selected using a process known as the qualifications-based system (QBS). Selection by design competition is used more rarely because the competition represents an added expense to the designers. Often owners will provide a fee for competing firms to help defray their costs, but owners perceive that practice as adding to their project cost.

The A/Es will design the project in accordance with the applicable **building codes**. If the owner is a government agency, there is a legal limit on the amount of the design fee, based on a percentage of the programmed amount of the project. As soon as the designer and owner have agreed on the terms of the design contract, the designer will begin work. Designers do not, and should not, begin work until they have a contract.

building codes
Community or regional codes to establish minimum standards of design and construction.

DESIGN PHASE

The primary requirement for any facility is that it must be safe, and to ensure facilities are safe, local jurisdictions have instituted building codes. These codes are enforceable by law and reflect the best experience of the model code organizations that wrote the code. Cities either use model codes or pattern their own local code after one of the model codes. Cities also require that plans and specifications be submitted along with an application for a building permit so they can ensure the plans meet the code requirements.

There were three major organizations in the United States that developed model codes, but there has been a consolidation of services, products, and operations. The new single-service organization is the **International Code Council** (ICC), which now reproduces the International Building Code (IBC). The ICC

International Code Council
The ICC was established in 1994 as a nonprofit organization dedicated to developing a single set of comprehensive and coordinated national model construction codes.

is composed of the old Building Officials and Code Administrators (BOCA), the International Conference of Building Officials (ICBO), and the Southern Building Code Congress International (SBCCI). More than 97% of U.S. cities, counties, and states that adopt model codes choose building and fire codes created by the three building safety groups that make up the ICC. There are many other codes directed at specific work such as the National Electrical Code, Uniform Mechanical Code, Uniform Plumbing Code, and Life Safety Code.

As the design progresses, the owner and A/E will schedule design reviews. The purpose of these reviews is to ensure the A/E's design meets the owner's expectations. Designers refer to the design phases as schematic design, design development, and construction documents. The purpose of the three-phase system is to ensure that the owner and designers have a mutually understood scope and are progressing together. They do not want to discover late in the design process that they have differences of understanding regarding the scope of the project. Correction of their misunderstandings can cause extra expense and lost time.

The owner has a responsibility to ensure the drawings are actually reviewed by all affected elements of the owner's staff. It is key for the owner's operations and maintenance staff to be aware of the systems they are going to have to support in the facility. Many companies have reduced their engineering staffs in an effort to reduce overhead costs and outsourced their engineering work. This leaner organization may save overhead costs, but it makes coordination during the design phase more difficult because the engineers are more geographically diversified and are not devoted exclusively to the owner's project. As a result, design review coordination may be difficult. However, comprehensive design reviews are necessary, and management controls should be in place to ensure the company's staff properly completes the reviews.

BID PHASE

contractor
An individual or firm undertaking the execution of construction work under the terms of a contract.

The bid phase is discussed here from the point of view of the **contractor**. At this point, it is assumed that the owner is fully committed to this project, has the necessary funding, has the completed design, and is using the design-bid-build project delivery method.

The first step for a contractor is to decide whether or not to bid the job. Contractors are generally limited in their ability to bid by three factors: (1) their bonding capacity, (2) the policies of management, and (3) internal resources. A contractor's backlog of work will diminish over time unless new work is obtained. Backlog relates to bonding capacity because the **bonding company** (surety) may not be willing to provide additional payment and performance bonds to a company with a large backlog. Bonding capacity is based on the company's financial stability, current backlog, management capability, and experience. If the company is too close to the upper limit of its bonding capacity, the surety will not write the necessary bonds for the contractor and the contractor will have to pass on bidding the job. Total volume, maximum job size, type of construction, and geographic location of the work define capacity.

bonding company
Provides bid, performance, and payment bonds to contractors. See Surety.

The second factor companies consider in making the bid/no bid decision is the guidance of upper management. Factors contractors consider in deciding whether or not to bid a particular project include

- Location of the work
- Identity of the owner
- Availability of key company personnel
- Experience in the type of work solicited
- Whether or not there is financing for the project
- Size of the project

Companies want to bid work in locations where they normally work because they have established relationships with subcontractors and the providers of materials and labor. They want to work for owners they know so they will not have to learn the nuances of unfamiliar specifications. They cannot bid work without the necessary staff to manage the work. Obviously companies want to bid the kind of work they know how to accomplish successfully. Last, they do not want to attempt projects that are drastically larger in scope than their previous building experiences dictate. If they violate one of the criteria of location, owner, key personnel availability, kind of work, or project size, the contractor's risk is increased.

If a construction company decides to bid the project, the process of estimating its own work and collecting quotations from subcontractors and material suppliers begins. Bid preparation is expensive and companies do not make a bid/no bid decision casually. The principles for preparing an estimate, described in Chapters 5 through 7, govern preparation of the bid. A general contractor will collect quotations from subcontractors and materials suppliers, and estimate the costs of the portion of the work it will build itself. The elements that make up the final amount the company will bid are shown in Table 3.1. In preparing a bid, contractors must consider the costs of equipment, labor, materials,

TABLE 3.1 Elements of a contractor's bid.

Cost category	
Direct costs	$ Materials
	$ Labor
	$ Equipment
	$ Subcontractors
	$ Job overhead
Indirect costs	$ Corporate overhead
	$ Contingency
	$ Profit
Bid amount	Sum of all direct and indirect costs

subcontractors, job and company overhead, contingency, and profit. They should also consider the number of competitor bidders and the bidding history of those competitors on similar projects. Any contractor can bid to get the job, but successful contractors bid to earn a profit.

New project engineers must remember that bids are *never* accepted late for public competitive bids. Late bids (1 second late is still late) are returned to the bidder unopened. An exception occurs in the case of owners who accept bids by mail. In that case, bids must be postmarked and time stamped by the post office before the time of bid opening. A few recent court cases have forced owners to accept late bids when the lateness was caused by a factor beyond the bidder's control, but it is seldom worth the time and expense to force an owner to accept a bid by going to court.

AWARD PHASE

The award phase consists of two kinds of activities, those that are owner requirements and those that the successful bidder needs to accomplish to prepare for the work. During the award phase, the owner will require the submission of properly executed *payment and performance bonds* and evidence of *worker's compensation and liability insurance.* The owner normally purchases the *builder's risk insurance.* Some owners require that the general contractor submit a *list of subcontractors,* either at bid opening time or within a designated number of days after the bid opening. The requirement to submit a subcontractor list is designed to prevent bid shopping and bid peddling, and to screen the list for subcontractors with unacceptable performance histories. A detailed *project schedule* is sometimes required, normally during the award phase or immediately after contract award. The project schedule will become the basis for control of the project, and for progress payments to the contractor.

Contractors use the time between bid opening and contract award for detailed preproject planning. This is the time to carefully plan as many of the job details as possible. Preplanning the job means planning the details of how work will proceed and in what sequence. Decisions are made on matters such as construction procedures, type of equipment to be used, job access, location of the field office, storage areas, and final selection of subcontractors and suppliers. A **cash flow analysis** should be completed to determine how much cash is needed and on what schedule. A detailed project schedule is prepared, usually using commercial scheduling software, and the schedule must be submitted to the owner for approval. The work breakdown and pay schedule are also planned.

The award phase does not end until both parties have signed the construction contract and the owner issues the Notice to Proceed. The contractor cannot begin the work until the **Notice to Proceed** is received. Notice to Proceed documents the contractor's legal access to the site and establishes the official start date of the work. After the construction contract is executed, the general contractor enters into subcontracts with subcontractors and suppliers. These sub-

cash flow analysis
Process for analyzing a project to determine the amount of working capital that will be required each month.

Notice to Proceed
Document issued to the contractor after award of the contract. Formally authorizes access to the site and establishes the project start date.

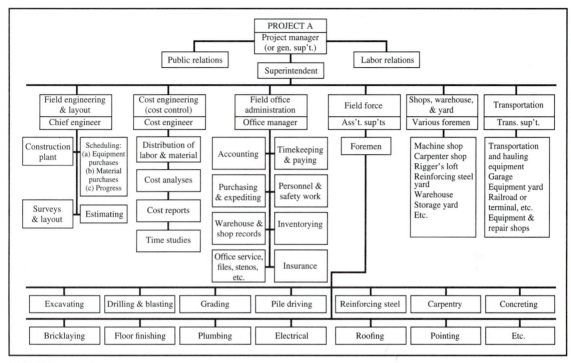

FIGURE 3.2 Construction project management organization for a large project.

contracts should incorporate the same terms as the contract between the owner and the general contractor.

The owner has an obligation to provide the contractor with plans and specifications that describe the project in such detail that completion of the project in accordance with the plans and specifications will produce a project that meets the owner's needs and expectations.

CONSTRUCTION PHASE

The size of the contractor's on-site project management organization is a function of the size and complexity of the project. A small, uncomplicated project, such as a sewer line installation or a small bridge repair project with a very limited number of subcontractors, may have only a project superintendent and one **foreman**. A large project, such as a multistory office building or a large shopping mall, may have a large management team consisting of a project manager, assistant project managers, field engineer, several **project engineers**, scheduler, materials expediter, estimator, and clerical staff. In addition, the contractor will employ subcontractors, material fabricators, and suppliers who have their own management teams. Figure 3.2 depicts a construction project management organization for a large project.

foreman
Supervisor of a specific trade group of workers.

project engineer
Entry-level contract management position at the site. Many times the first assignment for a new graduate is as a project engineer.

REQUEST FOR
INFORMATION/INTERPRETATION

PROJECT: The Big Building   **RFI NO.**   67

               Downtown, Anywhere, USA   **FROM:**   The Low Bid Contractor

TO: The Engineer   **DATE:**   09/10/03

               Suite 400, Engineer Ave., Downtown   **A/E PROJECT NUMBER:** 00101

               Telephone: 504-6300   **Fax:** 504-6331

SUBJECT: Elevators 11 & 12

IMPORTANCE: ☐ High   ☐ Low   ■ Normal   ☐ Urgent

Specification Section: Paragraph: Drawing Reference: A403D   Detail: 16/S505

Request:

Please Clarify:
On Sheet A403D the clear wall-to-wall dimension is called out as 23'4"for elevators 11 & 12.
On detail 16/S505 it is called out as 22'6". Which dimension is correct. This is an urgent matter.

---

*Request Date/Time for Response: _____

(The undersigned acknowledges review of Section 01044 in its entirety.)

Signed by: Mr. Lots of Questions

Response: 23'4" is the correct dimension. Change detail 16/S505 to reflect this directive.

☐ Attachments

Drawing Impact: ☐ Yes   ☐ No

Response From: Mr. Answer   To: Low Bid   *Date Rec'd: 9/11/03   *Date Ret'd: 9/11/03

Signed by: Mr. Answer

Copies: ☐ Owner   ☐ Consultants   ☐   ☐   ☐   ☑ File

*Contractor shall allow up to 7 calendar days review and response time for RFIs. (See section 01044.)

FIGURE 3.3 Project request for information example.

The job of the construction project team is to establish the means by which the team will construct the facility in accordance with the terms of the contract and any **change orders;** order the right materials and equipment; communicate clearly with the owner, and all materials suppliers and subcontractors; maintain a safe construction site; provide timely decisions; coordinate the activities of the subcontractors; accomplish the job administrative requirements such as requests for information (RFIs) (see Fig. 3.3), **shop drawings**, and change orders; prepare the monthly pay requisitions; and accept responsibility for the performance of the contract. Above all else, the mission is to complete a high-quality project, on time, safely, at a profit.

Requests for information are used by the contractor to document guidance from the architect, engineer, or project manager regarding issues that may not be clear in the construction documents—usually on the drawings or in the specifications. The contractor submits the request for information in written form and receives a written response. Responses to RFIs normally are incorporated in the **as-built drawings**. The as-built drawings are completed by the contractor as the work progresses, and when submitted to the owner at the completion of the work, they serve as a permanent record of the as-built conditions and dimensions.

Shop drawings of the materials to be installed are reviewed by the project designer. For mechanical equipment, this may require review of the manufacturer's make and model. For items such as structural steel, the engineer designs the beams and the columns, but the fabricator is responsible for the connections. These connections require the review of the structural engineer.

One of the first job assignments of an engineering or construction management graduate on a construction project will most likely be as a project engineer, who is assigned responsibility for such documents as daily reports, progress reports, concrete quantity calculations, RFIs, and shop drawings. New project engineers will spend most of their time and effort learning how a facility is built. The new engineer must learn the sequence of work, the working relationships on the site, the capabilities of the tools and equipment, the methods of controlling schedule and cost, the flow of information, and the functions of subcontractors. Often recent graduates bring new skills, such as computer expertise, that the rest of the team needs. Such skills offer an opportunity for a positive contribution to the project management team. Large projects may have several project engineers, each with specific responsibilities.

Construction Company Team Functions

The **project manager** (PM) is in overall charge of the project, responsible for the rate of progress, financial control, safety, and ultimate profitability of the job. The PM is the **superintendent's** supervisor and is the person responsible for ensuring the superintendent has the materials, supplies, equipment, labor, and subcontractors available when they are needed. Project managers select

change orders
Directives issued by the owner to change the contract by adding or subtracting features within the scope of work.

shop drawings
Drawings that supplement the contract drawings to identify fabrication details, makes and models, colors, and other details as required by the contract.

as-built drawings
Completed by the contractor as the work progresses, and submitted to the owner at the completion of the work as a permanent record of the as-built conditions and dimensions.

project manager
Senior construction company manager at the site, in overall charge of the project.

superintendent
Person in overall charge of the day-to-day detailed management of the construction process. Supervises the foremen.

equipment needed on the site, they negotiate the subcontracts, and they submit the progress and financial reports to the company; they are responsible for all communications with the owner.

Superintendents often come up through the trades and have many years of experience. Their primary function is to coordinate the fieldwork and supervise the trade foremen. Their greatest strength is their complete knowledge of construction practice and the requirements of the project. They do not prepare shop drawings, but they need to know the status of all critical shop drawings so they know when materials and subcontractors can be expected on site. They submit weekly cost reports to be used in the control of the job costs. They work with the project schedulers every day. They inform the project manager when a change order is required and what the change order must accomplish. Superintendents bring experience to the project. They are the planner of next week's activities, the referee of subcontractor conflicts, and the controller of the site layout. At the working level, they are in charge. They set the standard for performance.

Large projects also may have schedulers, estimators, and materials expediters. These are specialized jobs that require experience and project knowledge. Scheduling and estimating are also home office functions, which take place during the bidding and award stages of the work, but on large projects they may be needed full-time on site.

Owner's Project Team

The owner will be represented on the site, and, like the contractor, the size of the owner's project team will depend on the size and complexity of the project. The type of contract will also impact the size of the owner's team. A **unit-price contract**, such as is used for highways, would require a larger team because it is necessary to measure the quantities of all work items. For example, the borrow pit may be surveyed before and after hauling material to calculate the quantity of material removed. A **lump sum contract**, such as would be used for the construction of a building, has no such requirement.

For a small building project, the architect might visit the site and consult with the contractor weekly. The architect would report progress to the owner, review shop drawings, negotiate change orders, respond to RFIs, and approve progress payments. At the other end of the spectrum, on a large highway project, a state department of transportation might assign a complete team consisting of a resident engineer, inspectors, surveyors, and quality assurance technicians. Highway projects are unit-price contracts, which have a bid price and estimated quantity for each bid item, and therefore require a larger owner's presence in the field because the installed quantities must be measured and counted. The **resident engineer** (RE) is in overall charge of the owner's field team, and must be a registered professional engineer. The RE will have some authority to make engineering decisions, stop the work in event of unsafe conditions, order and approve change orders, approve progress payments, negotiate

---

unit-price contract
Contract used for heavy/highway projects that do not allow the contractor or the owner to know the exact quantities of materials before the start of work. Bidding is based on the engineer's estimated quantities.

---

lump sum contract
Type of contract commonly used for buildings where the contractor can accurately estimate the costs.

---

resident engineer
Senior representative of the owner, usually on public works projects.

change orders, respond to RFIs, and report to the owner. **Inspectors** are the eyes and ears of the resident engineer. They have no authority to make changes or interpret the contract. They measure quantities of materials, and observe and report the quality of work. Some public agency owners have in-house surveyors to provide elevations and boundary markers, but this requirement is increasingly included in the construction contract. Surveyors perform initial and final profiles to calculate soil or rock volume. They also provide the survey data for the final as-built drawings. Quality assurance technicians perform tests on materials such as soils, asphalt, and concrete. In some cases, the technicians take the materials samples. In other cases, the inspectors take the material samples for the technicians. Large projects may have a materials testing laboratory established in the field, but for most projects, the samples are sent to a commercial laboratory.

inspector
Employee of the owner whose function is to inspect the work as it progresses, calculate the quantity of materials used, and assess the quality of the work.

Managing Critical Activities

Projects are broken down into activities for purposes of scheduling, estimating, progress control, and cost control. Large projects can have several hundred activities. The project manager cannot possibly manage all of them, so the knowledgeable manager knows which activities are on the **critical path** and which activities have the most potential impact on profit. Items on the critical path are critical-time activities and have the potential to delay the project on a day-by-day basis. These activities must be intensively managed.

critical path
Path of activities on a critical path diagram that defines the duration of the project. Critical path activities have no float.

There are also **critical-cost activities**, which can be identified by their potential impact on cost and profit. The cost-critical activities that the project manager must watch are those that could impact the cost of the work by at least 0.5% of the bid price. For example, on a $1,000,000 project, any activity with a potential for cost overrun or underrun of $5,000 or more is by definition a cost-critical activity. Usually about 20% of all activities are cost critical. The project manager should give special attention to those activities. The remainder of the activities should be managed by the project management staff, and should come to the project manager's personal attention only on an exception basis. Some activities will exceed their estimated cost, and some will be under their estimated cost, but the net impact should be minimal. This is a practical application of Pareto's 80–20 rule (Vilfredo Pareto, 1848–1923), which states that 20% of the activities are critical and should be managed carefully. The other 80% will average out.

critical-cost activity
Any single activity that has the potential to impact the estimated cost of the project by more than 1/2 of 1%.

Many people have difficulty delegating work. Often that difficulty comes from a sense of responsibility for everything (a good thing) combined with a lack of ability to know what to delegate (a bad thing). Pareto's law can help. It suggests that the PM has the tools to determine which activities to delegate to the staff, and thereby to create time for leadership and overall management. Keep the structural steel, mechanical systems, and electrical systems. Delegate the kitchen cabinets and plumbing fixtures.

PROJECT COST AND SCHEDULE CONTROL

The first project control principle is that watching and controlling are not the same. Watching implies observing events and taking notes. Controlling requires that the PM team establish some standards of quality, minimum acceptable rates of production, rates of planned expenditures, or other required metrics. If the metric standards are not met, corrective action is taken. Possible corrective actions could include

- Adding additional trades workers or crews
- Adding or removing equipment
- Working overtime
- Bringing in additional subcontractors
- Making the job more efficient
- Eliminating factors that cause subcontractors to interfere with one another's operations or otherwise create inefficiencies

productivity
A measure of the progress of a crew or contractor against a standard such as the project schedule.

The most important job factor the project team must control is **productivity**. There is no more important factor in the equation of job cost that the project team can impact than productivity. The cost of labor, equipment (rental or ownership), and materials is essentially fixed and cannot be changed significantly by the project management team. Productivity is different and must be managed. The effect of production on activity time is given by Eq. [3.1].

$$T = \frac{Q}{R} \qquad [3.1]$$

where

T = total time
Q = the total quantity to be installed
R = production rate

The total cost is determined by the equation:

$$C_t = C_h \times T \qquad [3.2]$$

where C_t = total cost and C_h = cost per hour, substituting

$$C_t = C_h \times \frac{Q}{R} \qquad [3.3]$$

Therefore, if R is increased, the total cost, C_t, is reduced. If R is decreased, C_t is increased. Consider that the cost per hour is $1 and production is varied for producing 100 units:

$$C_0 = \$1 \text{ per hr} \times \frac{100 \text{ units}}{1 \text{ units/hr}} = \$100$$

$$C_1 = \$1 \text{ per hr} \times \frac{100 \text{ units}}{10 \text{ units/hr}} = \$10$$

$$C_2 = \$1 \text{ per hr} \times \frac{100 \text{ units}}{20 \text{ units/hr}} = \$5$$

$$C_3 = \$1 \text{ per hr} \times \frac{100 \text{ units}}{30 \text{ units/hr}} = \$3$$

The relationship between C_t and R is not a straight line. The relationship is inverse exponential, as shown in Fig. 3.4. As the curve approaches the vertical and horizontal axes, it is asymptotic (approaches zero at an infinite distance from the origin).

EXAMPLE 3.1

Let the total cost of an activity be C_i. That cost can be estimated by the expression

$$C_i = C_h \times \frac{Q_i}{R}$$

where C_i is the estimated cost of activity i, and Q_i is the quantity of work units of activity i. If total daily quantity $Q_i = 100{,}000$ and productivity $R_i = 100$ per hr, and the hourly cost $C_h = \$150$ per hr,

$$C_i = \$150 \text{ per hr} \times \frac{100{,}000 \text{ units}}{100 \text{ units/hr}} = \$150{,}000$$

If production is reduced to half the previous rate, to 50 units per hr, the cost will double:

$$C_i = \$150 \text{ per hr} \times \frac{100{,}000 \text{ units}}{50 \text{ units/hr}} = \$300{,}000$$

If the productivity is reduced to half again, the cost will double again:

$$C_i = \$150 \text{ per hr} \times \frac{100{,}000 \text{ units}}{25 \text{ units/hr}} = \$600{,}000$$

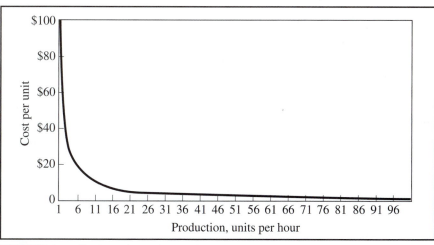

FIGURE 3.4 Total cost/productivity relationship.

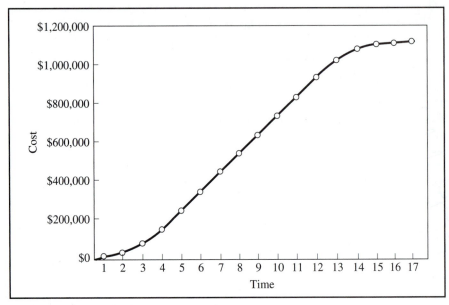

FIGURE 3.5 Budgeted cost of work schedule curve.

Cost control requires monitoring and control of productivity, on an hourly, daily, and weekly basis. The control of production rates is of paramount importance to controlling job costs.

Cash Flow Analysis

An S-curve, such as is shown in Fig. 3.5, can be used to predict project cash flow requirements. It is a picture of how the project budget will be spent if the work is constructed exactly as estimated and scheduled. Cash flow and costs incurred are different as will be discussed in this section. This curve is developed by cumulatively adding the costs from the **work breakdown structure** submitted to the owner according to the planned completion time on the original master project. It is known as the **budgeted cost of work schedule** (BCWS). Costs will usually start slowly because the contractor is using the early days of the project to mobilize and organize. Project costs tend to slow down as the project nears completion when all that remains is finishing the last remaining tasks and making the punch list corrections. During the middle portion of the work, a project will accumulate costs at a more or less steady rate, thereby producing the S-shape observed in Fig. 3.5.

The S-curve can be used to predict the monthly cash flow requirements for the contractor and the monthly payment requirements for the owner. Contractors should anticipate that the owner will plot the contractor's projected monthly earnings from the critical path method (CPM) schedule onto a curve such as is depicted in Fig. 3.5. An excessively front-loaded schedule will look more like Fig. 3.6 and will be rejected. This does not mean that permissible levels of front

work breakdown structure
A hierarchical breakdown of the scope of work for a project.

budgeted cost of work schedule
A graph showing how the project is budgeted to be accomplished according to the original master schedule.

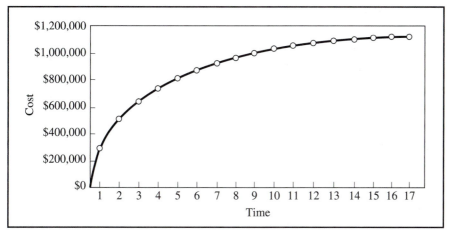

FIGURE 3.6 Front-loaded cost of work schedule curve.

loading (inflating unit prices of items completed earlier in the project) are discouraged; indeed they are encouraged.

Measuring Project Progress   The BCWS curve is a plot of the estimated cost
assuming that the work is performed at its scheduled point in time; on the same
diagram, a curve of **actual costs of work performed** (ACWP) can be plotted to
provide an assessment of project progress (see Fig. 3.7). A third curve

**actual costs of work
performed**
*A graph showing the
real costs of work
accomplished plotted
against project time.*

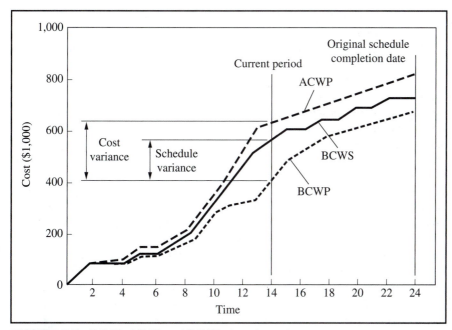

FIGURE 3.7 BCWS, ACWP, and BCWP plotted together.

**budgeted cost of
work performed**

*A graph showing the
planned (estimated)
cost of the work
allocated to completed
activities.*

representing the **budgeted cost of work performed** (BCWP) can also be plotted on the diagram. The BCWP is computed from the measured accomplished work and the budgeted costs for the work. By extrapolating from the curves, the estimated cost to complete the project and the probable completion date can be obtained.

The difference at any point in time between the budgeted cost of work scheduled and budgeted cost of work performed is the schedule variance of the project, Eq. [3.4] (see Fig. 3.7). The project's cost variance is the difference between the budgeted cost of work scheduled and actual cost of work performed, Eq. [3.5] (see Fig. 3.7).

$$\text{Schedule variance} = \text{BCWP} - \text{BCWS} \qquad \textbf{[3.4]}$$

$$\text{Cost variance} = \text{BCWP} - \text{ACWP} \qquad \textbf{[3.5]}$$

These two variances assist in evaluating and controlling project risk by measuring progress in monetary terms. This method of tracking and analyzing construction work is termed the **earned value approach**. Earned value is discussed in greater detail in Chapter 9.

Front loading is achieved by assigning higher costs to early activities, and lower costs to late activities, but keeping the total project cost unchanged. Many people consider front loading unethical. Contractors consider it essential if they are going to manage their cash flow effectively. It is necessary for cash flow and their least expensive source of borrowed money. Subcontractors, particularly, find it necessary to overbill as they must pay their labor and materials bills prior to even submitting their invoices, and then wait 60 days or more after that to receive their progress billing.

Cash flow analysis is a function that is done by both the owner and the contractor separately. The owner will be required to make regular progress payments to the contractor. The timing of those payments is a matter of contractual agreement, or an agreement negotiated at the preconstruction conference. In some states, it is a matter of law. Either way, the due date is an obligation and the owner should never be late. For that reason, the owner needs to plan the costs of the progress payments before the project starts and ensure the funding is available to make the payments.

The contractor has a slightly different reason for making a cash flow analysis. The contractor will need working capital during the construction, and must know how much cash flow is needed before the project starts. Halfway through the project is not the proper time for the contractor to learn the magnitude of the working capital requirement. By a cash flow analysis made before the work starts, the amount of working capital needed each month is determined. The cash flow analysis is a key management tool for the contractor because it is the score card against which profitability is measured during the project.

Example 3.2 is provided to illustrate the principles of cash flow analysis, as done by the contractor. The owner would use the same principles.

**earned value
approach**

*Analysis of the
contract-completed
activities to determine
the cost to date and
compare with the
revenue to date. The
difference is the earned
value.*

front loading

*Practice of assigning
high values to
activities that are
completed early, and
compensated for by
assigning low values
to activities that are
completed late in the
project. The result is
higher initial monthly
pay estimates.*

EXAMPLE 3.2

Assume the contract has a 6-month duration. The cost of borrowing money is 0.75% per month. Monthly pay estimates will be submitted on the fifth of the month, for work completed as of the last day of the preceding month, with payment anticipated 30 days later. Under the terms of the contract, the owner will retain 10% of the first 50% of the contract amount. Estimated cost is $500,000. Using a 10% markup, the contractor bid $550,000. Expected monthly costs are as shown:

Month	1	2	3	4	5	6
Estimated Cost	$50,000	$100,000	$100,000	$100,000	$100,000	$50,000

The working capital requirements are

Month 1.   Costs = $50,000
         Interest = $50,000 × 0.0075 = $375
         Cash requirement = $50,000 + $375 = $50,375
Month 2.   Costs = $100,000
         Invoice 1 = $55,000 − $5,500 = $49,500
         Interest = ($50,375 + $100,000) × 0.0075 = $1,128
         Cash requirement = $50,375 + $100,000 + $1,128 = $151,503
Month 3.   Costs = $100,000
         Invoice 2 = $110,000 − $11,000 = $99,000
         Cash received from owner = $49,500
         Interest = ($151,503 + $100,000 − $49,500) × 0.0075 = $1,515
         Cash requirement = $151,503 + $100,000 − $49,500 + $1,515
         = $203,518
Month 4.   Costs = $100,000
         Invoice 3 = $110,000 − $11,000 = $99,000
         Cash received from owner = $99,000
         Interest = ($203,518 + $100,000 − $99,000) × 0.0075 = $1,534
         Cash requirement = $203,518 + $100,000 − $99,000 + $1,534
         = $206,052
Month 5.   Costs = $100,000
         Invoice 4 = $110,000
         Cash received from owner = $99,000
         Interest = ($206,052 + $100,000 − $99,000) × 0.0075 = $1,553
         Cash required = $206,052 + $100,000 − $99,000 + $1,553
         = $208,605
Month 6.   Costs = $50,000
         Invoice 5 = $110,000
         Cash received from the owner = $110,000
         Interest = ($208,605 + $50,000 − $110,000) × 0.0075 = $1,115
         Cash requirement = $208,605 + $50,000 − $110,000 + $1,115
         = $149,719

Month 7.   Costs = $0.00
Invoice 6 = $82,500 (the unbilled balance of the bid amount)
Cash received from owner = $110,000
Interest = ($149,719 − $110,000) × 0.0075 = $298
Cash requirement = $149,719 − $110,000 + $298 = $40,017

Month 8.   Costs = $0.00
Cash received from owner = $82,500
Project profit = $82,500 − $40,017 = $42,483

These calculations are shown in Fig. 3.8. Note that the maximum working capital requirement, which occurs at the end of month 5, is $208,605. To perform the contract, the contractor must have at least that amount available. Change orders will increase the working capital required. Also note that the estimated costs were $500,000 and the bid was $550,000 for an estimated profit of $50,000, but the cost of borrowed money reduced the profit to $42,483. That profit reduction is real whether the company borrows the money or uses its own assets.

A schematic of the cash flow requirements is shown in Fig. 3.9.

profit (loss) to date
Calculation of the profit to date based on the progress and expenditures to date, compared to the expected progress and expenditures to date.

The cash flow analysis shown in Example 3.2 should be accomplished before the project begins so the construction contractor will know there is sufficient working capital available. During the project, the project manager must calculate **profit (loss) to date** on a regular, weekly basis. The four items that make up the financial statement and come from the project manager are

■ Cost to date

■ Reestimated cost to complete

Project Cash Flow Calculation
Estimated cost $500,000
Bid price $550,000
Cost of money = 0.0075
Invoice date = 5th of the month
Payment received 30 days later
Retainage = 10% of the first 50% (of the contract amount)

Month	1	2	3	4	5	6	7	8	Total
Est. cost	50,000	100,000	100,000	100,000	100,000	50,000			500,000
Invoice amt.		49,500	99,000	99,000	110,000	110,000	82,500		550,000
Amt. received	0	0	49,500	99,000	99,000	110,000	110,000	82,500	550,000
Cash req'd.	50,000	150,375	202,003	204,518	207,052	148,605	39,719		
Interest	375	1,128	1,515	1,534	1,553	1,115	298		7,517
Total	50,375	151,503	203,518	206,052	208,605	149,719	40,017	0	
Profit									42,483

FIGURE 3.8   Excel spreadsheet of cash flow requirements (Example 3.2).

- Amount billed
- Contract amount (once the job has started and there are change orders)

Therefore, it is not the accounting department for whom we have to wait, but instead the project manager, who has all of the information in the glove box and above the visor of the pickup truck.

It does not make sense to wait for the accounting department to figure out that the project lost money 3 months after the project ends, when all the bills have been paid, and there is no opportunity to change construction methods. Efficient construction companies have weekly job cost reports and monthly work-in-progress reports that are invaluable to the management of the project.

Most project managers calculate percentage complete and profit to date by the *cost to cost method.* Using this method, the percentage complete is calculated as the ratio of *total cost to date* (including indirect costs) to *estimated total cost.* There is some argument about counting the costs of materials that have been

Month	1	2	3	4	5	6	7	8
Cost	$50,000	$100,000	$100,000	$100,000	$100,000	$50,000		
Invoice amount		49,500	99,000	99,000	110,000	110,000	$82,500	
Amount paid	0	0	49,500	99,000	99,000	110,000	110,000	$82,500
Working capital required	50,375	151,503	203,518	206,052	208,605	149,719	40,117	

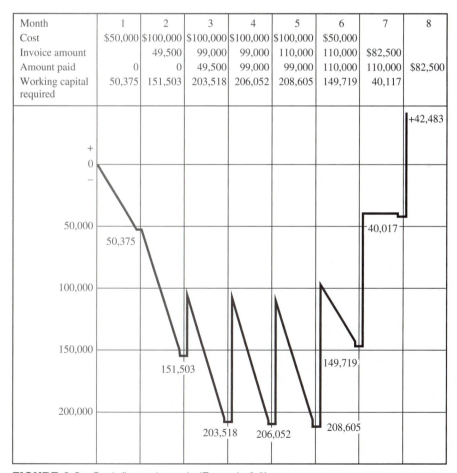

FIGURE 3.9 Cash flow schematic (Example 3.2).

paid for and are on the site but not yet incorporated in the facility. Most contractors would argue that to not count uninstalled materials leads to a low progress estimate.

EXAMPLE 3.3

Given these data, calculate the percentage complete and profit earned to date:

Original estimated total cost	$500,000
Cost to date, including indirect costs	$ 75,000
Current reestimated cost to complete	$425,000
Percentage complete	$ 75,000 / $500,000
	= 0.15 or 15%
Bid amount	$550,000
Profit earned to date	($550,000 − $500,000)
	× 0.15 = $7,500

A quick analysis of the cash flow schematic diagram in Fig. 3.9 will confirm that the project cash flow is negative, by around $75,000 at 15% complete, yet we show a profit of $7,500. There is a difference between cash flow and profit. In this case, the cash flow is negative, but the company has made a profit. That profit will not be available to be spent until the cash flow turns positive.

As the project progresses, the *estimated cost to complete* will change, which will change the total estimated direct costs. Some of the costs of activities that have been completed will have been above or below the budget. Some of the future costs will also change. Change orders will change the *cost to complete* and the *percentage of completion.* Progress and profit estimates should always be based on the best available data. When the total estimated cost to complete changes, use the reestimated amount. Above all else, calculate *profit to date* and *reestimated cost to complete* every week.

Schedule Control

Scheduling is addressed in detail in Chapter 4. Schedules should be developed using the simplest method that will satisfy the requirement to control the job. The owner will usually require the contractor to submit a project schedule. Many times the project schedule must be prepared using specific commercial scheduling software. The advantage of using software such as Primavera is that the program automatically identifies the activities on the critical path and allows for resource allocation.

The cost of each activity is estimated when the job estimate is computed. In that way, the project manager has a schedule of both the physical progress and the estimated costs. Progress can be compared to the schedule to determine if it is on time, late, over budget, or under budget. It is not possible to compare financial status and progress without a cost-loaded schedule.

the temporary structure design. The design must be reviewed and stamped by a registered professional engineer and must be submitted to the project A/E for review.

The design of temporary structures is not included in the project design professional's contract, but review of the design is usually part of the A/E's duties. The requirements for design review of temporary structures are eliminated in design-build contracts.

FIGURE 3.11 Temporary shoring to support concrete slab placement.

FIGURE 3.12 Temporary sheet pile cofferdam for bridge foundation work.

There may be cases of design requirements for the permanent facility that must be completed by the contractor's engineer. For example, the steel girders of a highway bridge will "sag" when the concrete deck is constructed. It is the contractor's responsibility to calculate the amount the camber of the steel girders will change (flatten) under the weight of the concrete, because it is the contractor's responsibility to procure the correct amount of concrete and to build the bridge deck to a predetermined finish contour. Failure to properly account for the loss of camber in the girders will result in a bridge deck that is either too flat or too humped. Either way, the profile would be incorrect. The construction company's engineer or the constructor's consulting engineer accomplishes these and other design tasks.

In many ways, the design of temporary structures is more difficult than the design of the permanent structure. Soils and foundation problems are less defined. Additionally, the materials used for temporary structures must be removable. However, the temporary structures must be safe for their intended purpose.

RISK MANAGEMENT

There are many risks inherent in the construction industry. For many years construction contractors were expected to bear all the risks during the construction process. Currently, the industry is moving toward allocating risks to the party most able to deal with the specific risk. Managing risks means minimizing the risks, insuring against the risks, and sharing risks. To a construction contractor, a risk is an event that will cause costs that were not planned and from which no profit will result. Subsurface conditions that are different than anticipated can lead to excavation costs that are greater than expected. A construction company should make provisions to protect its interest if unanticipated conditions are encountered. The loss of a key piece of equipment can cause a contractor additional unplanned costs. Such losses can be covered by insurance.

A partial list of the risks contractors face includes

- Construction-related risks, such as the inability of a subcontractor to perform
- Labor availability and labor productivity
- Strikes
- Economic risks, such as cost escalation
- Political and public risks, such as disapproval of the required project permits
- Physical risks, such as subsurface conditions
- Weather
- Contractual and legal risks, such as risks assigned by contract over which the contractor has no control
- Design risks, such as a project design that is not constructible

■ Safety risks, such as worker injury or an injury to a member of the public

■ Construction vehicle accidents

The concept of risk management is that we want to minimize risks, no matter whose risks they are, and we want an equitable sharing of risks. The question is, who assumes which risks, and how are risks shared? Risks are best assumed by the party with the greatest ability to control the risk. Contractors need a formal process to apply to all projects at the start and throughout the work to identify, quantify, and allocate risks. One cannot be involved in the construction process and avoid all risks.

The primary way to manage risks is by purchasing insurance to cover specific events that would result in a loss if they occur. The loss of a bulldozer to a contractor who owns and operates only one bulldozer would be a risk that would put the contractor out of business. The contractor manages that risk by insuring the bulldozer. The site subsurface conditions might not be exactly what the contractor anticipated. The contractor would protect against that risk by examining the contract language addressing changed conditions. Contractor safety programs are another example of risk management. Companies with high accident rates have a high **experience modifier rating (EMR)**, which increases the cost of **worker's compensation insurance** and therefore makes the contractors less competitive. Good contractors work very hard at reducing the occurrence and severity of accidents.

Subcontracting is also a form of risk management. There is great risk in performing work for which a company is not qualified. A contractor can reduce project risk by hiring qualified subcontractors. The cost may be increased, but the return justifies the cost by reducing the contractor's exposure to risk. The prime contractor can also require performance and payment bonds from subcontractors.

The various kinds of contractor insurance are discussed in Chapter 8.

VALUE ENGINEERING

Value engineering (VE) is also known as function analysis or **value analysis**. Ideally, it is performed as early as possible during the design process to maximize results. As an engineering function, it uses a process of function cost analysis to identify features that can be changed to increase the project's value.

The main objective of value engineering is to increase the project's value without reducing the quality, reliability, performance or other critical factors important in meeting the owner's requirements. VE is a rigorous, systematic effort to improve the value and optimize the life cycle costs of a project. VE generates these cost improvements without sacrificing needed performance levels [11]. According to Dell'Isola [11], three basic elements provide a measure of value to the user: function, quality, and cost. These elements can be interpreted by the following relationship:

experience modifier rating (EMR)
Factor used to calculate a contractor's cost of worker's compensation insurance, based on the accident experience of the company.

worker's compensation insurance
Insurance that pays employee medical bills and a portion of hourly wage in case of job-related injury.

value engineering
Process of evaluating projects to find methods and materials that are less expensive than the specified requirement, but do not decrease the value of the completed project.

value analysis
Used during construction to estimate the contractor earnings to date. Construction activities are assigned a value at the beginning of the project. Completed activities are added up.

$$\text{Value} = (\text{Function} + \text{Quality})/\text{Cost}$$

Where

Function = The specific work the project or design must perform

Quality = The owner's needs, desires, and expectations

Cost = The life cycle cost of the project

Therefore, we can say that:

Value = The most cost-effective way to reliably accomplish a function that will meet the owner's needs, desires, and expectations

Examples of VE cost reduction can be achieved through intelligent siting of the building to take advantage of the existing streets and utilities or the prevailing winds and available solar heat. There may be cost savings in removing unneeded features such as insulation in walls and roofs in unheated spaces. There may be value in simply changing the shape of a sewer pipe from round to oval, based on availability of pipe, or the required depth of excavation may be reduced as a result of the change. Value engineering changes should produce a facility that still performs the same functions as it did before the change was made but has increased value due to the change.

Many construction contracts offer to share the savings of value engineering suggestions submitted by the contractor. Such provisions are beneficial to both the contractor and the owner. The owner's benefit is obviously the cost savings. The contractor has an opportunity to earn money based on the contractor's knowledge and experience. The typical contract offers to share the savings on a 50–50 basis. For a VE proposal worth $250,000, the contractor would earn $125,000 just for suggesting it.

According to Dell'Isola [11], large construction agencies can expect to spend 0.1–0.3% of total project costs for an effective VE program. This investment should result in a minimum of 5–10% savings in initial costs and 5–10% savings in annual maintenance and operations costs. It is important to repeat that VE efforts are most effective when performed as early as possible.

SUMMARY

It is the owner's responsibility to work with the design professionals to define a project that meets the owner's requirements for scope, quality, time, and cost. The planning phase is used to define the project goals, identify the source of financing and the location of the facility, decide what delivery system will be used, and select the designer.

In the bid phase, general contractors decide whether or not to bid on the project. Bids consist of the total cost estimate of the work the general contractor will self-perform, quotes from subcontractors, contingency, overhead, and profit. The award phase is used by the low bidder to prepare for the work by developing a detailed schedule, finalizing contracts with subcontractors, issuing purchase orders to suppliers, and submitting information required by the owner.

The job of the project manager is to complete the project on time, under budget, and safely, so that it meets the quality expectations of the owner. The contractor's project management team will control the sequence of the work; coordinate the activities of the subcontractors; process RFIs, monthly pay estimates, and change orders; and maintain a complete project file. They are responsible for quality on the project, which means they will ensure all materials are properly installed in accordance with the contract.

Project cost control and project schedule control are two essential functions of the project management team. Cost control is achieved by preplanning the cash flow requirements for the project, and then monitoring the cash requirements during the project. Schedule control is analogous to cost control. A detailed project schedule, with costs for each activity, is developed prior to the start of work. Changes resulting from change orders or sequence changes are reflected in the updated schedule as work progresses. Risk should be managed by assigning the risks to the owner, contractor, or subcontractors based on which team partner is best able to control the risks.

REVIEW QUESTIONS

3.1 Give an example of a company or government organization that has a master plan. Contact the owner or agency to determine how they actually develop and update their master plan, and how it impacts their capital budget.

3.2 Whose responsibility is scope definition? What kind of questions must be answered in the process of scope definition?

3.3 What is the difference between a construction mortgage and a permanent mortgage? If a banker would charge 7.5% per year for a mortgage, can the owner discount the cost of financing the project if it is self-financed? Why?

3.4 Why are designers selected using qualifications-based selection (QBS)?

3.5 What is the purpose of model codes? What code is used in your local area? Contact your local municipality to determine if it makes changes to the model code.

3.6 What are the stages at which the designer will schedule design reviews? What is the purpose of the design reviews? Who performs the reviews? Is there any impact from ignoring design reviews? What should a designer do when the owner is clearly incapable, through lack of staff or experience, of completing a competent design review?

3.7 What will the general contractor accomplish during the bid phase? How do contractors decide whether or not to bid a project?

3.8 What will a contractor accomplish during the award phase? What will the owner require during this phase?

3.9 What are the functions of

    a. Project manager

    b. Project superintendent

    c. Project engineer

    d. Foreman

    e. Scheduler

    f. Purchasing agent/expediter

3.10 Why are most buildings bid lump sum contracts, while highway projects are bid unit price?

3.11 What are the goals of partnering? Is partnering common in your area? Try to find a construction company that has worked on a partnered project and determine its impression of partnering.

3.12 Why is there an inverse exponential relationship between productivity and total cost, instead of a straight-line relationship? What implications does this relationship have for project managers and project engineers?

3.13 Explain why a markup of 10% on a $600,000 cost estimate (bid = $660,000) will produce a profit less than $60,000.

3.14 What is the purpose of value engineering? How does VE offer a profit opportunity for contractors?

3.15 Develop a list of contractor risks to supplement the list provided in the reading.

REFERENCES

1. Barrie, Donald S., and Boyd C. Paulson, Jr. (1984). *Professional Construction Management,* 2nd ed., McGraw-Hill, New York.

2. Fisk, Edward R. (2003). *Construction Project Administration,* 7th ed., Prentice Hall, Upper Saddle River, NJ.

3. Gibson, G. Edward, Jr., and Peter R. Dumont (1996). *Project Definition Rating Index (PDRI) for Industrial Projects,* Construction Industry Institute Implementation Resource 113-2. Construction Industry Institute, Austin, TX, July.

4. Gould, Frederick (1997). *Managing the Construction Process,* Prentice Hall, Upper Saddle River, NJ.

5. Griffis, Fletcher H., and John V. Farr (2000). *Construction Planning for Engineers,* McGraw-Hill, New York.

6. *Management of Project Risks and Uncertainties* (October 1989). CII Publication 6-8, Construction Industry Institute, Bureau of Engineering Research, University of Texas at Austin.

7. Oberlender, Garold (2001). *Project Management for Construction,* 2nd ed., McGraw-Hill, New York.

8. Paulson, Boyd C., Jr. (1976). "Designing to Reduce Construction Costs," *Journal of the Construction Division,* 102 (CO4), December, American Society of Civil Engineers, New York.

9. *Quality in the Constructed Project* (2000). Manuals and Reports on Engineering Practices No. 73, American Society of Civil Engineers, Reston, VA.

10. Shilito, M. L., and D. J. DeMarle (1992). *Value, Its Measurement, Design, and Management,* John Wiley & Sons Inc., New York.

11. Dell'Isola, Alphonse (1997). *Value Engineering: Practical Applications for Design, Construction, Maintenance and Operations.* R.S. Means Company, Inc., Kingston, MA.

WEBSITE RESOURCES

1. www.construction.about.com   Contains various features such as construction newsletter, project management papers, links to multiple construction-related sites, and a contract guide.

2. www.B4UBUILD.com   This site is devoted to home building for the inexperienced. It contains everything from design sources and code explanations to descriptions of tools and mortgage advice.

3. www.uniteddesign.com   Offers plans, scheduling and estimating software, and project management references.

4. www.constructionweblinks.com   Contains current legal news relating to the construction industry in areas such as codes, prompt payment, and bid issues.

CHAPTER

Scheduling Techniques for Construction Projects

The critical path method (CPM) is a planning and control technique that provides an accurate, timely, and easily understood graphic depiction of project task sequence. Its purpose is to allocate resources over time in an optimal manner and in a way that allows effective reallocation and schedule control after the project starts. One of the most important features of the CPM is the logic diagram. The logic diagram graphically portrays the relationships between project activities. Another method of scheduling projects that is particularly useful if the major activities must be conducted sequentially is linear scheduling. Linear scheduling is a planning method that seeks to ensure that the work can proceed continuously, that there are no idle resources, and that activities do not interfere with one another. A linear schedule helps management ensure that equipment and labor are not trying to occupy the same physical space at the same time.

INTRODUCTION

Project planning is the process of considering alternatives and methods to complete a task. Planning creates an orderly sequence of events, defines the principles to be followed in carrying forth the plan, and describes the ultimate disposition of the results. It serves the manager by pointing out the things to be done, their sequence, how long each task should take, and who is responsible for which tasks or actions.

The goal of planning is to minimize resource expenditures while satisfactorily completing a given task. Planning aims at producing an efficient use of equipment, materials, and labor, and ensuring coordinated effort. Effective project management requires continual monitoring of task accomplishment—progress. A comparison of actual progress to scheduled progress helps the manager identify problems early and permits development of revised plans to maintain the proper course toward the objective.

Historical Perspective

Before King Cheops planned the construction of his pyramid, humans had by necessity developed a schedule of planting each year's crops. A schedule based on the phases of the moon and knowledge of the annual seasons ensured that planting was accomplished at the right time each year. The start date for the work was defined by nature's schedule.

Construction managers employ practices, principles, and techniques that are derived from earlier concepts and experiences. Frederick Taylor (1856–1915) conducted studies aimed at improving productivity at the Bethlehem Steel Company's plant in Pittsburgh and in 1911 he published *Principles of Scientific Management.* Today we utilize Taylor's concept of breaking work into elementary parts: the work breakdown structure used for scheduling and cost control. Henry L. Gantt, an associate of Taylor's, developed the Gantt chart (**bar chart**). This is a visual tool for scheduling and planning work tasks. Gantt did his work in response to the need for better industrial logistical planning during World War I. The Gantt chart is used for scheduling multiple overlapping tasks.

In the 1950s, two major scheduling methods were developed: (1) the program evaluation and review technique (PERT), by Willard Frazar, came out of the Navy's need in managing the Polaris Missile Program, and (2) the **critical path method** (CPM) by Morgan R. Walker of E. I. Du Pont de Nemours and James E. Kelly of Remington Rand Univac.

bar chart
A graphical representation of planned construction activities, the estimated activity durations, and the planned sequence of activity performance

The Planning Process

Assumptions Based on Facts When planning construction operations, the manager must carefully assess all factors that impact the work. Many of these factors relate to the environment within which the work will be undertaken. An example is the effect of weather on operations. The effect of climate on construction operations is so great that the evaluation of this item alone can be as important as all other factors combined. The diversion of a river for dam construction must be planned to take place during low-runoff periods. Contractors try to have buildings "closed-in" before winter weather (Fig. 4.1) brings cold temperatures and moisture.

critical path method
A project scheduling method where activities are arranged based on interrelationships and the longest time path through the network called the critical path is determined.

Alternative Courses of Action Various courses of action are compared in terms of personnel, material, equipment, and time. This is often difficult because, typically, planning considers multiple sources of uncertainties.

Select the Course of Action The plan addresses all aspects of the project including administration and logistics and provides a "roadmap" for successful completion of the project. Important project aspects that should be addressed are

- Availability of labor, equipment, and materials
- Moving onto the job site
- Bringing supplies and equipment into the job site

FIGURE 4.1 Closing in a building before winter weather.

- Obtaining and using natural resources
- Planning for inclement weather
- Providing for adequate construction site drainage

ACTIVITIES

A common technique used to understand and organize a complex and multidimensional undertaking is to break the project into smaller pieces (divide into subparts), Taylor's concept. In construction, this technique is applied in both planning and estimating. To create a construction plan, all of the work tasks necessary to accomplish the project are first identified. In the terminology of scheduling, tasks or work items are referred to as activities. Each activity is a discrete task. It is also necessary to think about critical success factors and key milestone dates for each activity.

The manager must carefully study all of the contract documents, construct the project mentally, and break it down into its component activities. The activities to build the bridge depicted in Fig. 4.2 are

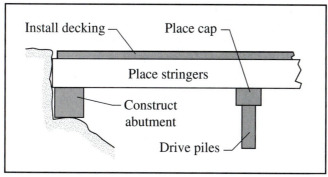

FIGURE 4.2 An activity is a time-consuming, definable task.

- Drive piles
- Construct abutment
- Place cap
- Place stringers
- Install decking

Activities definition should be only as specific as the level of management control. When developing an activity list, always consider the purpose of the schedule and who will use it. The number and detail of the listed activities will vary from job to job and will depend on the intended level of control.

Planning involves the scheduling of activities, but management includes monitoring and controlling the execution of the work; therefore, the activities provide the building blocks for focusing management attention.

> Activity descriptions should be concise and unambiguous. The description communicates the scope and location of the activity. A verb should be used in the description of production-type activities.

Guidance from the Associated General Contractors of America (AGC) asserts that activities have five characteristics [3]:

1. *Activities consume time.* To be classified as an activity, a project task must consume time.

2. *Activities* usually *consume physical resources.* A project activity usually consumes labor, material, or equipment resources. However, there are a few construction activities, such as curing for concrete, that do not consume physical resources.

3. *Activities have definable start and finish points.* An activity represents a definable scope of work, that work has a starting point in time and, when all of the work is completed, an ending point in time.

4. *Activities are assignable.* Each activity is to be accomplished by a particular member of the construction team. This characteristic of an activity facilitates management of the work.

5. *Activities are measurable.* The progress toward completion of the activity's scope of work must be measurable.

An activity is therefore a task, function, or decision that *requires a time duration.*

Activity Duration

One of the most important steps in planning a project is estimating the time required to complete each activity. The duration of an activity is a function of the quantity of work to be accomplished and the work production rate (see Eq. [5.4]). Work production rates are based on the planned composition of labor and equipment used to perform the task. Carelessly made estimates of production rates may cause uneconomical use of personnel, materials, equipment, and time.

When a contractor prepares a project estimate for bidding, the estimator calculates the quantities of material that must be put in place and assumes a production rate to use in arriving at the cost for each work task. In estimating production rates, the estimator considers construction methods and techniques. The selected construction method establishes the task duration. This bid preparation information can serve as the base for calculating an activity duration.

EXAMPLE 4.1

In estimating a three-story building project, it was determined that 480 fluorescent light fixtures would be installed on each floor. In developing the bid estimate, the estimator used a production rate of two fixtures per man-hour. This production rate was based on a crew size of five electricians. The normal workday will be 8 hr. What is the duration (in days) for an install fluorescent-light fixtures first-floor activity?

Using the number of fluorescent light fixtures, the production rate from the estimate, and Eq. [3.1]:

Production rate $\dfrac{2 \text{ fixtures}}{1 \text{ man} - \text{hour}} \times 5 \text{ electricians} = 10 \text{ fixtures per hour}$

Time to perform the work (Eq. [3.1]) $T = \dfrac{Q}{R}$

Per-floor activity duration hours $\dfrac{480 \text{ fixtures}}{\text{floor}} \times \dfrac{1 \text{ hr}}{10 \text{ fixtures}} = 48 \text{ hr per floor}$

Per-floor activity duration days $\dfrac{48 \text{ hr per floor}}{8 \text{ hr per day}} = 6 \text{ days per floor}$

The time unit used to express an activity's duration should match the objective of the scheduling effort. In the early stages of project development, a milestone schedule based on activity durations expressed in months might be appropriate. During the construction stage of a project, it is more common to use weeks, days, or hours as the activity time unit. In the case of a long-duration project, the use of a time unit expressed in weeks might be appropriate. For the majority of projects, the activity time unit is days. This use of days usually matches the terms in the construction contract, as it is common practice to express contractual time in either work or calendar days. In the case of a process plant shutdown or short-duration complex projects, the use of an hour time unit is often necessary; a good discussion of this is presented in Earl Tabler's paper "Installation Procedure for Hot Metal Facility" [11].

> All activities in the schedule must have their duration expressed in the same unit of time.

BAR CHARTS

In 1917, Gantt invented a chart scheduling method. A Gantt chart presents planned activities as stacked horizontal bands against a background of dates (along the horizontal axis). This Gantt, or bar, chart is a commonly used project planning and control tool. Almost all of today's commercially available scheduling software presents information in bar chart formats (see Fig. 4.3). It is a simple and concise graphical picture for managing a project. It is easy with a bar chart to compare planned production against actual production.

The bar chart is widely used as a construction-scheduling tool because of its simplicity, ease of preparation, and graphical format. Normally, the activities are listed in chronological order according to their start date. Discontinuous bars are sometimes used on hand-drawn bar charts to represent interruptions to activity work. The better practice is to have individual activities with their own bars.

The advantage of using a bar chart is that field personnel can easily understand the information. Additionally, it is a very useful tool for preliminary planning and scheduling. If cost, equipment, or personnel requirements are superimposed on the activities, the total resources required at any time can be computed. The cost versus time S-curve discussed in Chapter 3 can easily be prepared by cost loading the bar chart.

In the case of a stand-alone bar chart that is not generated from a critical path method diagram, the major disadvantage is that the user must have detailed knowledge of the particular project and of construction techniques.

Other disadvantages of planning solely with a bar chart are

- It does not clearly show the detailed sequence of the activities.
- It does not show which activities are *critical* to the successful, timely completion of the project.

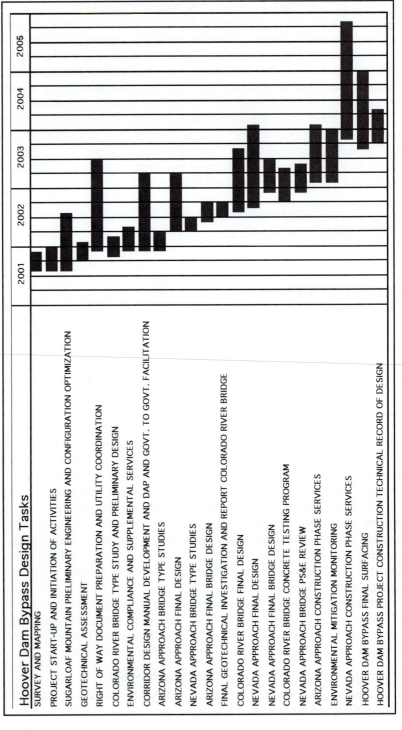

FIGURE 4.3 Bar chart for the design activities to support the Hoover Dam bypass project.

FIGURE 4.4   Pier construction San Francisco-Oakland
Bay Bridge East Span project.

CRITICAL PATH METHOD

The critical path method focuses management's attention on the relationships
between critical activities. It is an activity relationship representation of the project. The evaluation of critical tasks, those that control project duration, allows
for the determination of project duration. Because of the size and complexity
of major construction projects (Fig. 4.4), CPM scheduling is most often applied
using computer software to make the calculation (see Fig. 4.5).

> The CPM calls attention to which activities must be completed before
> other activities can begin.

The critical path method overcomes the disadvantages of a bar chart and
provides an accurate, timely, and easily understood graphic of the project. One
of the most important features of the CPM is the logic diagram. The logic diagram graphically portrays the relationships between project activities. With this
additional information, it is easier to plan, schedule, and control the project.

> CPM calculations define a time window within which an activity can be
> performed without delaying the project.

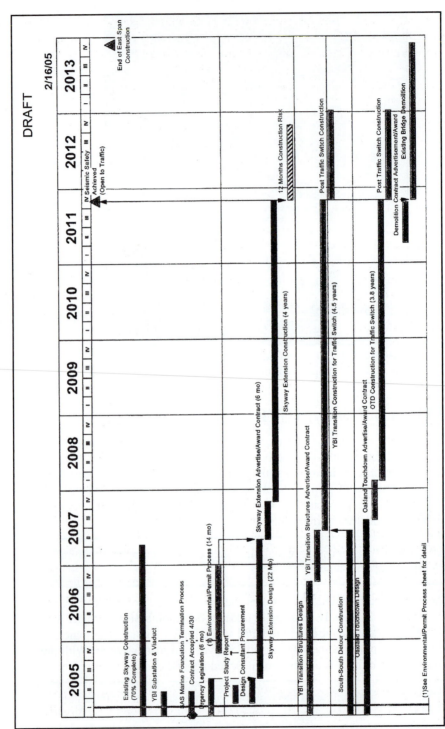

FIGURE 4.5 Computer-generated schedule used in evaluating alternative construction processes for the San Francisco-Oakland Bay Bridge East Span project.

Using the CPM to develop a schedule requires detailed investigation into all identifiable tasks that make up a project. This means that the manager must visualize the project from start to finish and must estimate time and resource requirements for each task. It is good practice to also obtain input from experienced superintendents and subcontractors. The advantages of using a CPM logic network include

- Reduces the risk of overlooking essential tasks and provides a blueprint for long-range planning and coordination of the project
- Provides a clear graphic of the interrelationships between project activities (tasks)
- Focuses the project manager's attention by identifying the critical tasks that control project duration
- Generates information about the project so that the manager can make rational and timely decisions if complications develop during the progression of the work
- Enables the project manager to easily determine what resources are needed to accomplish the project and when these resources should be made available
- Allows the project manager to quickly determine what additional resources will be needed if the project must be completed earlier than originally planned
- Provides feedback about a completed project that empowers the project manager and the estimator to improve techniques and assure the best use of resources on future projects

A CPM is not a substitute for appropriate construction knowledge and planning. It does not make decisions for the project manager, nor can it contribute anything tangible to the actual construction. The greatest danger or disadvantage of a CPM is that with modern scheduling software it is possible for anyone with computer proficiency to construct a schedule that appears to be reasonable. It is easy to input various project activities into the software and to create relationships in such a way that when looking at a bar chart of the schedule, the work seems to flow in an entirely sensible manner. Unfortunately, there is no way to tell by looking at the plotted schedule whether this is a true picture of physical construction relationships or simply an attractive bar chart. Very often, there is no internal schedule logic, and, therefore, the schedule is essentially useless from the standpoint of monitoring and analyzing progress. The creation of these schedules can be the result of a lack of construction knowledge or inexperience, or they can result from premeditated abuse.

Schedules are produced to aid in managing the work, and the information contained in the schedule must be conveyed to many project participants (e.g., subcontractors, suppliers, and crew foremen). In the case of a large complex project, this is a huge information flow. Besides that difficulty, it must be recognized that for a participant receiving only part of the information, it is virtually impossible to check the information.

Activity Logic Network

The activity logic network supports the project manager by providing a graphical picture of the sequence of construction tasks. Before the diagram can be developed, the project must first be constructed mentally to determine activity relationships. The manager does this by asking the following questions for each activity on the activity list:

predecessor activity
An activity that must be completed before a given activity can be started.

- Can this activity start at the beginning of the project? (Start activities)
- Which activities must be finished before this one begins—**predecessor activity**(ies)? (Precedence)
- Which activities may either start or finish at the same time this one does? (Concurrence)
- Which activities cannot begin until this one is finished—**successor activity**(ies)? (Succession)

successor activity
An activity that cannot start until a given activity is completed.

In some cases, the sequencing of activities is clear, that is, this can be dictated by a spatial sequence—drive the bridge piles before the caps can be set in place (see Fig. 4.2). However, in other cases, the sequencing may be dictated by structural reasons that are not always clear by studying *only* the project plans issued by the owner. In many cases, the manager must carefully study other project documents, such as the shop drawings submitted by fabricators or subcontractors. Sometimes when estimators develop a schedule they build the sequencing of activities based on an assumed equipment constraint, such as having only one crane available. Equipment and crew resource constraints will be discussed later in the scheduling activities section of this chapter.

> In July 2001, a steel truss designated TR13 collapsed during construction of the Washington, DC, convention center. The shop drawing for the work indicated that truss TR13 should have been erected after truss TG109. The engineer of record stated, "erection directions were not followed" as TG109 was not in place [13].

One way to determine these relationships is to list all activities in a columnar format with a second column to the right of the activities list titled "Preceded Immediately By (PIB)." For each activity, use the second column to list all other activity numbers (identifier codes) that must *immediately* precede the activity in question. If the activity can begin at the beginning of the entire project, write "None."

It is common practice and a necessity for the CPM algorithm to work that a network have only *one start* and *one finish* activity. Therefore, in the case of a network with multiple activities that have no predecessor, a dummy "start" activity can be inserted at the beginning of the network. Similarly, a dummy "finish" activity is placed at the end of a network and all activities that have no successor activity are tied to this final activity.

EXAMPLE 4.2

Our schedule to build a concrete slab on grade has seven activities: excavate, build forms, procure reinforcing steel, fine grade, set forms, place reinforcing steel, and place and finish concrete. Use the preceded immediately by (PIB) method to organize these activities for a CPM schedule.

Activity	PIB
A. Excavate	None
B. Build forms	None
C. Procure reinforcing steel	None
D. Fine grade	A
E. Set forms	A and B
F. Place reinforcing steel	C, D, and E
G. Place and finish concrete	F

There are two CPM logic-diagramming formats: (1) activity-on-the-arrow and (2) activity-on-the-node. With the activity-on-the-arrow (AOA) format (see Fig. 4.6), the arrows of the diagram represent the activities of the logic diagram. Unless the diagram is time scaled, the length of the arrows has no relationship to activity time durations. The **nodes** (events) on an AOA diagram serve only as connecting points for activities. When using the AOA diagramming format, it is often necessary to use **dummy activities** (arrows) to denote dependencies. Dummy activities consume no resources. AOA diagramming is sometimes referred to as I–J diagramming, with the I and J referencing to the beginning and ending nodes, respectively, of an activity arrow. The logic of the diagram is that activities (arrows) leaving a node cannot begin until all of the activities heading into that node have been completed.

Activity-on-the-arrow logic diagramming was once the most popular CPM method. Today, however, most CPM users employ the activity-on-the-

node
On an activity-on-arrow schedule, the nodes mark the start and end point of an activity. With precedence diagramming the nodes represent the activities of the schedule.

dummy activities
A pseudoactivity with duration of zero. A dummy is a dotted line arrow and used solely to indicate sequence.

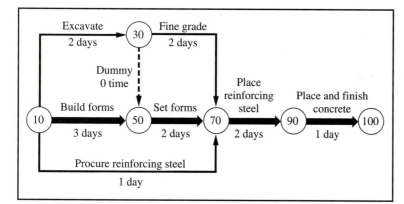

FIGURE 4.6 Example of an activity-on-the-arrow network diagram.

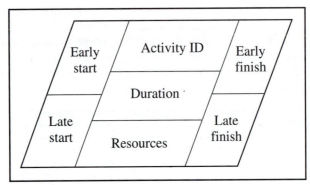

FIGURE 4.7 Example of activity node information; the specific arrangement will depend on the scheduling software used.

node (AON) format, where each node represents an activity. This is sometimes referred to as "precedence diagramming." The two basic logic symbols on the precedence diagram are the *node* and precedence *arrow,* but what they represent is reversed from the AOA case.

Nodes   A node is simply a parallelogram that represents an activity, and each activity on the activities list is represented by a node on the logic diagram. The node is of a standard shape and format, and contains all the necessary activity information. Notations on each node indicate the activity's alphanumeric identifier (ID code) and duration. Sometimes additional information, such as early and late start times, early and late finish times, and required resources, is also included on the node (Fig. 4.7). Each activity in a network should have a unique identifier.

Activities are identified by user-defined codes that aid in describing activities. These codes are essential for sorting and categorizing activities when using computer-based scheduling. Each activity can have one or more activity codes. Common code categories are

- Responsibility
- Area of work
- Phase of project
- Subphase
- Type of work
- Specification section
- Bid item

When creating activity identifiers, the practice is to use a numbering system that increases in magnitude when moving from project start to finish. Reading of project status reports generated from the schedule is much easier when directional numbering is used. It is also good practice when first numbering activities

TABLE 4.1 Activity logic relationships.

Logic relationships	Diagram example
SEQUENTIAL LOGIC Activity 20 cannot start until activity 10 is completed	10 → 20
CONCURRENT LOGIC Activities 5 and 10 can proceed concurrently. **Multiple predecessor logic** Activity 20 cannot start until both activities 5 and 10 are completed.	5, 10 → 20
Activities 30 and 40 cannot start until both activities 10 and 20 are completed.	10, 20 → 30, 40
Multiple successor logic Activity 20 must be completed before either 30 or 40 can start, 30 can start only after 10 and 20 are completed, 40 can start immediately after 20 is completed.	10, 20 → 30, 40

to leave numbering gaps (e.g., 5, 10, 15, etc.). This permits easy inclusion of additional activities as planning progresses or when change orders are issued.

Precedence Arrow The precedence arrows show the order, sequence, and relationship between activities (such as what activities must precede and follow another activity). The configuration of the diagram's nodes and arrows is the result of the PIB list (or the answers to the five questions that were previously asked concerning each activity). The logic behind the diagram is such that an activity cannot begin until all preceding activities are complete. Table 4.1 presents several common logic relationships.

The logic network is constructed without regard to how long an activity will last or whether all required resources are available. It simply displays the relationships among activities, provides project understanding, and improves communications. Once the logic network has been developed, the manager then assigns activity duration and resource requirements to each activity.

Schedule Calculations

The next step in the CPM development process is to calculate the earliest and latest times for each activity. These times are computed such that the network logic is not violated and the project's overall duration is held to a minimum. This

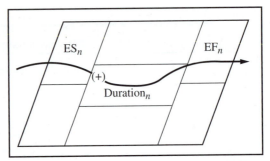

FIGURE 4.8 Calculation of an activity's early finish.

provides the manager with a *time frame* for each activity. Within this time frame, the activity must be completed, or other activities will be delayed and then there is the possibility of creating a ripple effect that delays the entire project. Once the activity times are calculated, the manager will be able to identify those tasks that are on the critical path—the sequence of linked activities that controls the project duration. By carefully managing these critical activities, it is possible to reduce the time to complete the project. An activity cannot begin until all activities previous to it (arrows leading to it in the logic node) are completed.

Forward Pass

A forward computational sequence through the logic network will yield this information:

1. The earliest time each activity in the network can start and finish.
2. The minimum overall duration of the project.

forward pass

A schedule calculation that determines the earliest start and finish time of the precedence diagram activities and the minimum project duration.

In performing the **forward pass** calculations, all successor activities are started as early as possible. The calculations maintain the rigor of the network logic; therefore, the earliest an activity can start is when all predecessors to that activity are completed. Furthermore, each activity is postulated to finish as soon as possible. This finish requirement yields Eq. [4.1] (see Fig. 4.8).

$$\text{Early finish}_n \text{ (EF)} = \text{early start}_n \text{ (ES)} + \text{duration}_n \qquad \textbf{[4.1]}$$

where n denotes the nth activity.

Early Start/Early Finish   The *early start time* (ES) of an activity is the earliest point in time, taking into account the network logic, that an activity may start. For this text, the early start time is positioned in the upper left corner of the activity node box. Some scheduling software may position or display this information in another location on the node.

The starting point for performing a forward pass is the first activity in the network. In the case of the first activity of a project, the earliest time it may start is zero (the end of day 0 or the beginning of day 1).

> The event-time numbers calculated for an activity represent the end of the time period. Thus, a start or finish time of day 5 would mean the end of the fifth day (or the beginning of the sixth day).

If only one precedence arrow leads into an activity, then that activity's early start time is the same as the previous activity's (at the tail of an arrow) early finish time. To determine an activity's early start time when more than one arrow leads into its node, select the *largest* early finish time of all activities at the tail of the arrows (Eq. [4.2]). Logically, an activity cannot begin until *all* preceding activities are complete (see Fig. 4.9).

$$\text{Early start}_n = \text{Maximum early finish of all predecessors activities} \quad \text{[4.2]}$$

Add the duration of each activity (center of the node in our nomenclature) to the early start time to compute the *early finish time* (EF) (Eq. [4.1]), position the EF in the upper right corner of the activity node. The early finish time is the earliest time the activity may finish.

Using this systematic process, work through the entire logic diagram computing all early start and early finish times from the beginning activity to the finish of the project. This computational sequence through the logic diagram completes the forward pass. The overall duration for the project will be the EF of the last activity in the network. For the network shown in Fig. 4.10, the forward pass yields a project duration of 22 days, as determined by the sequence of construction and the time duration assigned each activity.

Backward Pass

A backward computational sequence through the logic network will produce the latest point in time that each network activity can start and finish, and still

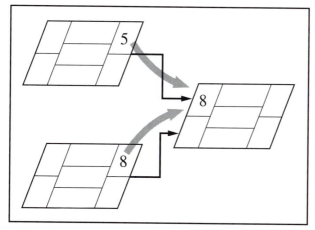

FIGURE 4.9 Calculation of early start; use largest preceding early finish.

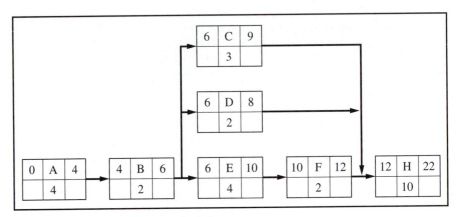

FIGURE 4.10 Forward pass calculations.

backward pass

A schedule calculation that determines the late start and late finish times of the precedence diagram activities under the condition that the project's minimum duration be maintained.

maintain the minimum overall project duration. The **backward pass** calculation starts with the last activity in the network. This last activity is assigned a late finish time equal to its early finish time as calculated by the forward pass.

Late Finish/Late Start   To calculate *late finish time* and *late start time* of an activity, follow the precedence arrows backward through the logic diagram (right to left). Taking into account the network logic, the late finish time (LF) of an activity is the latest point in time that an activity may finish without delaying the entire project. For this text, the late finish time is positioned in the lower right corner of the activity node box. Since the last activity in the Fig. 4.10 network (activity H) had an early finish of 22 days, the latest time that this activity can finish and not lengthen the project duration is the end of the 22nd day. The number 22, therefore, should be assigned as the LF for this network and placed in the lower right corner of the H node.

The preceding activity's late time (at the tail of the precedent arrow) is the succeeding activity's late start time (at the head of the precedent arrow). To determine an activity's late finish time when more than one arrow tail leads away from its node, choose the *smallest* late start time of all activities at the arrows' heads (see Fig. 4.11). Logically, an activity must finish before *all* follow-on activities may begin.

To compute the late start time (LS) of an activity, use Eq. [4.3], which states, subtract the activity's duration (center of the node) from its late finish time (see Fig. 4.12). Position the LS in the lower left corner of the activity node. The late start time is the latest time the activity may start without delaying the entire project.

$$\text{Late start}_n \ (\text{LS}) = \text{late finish}_n \ (\text{LF}) - \text{duration}_n \qquad \textbf{[4.3]}$$

Using this backward systematic process, work through the entire logic diagram (against the arrows) computing all late finish and late start times. This computational movement back through the logic diagram is known as the

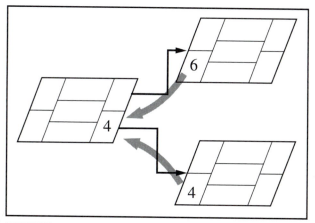

FIGURE 4.11 Calculation of an activity's late finish; use the *smallest* late start time of all succeeding activities.

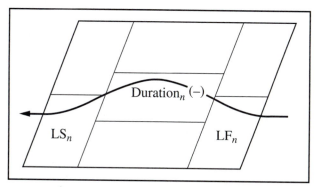

FIGURE 4.12 Calculation of an activity's late start.

backward pass. The late start time of the first activity must be zero. For the network shown in Fig. 4.13, the late start time of activity A is zero.

Critical Path and Critical Activities

The critical path through a schedule network is the longest time duration path through the network. It establishes the minimum project time duration. All activities on the critical path are by definition critical. A critical activity can be determined from the logic network by applying either of these rules:

1. The early start and late start times for a particular activity are the same.
2. The early finish and late finish times for a particular activity are the same.

For the network shown in Fig. 4.13, activities A, B, E, F, and H meet the listed rules, thus they are all *critical activities.* A critical activity, if delayed by any amount of time, will delay the project's completion by the same amount of

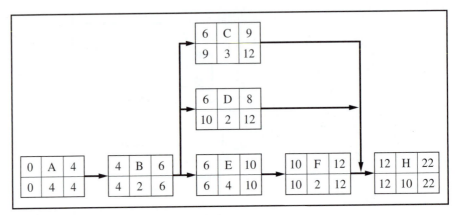

FIGURE 4.13 Backward pass calculations.

time. Critical activities are linked together forming a path from the start activity to the finish activity called a *critical path*. There can be more than one critical path through the network and the critical path may branch out or come back together at any point. All critical paths must be continuous; any assumed critical path that does not start at the start node and end at the finish node indicates a logic mistake.

Critical paths are indicated on the logic diagram by methods such as double lines, bold lines, or color highlighted lines (see Fig. 4.14). Any activity node not on the critical path will contain **float**. *Float* is duration of time that is available to complete an activity beyond the activity's work duration, such as having 6 days to do a task that requires only 4 work days. Activities on the critical path have no float.

Total Float *Total float* (TF) is the amount of time that an activity can be delayed without delaying project completion. Total float assumes that all preceding activities are finished as early as possible and all succeeding activities are started as late as possible.

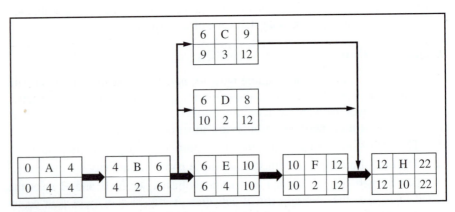

FIGURE 4.14 The critical path through a network.

Total float for an activity can be determined by either Eq. [4.4] or Eq. [4.5].

$$\text{Total float activity}_n = \text{late start activity}_n - \text{early start activity}_n \quad \textbf{[4.4]}$$

$$\text{Total float activity}_n = \text{late finish activity}_n - \text{early finish activity}_n \quad \textbf{[4.5]}$$

where n denotes the nth activity.

Both equations yield the same result.

Total float is calculated and reported individually for each activity in a network. However, total float is a *network path attribute* and not associated with any one specific activity along the path.

Free Float *Free float* is the time duration that an activity can be delayed without delaying the project's completion and *without delaying the start of any succeeding activity*. Free float is the property of an activity not a network path. Free float for an activity is determined by the equation

$$\text{Free float}_n = \frac{\text{minimum early start of}}{\text{all successor activities}} - \text{early finish}_n \quad \textbf{[4.6]}$$

where n denotes the nth activity.

Activity C of the network shown in Fig 4.14 is not on the critical path. Therefore, the activity has float. Free float would be 3 (minimum early start of successor, 12–EF of 9).

free float
The time duration that an activity can be delayed without delaying the project's completion and without delaying the start of any succeeding activity.

Interfering Float *Interfering float* is the delay time available for an activity that will not delay the project's completion, but such activity delay into interfering float will delay the start of one or more following noncritical activities. Interfering float for an activity is determined by Eq. [4.7].

$$\text{Interfering float}_n = \text{late finish}_n - \frac{\text{smallest early start of}}{\text{succeeding activity(ies)}} \quad \textbf{[4.7]}$$

where n denotes the nth activity.

If by the network logic, the activity in question has more than one successor, use the smallest successor ES in Eq. [4.7].

The aggregate of *free float* (FF) and *interfering float* (IF) equals the total float, Eq. [4.8].

$$\text{Total float}_n = \text{free float}_n + \text{interfering float}_n \quad \textbf{[4.8]}$$

interfering float
The time available to delay an activity without delaying the project's completion but by delaying an activity into interfering float causes the delayed start of one or more following noncritical activities.

EXAMPLE 4.3

You and the project manager are reviewing the group of activities on the project schedule that is shown in Fig. 4.15 (activities RPB, TIB, RPW, FIB, and CWW). The time durations are in days.

a. What is the total float for each of the activities shown?
    Using Eq. [4.4] (Total float$_n$ = Late start$_n$ – Early start$_n$) the total float for each activity is

Activity	Activity ID	Total float
Rough plumbing bath rooms	RPB	3 − 0 = **3**
Tile bath rooms	TIB	25 − 13 = **12**
Rough plumbing wash rooms	RPW	16 − 13 = **3**
Finish bath rooms	FIB	32 − 20 = **12**
Cabinet work wash rooms	CWW	31 − 31 = **0**

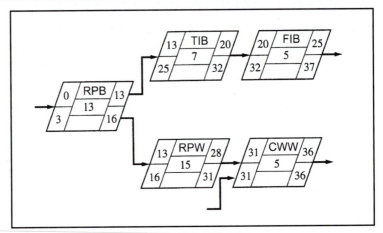

FIGURE 4.15 Portion of a CPM diagram.

b. By inspection, can you determine the free float for any of these activities?

The definition of *free float* states that it is the time an activity can be delayed without delaying the project's estimated completion time and not delaying the start of any succeeding activity.

Activity	Free float	
RPB	0	If we would delay RPB, the result would be to also delay TIB and RPW, as RPB is the only predecessor of each and therefore controls their start time.
TIB	0	If we would delay TIB, the result would be to also delay FIB, as TIB is the only predecessor of FIB and therefore controls FIB's start time.
RPW	3	We can delay RPW for 3 days and not delay the start of CWW, since the other predecessor of CWW sets RPW's start date.
FIB	Cannot determine	We do not have the necessary information about down-stream activities for which FIB is a predecessor.
CWW	0	ES and LS are equal; this is a critical activity with no float of any type.

PRECEDENCE LOGIC DIAGRAMS

Precedence diagramming is an extension of the activity on the node scheduling method. The precedence method allows the overlapping of concurrent activities. Most scheduling software allows the use of the precedence method. The advantages of precedence modeling are

■ Concurrent activities can be easily modeled without dividing the work task into a number of discrete activities.
■ The workflow of continuous operations is better represented.

The terms *lag* and *lead* are used to describe the amount of time between the start or finish of an activity and that of the activity's predecessor or successor. Lag is the time an activity must be delayed from the start or finish of a predecessor. Lead is the amount of time by which an activity precedes the start or finish of a successor. The four precedence relationships are

■ Finish-to-start (FS)
■ Start-to-start (SS)
■ Finish-to-finish (FF)
■ Start-to-finish (SF)

Finish-to-Start

With the finish-to-start relationship, it is implied that the successor activity cannot start until the predecessor activity has been completely accomplished (Fig. 4.16). When using precedence diagramming, a lag notation can be added to the finish-to-start relationship to denote a required delay between the finish of the predecessor activity and the start of the succeeding activity (Fig. 4.17).

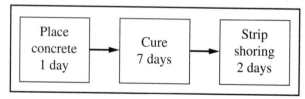

FIGURE 4.16 AON network finish-to-start relationship without lag.

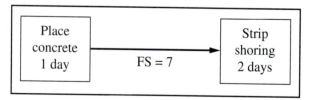

FIGURE 4.17 AON network finish-to-start relationship with lag.

Assuming that activities are performed in a *continuous* manner, the early start of the succeeding activity and late finish of the preceding activity can be calculated using Eqs. [4.9] and [4.10].

$$\text{Early start}_{n+1} = \text{early finish}_n + \text{lag} \qquad [4.9]$$

$$\text{Late finish}_n = \text{late start}_{n+1} - \text{lag} \qquad [4.10]$$

Start-to-Start

There are many construction activities that can start after a certain amount of work has been accomplished on a preceding activity. If only finish-to-start diagramming notation is used, the preceding activity would have to be broken into incremental units of work to represent this condition (Fig. 4.18). Using start-to-start notation with a lag, the same relationship can be described using fewer activities (Fig. 4.19).

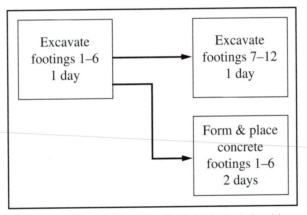

FIGURE 4.18 AON network finish-to-start relationship.

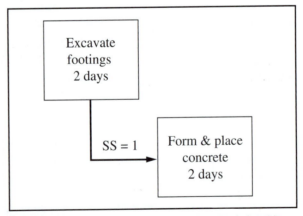

FIGURE 4.19 AON network start-to-start relationship with lag.

The early start of the succeeding activity and late finish of preceding activities can be calculated by Eqs. [4.11] and [4.12]. The equations are valid, however, only if the activities are performed in a continuous manner.

$$\text{Early start}_{n+1} = \text{early start}_n + \text{lag} \qquad \textbf{[4.11]}$$

$$\text{Late finish}_n = \text{late start}_{n+1} - \text{lag} \qquad \textbf{[4.12]}$$

Finish-to-Finish

There are many construction activities that take place concurrently but with the succeeding activity lagging some time interval behind the accomplishment of a certain amount of work on a preceding activity. When finish-to-start diagramming notation is used, the trailing activity must be broken into incremental units of work to represent this condition (Fig. 4.20). By use of finish-to-finish notation with a lag, the same relationship can be described using fewer activities (Fig. 4.21).

The early start of the succeeding activity and late finish of preceding activities can be calculated by Eqs. [4.13] and [4.14]. Again, the assumption is made that the activities are performed in a *continuous* manner.

$$\text{Early start}_{n+1} = \text{early finish}_n + \text{lag} - \text{duration}_{n+1} \qquad \textbf{[4.13]}$$

$$\text{Late finish}_n = \text{late start}_{n+1} - \text{lag} \qquad \textbf{[4.14]}$$

Start-to-Finish

The start-to-finish relationship is not supported by most scheduling software. This is probably caused by the fact that the other relationships allow appropriate modeling of construction processes.

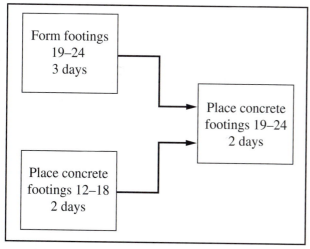

FIGURE 4.20 AON network finish-to-start.

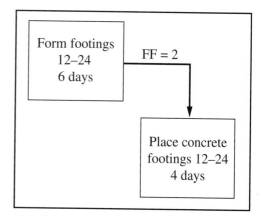

FIGURE 4.21 AON network finish-to-finish relationship with lag.

CALENDAR DATE SCHEDULE

If all activities are scheduled to begin on their ES dates, the schedule is known as an *early start schedule*. This logic diagram schedule provides the manager with considerable planning information. But everything to this point was done based on workday time durations. To communicate schedule information in a meaningful manner, the workday schedule needs to be converted to a calendar schedule with points in time, such as an activity start expressed as a specific date. If done manually, this is a simple counting procedure, and if the schedule was created using a software package, the conversion can be done automatically.

Project start dates are usually specified in the contract documents. Other decisions or contract requirements that will affect the calendar conversion are

1. *Workdays per week.* This may be specified in the contract or it can be a contractor decision.

2. *Holidays that will be observed.* Again, this may be specified in the contract, be covered by union rules, or be a contractor decision.

3. *Weather days.* These are a function of specific work activities, the climatic conditions at the project location, and when during the seasons the project begins.

SCHEDULING ACTIVITIES

It is usually assumed that the activities will start on their early start dates and typically the first CPM network is constructed under this assumption. But this does not have to be the case, and what the CPM really does is provide project management with the information needed to schedule the project activities in a manner that satisfies multiple project constraints. If the resource information is included with each activity, it is possible to determine the magnitude of resource requirements for each day of the project.

An activity can take place at any time within the interval defined by its early start- and late-finish dates, and still not delay the project. This provides flexibility in scheduling activity accomplishment. Using this flexibility, a manager can exercise control over the magnitude of *resource requirements* needed on a given day. A simple scheduling process can be illustrated graphically using the activities in the Fig. 4.22 network. The critical path for this network is emphasized by heavier activity-connecting arrows.

The first step is to create a bar chart of the network with all activities listed in identifier order. Ordering the activities by ID sequence makes it easier to track successor activities. After each activity, place a reminder note in parentheses of all immediately dependent activities (see the activity column of Fig. 4.23). For example, since activities P and Q cannot begin until activity O is completed, annotate activity O in the schedule like this: O (P, Q). The next step is to mark on the bar chart the time frame for each activity during which it may be performed without delaying the project or violating any of the diagram logic (sequence) relationships (see Fig. 4.23).

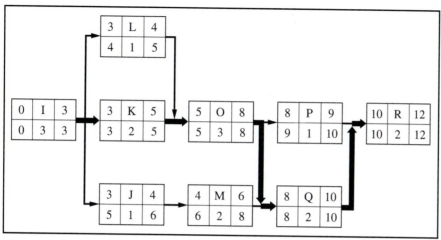

FIGURE 4.22 Scheduling network activities.

Consider activity J in Fig. 4.22; the ES shows that the earliest this activity can begin is the end of day 3 (or the beginning of day 4). Thus, the beginning of day 4 to the end of day 6 (as determined from the LF) is the available time span in which to complete this activity (Fig. 4.23). Because of the nature of the logic diagram, this activity cannot be scheduled earlier, since activity I must be completed first. It cannot be scheduled later, as that would delay the project. Considering the end of day convention, where a start date means the end of the day or beginning of the following day, the left bracket is placed at the beginning (morning) of the following day (e.g., "ES+1").

Activity	Workdays											
	1	2	3	4	5	6	7	8	9	10	11	12
I (J, K, L)	2	2	2									
J (M)				1	✕	✕						
K (O)				3	3							
L (O)				2								
M (Q)					1	1						
O (P, Q)						4	4	4				
P (R)									2			
Q (R)									3	3		
R											2	2
Sum	2	2	2	6	4	5	4	4	5	3	2	2

FIGURE 4.23 Bar chart for scheduling network activities.

Once the brackets are placed correctly, the next step is to formulate a trial schedule, scheduling each activity as soon as possible within the time frame, or flush with the left bracket. To schedule a particular activity, place the number of resources required inside each box along the activity line. Do not exceed the activity's duration; stop at the end of the early finish day. The remaining boxes within the brackets are left blank at first and will become either free or interfering float. In Fig. 4.23, the boxes of the critical activities have been darkened.

The diagonal lines forming an X on the last two days of the activity J bracketed area denote days of interfering float. Some activities have free float (activities L, M, and P), and some have all interfering float (activity J). The placement of the Xs serves as a reminder that when the actual accomplishment of an activity is moved into that time frame, there is a ripple effect with succeeding activities.

Resources

If carpenter crews are identified as a controlling resource, the numbers within the activity data boxes would indicate how many carpenter crews are required for each of the activities. It is necessary to consider only major controlling resources when developing the schedule. Each project has its own controlling resources depending on the nature of the work. On a building job, the controlling resource might be the cranes. On a mass concrete project, the controlling resources might be forming crews and placement crews. There are usually several controlling resources, and all would have to be considered when scheduling the work. In cases where many different kinds of resources are necessary for an activity, managers may choose to use several lines contained within one set of brackets and use each line for a different type of resource. The resources required on any particular day are the sum of those required for the individual activities taking place on that day. The last row of Fig. 4.23 shows the resource requirement sum for each day of the project. There are computer-scheduling algorithms available to aid in scheduling resources on large projects.

The resource required to complete activity M is one carpenter crew for 2 days. To show this activity scheduled as soon as possible, place the number 1 (number of carpenter crews) in only the first two boxes of the bracketed duration (Fig. 4.23). Scheduling all the activities as soon as possible yields the early start schedule as shown in Fig. 4.23. All activities are scheduled to begin at their ES times.

The number of available resources is often less than the needed resources for a given day, and activities must be delayed (into float whenever possible) to spread resource use across the time frame of the project. When daily resource requirements exceed what is available, the manager must resource constrain the project in such a manner as to have as little effect as possible on the duration of the project. The terminology is not universal, but constraining usually means that because of a resource deficiency, the project will have to be extended. However, sometimes resources can be shifted in such a way as to avoid delaying the

Activity	Workdays											
	1	2	3	4	5	6	7	8	9	10	11	12
I (J, K, L)	2	2	2									
J (M)				☒	☒	☒						
K (O)				3	3							
L (O)					2							
M (Q)							1	1				
O (P, Q)						4	4	4				
P (R)									2			
Q (R)									3	3		
R											2	2
Sum	2	2	2	3	5	5	5	5	5	3	2	2

FIGURE 4.24   A smooth resource constrained schedule, maximum five crews.

overall project duration. The term *leveling,* or *leveling project resources,* has the connotation that the project duration will not be extended.

If the objective, considering the Fig. 4.22 network, is to limit the number of crews to five or less and to avoid a situation where there are peaks and valleys in the number of crews required over the duration of the project, Fig. 4.24 illustrates one solution. In Fig. 4.24 activity J has been shifted 2 days into its interfering float. The result of that action was that activity M also had to be shifted 2 days, as it is dependent on J. Additionally, activity L was shifted 1 day but this was into free float so no other activity was affected. If any of the critical activities are shifted to achieve the desired leveling of resources, the result will be to lengthen the project duration.

Scheduling the work according to the Fig. 4.24 task timing results in a uniform increase in the number of required crews, a level five crew requirement from day 5 through day 9, and a uniform decrease of crew requirements at the end of the project. This is a good situation because it is difficult to manage the work when the required number of resources is constantly changing. The problem with this schedule is that, with the exception of only activity P, we have now made every activity critical. If anything should delay a critical activity, the project will be delayed.

Another possible schedule that holds the maximum number of crews to five or less is presented in Fig. 4.25. With this schedule, some float has been retained for activities J, M, and P. The problem, however, is that on day 8 it is necessary to lay off one crew and then on day 9 you need one more crew for only 1 day. If the resource were a crane instead of a carpenter crew, there would be the added consideration of the cost to mobilize and demobilize the machine. In that case,

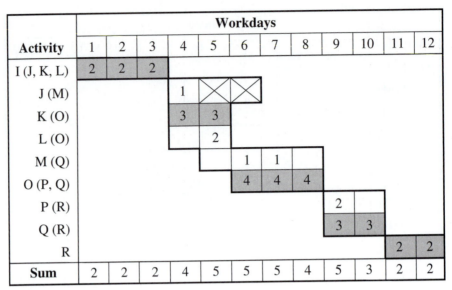

Activity	\multicolumn{12}{c}{Workdays}
	1	2	3	4	5	6	7	8	9	10	11	12
I (J, K, L)	2	2	2									
J (M)				1	✕	✕						
K (O)				3	3							
L (O)					2							
M (Q)							1	1				
O (P, Q)						4	4	4				
P (R)									2			
Q (R)									3	3		
R											2	2
Sum	2	2	2	4	5	5	5	4	5	3	2	2

FIGURE 4.25 Resource constrained schedule, maximum five crews.

the manager may decide to pay the extra day of machine rent and retain schedule flexibility. Leveling of crew resources is always a challenge when trying to keep productive crews together.

Crashing

If the CPM indicates that the project's duration exceeds the desired completion date, the manager should examine the logic diagram's critical path to find activity durations that may be shortened. This is known as *expediting, compressing,* or *crashing* the project. The only way to reduce the duration of the project is to shorten the duration of the critical activities. Shortening a noncritical activity will not shorten the project duration, but it will increase cost. By increasing the allocation of resources to critical path activities, it may be possible to reduce the duration of the project. Additional equipment and personnel can be committed or the same equipment and personnel can be used for longer hours. Normally, a moderately extended workday is the most economical and productive solution. Managers may choose to work double shifts or weekends. When expediting activities consider the long-term effects on safety, morale, equipment, and consequential decreases in efficiency.

velocity diagram
A graphic for monitoring the relationship between time and the accomplishment of an activity.

VELOCITY DIAGRAM

A **velocity diagram** presents a graphical picture of the relationship between time and the accomplishment of an activity (Fig. 4.26). The vertical axis of the diagram represents accomplishment of a work task: cubic yards excavated, miles

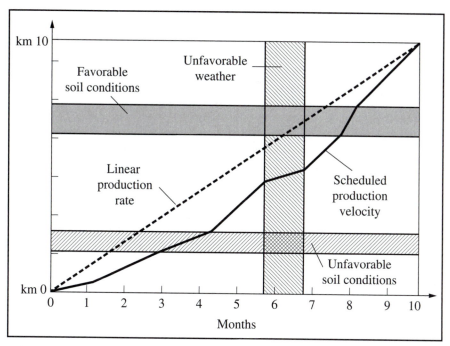

FIGURE 4.26 Velocity diagram as a scheduling tool. (Dressler, Joachim. 1980. "Construction Management in West Germany," *Journal of the Construction Division*, ASCE, 106 (CO4). Reproduced by permission of ASCE.)

of road built, kilometers of pipeline constructed. The horizontal axis presents construction time. The slope of the production line gives the activity production rate. When the vertical axis represents activity completion in terms of advancement along a linear dimension, such as stations for a highway project or floors of a building job, the velocity diagram delineates an activity's rate of progress in time and space. The tool focuses management attention on the rate of accomplished work.

In the section on activity duration, the process of converting bid estimate production rates into an activity duration was demonstrated. If the bid estimate production rates are based on the same time unit as that of the proposed velocity diagram, the manager can plot the diagram directly. If the time units are different, a conversion will be necessary. In the case where a CPM was prepared with activity's duration, the production rate for a velocity diagram can easily be calculated by rearranging Eq. [3.1].

$$T = \frac{Q}{R}$$

$$\text{Planned production rate} = \frac{\text{Quantity of work}}{\text{Estimated duration}}$$

EXAMPLE 4.4

The construction CPM allows 8 months to construct 16 kilometers of pipeline. Assuming uniform conditions, what would be the average planned production rate?

Using Eq. [3.1]:

$$\text{Planned production rate: } \frac{16 \text{ km}}{8 \text{ months}} = 2 \text{ km per month}$$

This average or linear production rate is the diagonal line in Fig. 4.27.

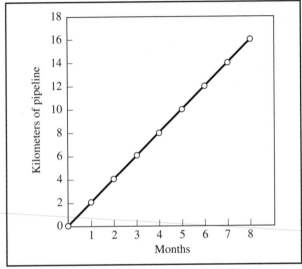

FIGURE 4.27   Graphical display of Example 4.4 product.

For simplicity, the planned pipeline production rate was assumed to be constant, resulting in a linear plot. If it is anticipated the activity production rate will vary significantly during certain portions of the work, then the velocity line can be adjusted to reflect the expected situation. In the case of the Fig. 4.26 project, there are data indicating changing ground conditions and inclement weather are considerations. As we can see, the production rates in the diagram are not linear but fully consider the impact of both physical and environmental conditions affecting the work.

By plotting actual progress on the diagram as the work progresses, the manager can measure performance against a baseline needed to achieve timely completion. The velocity diagram lends itself to efficiently managing projects that require control of time and space between construction activities.

LINEAR SCHEDULING

linear schedule
A simple diagram that shows the relationships between activities, their duration, and the space where they take place at a given time.

A **linear schedule** is simply a series of individual velocity diagrams. They provide a graphical display of the movement of labor and equipment in time and space. Many construction projects involve groups of activities performed *consecutively* by the same crew. This is especially true in the case of linear projects such as highway work, but it is also true in the case of multistory commercial building work. On such projects, it is advantageous to arrange the schedule so that the crews can work continuously. By scheduling the project so that the work can proceed continuously, costs are reduced because intervals of idle equipment and manpower are eliminated. Another important property of linear projects is that numerous activities can be carried out concurrently at different locations.

Linear scheduling is a planning method that seeks to ensure that the work can proceed continuously, that there are no idle resources, and that activities do not interfere with one another. In the scheduling literature, this method has been referred to as repetitive-unit construction, time-space scheduling, and as the vertical production method. It is an effective tool for relating activities and production rates.

As with CPM scheduling, the process of developing a linear schedule consists of four steps:

1. Identify the activities.
2. Estimate activity production rates.
3. Develop activity sequence.
4. Draft the linear schedule.

Identify Activities

The process of identifying activities was discussed in the earlier activities section of this chapter, though activity detail in linear schedules is usually not as extensive as in the case of a CPM schedule.

Estimate Activity Production Rates

Like velocity diagrams, linear schedules show the time and space relationship of activities. The production rate of the individual activities must be expressed in the same units. Because of how the linear schedule diagram is drafted, production of an activity (all activities) is usually expressed in terms of position.

As discussed concerning velocity diagrams, the production rate is often assumed to be linear. But the average production can be adjusted to reflect the effect of physical conditions or environmental conditions as shown in Fig. 4.26.

Develop Activity Sequence

Even though the activities in a linear construction project can be carried out concurrently, there is a sequence of activities that is determined by the activity relationships. Due to the nature of the construction operations, activities are normally sequenced based on physical relationships. An activity starts, and after a certain amount of work is accomplished, a following activity can begin and continue

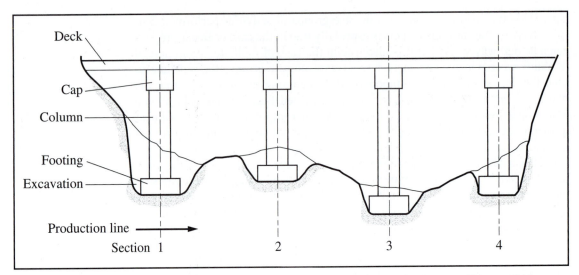

FIGURE 4.28 Linear schedule for a bridge project with sequence of activities. (Selinger, Shlomo. 1980. "Construction Planning for Linear Projects," *Journal of the Construction Division*, ASCE, 106 (CO2). Reproduced by permission of ASCE.)

concurrently. Consider the bridge work shown in Fig. 4.28. Section 1 excavation must be finished before the section 1 footing activity can begin. Once the excavation for the first footing is complete, the excavation crew moves to the

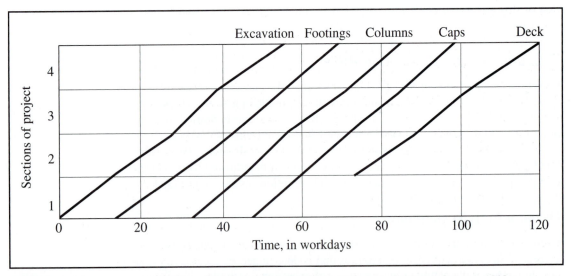

FIGURE 4.29 Linear schedule for a bridge project with velocity diagrams. (Selinger, Shlomo. 1980. "Construction Planning for Linear Projects," *Journal of the Construction Division*, ASCE, 106 (CO2), Reproduced by permission of ASCE.)

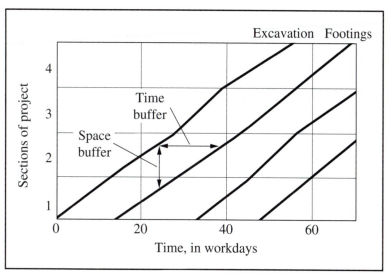

FIGURE 4.30 Linear schedule for a bridge project showing buffers. (Selinger, Shlomo. 1980. "Construction Planning for Linear Projects," *Journal of the Construction Division*, ASCE, 106 (CO2). Reproduced by permission of ASCE.)

second footing and the carpenter crew building the footings starts work on the first footing. This process continues with the activities that follow: column, cap, and deck.

Draft the Linear Schedule

A linear schedule is a series of individual velocity diagrams (Fig. 4.29). Time is plotted on the horizontal scale and the common production measure is plotted on the vertical axis. The individual activity velocity diagrams are plotted in order of occurrence allowing for start and/or finish lags that ensure that the activity production lines do not overlap. An overlap would indicate a work conflict and lost productivity.

To avoid interferences, buffers are established between activities based on expected variations in production rates. The buffers can be described in terms of either time or space (Fig. 4.30).

COMPUTER SUPPORT

Construction scheduling software has transitioned from programs on large, mainframe computers using keypunched cards and activity on arrow sequencing in the 1970s to the current stand-alone programs found on personal computers and notebooks with a variety of input devices relying on precedence sequencing. Today's computerized scheduling programs range in price from a few hundred dollars to a few thousand dollars, based in part on their flexibility, reporting

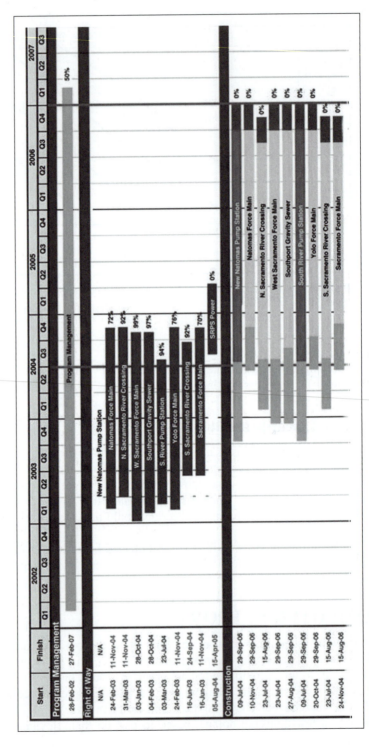

FIGURE 4.31 Monthly progress report schedule for a large sewer project.

capabilities (Fig. 4.31), graphics, and resource allocation characteristics. There are even programs that are four-dimensional (4D) CAD systems where the fourth dimension is time and the three-dimensional (3D) CAD drawings of the project components are linked to the schedule to give a graphical representation of the work sequences. These systems are much more expensive than traditional scheduling software.

Some estimating programs have limited scheduling components built in and/or facilities to link their estimate bid items (activities) to the most popular scheduling software packages. Some project management software packages can also be linked to scheduling packages. With the advent of project websites and the pervasive nature of e-mail, project data can be shared with as many or as few people as desired around the world. Online project collaboration has enabled remote schedule review via PDAs (personal data assistants) and likely soon via cell phone, in addition to all typical computer platforms.

Most contractors use commercial scheduling software such as Primavera's SureTrak or Microsoft Project. Either of these two software packages has good capability for planning and controlling simple to medium-sized (complex) projects. Primavera's more sophisticated project planner, P3, is designed for large-scale complex projects and offers more features for developing schedules.

SUMMARY

Scheduling a project is analogous to preparing a bid estimate. Both attempt to model future project events based on a set of plans and specifications. Scheduling is essentially time estimating. The critical path through a logic network is the longest time path through the network. It establishes the minimum overall project duration. All activities on the critical path are by definition critical.

Velocity diagrams present a graphical picture of the relationship between time and the accomplishment of an activity. Linear scheduling is a planning method that seeks to ensure that the work can proceed continuously, that there are no idle resources, and that activities do not interfere with one another.

This discussion has addressed the mechanics of scheduling construction projects. To assist contractors and others involved in the construction process, there are many good scheduling software packages available. Information about such programs can be found at the websites listed at the end of the chapter.

Critical learning objectives include

- The ability to create a simple bar chart and a simple CPM logic network
- An understanding of how to use a CPM for determining project status and identifying float
- The ability to calculate resource needs and an understanding of resource leveling strategies
- The ability to create a velocity diagram
- The ability to develop a linear schedule

These objectives are the basis for the review questions in this chapter.

REVIEW QUESTIONS

4.1 The estimate for a five-story office building included 72 doors on each of the upper three floors. In developing the bid, the estimator used a production rate of 2.0 carpenter man-hours to hang a door. The project superintendent is organizing the carpenter crews to include three carpenters per crew. What is the duration (in days) of the activity to hang all the doors on the upper three floors? Assume an 8-hr workday.

4.2 The estimate for a three-story dormitory included 30 plumbing fixtures on each of the floors. In developing the bid, the estimator used a production rate of 0.625 man-hours to install a fixture. The project superintendent is organizing the plumbing crews to include two plumbers per crew. What is the duration (in days) of the activity to install all the fixtures in the dormitory? Assume an 8-hr workday.

4.3 Construct a bar chart to use in managing the construction of a single-span bridge. The identified activities are
■ Drive piling west abutment (WA), 2 days
■ Drive piling east abutment (EA), 2 days
■ Construct WA cast in place pile cap, 3 days
■ Construct EA cast in place pile cap, 3 days

Each pile cap requires 5 days curing time before the precast girders can be placed.
■ Set girders, 1 day
■ Construction of cast-in-place concrete deck, 7 days
■ Cure deck, 14 days
■ West and east concrete approach slabs, 5 days each

There is only one crane available for pile driving and setting girders. The company plans to use only one crew of carpenters and finishers for the concrete work.
a. What is the project duration?
b. Are the two "construct approach slab" activities critical to the project completion?

4.4 Construct a bar chart to use in managing the construction of a single-span bridge as described in Question 4.3. Again, there is only one crane available for pile driving and setting girders, but the company now plans to use two carpenter crews for the concrete work. The deck construction will take only 5 days using two crews.
a. What is the project duration?
b. If two cranes were available for pile driving, what would be the project duration?

4.5 Develop a logic network for the sequence of activities listed in the table.

Activity	PIB
5	None
10	None
15	5
20	5
25	10
30	15
35	25
40	20, 30, 35

4.6 Develop a logic network for the sequence of activities listed in the table.

Activity	PIB
5	None
10	5
15	5
20	10, 15
25	15
30	20
35	20, 25
40	25
45	30, 35, 40

4.7 Develop a logic network for the sequence of activities listed in the table. Calculate the ES, EF, LS, and LF times for all activities.
   a.   What is the project duration?
   b.   What are the critical activities?

Activity	Duration	PIB
5	2	None
10	3	None
15	3	5
20	4	5
25	1	10
30	2	15
35	2	25
40	5	20, 30, 35

4.8 Develop a logic network for the sequence of activities listed in the table. Calculate the ES, EF, LS, and LF times for all activities.

a. What is the project duration?

b. What are the critical activities?

Activity	Duration	PIB
5	4	None
10	3	5
15	4	5
20	2	10, 15
25	3	15
30	1	20
35	4	20, 25
40	3	25
45	2	30, 35, 40

4.9 Calculate the total, free, and interfering float for the Question 4.7 activities.

4.10 Calculate the total, free, and interfering float for the Question 4.8 activities.

4.11 Determine the float for the activities in the partial network shown here.

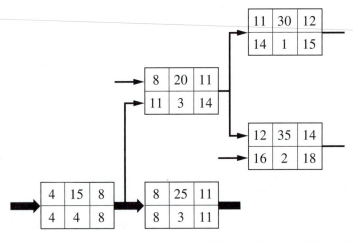

4.12 What type of float (TF, FF, IF) can activities B, E, F, and H have? Activity H is a critical activity on the critical path.

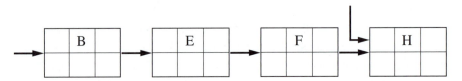

4.13 The bid estimate for a five-story building having a basement had the concrete quantities and durations for the individual floors as shown here.

Floor	Quantity of concrete (cy)	Duration, workdays
Basement	4,000	20
First	4,000	25
Second	3,000	15
Third	3,000	15
Fourth	2,000	10
Fifth	2,000	10

Develop a velocity diagram for the concrete placement activity.

4.14. The bid estimate for a highway had the excavation quantities and durations for the roadway stations as shown here.

Station	Quantity of excavation (cy)	Duration, workdays
0+00	0	0
5+00	36,000	5
10+00	43,000	6
15+00	20,000	3
20+00	5,000	1
25+00	4,000	1

Develop a velocity diagram for the concrete placement activity.

REFERENCES

1. Bubshair, Abdulaziz A., and Michael J. Cummingham (1998). "Comparison of Delay Analysis Methodologies," *Journal of Construction Engineering and Management,* ASCE, 124,(4), July–August.

2. Callahan, Michael, Jim Rowings, and Daniel Quackenbush (1991). *Construction Project Scheduling,* The McGraw-Hill Companies, New York.

3. *Construction Planning & Scheduling* (1997). The Associated General Contractors of America, 333 John Carlye Street, Suite 200, Alexandria, VA.

4. Dressler, Joachim (1980). "Construction Management in West Germany," *Journal of the Construction Division*, ASCE, 106(CO4), December.

5. Finke, Michael R. (1999). "Window Analyses of Compensable Delays," *Journal of Construction Engineering and Management*, ASCE, 125(2), March–April.

6. Griffis, Fletcher H., and John V. Farr (1999). *Construction Planning for Engineers,* The McGraw-Hill Companies, New York.

7. Harris, Robert B., and Ioannou, Photios G. (1998). "Scheduling Projects with Repeating Activities," *Journal of Construction Engineering and Management,* ASCE, 124(4), 269–278.

8. Hunkele, Lester, and Cliff J. Schexnayder (2001). "Contracting for the Pentagon," *29th Annual Conference of the Canadian Society for Civil Engineering (CSCE),* Canadian Society for Civil Engineering, Victoria, June, C50.

9. Kartam, Saied (1999). "Generic Methodology for Analyzing Delays Claims," *Journal of Construction Engineering and Management,* ASCE, 125(6), November–December.

10. Selinger, Shlomo (1980). "Construction Planning for Linear Projects," *Journal of the Construction Division,* ASCE, 106(CO2), June.

11. Tabler, L. Earl, Jr. (1965). "Installation Procedure for Hot Metal Facility," *Journal of the Construction Division,* Proceedings ASCE, 91(CO1), May.

12. Thabet, W.Y., and Y. J. Beliveau, (Dec 1994). "HVLS Horizontal and Vertical Logic Scheduling for Multistory Buildings," *ASCE Journal of Construction Engineering and Management,* 120(4).

13. "Wind and Steel Erection Cited as Cause of D.C. Truss Collapse" (2001). *ENR,* May 21, p. 20.

14. Vorster, M. C., Y. J. Beliveau, and T. Bafna (1992). "Linear Scheduling and Visualization," *Transportation Research Record No. 1351,* 32–39.

WEBSITE RESOURCES

1. http://rdl.train.army.mil/soldierPortal/atia/adlsc/view/public/10898-1/fm/5-412/toc.htm   FM 5-412, Project Management. Department of the Army. Provides an online copy of the Army Corps of Engineers' project management field manual.

2. www.primavera.com   Primavera Systems, Inc. Primavera is a leading provider of comprehensive project management, control, and execution software.

3. http://office.microsoft.com/en-us/project/FX100649011033.aspx   Microsoft Project 2007. Microsoft Project is a tool to schedule, organize, and analyze tasks, deadlines, and resources.

4. www.icivilengineer.com/Construction/Construction_Management/Construction_Scheduling   iCivilEngineer, the construction scheduling subsection of the construction management section carries articles on construction scheduling.

5

Construction Cost Estimates

Estimating is determining how to construct the specified work in the most economical manner and within the time allowed by the contract. The format of all estimates should be as consistent as possible. A work breakdown structure should be established for this purpose. An alternate method to detailed task-by-task estimate preparation, especially in the early stages of project development when details are not available, is parametric estimating. Direct costs are those that can be attributed to a single task of construction work. Indirect costs are those that cannot be attributed to a single task of construction work.

INTRODUCTION

One of the most consequential elements that govern the success of a construction company is the ability to prepare fully detailed and accurate estimates for the cost of performing work (Fig. 5.1). For a typical facility, the **estimator** prepares a series of different estimates throughout the project's development and life (Fig. 5.2). The types of estimates are

1. *Estimates for conceptual planning.* An estimate at this stage in a project's development is, by necessity, based on only very general parameters of facility size, anticipated quality of construction, and use of the facility. A **conceptual estimate** is prepared based on a defined project scope from the facility owner. The owner should approve the scope document that serves as the basis of the estimate as the scope identifies the baseline from which the estimate is derived. The expected accuracy of such an estimate is in the order of plus or minus 15% to 25%. These estimates are prepared for early budgeting to determine project feasibility and to develop project financing.

2. *Estimates for feasibility.* Using preliminary design information and after project scope is completely defined, a broad budget **feasibility estimate** can be prepared. Major items of equipment can be priced and their costs

estimator
One whose primary assignment is to estimate the cost of projects and change orders for bids and negotiations.

conceptual estimate
A cost estimate prepared from only a conceptual description of a project. This is done before plans, specifications, and other project details have been fully developed.

feasibility estimate
An estimate of project cost that is prepared before complete construction documents are available.

FIGURE 5.1 Construction of a concrete frame building.

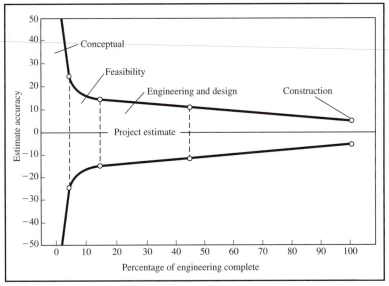

FIGURE 5.2 Estimate accuracy versus percentage of engineering complete.

input into the estimate. The local labor market should be checked for qualified craftsmen. With better scope definition, the expected accuracy is in the order of plus or minus 10% to 15%.

3. *Estimates during engineering and design.* Based on schematic level design documents, it is probable that major quantities can be quantified (tons of steel) and types of construction (steel vs. concrete) identified. Typically, organizations update an estimate at specified design milestone points (15%, 30%, and 60% design) or when there is a scope change. Therefore, as design progresses, estimated accuracy improves into a range of plus or minus 5% to 10%. The designer and project owner may use the early estimates not only to review expected total project costs but for the purpose of reviewing design cost drivers and performing **value engineering** (VE) analyses.

4. *Estimates for construction.* These are itemized cost computations based on a complete set of contract documents. Construction estimates can be prepared using historical average costs or by creating crews and assigning a production rate to crew and equipment. The method employed depends on the type of work. Building work contractors make greater use of historical data whereas heavy/highway contractors typically estimate using crew production. In either case, the expected accuracy is plus or minus 5%.

5. *Estimates for change orders.* These are generally performed on the project site in response to an owner-directed change in project scope.

To provide a consistent set of elements that are applicable to bidding, budgeting, and change order preparation, uniform methods must be employed in the assembly of the estimate. Therefore, this chapter describes methods, procedures, and formats for the preparation of construction cost estimates. All expected costs should be included in the cost estimate. An **engineer's estimate** should be developed as accurately as possible, and should be based on the best information available.

Estimating Methodology

Whenever adequate design information is known, detailed estimating methods should be employed. When details cannot be reasonably assumed, historical unit prices can be used. The methodologies and procedures described here are applicable whether the cost estimate is prepared manually or by using computer-estimating software systems. There are database-driven estimating software systems for all Construction Specifications Institute (**CSI**) MasterFormat divisions that are universally accepted and supported on a national level [web 3].

"MasterFormat is a master list of numbers and titles for organizing information about construction requirements, products, and activities into a standard sequence" [web 3]. Until 2004 the CSI MasterFormat organized construction work activities into 16, level 1 titles (Table 5.1) called divisions. The MasterFormat 2004 Edition, which replaced MasterFormat 1995, expanded the original 16 divisions into 50 divisions. The expanded MasterFormat 2004 includes all aspects of a construction project—from the owner's concept to design, facility management, and finally demolition. The system goes two levels further in

value engineering
A evaluation of a project's objective function for the purpose of bettering the objective in terms of cost and functional parameters.

engineer's estimate
A forecast of construction costs prepared on the basis of a detailed analysis of materials, equipment, and labor for all items of work.

CSI
The Construction Specifications Institute provides technical information to enhance communication among all disciplines of nonresidential building design and construction.

TABLE 5.1   CSI MasterFormat level 1 titles.

MasterFormat 1995	MasterFormat 2004
	Procurement and Contracting Requirements Group
	00 Procurement and Contracting Requirements
	Specifications Group
	General Requirements Subgroup
01 General requirements	01 General requirements
	Facility Construction Subgroup
02 Site construction	02 Existing conditions
03 Concrete	03 Concrete
04 Masonry	04 Masonry
05 Metals	05 Metals
06 Wood and Plastic	06 Wood, Plastics, and Composites
07 Thermal and Moisture Protection	07 Thermal and Moisture Protection
08 Doors and Windows	08 Openings
09 Finishes	09 Finishes
10 Specialties	10 Specialties
11 Equipment	11 Equipment
12 Furnishings	12 Furnishings
13 Special Construction	13 Special Construction
14 Conveying Systems	14 Conveying Systems
15 Mechanical	15 RESERVED FOR FUTURE EXPANSION
16 Electrical	16 RESERVED FOR FUTURE EXPANSION
	Facility Services Subgroup
	20 RESERVED FOR FUTURE EXPANSION
	21 Fire Suppression
	22 Plumbing
	23 Heating, Ventilation and Air Conditioning
	24 RESERVED FOR FUTURE EXPANSION
	25 Integrated Automation
	26 Electrical
	27 Communications
	28 Electronic Safety and Security
	29 RESERVED FOR FUTURE EXPANSION
	Site and Infrastructure Subgroup
	30 RESERVED FOR FUTURE EXPANSION
	31 Earthwork
	32 Exterior Improvements
	33 Utilities
	34 Transportation
	35 Waterway and Marine
	36 RESERVED FOR FUTURE EXPANSION
	37 RESERVED FOR FUTURE EXPANSION
	38 RESERVED FOR FUTURE EXPANSION
	39 RESERVED FOR FUTURE EXPANSION
	Process Equipment Subgroup
	40 Process Integration
	41 Materials Processing and Handling
	42 Process Heating, Cooling, and Drying Equipment
	43 Process Gas and Liquid Handling, Purification and Storage Equipment
	44 Pollution Control Equipment
	45 Industry-Specific Manufacturing Equipment
	46 RESERVED FOR FUTURE EXPANSION
	47 RESERVED FOR FUTURE EXPANSION
	48 Electrical Power Generation
	49 RESERVED FOR FUTURE EXPANSION

TABLE 5.2 One company's cost experience for high-tech facilities.

Task type	Cost range (%)	Cost percent average
Site work	1–3	2.0
Concrete	5–10	6.5
Structural	5–8	6.0
Architectural finishes	6–9	6.0
Tech-specific rooms	6–13	8.0
Mechanical	36–48	42.0
Electrical	11–14	12.0
Indirect costs	14–22	17.5
		100.0

detailing work activities. There are also suggested level 4 titles. This format provides the building construction industry with a uniform system for organizing information.

When analyzing construction tasks in an estimate, the estimator should identify those tasks that account for the major portion of the project costs (see Table 5.2). These tasks can be identified by applying the 80/20 rule, which states that approximately 80% of the project cost is contained in 20% of the tasks. Because these significant tasks account for most of the project cost, they should receive prime emphasis and effort in both estimate preparation and review.

In the case of the information shown in Table 5.2, which is only one company's experience with high-tech facilities, it is clear that when estimating those types of facilities, emphasis must be placed on costing the mechanical electrical work. Additionally, **indirect costs** (overhead expense) are a significant portion of the total cost, so they must be prudently considered. Indirect costs include expenses such as job supervision, office staff, project office facilities, and temporary utilities. The three tasks: mechanical, electrical, and work in the "tech-specific rooms" plus indirect costs typically account for 80% of the project cost when building high-tech facilities.

indirect costs
Overhead or auxiliary costs incurred in completing a project but not attributable to any specific tasks.

Work Breakdown Structure

The format of all estimates should be as consistent as possible. A work breakdown structure (WBS) should be established for this purpose. The WBS is a hierarchical breakdown of the scope of work (Fig. 5.3). It provides a common, ordered hierarchy framework for summarizing information and for quantitative reporting to customers (the client/owner) and management. The purpose of the WBS is to (1) provide an organized manner of collecting project cost data in a standard format for estimating, cost reporting, and cost tracking; (2) provide a checklist for categorizing costs; and (3) provide a means to maintain historical cost data in a standard format. Separate unique WBSs can be developed for differing types of work (i.e., building construction or heavy/highway). The partial estimate shown in Fig. 5.4 follows the WBS approach. For the Approach Forebay Walls concrete task, there are subtasks for forming, reinforcing steel, and concrete.

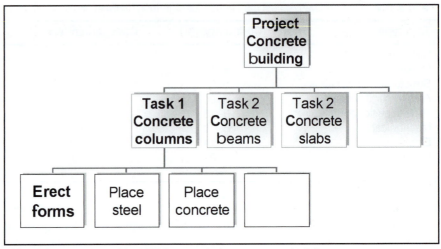

FIGURE 5.3 Hierarchical WBS example for a concrete building.

The WBS is a standard product-oriented format that identifies all project-required work. The WBS provides a common framework for preparing cost estimates and collecting construction project cost data. A WBS model could consist of stepped levels of detail. At each level there could be many parts.

Level 1 Facility (building)

Level 2 (System) 02—Superstructure (columns, beams, floors)

Level 3 (Subsystem) 01—Construction steps

The WBS elements must be:

■ Definable—can be described and easily understood by all project participants

■ Manageable—a meaningful unit of work where specific responsibility and authority can be assigned to a responsible individual

■ Estimateable—time duration and cost can be estimated.

■ Independent—minimum interface with or dependence on other ongoing elements (i.e., assignable to a single control account and clearly distinguishable from other work packages)

■ Integratable—integrates with other project work elements and with higher level cost estimates and schedules to include the entire project

■ Measurable—can be used to measure progress; has start and completion dates and measurable interim milestones

■ Adaptable—sufficiently flexible so the addition/elimination of work scope can be readily accommodated in the WBS framework

Degree of Detail

All cost estimates should be prepared on the basis of calculated quantities and production costs that are commensurate with the degree of design detail known

Bid item	Description	Units	Bid quantity	Labor	Burden	Permanent material	Construction material	Direct total	
270000	APPROACH FOREBAY WALLS	LS	1	3,855,323	5,697,439	9,379,140	2,481,735	38,799,408	
272000	CONCRETE WORKS	M3	79,222	1,974,779	2,864,424	7,801,610	2,057,868	17,622,021	$222
									$221
272200	WALL TOE	M3	39,883	523,942	749,520	3,668,550	152,206	5,751,959	
272300	WALL & COUNTERFORTS	M3	29,435	1,109,069	1,617,520	3,002,580	1,484,961	8,954,037	
272400	CABLE DUCT	M3	9,920	337,618	491,327	1,008,480	420,003	2,765,425	
			79,238	1,970,629	2,858,367	7,679,610	2,057,170	17,471,422	$221
272210	FORMS	M2	3,124	55,711	81,792	-	116,780	351,301	
272310	FORMS	M2	38,520	686,930	1,008,516	-	1,439,934	4,331,657	
272410	FORMS	M2	10,811	192,793	283,050	-	404,131	1,215,720	
			52,455	935,434	1,373,357	-	1,960,845	5,898,678	$74
272220	REINFORCEMENT STEEL	TONS	2,990	374,307	539,246	1,674,400	16,130	3,060,242	
272320	REINFORCEMENT STEEL	TONS	2,208	372,713	540,143	1,236,480	35,735	2,625,181	
272420	REINFORCEMENT STEEL	TONS	738	118,621	172,363	413,280	10,531	851,543	
			5,936	865,642	1,251,752	3,324,160	62,396	6,536,966	$83
272230	CONCRETE	M3	39,883	93,924	128,483	1,994,150	19,296	2,340,416	
272330	CONCRETE	M3	29,435	49,426	68,861	1,766,100	9,292	1,997,200	
272430	CONCRETE	M3	9,920	26,204	35,914	595,200	5,341	698,161	
			79,238	169,553	233,258	4,355,450	33,929	5,035,777	$64

FIGURE 5.4 Hierarchical WBS estimate example for a concrete structure.

or assumed. This is accomplished by separating construction into its incremental parts. These parts are commonly referred to as construction tasks and are the line-by-line listings of every estimate. Each task is then defined and priced as accurately as possible.

> Tasks are seldom spelled out in the contract documents but are necessary for evaluating the requirements and developing their cost.

At the most detailed level, each task is usually related to and performed by a crew. The estimator develops the task description by defining the type of effort required to construct a work item. To complete a work item, multiple tasks may be necessary. Task descriptions should be as complete and accurate as possible to lend credibility to the estimate and aid in later review and analysis. Whenever a significant number of design assumptions are necessary, such as in design-build projects, the estimator may have to use historical cost data from previous similarly designed projects and/or use parametric estimating techniques.

Parametric Estimates

During the earliest stages of project development, prior to any design work, there is limited information about the project. However, there is the need to establish the approximate cost in order to evaluate options and to make choices between needs and feasibility. Because there is very little project definition at this point in time, conceptual estimates usually rely on parametric techniques to extrapolate from past experience the cost of future projects.

Parametric estimating is the process of using various factors to develop an estimate. The factors are based on engineering parameters and developed from historical databases, construction practices, and engineering/construction technology. The parameters include physical measures that describe project definition characteristics (e.g., size, square feet; building type, hospital; foundation type, piling; exterior closure material, concrete block; roof type and material, composition; and number of floors). The appropriateness of selecting the parametric method depends on the extent of project definition available, the similarity between the project and other historical data models, and the ability to calculate details and known parameters or factors for the project.

EXAMPLE 5.1

Estimate the cost of a 250,000-sf warehouse building. Several similar buildings have been constructed in the last several years. The first one, built 6 years ago, cost $2,250,000 and had 150,000 sf of floor space. Two years ago, a small 80,000-sf warehouse was completed for $1,680,000, and just last year a 200,000-sf structure was completed for $3,400,000.

Using the historical data, a sf cost can be deduced and applied to the proposed warehouse to arrive at a parametric estimate of the cost.

$$\frac{\$2,250,000}{150,000 \text{ sf}} = \$15.00 \text{ per sf}$$

$$\frac{\$1,680,000}{80,000 \text{ sf}} = \$21.00 \text{ per sf}$$

$$\frac{\$3,400,000}{200,000 \text{ sf}} = \$17.00 \text{ per sf}$$

Therefore, the average cost per sf has been $17.666 per sf.

The estimated cost for the new warehouse is $4,416,500 (250,000 sf × $17.666 per sf).

The procedure could be given added refinement by making an inflation adjustment to the average sf cost. Another refinement that is sometimes used is a scale or project size factor to account for the size of the building.

There are multiple mathematical methods to arrive at specific indices. In Example 5.1, a simple average of the individual costs was used. Three common methods of deriving an index value are

$$\text{Index value (average)} = \frac{\sum\limits_{i=1}^{n} C_i}{n} \qquad [5.1]$$

$$\text{Index value (inverse)} = \frac{n}{\sum\limits_{i=1}^{n} (1/C_i)} \qquad [5.2]$$

$$\text{Index value (root of the product)} = (C_1 \times C_2 \times C_3 \times \cdots \times C_n)^{\frac{1}{n}} \qquad [5.3]$$

Where C = the individual costs elements

    n = the numbers of cost elements

EXAMPLE 5.2

Considering the cost elements that were derived in Example 5.1 calculate the index values using the three different methods.

The cost elements in Example 5.1 were $15.00, $17.00, and $21.00 per sf.

Using Eq. [5.1]: $\dfrac{\$15.00 + \$17.00 + \$21.00}{3}$ = $17.666 per sf.

Using Eq. [5.2]: $\dfrac{3}{(1/\$15.00 + 1/\$17.00 + 1/\$21.00)}$ = $17.330 per sf

Using Eq. [5.3]: $(\$15.00 \times \$17.00 \times \$21.00)^{1/3}$ = $17.495 per sf

From the results shown in Example 5.2, it is clear that the different methods yield very similar results and the choice of which method to use is a matter of personal choice.

Cost Estimating Relationships   It is often desirable to refine parametric estimates by the use of cost estimating relationships (CERs). If the cost data presented in Example 5.1 had come from projects in several different locations, index values such as those published by RSMeans [2, 3, and 4] could be used to adjust the cost to a particular base city.

$$\frac{\text{Index for City A}}{\text{Index for City B}} \times \text{City B cost} = \text{City A cost}$$

Similarly, indices can be applied to adjust cost data for:

Inflation—Unit cost can be adjusted for the time difference between the historical projects and the estimated project.

Project size—Project size can affect cost, therefore, in developing a CER, size of the historical projects in relation to the estimated project must be factored into the cost relationship.

Unit cost—The costs of certain items are independent of project size; as a result, an estimator must have a clear understanding of the proposed project scope.

Parametric estimation should be performed only by experienced estimators because these estimates are based primarily upon the assumptions that are being made based on past experience.

Accuracy and Completeness

Accuracy and completeness are critical to all cost estimates. An accurate and complete estimate establishes accountability and enables management to place greater confidence in the estimate. It is always prudent to make a deliberate examination of the contract documents for confusion about project scope, for conflicting requirements within the specifications, and to identify the primary risk factors.

Review

All construction cost estimates must be given an independent review. In the case of a construction company, the estimate is usually reviewed by or with other estimators in the firm and by company management. Engineering firms may have their estimates reviewed and checked by an independent estimating service. The estimate should be reviewed for the purpose of confirming the validity of the assumptions and the logic used in estimating the cost of construction tasks. The review should include a check of the quantities, prices, basis for production values, and arithmetic. It is important that the reviewer develop and use a uniform checklist procedure in the review process to ensure that important considerations have not been overlooked.

unit cost
The estimated or actual cost of a single unit of work.

construction documents
The drawings and specifications setting forth the requirements for constructing the project.

general conditions
That part of the contract documents that defines the rights, responsibilities, and relationships of all parties to the construction contract.

BASICS FOR PREPARATION OF ESTIMATES

Construction cost estimates consist of

1. Descriptions of work elements to be accomplished (tasks)
2. A quantity of work required for each task
3. A cost for each task quantity
4. Time to complete each task

A **unit cost** for each task is developed to increase the accuracy of the estimating procedure and should provide a reference comparison to historic experience. Lump sum estimating, when used at the task level, must be documented.

Review Contract Documents

The first step in preparing a construction estimate is to thoroughly examine the **construction documents** starting with the general and supplemental conditions of the contract. Clauses in the **general conditions** of the specifications can affect project indirect cost and methods of construction. Additionally, the estimator should carefully review the supplemental (special) conditions as these describe the unique characteristics of the project and alter the general conditions. The estimator should also review any environmental impact profile and mitigation commitment statements if the work must be built under such agreements. During this examination, it is appropriate to make notes of anything that can affect construction duration and the cost to perform the work, or add indirect and overhead cost to the work.

If the owner conducts a **prebid conference**, the estimator should attend. Before attending such meetings, it is necessary to review thoroughly the project plans and specifications as the meeting will offer an opportunity to resolve questions.

Planning the Work

The estimator must thoroughly understand the scope of work and the project environment. To do this, the estimator must review the drawings (Fig. 5.5), specifications, and referenced documents.

> The detailed review enables the estimator to formulate a construction sequence and duration. A site visit is strongly recommended to enable the estimator to relate the physical site characteristics to the available design parameters and details.

prebid conference
A meeting held prior to bid opening for the purpose of explaining the project and answering questions that bidders have with respect to the contract documents and the work.

FIGURE 5.5 The estimator must carefully review the project drawings.

This is particularly important on projects with difficult site conditions, major maintenance and repair projects, alteration/addition projects, and environmental projects. The construction sequence must be developed as soon as possible and should be used to provide a checklist of construction requirements throughout the cost estimating process.

Quantities

Quantity take-off is an important part of cost estimation. It must be as accurate as possible and should be based on all available engineering and design data. After the scope has been analyzed and broken down into the construction tasks, each task must be quantified prior to pricing. Quantities should be shown in standard units of measure and should be consistent with design units (English or metric). The detail to which the quantities are prepared for each task is dependent on the level of design detail.

> Project notes should be added at the appropriate level in the estimate to explain the basis for the quantity calculations and to explicitly explain contingencies.

Types of Costs

Various types of cost elements must be evaluated in detail.

direct costs
All cost elements that can be associated with a specific item of project work.

profit
The amount of money, if any, that a contractor retains after completing a project and paying all costs for materials, equipment, labor, and overhead.

supplies
Materials used during the construction process that are not permanently incorporated into the completed facility.

Direct costs. The costs attributed to a single task of construction work are known as **direct costs**. These costs are usually associated with a construction crew performing a task using specific materials and equipment. The costs of labor to erect forms and of the concrete placed in the forms are direct task costs. Subcontracted costs should be considered direct costs to the prime contractor in estimates.

Indirect costs. The costs that cannot be attributed to a single task of construction work are classified as indirect costs. These costs include overhead, **profit**, and bond.

Estimates based on a detailed design will be developed by calculating the direct cost of labor, materials, construction equipment, **supplies**, and subcontractors. Applicable indirect costs will be added to reflect the total construction cost. Other pricing considerations, including escalation, construction contingencies, and profit, will be added to the construction costs to determine the total project cost.

Production

It is helpful to have a common method of recording and reporting production. The most convenient and useful *unit of work done* and *unit of time* to use in calculating productivity for a particular activity is a function of the specific work task being analyzed. To make accurate and meaningful comparisons and conclusions about production, it is best to use standardized terms. *Production rate* is the relationship of work done and the time required to accomplish that work (Eq. [5.4]). It can be cubic yards per hour, tons per shift (also indicate the duration of the shift), or feet of trench per hour.

$$\text{Production rate} = \frac{\text{Work done}}{\text{Unit of time}} \qquad [5.4]$$

- *Unit of work done.* This denotes the unit of production accomplished. It can be the volume or weight of the material moved, the number of pieces of material cut, the distance traveled, or any similar measurement of production.
- *Unit of time.* This denotes an arbitrary time unit such as a minute, an hour, a 10-hr shift, a day, or any other convenient duration in which the unit of work done is accomplished.

Time-Required Formula The inverse of the production-rate formula (Eq. [5.4]), is sometimes useful when scheduling a project because it defines the time required to accomplish an arbitrary amount of work.

The most convenient and useful units of *time* and *work done* for a particular task or piece of equipment are a function of the specific work task. To be able to make accurate and meaningful comparisons and conclusions, it is best to standardize the production data terms.

Productivity and Price Resources

Various productivity and pricing resources should be available to the estimator. When using the data from such resources, the estimator's experience and ability to relate the data to a specific circumstance is important. To do this, the estimator must visualize the building process. *Historical data* are the most commonly used productivity and pricing resources. Historical production and costs data from similar past work are excellent resources when adequate details have been saved and adjustments to project specifics can be defined. Portions of other estimates having similar work can be retrieved and repriced to the current project rates.

Development of Specific Tasks

The developed task descriptions must describe the scope and material requirement. The unit cost for each task is developed as a direct cost with separate costing for the labor, material, equipment, and supplies. Notes that explain key factors in the pricing and methodology should accompany the task development. Comparison with pricing guides, such as those published by R. S. Means [1, 2, and 3] or Frank R. Walker [10], is recommended.

Labor Unit Cost

This cost is based on a defined crew that performs the tasks at an assigned production rate. Hourly rates for each craft are applied to the crew labor to arrive at the hourly crew **labor cost**. The total crew labor cost/hour is divided by the expected production rate (units/hour) to derive the labor cost/unit.

Equipment Unit Cost

The only reason for purchasing equipment is to perform work, which will generate a profit for the company. The expense associated with productive machine work is commonly referred to as ownership and operating cost (O&O). O&O cost is stated on an hourly basis (i.e., $90/hr for a dozer) because it is used in calculating the cost per unit of machine production. If a dozer can push 300 cy per hour and it has a $90/hr O&O cost, production cost is $0.300/cy ($90/hr ÷ 300 cy/hr). The estimator/planner can use the cost per cubic yard figure directly.

> *Ownership cost* is the cumulative result of those cash flows an owner experiences whether or not the machine is productively employed on a job. Most of these cash flows are expenses (purchase price), but a few are cash inflows (salvage value).
>
> *Operating cost* is the sum of those expenses an owner experiences by working a machine on a project. Typical expenses include (1) fuel; (2) lubricants, filters, and grease; (3) repairs; (4) tires; and (5) replacement of high-wear items.

Operator wages are sometimes included under operating costs, but because of wage variance between jobs, the general practice is to keep operator wages as a separate cost category. The calculation procedures for equipment costs are developed in Chapter 12.

Material Unit Cost

This cost is developed using vendor quotes, historical costs, commercial pricing sources, or component calculations. The price should include taxes and delivery to the project site.

Unit Cost Reference Data

There are available, through subscription or purchase [2, 3, 4, and 10] estimating cost data books. The basis of the cost data is typically explained along with adjustment methodology. Such publications provide a valuable "second opinion" to the estimator.

Costs and Pricing

The cost for each task is developed by summing the direct cost elements for labor, equipment, materials, supplies, and subcontractors. The indirect costs and other markups associated with each task or work item should be identified and considered separately.

When using historical pricing, adjustments must be made for project location, work methodology, quantity of work, and other dissimilarities that affect prices.

For cost estimates prepared during preliminary or planning phases, when a detailed design is not available, predetermined unit prices for key systems adjusted to current pricing levels may be used. The estimator must use extreme care and sound judgment when using predetermined unit costs. The basis for the unit costs should be well documented and included in the supporting data of the estimate.

Supporting Documentation

The estimator should always remain mindful of the documentation necessary to support the estimate. Support documentation includes project work narratives and schedule, backup data, and drawings and sketches (Fig. 5.6).

Narrative of Contract Costs This part of the *estimate of construction cost* consists primarily of those sheets, with notes, that describe the tasks and costing. It contains discussions, considerations, and the developed construction plan. The types of items normally included are

CREWS COMPOSITION						
Overburden						
		PRODUCTIVITY =	475	M3/HR @ 80%	Crew E01110	
	1	HYD EXC 385BL OF	6	M3 BUCKET	1	Foreman
	1	LOADER 988FX OF	0	M3 BUCKET	1	Operator Cl 1
	4	CAT 773B CAPACITY	55	TONS	5	Operator
	1	DOZER D8 TO HELP THE MOVE THE MATERIAL			3	Labors
Rock						
		PRODUCTIVITY =	795	M3/HR @ 80%	Crew E01114	
	1	HYD EXC 385BL OF	6	M3 BUCKET	1	Foreman
	1	LOADER 988FX OF	9	M3 BUCKET	2	Operator Cl 1
	8	CAT 773B CAPACITY	55	TONS	12	Operator Cl 3
	3	DOZER D8 WITH SINGLE SHANK			4	Labors
	1	DOZER D8 TO HELP THE MOVE THE MATERIAL				
Rock - Structural						
		PRODUCTIVITY =	288	M3/HR @ 80%	Crew E01121	
	1	HYD EXC 345BL OF	4	M3 BUCKET	1	Foreman
	1	LOADER 988FX OF	0	M3 BUCKET	1	Operator Cl 2
	4	CAT 769D CAPACITY	36	TONS	6	Operator Cl 3
	1	DOZER D8 WITH SINGLE SHANK			3	Labors
	1	DOZER D8 TO HELP THE MOVE THE MATERIAL				

FIGURE 5.6 Crew documentation for an excavation project.

■ *Project narrative.* The project narrative provides general details of the project. The narrative defines the assumptions made during the preparation of the cost estimate. It describes the project requirements that must be performed in sufficient detail to give a clear understanding of the scope of work. It describes project details including length, width, height, and shape of primary features; special problems that will be encountered in performing the work; site conditions affecting the work; time for mobilization and demobilization of all equipment; and the reasons for unusually high or low unit prices. Other factors to be considered in the project narrative are use of overtime, shift work, limited working hours, phasing, and subcontracting. Project-related details include site access, borrow areas, construction methodology, unusual conditions (soil, water, or weather), unique techniques of construction, equipment/labor availability, and environmental concerns as appropriate.

EXAMPLE 5.3

Narrative—CSI Division 6 Wood, Plastics, and Composites

Rough carpentry includes wood blocking at the roof and interior blocking.

Finish carpentry includes vanity tops in public toilets, dressing room counters and cabinets, concession counters, ticket counters, projection room stands, wardrobe units, storage shelving, and suite cabinets and counters.

■ *Construction schedule.* The cost estimator will prepare a construction schedule (Fig. 5.7) to support the cost estimate that is consistent with the schedule for completion of the project. This schedule may be in the form of a bar chart or network analysis system, but it must identify the sequence and duration of the major tasks on which the cost estimate is developed. The **preliminary schedule** must be prepared in sufficient *detail to adequately develop the required labor, equipment, crew sizes, and production rates* required for each of the major construction tasks.

preliminary schedule
A schedule prepared to support development of the project cost estimate.

■ *Equipment and materials utilization.* On those projects involving considerable heavy construction equipment, it is necessary to sufficiently plan the equipment usage against the work schedule to identify the actual number by specific type, and allow for proper mobilization to ensure that demand for the equipment is not over- or understated. Materials that require long lead times and can become critical to the construction schedule should be noted, planned, and adequately considered.

■ *Labor discussion and utilization.* The estimate should clearly state the sources for the various labor classifications and rates. This would include the composition of skills in each crew and what those skills are doing, or can do, according to union rules if applicable. When extensive overtime beyond the normal workday is used in the estimate, an explanation should be included.

Description	Sept.	Oct.	Nov.	Dec.	Jan.	Feb.	Mar.	Apr.	May	June	July	Aug.	Sept.	Oct.	Nov.	Dec.
Clear & Grubb G.C.			▓	▓												
Construction Surveying G.C.																
① Oak Ridge Road Bridge over Bryan Blvd. Sta. 19+03.66-L-																
TPCB & Fills G.C.								▓	▓							
EB 1 & 2										▓	▓					
Close Airport Blvd.										?						
Bents 1, 2, 3											▓					
Erect 16 Girders												▓				
Deck (4 each)													▓	▓		
Appr. Slabs, Rail, Jt.															▓	
Groove Deck SUB																▓
② Inman Road SR2140 Bridge over Bryan Blvd. Sta. 59+56.53-L-																
Utility			▓													
Detour & Cut G.C.							▓	▓								
Substructure								▓	▓							
Erect 28 Girders											▓					
Deck (4 each)												▓	▓			
Appr. Slabs, Rail, Jt.														▓		
Groove Deck SUB																▓

FIGURE 5.7 Schedule prepared by a bridge estimator. Note clearing and grubbing item belongs to the general contractor. The bridge contractor is bidding as a subcontractor.

Estimate Backup Data This part of the estimate consists of all the support and backup documentation.

- *Quantity computations.* The quantity take-off computations for the tasks estimated should be organized by task for the bid items and kept as backup data.
- *Crew, labor, and equipment rates.* The details used to prepare and express the crew composition and associated rates for labor and equipment costs must be shown. The information contained on these sheets provides the backup support for the task unit labor and equipment costs.
- *Production rates.* The information used to develop and analyze crew production rates must be documented.
- *Mobilization, preparatory work, and demobilization.* These costs should be itemized and totaled separately. These costs may be combined at the summary level with overhead if they are not paid as a separate bid item.
- *Overhead costs.* The itemization and calculations of overhead costs, for both the job site and the home office, should be presented.
- *Bond costs.* Bond costs should be calculated.

> The take-off should reference the drawing and clearly explain the computation.

Quotations   Quotations should be collected and compiled by task or bid item into an organized reference. When quotations were not obtained for significant material and supply items, the basis for the cost used in the estimate should be fully described.

> Quotations should be considered proprietary information and should be kept confidential to protect the information entrusted to the estimator and company by suppliers and subcontractors.

Drawings and Sketches   Include all other information pertinent to the estimate, such as drawings and sketches that were used as the basis of the cost estimate. Drawings may include a project map showing the location of the work with respect to principal cities, roads, railways, and waterways; a site map showing the location of the work, borrow, quarry, spoil areas, and existing work access roads; any existing usable facilities; and a construction layout. Supporting documents such as sketches, soil boring, and material classifications are also important.

LABOR

The cost of labor is probably the most variable and difficult to estimate. The local labor market and conditions should be investigated to determine the available supply of all classes of labor and its competence. Local work practices must be studied to ascertain their effect on productivity.

Direct labor costs are defined as base wages plus labor cost fringes (additives) including payroll taxes, fringe benefits, travel, and overtime allowances paid by the contractor for personnel who perform a specific construction task. The various union crafts in construction usually negotiate their own wage rates and working conditions. Therefore, if working union, these must individually be examined and understood. In addition to the actual workers, there are generally working crew foremen who receive an hourly wage and are considered part of the direct labor costs.

Indirect labor costs are wages and labor cost fringes *paid to contractor personnel whose effort cannot be attributed to a specific construction task.* Personnel such as superintendents, engineers, clerks, and site cleanup laborers are usually included as indirect labor costs (project overhead).

Crews

Direct labor cost requirements are broken into tasks of work. A labor crew, including equipment, usually performs each work task. Therefore, the crew must be defined, its cost defined, and a production rate established for the task. Crews

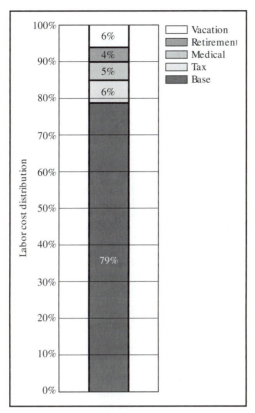

FIGURE 5.8 Typical distribution of total wage rate for a laborer.

may vary in size and mix of skills. The number and size of each crew should be based on such considerations as having sufficient workers to perform a task within the construction schedule, and the limitation of workspace. Once the crews have been developed, the task labor costs can be determined based on the production rate of the crew and the labor wage rates.

Wage Rates

A wage rate must be determined for each labor craft that will represent the total hourly cost rate to the construction contractor. This total rate will include the base wage rate plus taxes and insurance, fringe benefits, and travel or subsistence costs (Fig. 5.8).

Wage rates on government-funded work are generally well defined. The Davis-Bacon Act, PL 74-403, requires a contractor performing construction in the United States for the federal government to pay not less than the prevailing rates established by the Department of Labor. A schedule of minimum rates is

```
* PAIN0051M  06/16/2000
                                    Rates      Fringes
PAINTERS:
 Brush, Roller, Spray and
 Drywall Finishers                  20.23      5.47
--------------------------------------------------------
* PLAS0891A  05/01/2000
                                    Rates      Fringes
CEMENT MASONS                       19.80      3.895
--------------------------------------------------------
PLUM0005E  09/01/1999
                                    Rates      Fringes

PLUMBERS:
 Apartment Buildings over 4 stories
 (except hotels), schools, colleges,
 and speculative office buildings,
 strip shopping centers, churches,
 water coolers, room air conditioning
 units, appliances, packaged ice
 machines, and light commercial
 refrigeration and/or air conditioning
 systems serving a single business in
 a single story building and not to
 exceed 5 h.p. or tons, self-contained
 package unit up to and including 5
 h.p. or tons.                      16.54      4.835
 All other work                     24.85      7.735
--------------------------------------------------------
* PLUM0602A  08/01/2000
                                    Rates      Fringes
STEAMFITTERS, REFRIGERATION AND
 AIR CONDITIONING MECHANICS:
 Light commercial refrigeration
 and/or air conditioning systems
 serving a single business; the
 air conditioning systems shall
 not total more than 15 tons
 and the refrigeration system
 shall not total more than 7 1/2
 tons; apartment buildings over
 4 stories with individual units
 not to exceed 5 tons (excluding
 split units)                       13.75      7.505
 All other work                     25.71      7.505
```

FIGURE 5.9   A Department of Labor wage decision.

included in the project documents. Fig. 5.9 presents a sample Davis-Bacon wage determination. Where labor is in short supply for certain crafts, the work is in a remote area, or it is known that rates higher than the set rate scale will be paid, the higher wage rates should be used instead of the minimum wage since they will be required to attract labor to the job. The Bureau of Labor Statistics provides online data on unemployment by areas of the county [web 2]. The wage rate should be adjusted to include travel time or night differential where these are a customary requirement.

Overtime and Shift Differential

The cost estimator should carefully consider the available working time in the construction schedule to accomplish each task. The efficiency of a second or third shift should be adjusted to recognize that production will not be as high as the day shift for most types of construction operations. A three-shift operation is normally avoided because of lower labor efficiency and the requirement to allow time for equipment maintenance.

Overtime should be included in the labor cost computation when work in excess of 40 hours is required by the construction schedule or is the custom of labor in the local vicinity. Overtime is normally calculated as a percentage of the base wage rate. It is usually based on time and one-half, but may be double time, depending on the existing labor agreements. Tax and insurance costs are applied to overtime, but **fringe benefits** and travel and/or subsistence costs are not.

Example 5.4 shows overtime calculation for 40 hours regular time, plus 8 hours overtime at time and one-half.

fringe benefits
Employee benefits that an employer must pay in addition to an employee's base pay. Typically these may be paid directly to the employee or they may be paid to various agencies on behalf of the employee. They would include health insurance, pension plans, and certain taxes.

EXAMPLE 5.4

Calculate the percentage adjustment that should be used to adjust labor wages on a project that will normally work 8 hr of overtime per week.

Work hours	Equivalent straight time hours
40 hr at straight time =	40.0
8 hr at 1 1/2 times =	12.0
	52.0 hr paid equivalent straight time
52 hr paid/48 hr worked =	1.0833
	$(1.0833 - 1) \times 100\% = 8.33\%$

Taxes and Insurance

Wages paid to employees are subject to payroll taxes such as the Federal Insurance Contributions Act (FICA), Federal Unemployment Tax Act (FUTA), and state unemployment taxes. The FICA tax is for Social Security and Medicare, while FUTA is the federal employer tax used to fund state workforce agencies. Rates for all taxes and insurance should be verified prior to computation for these and any additional components.

Workman's compensation and *employer's liability insurance* costs applicable for the state in which the work is performed should be included in the composite wage rate. The project compensation rate is based on the classification of the major construction work and applies to all crafts employed.

Unemployment compensation taxes are composed of both state and federal taxes. Unemployment compensation tax will vary with each state, while the federal unemployment tax will be constant for all projects. Insurance rates can be obtained from the state unemployment office.

Social Security tax is paid by both the employee and the company. Half is paid out of the employee's base pay. That portion does not represent an extra cost to the employer, but the other half, which is paid by the employer, is an additional cost. The rates and the income ceilings on which social security taxes must be paid vary from year to year. Therefore, the estimator must verify the rate to be used in the cost estimate. Rates can be obtained from the Social Security Administration.

The total percentage of the preceding taxes and insurance is summed and then applied to the basic hourly wage rate plus overtime for the various crafts. Example 5.5 illustrates a method for deriving the total tax and insurance percentage. Since rates are subject to change and in some cases vary by region, the calculations shown are presented as an example only. Actual values must be determined by the estimator for the specific project being estimated.

EXAMPLE 5.5

Calculate the total tax and overhead percentage that should be used to adjust the basic labor rates.

Workman's compensation and employer's liability (this varies according to project location, state, and contractor)	7.60%
State unemployment compensation (varies with each state)	3.20%
Federal unemployment compensation (FUTA)	0.80%
Social Security and Medicaid (FICA)	7.63%
Total taxes and insurance	19.23%

Note that the rates for indirect and overhead labor must also include these applicable costs.

The calculation, however, is not as simple as shown in Example 5.5 because the percentages apply only to certain limited amounts of the employee's wages. As an example, the FUTA tax rate is 6.2% of taxable wages but the taxable wage base is only the first $7,000 in wages during a calendar year. Additionally, employers who pay the state unemployment tax, on a timely basis, receive an offset credit of up to 5.4% regardless of the rate of tax they pay the state. Therefore, the net federal tax rate is generally 0.8% (6.2% − 5.4%). This would equate to a maximum of $56.00 per employee, per year (.008 × $7,000. = $56.00) in federal FUTA tax.

Fringe Benefits

Fringe benefits may include health and welfare, pension, and apprentice training depending on the craft and the location of the work. These summed costs are usually expressed as an hourly cost with the possible exception of vacation, which may be easily converted to an hourly cost. On federally funded projects, the type of fringe benefits and the amount for the various crafts can usually be found with the Department of Labor wage determination in the specifications. Non-union contractors pay comparable fringe benefits directly to their employees.

Example 5.6 illustrates the calculations for fringe benefits. Since the values change and vary by region and contractor, the calculations shown are presented as an example only. The estimator must determine actual company values.

EXAMPLE 5.6

Calculate the total fringe benefit costs that must be added to the basic labor rates.

Health and welfare	$0.70/hr	
Pension	0.75/hr	
Apprentice training	0.00/hr	(N/A in this case)
Total fringe benefits	$1.45/hr	

Some fringe benefits and travel/subsistence costs are subject to payroll taxes. For example, vacation benefits are taxable and should be added to the basic wage rate.

LABOR PRODUCTIVITY

The unit of work one tradesman or one crew can accomplish per hour or per 8-hour shift is the labor productivity for a task. Labor productivity is subject to many diverse and unpredictable factors. But one primary factor that is often overlooked is *leadership*. The estimator must always inquire into who will actually build the project. Most project managers, superintendents, and project engineers are very ambitious. The project leaders who are ambitious first and foremost for the company, and not for themselves personally, usually achieve better results—their labor force is more productive. Leaders who help everyone to succeed set their projects up for success.

There is no substitution for the knowledge and experience of the estimator when estimating labor productivity. For some types of work, the task productivity of crew members such as equipment operators, helpers, or oilers is determined by the productivity of the equipment. For some labor-based crews, the task productivity of craftsmen such as carpenters, steelworkers, and masons may be based on the average of historical data tempered with the experience of the estimator.

Productivity Adjustment

The complexity of the variables affecting productivity makes it difficult to estimate. Production rates are usually based on averaging past production rates for the same or similar work. The estimator must incorporate particular job factors and conditions to adjust historical data to the project being estimated. Other sources for production rates include reference manuals, field office reports, construction logbooks, and observation of ongoing construction.

> Labor productivity is the critical risk factor affecting profitability.

The labor effort needed to perform a particular task varies with many factors, such as

- Relative experience, capability
- Morale of the workers
- Size and complexity of the job
- Availability of materials, tools, and equipment
- Climatic and topographic conditions
- Degree of mechanization
- Quality of job supervision to include planning and sequencing of work
- Amount of task repetition
- Existing labor-management agreements and/or trade practices

The effect of these labor efficiency factors and the work practices that exist in the project locality must be considered in selecting productivity rates.

Heavy civil projects are normally equipment intensive; therefore, productivity for this type of work should be based on equipment production rates.

CONSTRUCTION EQUIPMENT AND PLANT

Construction equipment and plant refers to the tools, instruments, machinery, and other mechanical implements required in the performance of construction work. Construction plant is defined as concrete batch plants, aggregate-processing plants, conveying systems, and any other processing plants that are erected in place at the job site and are essentially stationary or fixed in place. Equipment is defined as items that are portable or mobile ranging from small hand tools through tractors, cranes, and trucks. For estimating purposes, plant and equipment are grouped together as equipment costs.

Selection of Equipment

An important consideration in the preparation of an estimate is the selection of the proper equipment to perform the required tasks. The estimator should carefully consider number, size, and function of equipment to arrive at optimum equipment usage. Some factors to consider during the selection process are

- Job progress schedule (production rate)
- Availability of space, mobility and availability of equipment
- Equipment capabilities
- Distances material must be moved

- Steepness and direction of grades
- Weather conditions
- Hauling restrictions
- Mobilization and demobilization costs

The estimator preparing the estimate must be familiar with construction equipment and job-site conditions. The equipment selected should conform to contract requirements and be suitable for the materials to be handled and conditions that will exist on the project.

Equipment Production Rates

After determining the type of equipment to be employed, the estimator should select the specific equipment size that has a production rate suited to the efficient and economical performance of the work. The size and number of units required will be influenced by equipment production rate, job size, availability of space for equipment operations, and the project construction schedule for the various work tasks. Emphasis must be placed on establishing reasonable production rates (see Chapters 10 and 11).

Mobilization and Demobilization

Mobilization costs for equipment include the cost of loading at the contractor's yard; transportation to the construction site, including permits; unloading at the site; necessary assembly and testing; and standby costs during mobilization and demobilization. All labor, equipment, and supply costs required to mobilize the equipment should be included in the mobilization cost. Demobilization costs should be based on that portion of the equipment that would be expected to be returned to the contractor's storage yard and may be expressed as a percentage of mobilization costs.

Mobilization and demobilization costs for plant should be based on the delivered cost of the item, plus erection, taxes, and dismantling costs at the end of the project. Maintenance and repair are operating costs and should be distributed throughout work accomplishment.

Equipment Ownership and Operating Cost

Equipment ownership and operating costs are discussed in Chapter 12.

Plant Cost

In cases of highly specialized plant, 100% of the plant cost may have to be charged to the project. For less highly specialized plant, some salvage may be anticipated, depending on storage cost, resale value, and probability of sale or reuse in the immediate future. The total amount charged the project, including operation, maintenance, and repair, should be distributed in proportion to the time the plant is used on the various contract items. Cost of plant required for the production of concrete, aggregates, ice or heat for cooling or heating of concrete, etc., should normally be included in the estimate as part of the cost of these materials or supplies manufactured or produced at the site.

Small Tools

The costs of small power and hand tools and miscellaneous noncapitalized equipment and supplies are usually estimated as a percentage of the labor cost. The allowance must be determined by the estimator in each case, based on experience for the type of work involved. Unit prices based on historical data already include a small tools allowance. The small-tool cost will be considered as part of equipment cost. Such allowance can range as high as 12% of direct labor cost but is usually much lower. The cost estimator must ensure that this cost is not duplicated in the overhead rate percentages.

PERMANENT MATERIALS AND SUPPLIES

Permanent materials
Materials that are installed permanently in the completed project.

Permanent materials are those items that are *physically incorporated into and become part of the permanent structure.* Supplies are those items that are used in construction but *do not become physically incorporated into the project,* such as concrete forms. For the purpose of estimating, both can be considered materials unless they need to be separated because of different tax rates.

Sources of Pricing Data

Prices for materials and supplies may be acquired by obtaining quotations or from catalogs and historical data. The estimator should review the pricing and assess the reasonableness prior to use. Care should be taken to make proper allowances for quantity discounts, inflation, and other factors affecting cost.

quotes
To make an offer at a guaranteed price.

Quotes from Manufacturers and Suppliers   **Quotes** should be obtained for all significant materials and installed equipment, and for specialized or not readily available items. Quotations may be received either in writing, telephonically, or by fax. It is preferable to obtain quotes for each project to ensure that the cost is current and that the item meets specifications. If possible, more than one quote should be obtained to be reasonably sure the prices are competitive. The estimator should attempt to determine and ensure that contractor discounts are considered in the estimate. *Quotes should be kept proprietary to preserve the confidentiality entrusted.* A telephone quotation data sheet similar to that shown in Fig. 5.10 should be used for recording telephonic quotations.

Waste and loss considerations may be included in material unit price computations. This methodology results in a quantity take-off of work placement that is not altered to reflect material losses. However, the alternative methodology of increasing the measured quantity by waste and loss quantity is acceptable if the excess quantity will not be used for any other purpose. The methodology used by the estimator should consider the impact of charging labor on the excess quantity. In either case, a statement is required in the estimate explaining the methodology used.

TELEPHONE QUOTATION SHEET	CSI Number:		
Firm quoting:	Project:		
Address:			
Phone number:			
and FAX numbers:			

Person quoting: Estimator:

Item quoted: Date: Total quantity quoted:

Amount:

Description:

Includes sales tax: YES: _____   NO: _____

Per plans & specs: YES: _____   NO: _____

EXCEPTIONS:

FREIGHT INCLUDED TO:

_____ FAS PORT   _____ FOB FACTORY   _____OTHER   _____FOB JOB

		TOTAL QUOTATION	$ _____
ALLOWANCE FOR WASTE AND LOSS			$ _____
Weight	Export packing		$ _____
Volume	U.S. inland freight		$ _____
Quote valid for _____ days	TOTAL MATERIAL COST		$ _____
REMARKS:			

FIGURE 5.10 Example of a telephone quote sheet.

Forward Pricing

Sometimes quotes are requested in advance of the expected purchase date. However, suppliers are reluctant to guarantee future prices and often will only quote current prices. It may be necessary, therefore, to adjust current prices to reflect the cost expected at the actual purchase date. This cost adjustment, if required, should not be included as a contingency, but should be clearly and separately defined in each estimate. Adjust current pricing to future pricing using specific escalation factors. Computations of adjustment should be clear and should be maintained as cost estimate backup support.

Freight

The estimator should check the basis for the price quotes to determine if they include delivery. If they do not include delivery, freight costs to the project site must be determined and included. The supplier can usually furnish an approximate delivery cost. For delivery charge, free on board (FOB) refers to the point to which the seller will deliver goods without additional charge to the buyer. If the materials or supplies are FOB factory or warehouse, freight costs to the construction site should be added to the cost of the materials or supplies.

If the cost of materials or supplies includes partial delivery, FOB to the nearest rail station, the cost of unloading and transporting the materials or supplies should be included in the estimate. If the materials or supplies are a large quantity in bulk, which would require extensive equipment for unloading and hauling, it may be desirable to prepare a labor and equipment estimate for the material handling and delivery.

Handling and Storage

The contractor is usually required to off-load, handle, and stockpile, or warehouse materials on site. These costs should be included in the estimate. An item of electronic equipment requiring special low-humidity storage might have this special cost added to the direct cost of the equipment. For common items, such as construction materials or equipment needing secure storage, the cost for the security fencing, temporary building, and material handling should be considered as an indirect cost and be included in the job-site overhead cost.

Taxes

When applicable, state, and local sales taxes should be added to the cost of materials or supplies. Care should be taken that the sales tax rate is applied as required. The estimator should verify the tax rates and the applicability of these rates for the project location. Sales tax is considered to be a direct cost of the materials and supplies, and is included in the estimate.

SUBCONTRACTED WORK

In construction, specialty items such as plumbing, heating, electrical, roofing, and tile work are usually more effectively performed by subcontract. With so many specialties being performed, subcontract work becomes a very significant portion of the total costs of building construction. Since each estimate should be prepared as practically and as realistically as possible, subcontract costs become a necessary consideration.

The estimator must first determine those parts of the work that will probably be subcontracted. When the work to be subcontracted has been determined, those items will be identified in the estimate.

The cost of subcontracted work is the total cost to the prime contractor for the work performed. Subcontractor's costs include direct labor, materials, supplies, and equipment, second-tier subcontracts, mobilization and demobilization, transportation, setup, and charges for overhead and profit. The total subcontract cost is considered a direct cost to the prime contractor.

Use of Quotations

The estimator may use quotes for the expected subcontracted work when reviewed and verified as reasonable. In lieu of a quotation, each task of the subcontract should be priced as a direct cost with an appropriate rate of subcontractor's overhead and profit added.

OVERHEAD COSTS

Overhead costs are those costs that cannot be attributed to a single task of construction work. Costs that can be applied to a particular item of work should be considered a direct cost to that item and are not to be included in overhead costs. The overhead costs are customarily divided into two categories:

- Job office overhead, also referred to as general conditions or field office overhead
- General home office overhead, commonly referred to as **general and administrative** (G&A) **overhead**

The estimator must be sure that overhead costs are not duplicated between the two categories. Because of the nature of overhead costs, it is not practical to discuss all overhead items. Specific considerations must be carefully evaluated for each project. The estimator must use considerable care and judgment in estimating overhead costs. In the case of building construction, many indirect cost items are frequently described in the General Requirements Section (CSI Division 01) of the contract specifications. If not related to a specific work task, these costs must be identified and appropriately assigned as overhead costs.

Overhead will vary from project to project and may even vary from month to month within any given project. Job overhead items should be estimated in detail for all projects. Detailing of overhead costs for subcontract work is recommended when the impact of these costs is significant.

general and administrative overhead
The contractor's general operating expenses that are not related to a specific project but are incurred in operating the business.

Job Office Overhead

Job overhead costs are those costs at the project site that occur specifically as a result of the project. Examples of job overhead costs are

- Job supervision and office personnel
- Engineering and shop drawings/surveys
- Site security
- Temporary facilities, project office
- Temporary material storage
- Temporary utilities, such as electricity and water

- Preparatory work and laboratory testing
- Telephone and communications
- Permits and licenses
- Insurance (project coverage)
- Quality control
- Operation and maintenance of temporary job-site facilities

The costs of mobilization and preparatory work, including the setup and removal of construction facilities and equipment, are part of overhead costs, unless there is a specific bid item. For large projects, the cost for each part of this initial work should be estimated on a labor, materials, and equipment basis.

General and Administrative Overhead

Home office overhead expenses are those incurred by the contractor in the overall management of the business. Since they are not incurred for any one specific project, they must be apportioned to all the projects.

Each contracting company is organized differently. Each incurs costs differently from varying sources and manages operations of the home office by its own methodology. It is important to understand that home office costs are not standard and fixed. Even though the cost for a specific contractor varies from period to period, a rate is normally averaged as a computation of total home office costs over a sufficient period divided by the total volume of business during that specific period. This rate computation methodology allows distribution and projection to future project estimates. When more specific data are not available, the cost estimator may include empirical rates. Empirical G&A rates typically range from 3% for large contractors to 10% for small contractors. Home office costs are typically included in the estimate of overhead as the product of an average experienced percentage rate times the expected contract amount. Typical categories of home office overhead are

- Main office building, furniture, equipment
- Management and office staff, salary and expense
- Home office utilities
- General communications and travel
- Supplies
- Corporate vehicles
- General business insurance
- Taxes

Duration of Overhead Items

After the overhead items have been listed, a cost must be determined for each. Each item should be evaluated separately. Some items, such as erection of the project office may occur only once in the project. The estimator should use the developed job schedule in estimating overhead duration requirements. Costs reflective of each particular item during the scheduled period should then be applied. The product of duration and unit cost is the overhead cost for the item.

> If a job is not finished on time, the contractor loses dollars because project overhead costs exceed the amount in the estimate.

Distribution of Overhead

The contractor's overhead costs should be summed and distributed to the various bid items. A proportionate distribution is commonly made by percentage ratio of total overhead costs to those direct costs in each item. Regardless of the method of distribution, the estimates should clearly demonstrate the procedures and cost principles applied.

SUMMARY

Successful project owners and construction companies have the ability to prepare fully detailed and accurate estimates for the cost of performing work. Direct labor costs are defined as base wages plus labor cost fringes (additives) including payroll taxes, fringe benefits, travel, and overtime allowances paid by the contractor for personnel who perform a specific construction task. Labor productivity is subject to many diverse and unpredictable factors. There is no substitution for the knowledge and experience of the estimator when estimating labor productivity. An important consideration in the preparation of an estimate is the selection of the proper equipment to perform the required tasks. The estimator should carefully consider number, size, and function of equipment to arrive at optimum equipment usage. Overhead costs are those costs that cannot be attributed to a single task of construction work. Critical estimating learning objectives include

- An understanding of the process whereby various factors are used for parametric estimating
- An understanding of labor unit cost for construction
- An understanding of equipment unit cost for construction
- An understanding of material unit cost for construction
- An understanding of overhead cost for construction
- An ability to calculate the tax and insurance additives to labor cost

These objectives are the basis for the review questions in this chapter.

REVIEW QUESTIONS

5.1 Estimate the cost of a 50,000-sf office building. Similar buildings have been constructed recently. The first one, built 3 years ago, cost $4,100,000 and had 100,000 sf of floor space. Two years ago, a small 50,000-sf office was completed for $2,250,000, and just last year, a 70,000-sf structure was completed for $3,290,000. Use the historical data and a simple average price per sf to arrive at an estimated cost. ($2,216,650)

5.2 Estimate the cost of a 600,000-sf retail building. Similar buildings have been constructed recently. The first one, built 2 years ago, cost $9,100,000 and had 375,000 sf of floor space. Eighteen months ago, a 450,000-sf retail outlet building was completed for $12,250,000, and last year a 700,000-sf structure was completed for $17,290,000. Use the historical data and an inverse average price per sf to arrive at an estimated cost.

5.3 If, in estimating the building in Question 5.2, the root of square method of deriving an average price per sf had been used, what would be the estimated cost?

5.4 A project will require 100,000 gal of diesel fuel over a period of 3 years. The estimate has been prepared using a fuel price of $2.90 per gal. Perform a sensitivity analysis of the effect of rising fuel prices. Some have predicted that the price of diesel could rise to $3.30 per gal in two years. What fuel cost will you use in the estimate? If the estimated consumption is 50,000 gal in the first year, 35,000 gal in the second year, and 15,000 gal in the third year, how would that affect your estimated cost?

$2.90 per gal.		1	2	3	
$/gal	100,000 gal	50,000	35,000	15,000	
2.90	$290,000	$145,000	$101,500	$43,500	
2.95	$295,000	$147,500	$103,250	$44,250	
3.00	$300,000	$150,000	$105,000	$45,000	
3.05	$305,000	$152,500	$106,750	$45,750	
3.10	$310,000	$155,000	$108,500	$46,500	
3.15	$315,000	$157,500	$110,250	$47,250	
3.20	$320,000	$160,000	$112,000	$48,000	
3.25	$325,000	$162,500	$113,750	$48,750	
3.30	$330,000	$165,000	$115,500	$49,500	$309,750
3.35	$335,000	$167,500	$117,250	$50,250	310,000

5.5 Calculate the percentage adjustment that should be used to adjust labor wages on a project that will normally work 10 hours of overtime per week. (10%)

5.6 Calculate the percentage adjustment that should be used to adjust labor wages on a project that will normally work 9 hours of overtime per week.

5.7 Calculate the total tax and overhead percentage that should be used to adjust the basic labor rates. Based on the project location and the historical performance, the company's workman's compensation rate is 6.7%. State and federal unemployment compensation are 2.7% and 0.8%, respectively. The Social Security and Medicaid rate is 7.65%.

5.8 Using the Bureau of Labor Statistics website, determine the current unemployment rate for your area of the country.

REFERENCES

1. *Construction Estimating & Bidding Theory Principles Process* (1999). Publication No. 3505, The Associated General Contractors of America, 333 John Carlyle Street, Suite 200, Alexandria, VA.

2. *Means Assemblies Cost Data* (issued annually). (Data for preliminary or conceptual estimates.) R. S. Means Company, Inc., Construction Plaza, 63 Smiths Lane, Kingston, MA 02364-0800.

3. *Means Building Construction Cost Data* (issued annually). (Price data, reference tables, estimating aids, and illustrations.) R. S. Means Company, Inc., Construction Plaza, 63 Smiths Lane, Kingston, MA 02364-0800.

4. *Means Heavy Construction Cost Data* (issued annually). R. S. Means Company, Inc., Construction Plaza, 63 Smiths Lane, Kingston, MA 02364-0800.

5. Oberlender, Garold D., and Steven M. Trost (2001). "Predicting Accuracy of Early Cost Estimates Based on Estimate Quality," *Journal of Construction Engineering and Management,* ASCE, 127(3), May–June.

6. *OHSA Safety & Health Standards for Construction* (OSHA 29 CFR 1926 Construction Industry Standards) (2002). The Associated General Contractors of America, 333 John Carlyle Street, Suite 200, Alexandria, VA.

7. Peurifoy, Robert L., and Garold D. Oberlender (2001). *Estimating Construction Costs,* McGraw-Hill, Inc., New York.

8. Public Law: PL No. 74-403 Davis Bacon Act.

9. Schuette, Stephen D., and Robert W. Liska (1994). *Building Construction Estimating,* McGraw-Hill, New York.

10. *Walker's Building Estimator's Reference Book,* 26th ed. (a catalogue of material unit prices and labor rates), Frank R. Walker Company, P.O. Box 3180, Lisle, IL 60532.

WEBSITE RESOURCES

1. www.aspenational.org   American Society of Professional Estimators (ASPE). ASPE serves construction estimators by providing education and opportunity for professional development. ASPE, 2525 Perimeter Place Drive, Suite 103, Nashville, TN 37214.

2. www.bls.gov/home.htm   Bureau of Labor Statistics (BLS). BLS is the principal fact-finding agency for the federal government in the broad field of labor economics and statistics.

3. www.csinet.org   Construction Specifications Institute (CSI). CSI is an individual membership technical society whose purpose is to improve the process of creating and sustaining the built environment. The Institute provides technical information and products to enhance communication among all disciplines of nonresidential building design and construction.

4. www.hcss.com   Heavy Construction Systems Specialists Inc. (HCSS). HCSS specializes in construction estimating, bidding, and job cost software for contractors. HCSS, 6200 Savoy, Suite 1100, Houston, TX 77036.

5. www.sagetimberlineoffice.com   Sage Timberline. Timberline Software is an international supplier of accounting and estimating software for construction and property management. Timberline Software Corporation Headquarters, 15195 NW Greenbrier Parkway, Beaverton, OR 97006.

6. www.uscost.com/offices.asp   U.S. COST. U.S. COST provides project control services and software to facility owners, engineers, designers, and contractors throughout the world. SuccessEstimator is the flagship product. It is an integrated estimating and cost management system. U.S. COST, 600 Northpark Town Center, 1200 Abernathy Road, NE, Building 600, Suite 950, Atlanta, GA 30328.

7. www.frankrwalker.com   Frank R. Walker Company. *Walker's Building Estimator's Reference Book* is organized in CSI format and contains detailed construction cost data, reference tables, charts and diagrams for quick reference. Frank R. Walker Company, P.O. Box 3180, Lisle, IL 60532.

Estimating Heavy/ Civil Projects

When the engineer prepares a cost estimate for an earthwork project, the critical attributes that must be determined are (1) the quantities involved, basically volume or weight; (2) the haul distances; and (3) the grades for all segments of the haul roads. The mass diagram is an excellent method of analyzing linear earthmoving operations. The analysis objective is to predict the production rate for a group of machines (linked-system production rate) and the cost per unit of production. Typically, on a unit-price job, an owner pays for concrete work under only one or two bid items such as concrete by volume and reinforcement steel by weight.

INTRODUCTION

A contractor's estimating process begins with the decision to submit a bid. After reviewing the project drawings and specifications, the estimator visits the site of the proposed project and gathers information about location-specific conditions such as surface topography, drainage, and access. The information developed during the site visit is recorded in a report that is made part of the final project estimate.

After the site visit, the estimator determines the quantity of materials that must be furnished or moved. This process is called the take-off or **quantity survey**. During the take-off process, the estimator must make decisions concerning equipment needs, sequence of operations, and crew size. The type of excavation that must be performed to complete the project will influence the estimator's decisions concerning construction equipment and methods. Some projects such as dams require mass excavation.

Mass excavation involves moving a substantial volume of material and the excavation work is a primary part of the project. On the Eastside Project in California, the contractor building the West Dam moved over 68 million cubic yards

quantity survey
The process of calculating the quantity of materials required to build a project.

mass excavation
The requirement to excavate substantial volumes of material, usually at considerable depth or over a large area.

of material. Mass excavations are typically operations of considerable excavation depth and horizontal extent, and may include requirements to drill and blast rock (e.g., the movement of consolidated materials). **Structural excavation** is a different type of undertaking. The excavation work is performed to support the construction of other structural elements. This work is usually done in a confined area, it is typically vertical in extent, and the banks of the work may require support systems. The volume of excavated material is not a decisive factor as much as dealing with limited work space and vertical movement of the material. With either type of work, allowances for the waste of materials, inclement weather, delays, and other factors that may increase costs must also be incorporated in the estimate. The project plans provide graphical information necessary to calculate work quantities.

structural excavation
Excavation undertaken in support of structural element construction, usually involves removing materials from a limited area.

GRAPHICAL PRESENTATION OF EARTHWORK

station
A horizontal distance of 100 feet.

When the engineer prepares a cost estimate for an earthwork project, the critical attributes that must be determined are (1) the quantities involved, measured by volume or weight; (2) the haul distances; and (3) the grades for all segments of the haul roads.

Horizontal distances along a project are referenced in **stations**. The term *station* refers to locations on a 100-base numbering system. Therefore, the distance between two adjacent stations is 100 feet. Station 1 is written 1 + 00. The plus sign is used in this system of referencing points. The term refers to the surveyor notation for laying out a project in the field and is used on the plans to denote locations along the length of the project.

plan view
A construction drawing representing the horizontal alignment of the work.

Three kinds of views are presented in the contract documents to show earthwork construction features.

1. *Plan view.* The **plan view** is looking down from above on the proposed work and presents the horizontal alignment of features. Figure 6.1 is a plan view of a highway project; it shows the project centerline with stationing noted and the project limits are the two dark exterior lines.

profile view
A construction drawing depicting a vertical plane cut through the centerline of the work. It shows the vertical relationship of the ground surface and the finished work.

2. *Profile view.* The **profile view** is a side view, typically along the centerline of the work. It presents the vertical alignment of features. Figure 6.2 is a profile view; the bottom horizontal scale shows the centerline stationing, the vertical scale gives elevation, the dashed line is the existing ground line, and the solid line is the proposed final grade of the work.

cross section view
A construction drawing depicting a vertical section of earthwork at right angles to the centerline of the work. Used together with centerline distances to calculate earthwork quantities.

3. *Cross section view.* The **cross section view** is formed by a plane cutting the work vertically and at right angles to its long axis. Figures 6.3 and 6.4 present fill and cut cross sections, respectively. The straight heavy lines denote the final grade of the work and the existing ground is shown by the thin line with shading.

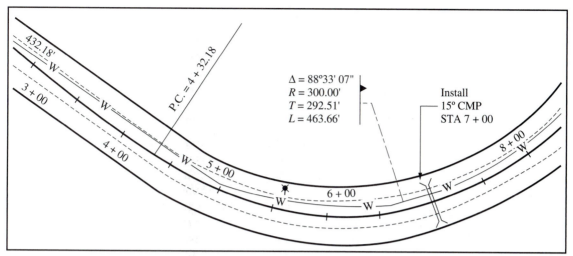

FIGURE 6.1 Plan view of a highway project.

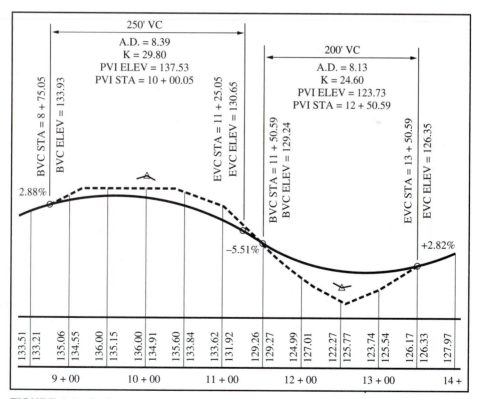

FIGURE 6.2 Profile view of a highway project.

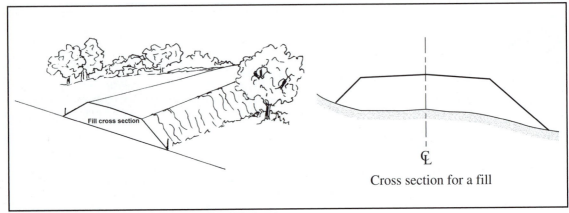

Fill cross section

Cross section for a fill

FIGURE 6.3   Earthwork cross section for a fill situation.

Cut cross section

Cross section for a cut

FIGURE 6.4   Earthwork cross section for a cut situation.

Cross Sections

cross sections

Earthwork drawings created by combining the project design with field measurements of existing conditions. These are typically views at right angles to the centerline of the project or a major project feature.

In the case of a project that is linear in extent, material volumes are usually determined from **cross sections**. Cross sections are pictorial drawings produced from a combination of the designed project layout and measurements taken in the field at right angles to the project centerline or the centerline of a project feature, such as a drainage ditch. When the ground surface is regular, field measurements are typically taken at every full station (100 ft). When the ground is irregular, measurements must be taken at closer intervals and particularly at points of change. Typical cross sections are shown in Figs. 6.3 and 6.4.

The cross sections usually depict the finished subgrade elevations; this fact should always be verified. If the grades depicted are the top of pavement, profile grade Fig. 6.5, the thickness of the pavement section must be used to adjust the earthwork computations. From the cross section drawings, *end areas* can be computed using any of several methods.

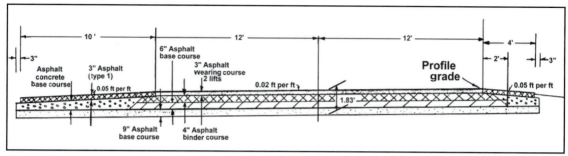

FIGURE 6.5   Cross section view of a pavement section for an interstate highway.

EARTHWORK QUANTITIES

Earthwork computations involve the calculation of earthwork volumes, the balancing of cuts and fills, and the planning of the most economical material hauls. The first step in estimating an earthmoving operation is calculation of the quantities involved in the project. The exactness with which earthwork computations can be made depends on the extent and accuracy of field measurements represented on the drawings.

End-Area Determination

The method chosen to compute a cross section end area will depend on the time available and the available computer or mechanical aids. Most companies use commercial computer software and digitizing tablets to determine cross section end areas. Other methods include the use of a planimeter, subdivision of the area into geometric figures with definite formulas for areas (rectangles, triangles, parallelograms, and trapezoids), and the use of the trapezoidal formula.

Digitizing Tablet   A digitizing tablet is a board with an imbedded wire mesh grid. When the cursor is traced over the board, a current is picked up by the board's grid, and the coordinates of the tracing device are passed on to the computer. When a plan is taped to the board and a scale is entered, the traced plan is converted to measurements, and then a software program computes the area, length, and volume of the traced data.

Planimeter   A planimeter is a mechanical device that is used to move a tracing point around the perimeter of the plotted area. It provides a value that is then multiplied by the square scale of the figure to calculate the figure's area. A planimeter can be used on any figure no matter how irregular the figure's shape might be.

Trapezoidal Computations   The mathematics of the area computations is often based on dividing the cross section drawing into parts. The computer can easily subdivide a cross section drawing into a large number of strips, calculate the area of each strip, and then sum the individual areas to arrive at the area of a section. If the calculations must be made by hand, the area formula for a triangle and a trapezoid are used to compute the volume.

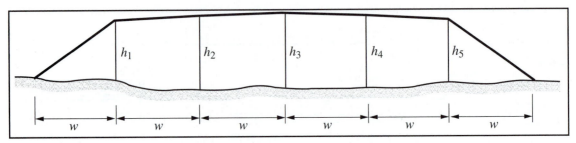

FIGURE 6.6　Division of a cross section drawing into triangles and trapezoids.

$$\text{Area of a triangle} = \frac{1}{2}hw \qquad [6.1]$$

where　h = height of the triangle
　　　　w = base of the triangle

$$\text{Area of a trapezoid} = \frac{(h_1 + h_2)}{2} \times w \qquad [6.2]$$

where (see Fig. 6.6)　w = distance between the two parallel sides
　　　　h_1 and h_2 = the lengths of the two parallel sides

The general trapezoidal formula for calculating area is

$$\text{Area} = \left(\frac{h_0}{2} + h_1 + h_2 + \cdots + h_{(n-1)} + \frac{h_n}{2} \right) \times w \qquad [6.3]$$

where (see Fig. 6.6)　w = distance between the two parallel sides
　　　　$h_0 \ldots h_n$ = the lengths of the individual adjacent parallel sides

The precision achieved using this formula depends on the number of subdivisions, but is about ±0.5%.

In the case of hillside construction, there can be both a cut area and a fill area in the same cross section (Fig. 6.7). When making area computations, it is always necessary to calculate cut and fill areas separately.

Average End Area

average end area
A calculation method for determining the volume of material bounded by two cross sections or end areas.

The **average-end-area** method is most commonly used to determine the volume bounded by two cross sections or end areas. The principle is that the volume of the solid bounded by two parallel, or nearly parallel, cross sections is equal to the average of the two end areas times the distance between the cross sections along their centerline (see Fig. 6.8). The average-end-area formula is

$$\text{Volume [net cubic yards (cy)]} = \frac{(A_1 + A_2)}{2} \times \frac{L}{27} \qquad [6.4]$$

where (see Fig. 6.8)
　　A_1 and A_2 = area in square feet (sf) of the respective end areas
　　　　L = the length in feet between the end areas

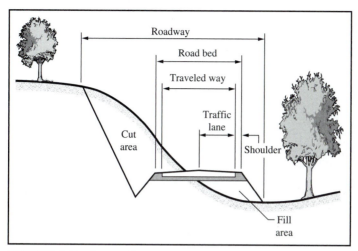

FIGURE 6.7 Cut and fill areas in the same section.

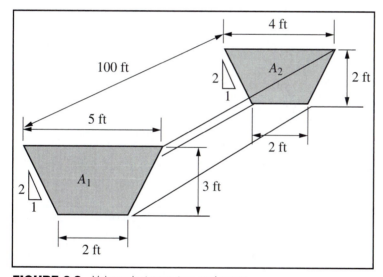

FIGURE 6.8 Volume between two end areas.

EXAMPLE 6.1

Calculate the volume between two end areas 100 ft apart (see Fig. 6.8). End area 1 (A_1) equals 10.5 sf and end area 2 (A_2) equals 6 sf.

$$\text{Volume} = \frac{(10.5 \text{ sf} + 6 \text{ sf})}{2} \times \frac{100 \text{ ft}}{27 \text{ cf/cy}} \Rightarrow 30.6 \text{ cy}^*$$

where cf means cubic feet

* If the cross sections (the two end areas) represented a cut situation, the units of the calculation would be bank cubic yards (1 cubic yard in its natural state, bcy). If the situation was a fill portion of the work the units of the calculation would be compacted cubic yards (ccy), as a section of an embankment represents a compacted condition.

The principle is not altogether true because the average of the two end areas is not the arithmetic mean of many intermediate areas. The method gives volumes generally slightly in excess of the actual volumes. The precision is about ±1.0%.

EXAMPLE 6.2

The volume between the embankment (fill) end areas given in the table here is calculated using the average-end-area method. The total volume of the section (the sum of the individual volumes) is also given.

Station	End area (sf)	Distance (ft)	Volume (ccy)
150 + 00	360	—	—
150 + 50	3,700	50	3,759
151 + 00	10,200	50	12,870
152 + 00	18,000	100	52,222
153 + 00	23,500	100	76,852
154 + 00	12,600	100	66,852
155 + 00	5,940	100	34,333
155 + 50	2,300	50	7,630
156 + 00	400	50	2,500
			Total volume 257,018

In this example, the interval between sections is 100 ft except between stations 150 + 00 and 151 + 00 and between stations 155 + 00 and 156 + 00, where the interval is 50 ft. The calculated total volume is 257,018 ccy.

If the 50-ft intervals are omitted (i.e., if we assume sections were taken only on full stations), the calculations would be as shown in this table.

Station	End area (sf)	Distance (ft)	Volume (ccy)
150 + 00	360	—	—
151 + 00	10,200	100	19,556
152 + 00	18,000	100	52,222
153 + 00	23,500	100	76,852
154 + 00	12,600	100	66,852
155 + 00	5,940	100	34,333
156 + 00	400	100	741
			Total volume 250,556

The difference is 6,462 ccy. This example emphasizes the importance of cross section spacing and the possible introduction of volume calculation error. The error in this case is 2.5%, based on an actual 257,018 ccy volume.

Although cross sections can be taken at any interval along the centerline, judgment should be exercised, depending particularly on the irregularity of the ground and the tightness of curves. In the case of tight curves, a spacing of 25 feet is often appropriate.

Stripping

The upper layer of material encountered in an excavation is often topsoil (organic material), resulting from decomposition of vegetative matter. Such organic material is commonly referred to as **stripping**. This material is unsuitable for use in an embankment, and usually it must be handled in a separate excavation operation. It can be collected and disposed of off the project, or stockpiled for later use on the project to plate slopes. If the structural embankments are of limited height, the organic material below the footprint of the fill sections must be stripped before embankment placement can commence (see Fig. 6.9). In the case of embankments over 5 feet in height, most specifications allow the organic material to remain if its thickness is only a few inches. When calculating the volume of cut sections, this stripping quantity must be subtracted from the net volume, as it cannot be used for embankment construction (Fig. 6.10). In the case of fill sections, the quantity must be added to the calculated fill volume (Fig. 6.10).

stripping
The upper layer of organic material that must be removed before beginning an excavation or embankment.

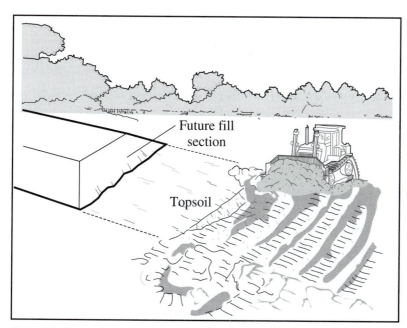

FIGURE 6.9 Stripping topsoil before building a fill.

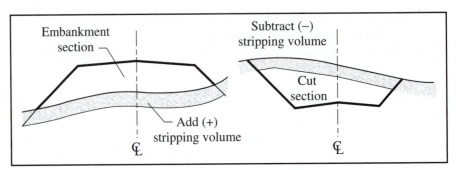

FIGURE 6.10 Effect of stripping on embankment and cut volume calculations.

Net Volume

The computed volumes from the cross sections represent two different material states. The volumes from the fill cross sections represent compacted volume. If the volume is expressed in cubic yards, the notation is compacted cubic yards. In the case of cut sections, the volume is a natural in situ volume. The term *bank volume* is used to denote this in situ volume; if the volume is expressed in cubic yards, the notation is bank cubic yards. If the cut and fill volumes are to be combined, they must be converted into compatible volumes. In Table 6.1, the conversion from compacted to bank cubic yards is made by dividing the compacted volume by 0.90. This value should be determined by soil testing of the actual material that will be handled.

Earthwork Volume Sheet

An earthwork volume sheet, which can easily be constructed using a spreadsheet program, allows for the systematic recording of information and making the necessary earthwork calculations (see Table 6.1).

> *Stations.* Column 1 is a listing of all stations at which cross-sectional areas have been recorded.
>
> *Area of cut.* Column 2 is the cross-sectional area of the cut at each station. Usually this area must be computed from the project cross sections.
>
> *Area of fill.* Column 3 is the cross-sectional area of the fill at each station. Usually this area must be computed from the project cross sections. Note there can be both cut and fill at a station (see row 5, Table 6.1).
>
> *Volume of cut.* Column 4 is the volume of cut between the adjacent preceding station and the station. The average-end-area formula, Eq. [6.4], is usually used to calculate this volume. This is a *bank* volume (bcy).
>
> *Volume of fill.* Column 5 is the volume of fill between the adjacent preceding station and the station. The average-end-area formula, Eq. [6.4], is usually used to calculate this volume. This is a *compacted* volume (ccy).

TABLE 6.1 Earthwork volume calculation sheet.

	Station (1)	End-area cut (sf) (2)	End-area fill (sf) (3)	Volume of cut (bcy) (4)	Volume of fill (ccy) (5)	Stripping cut (bcy) (6)	Stripping fill (ccy) (7)	Total cut (bcy) (8)	Total fill (ccy) (9)	Adj. fill (bcy) (10)	Algebraic sum (bcy) (11)	Mass ordinate (12)
(1)	0 + 00	0	0									
(2)	0 + 50	0	115	0	106	0	18	0	124	138	−138	−138
(3)	1 + 00	0	112	0	210	0	30	0	240	267	−267	−405
(4)	2 + 00	0	54	0	307	0	44	0	351	390	−390	−796
(5)	2 + 50	64	30	59	78	0	22	59	100	111	−52	−847
(6)	3 + 00	120	0	170	28	26	0	144	28	31	114	−734
(7)	4 + 00	160	0	519	0	76	0	443	0	0	443	−291
(8)	5 + 00	317	0	883	0	74	0	809	0	0	809	518
(9)	6 + 00	51	0	681	0	60	0	621	0	0	621	1,140
(10)	6 + 50	46	6	90	6	21	0	69	6	6	63	1,202
(11)	7 + 00	0	125	43	121	0	25	43	146	163	−120	1,082
(12)	8 + 00	0	186	0	576	0	81	0	657	730	−730	352
(13)	8 + 50	0	332	0	480	0	69	0	549	610	−160	−257

Stripping volume in the cut. Column 6 is the stripping volume of topsoil over the cut between the adjacent preceding station and the station. This volume is commonly calculated by multiplying the distance between stations or fractions of stations by the width of the cut. This provides the area of the cut footprint. The footprint area is then multiplied by an average depth of topsoil to derive the stripping volume. This represents a bank volume of cut material. Usually topsoil material is not suitable for use in the embankment. The average depth of topsoil must be determined by field investigation.

Stripping volume in the fill. Column 7 is the stripping volume of topsoil under the fill between the adjacent preceding station and the station. This volume is commonly calculated by multiplying the distance between stations or fractions of stations by the width of the fill. This provides the area of the fill footprint. To derive the stripping volume, the area of the embankment footprint is multiplied by an average depth of topsoil. The stripping is a *bank* volume, but it also represents an additional requirement for fill material, ccy of fill.

Total volume of cut. Column 8 is the volume of cut material available for use in embankment construction. It is derived by subtracting the cut stripping (col. 6) from the cut volume (col. 4) (see Fig. 6.10).

Total volume of fill. Column 9 is the total volume of fill required. It is derived by adding the fill stripping (col. 7) to the fill volume (col. 5) (see Fig. 6.10).

Adjusted fill. Column 10 is the total fill volume converted from compacted volume to bank volume.

Algebraic sum. Column 11 is the difference between column 10 and column 8. This indicates the volume of material that is available (cut is positive) or required (fill is negative) within station increments after intrastation balancing.

Mass ordinate. Column 12 is the running total of column 11 values from some point of beginning on the project profile. When the stations being summed are excavation sections, the value of this column will increase, while summing fill sections will result in a decrease of the column 12 values. Note that any material that could be used within a station length is not accounted for in the mass ordinate and therefore it is not accounted for in the **mass diagram**.

The mass diagram accounts for material that must be transported beyond the limits of the two cross sections that define the volume of material. Where there is both cut and fill between a set of stations, only the excess of one over the other is used in computing the mass ordinate. Cut material between two successive stations is first used to satisfy fill requirements between those same two successive stations before there is a contribution to the mass ordinate value. Likewise,

mass ordinate
The cumulative material mass differential as calculated from the start through the end of the work.

mass diagram
In earthwork calculations, a graphical representation of the algebraic cumulative quantities of cut and fill along the centerline, where cut is positive and fill is negative. Used to calculate haul distance.

FIGURE 6.11 The mass diagram provides the information for deciding in which direction material should be hauled.

if there is a greater fill requirement between two successive stations than there is cut available, the cut contribution is accounted for first. Only after all of the cut material is used will the remaining fill requirement contribute to the mass ordinate value. The material used between two successive stations is considered to move at right angles to the centerline of the project and therefore is often termed *crosshaul*. The remaining material in either case represents a longitudinal haul along the length of the project (Fig. 6.11).

MASS DIAGRAM

Earthmoving is basically an operation where material is removed from high spots (hills) and deposited in low spots (valleys) (Fig. 6.11). If there is a shortage of material, borrowing is required, and if there is any excess cut material, the excess must be wasted. The *mass diagram* is an excellent method of analyzing linear earthmoving operations. It is a graphical means for measuring haul distance (stations) in terms of earthwork volume (cubic yards). A station-yard is a measure of work: the movement of 1 cy through a distance of one station, or 100 linear feet.

On a mass diagram graph, the horizontal dimension represents the stations of a project (col. 1, Table 6.1), and the vertical dimension (col. 12, Table 6.1) represents the cumulative sum of excavation and embankment from some point of beginning on the project profile. The diagram provides information concerning

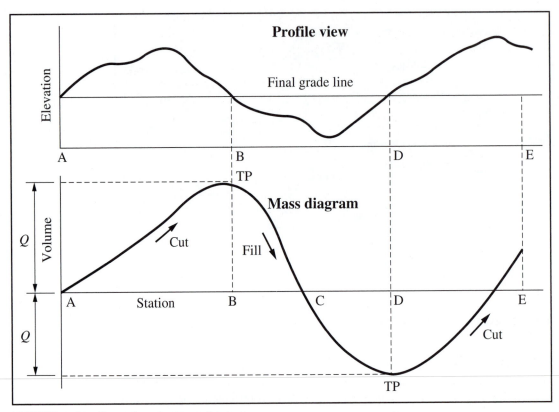

FIGURE 6.12 Properties of a mass diagram.

Quantities of materials

Average haul distances

Types of equipment that should be considered

When combined with a ground profile, the average slope of haul segments can be estimated. The mass diagram is one of the most effective tools for planning the movement of material on any project of linear extent.

Using column 1 (as the horizontal scale) and column 12 (as the vertical scale) of an earthwork volume calculation sheet, a mass diagram can be plotted. See the bottom portion of Fig. 6.12. Positive mass ordinate values are plotted above the zero datum line and negative values below. The top portion of Fig. 6.12 is the profile view of the same project.

Mass Diagram Properties

A mass diagram is a running total of the quantity of material that is surplus or deficient along the project profile. An excavation operation produces an *ascending* mass diagram curve; the excavation quantity exceeds the embankment quantity requirements. Excavation is occurring between stations A and B and stations D and E in Fig. 6.12, and the curve is ascending. The total volume of excavation between stations A and B is obtained by projecting horizontally to the vertical axis *the mass diagram line points* at stations A and B and reading the difference of the two volumes. Conversely, if the operation is a fill situation, there is a deficiency of material, and a *descending* curve is produced; the embankment requirements exceed the excavation quantity being generated. Filling is occurring between stations B and D, and the curve is descending. The volume of fill can be determined in a manner similar to that for the excavation.

The *maximum* or *minimum points* on the mass diagram, where the curve transitions from rising to falling or from falling to rising, indicate a change from an excavation to fill situation or vice versa. These points are referred to as *transition points* (TPs). On the ground profile, the grade line is crossing the ground line (see Fig. 6.12 at stations B and D).

When the mass diagram curve crosses the datum (or zero volume) line (as at station C) exactly as much material is being excavated (between stations A and B) as is required for fill between B and C. There is no excess or deficit of material at point C in the project. The final position of the mass diagram curve above or below the datum line indicates whether the project has surplus material that must be wasted or if there is a deficiency that must be made up by borrowing material from outside the project limits. Figure 6.12, station E, indicates a waste situation and excess material will have to be removed from the project.

USING THE MASS DIAGRAM

The mass diagram is an analysis tool for selecting the appropriate equipment for excavating and hauling material. The analysis is accomplished using balance lines and calculating average haul distances.

Balance Lines

A **balance line** is a horizontal line of specific length that intersects the mass diagram in two places. The balance line can be constructed so that its length is the maximum haul distance for different types of equipment. The maximum haul distance is the limiting economical haul distance for a particular type of equipment (see Table 6.2).

balance line
A horizontal line of specific length that intersects the mass diagram in two places.

TABLE 6.2 Economical haul distances based on basic machine types.

Machine type		Economical haul distance
Large dozers, pushing material		Up to 300 ft*
Push-loaded scrapers		300 to 5,000 ft*
Trucks		Hauls greater than 5,000 ft

*The specific distance will depend on the size of the dozer or scraper.

Figure 6.13 shows a balance line drawn on a portion of a mass diagram. If this line were constructed for a large push-loaded scraper, the distance between stations A and C would be 5,000 feet. Between the ends of the balance line, the cut volume generated equals the fill volume required. Between stations A and C, the amount of material the scrapers will haul is dimensioned on the vertical scale depicted by the vertical line Q. By examining either the profile view or the mass diagram, it is easy to determine the direction of haul that the cut must go to a fill location. Note the arrow on the profile view in Fig. 6.13.

In accomplishing the balanced earthwork operation between stations A and C, some of the hauls will be short, and some will approach the maximum haul distance (the distance between point A and point C). The *average haul distance* is approximately the length of a horizontal line placed one-third of the distance from the balance line in the direction of the high or low point of the curve. This is true in the case of a situation like that shown in Fig. 6.13 when the general shape of the mass diagram curve is a triangle. If the situation is similar to that shown in Fig. 6.14, where there are multiple balance lines and the area depicted is basically rectangular, the *average haul distance* is the length of a horizontal line placed midway between the balance lines. See the average haul for scrapers line shown in Fig. 6.14.

If the curve is above the balance line, the direction of haul is from left to right (i.e., up stationing). When the curve is below the balance line, the haul is from right to left (i.e., down stationing).

Because the lengths of balance lines on a mass diagram are equal to the maximum or minimum haul distances for the balanced earthmoving operation, they should be drawn to conform to the capabilities of the particular equipment that will be used. The equipment will, therefore, operate at haul distances that

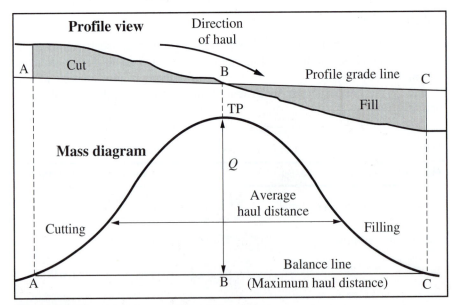

FIGURE 6.13   Mass diagram with a balance line.

are within its range of efficiency. Figure 6.14 illustrates a portion of a mass diagram on which two balance lines have been drawn. In this situation, it is planned that dozers will be used to push the short-haul material. Using dozers, the excavation between stations C and D will be placed between stations D and E. Then there will be a scraper operation to excavate the material between stations A and C and haul it to fill between stations E and G.

Average Grade

The project plans will provide a profile drawing of the work. If the mass diagram is plotted under the project profile, as shown in Figs. 6.12 and 6.15, the average haul grades of the earthmoving operations can be approximated. On the profile view, draw a horizontal line that roughly divides the cut area in half in the vertical dimension (see Fig. 6.15, line DE). Do the same for the fill area (line FG). This is a division of only the portion of the cut or fill areas defined by the balance line in question. The difference in elevation between these two lines provides the vertical distance to use in calculating the average grade for the haul involving the material in the balance. The average haul distance, as determined by construction of a horizontal line on the mass diagram, is the denominator in the grade calculation.

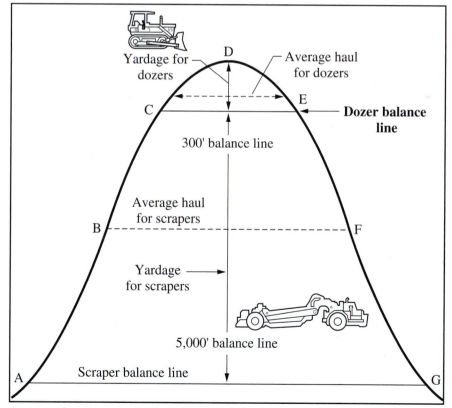

FIGURE 6.14 Two balance lines on a mass diagram.

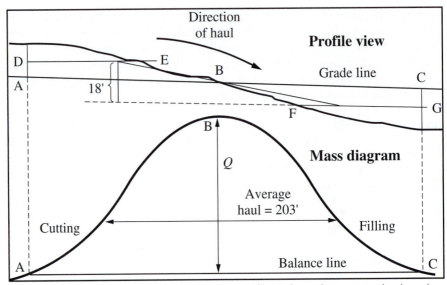

FIGURE 6.15   Using the mass diagram and profile to determine average haul grades.

$$\text{Average grade (percent)} = \frac{\text{Change in elevation}}{\text{Average haul distance}} \times 100 \qquad \textbf{[6.5]}$$

EXAMPLE 6.3

Calculate the average grade for the balance haul shown in Fig. 6.15.

$$\text{Average grade going from the cut to the fill} = \frac{-18 \text{ ft}}{203 \text{ ft}} \times 100 \Rightarrow -8.9\%$$

The return trip will be up an 8.9% grade.

Haul Distances

The mass diagram can be used to determine average haul distances. If the values in column 8 of Table 6.1 (volume of cut) from station 0 + 00 to 8 + 50 are summed, the total is 2,188 bcy. This is the total volume of excavation within the project limits. If the positive values in column 11 (algebraic sum) are summed, the total is 2,049 bcy, which is the total volume of excavation that must be moved longitudinally. The difference between these two values is the crosshaul for the project, 139 bcy. The crosshaul material is used between the two adjacent stations where it is generated and does not move along the length of the project. Many contractors treat the crosshaul as dozer work.

Reviewing the values in column 12, Table 6.1, we see there is a low point or valley of −847 bcy at station 2 + 50 and a high point or crest of 1,202 bcy at station 6 + 50. The sum of the absolute values of the peaks and low points equals the total excavation that must be moved longitudinally (847 + 1,202 = 2,049 bcy).

The curve can have intermediate peaks and low points (see Table 6.3), and all must be accounted for when calculating the amount of material that must be moved longitudinally.

Reviewing the values in column 12, Table 6.3, there is a low point or valley of $-28,539$ bcy at station $5 + 00$, a high point or crest of $-17,080$ bcy at $8 + 00$, and a second low point of $-22,670$ bcy at station $10 + 00$. The sum of the absolute values of the peaks and low points equals the total excavation that must be moved longitudinally [$(28,539 - 17,080) + (22,670 - 17,080) + (17,080 - 0) = 34,120$ bcy]. Haul 1 involves 11,459 bcy, haul 2 is 5,590 bcy, and haul 3 is 17,080 bcy.

Calculating Haul Distance   The average haul distance for any of the individual hauls can also be determined by calculation. Dividing the area (stations-cubic yards) that is enclosed by the balance line and the mass diagram curve by the amount of material hauled (cubic yards) will yield the average haul distance (stations).

The area (stations-cubic yards, usually referred to as simply station-yards) that is enclosed by the balance line, the mass diagram curve, and the zero datum line, which is haul 3 (Fig. 6.16), can be calculated using the trapezoidal formula, Eq. [6.3]. The vertical value at h_0 (station $0 + 00$) is 0 bcy, at h_1 is 3,631 bcy, at h_2 is 13,641 bcy, at h_3 through h_{13} is 17,080 bcy, at h_{14} is 8,502 bcy, and at h_{15} is 0 bcy. For stations $0 + 00$ and $15 + 00$, the beginning and ending stations in the equation (h_0 and h_n), the value is divided by 2. But 0 divided by 2 is still 0, so the area equals the sum of the verticals times the distance between the verticals, which is 1 (one station). The sum of the verticals is 213,654 bcy, and multiplying by 1 station produces an area of 213,645 station-bcy (sta.-yd). There is a slight error in the computation because the mass diagram curve crosses the zero balance line somewhere between station $14 + 00$ and $15 + 00$ and that last distance is not 100 feet. But there are no data to determine exactly where the curve and line cross, other than possibly by scaling the mass diagram.

The amount of material hauled is 17,080 bcy. Therefore, the calculated average haul for haul 3 is 12.5 stations or

$$12,500 \text{ feet} \left(\frac{213,654 \text{ station-bcy}}{17,000 \text{ bcy}} = 12.5 \text{ stations} \right)$$

Consolidated Average Hauls

Using the individual average hauls and the quantity associated with each, a project average haul can be calculated. The calculation process is similar to that used to calculate an averaged haul for an individual haul. Consider the three hauls depicted in Fig. 6.16 and their graphically determined average haul distances. Haul 1 is 11,459 bcy with an average haul distance of 400 feet or 4 stations. Haul 2 is 5,590 bcy with an average haul of 350 feet, and haul 3 is 17,080 bcy with an average haul of 12,510 feet. By multiplying each haul quantity by its respective haul distance, a station-yard value can be determined.

TABLE 6.3 Earthwork volume calculation sheet for the mass diagram in Fig. 6.16.

Station (1)	End-area cut (sf) (2)	End-area fill (sf) (3)	Volume of cut (bcy) (4)	Volume of fill (ccy) (5)	Stripping cut (bcy) (6)	Stripping fill (ccy) (7)	Total cut (bcy) (8)	Total fill (ccy) (9)	Adj. fill (bcy) (10)	Algebraic sum (bcy) (11) [0.90]	Mass ordinate (12)
0 + 00	0	0		0							
1 + 00	0	1,700	0	3,148	0	120	0	3,268	3,631	−3,631	−3,631
2 + 00	0	3,100	0	8,889	0	120	0	9,009	10,010	−10,010	−13,641
3 + 00	0	1,500	0	8,519	0	120	0	8,639	9,598	−9,598	−23,240
4 + 00	60	600	111	3,889	60	80	51	3,969	4,410	−4,359	−27,598
5 + 00	400	200	852	1,481	80	60	772	1,541	1,713	−941	−28,539 ↓
6 + 00	1,300	30	3,148	426	110	10	3,038	436	484	2,554	−25,985
7 + 00	2,400	400	6,852	796	120	85	6,732	881	979	5,753	−20,223
8 + 00	800	850	5,926	2,315	90	100	5,836	2,415	2,683	3,153	−17,080 ↓
9 + 00	50	1,250	1,574	3,889	5	120	1,569	4,009	4,454	−2,885	−19,965
10 + 00	95	180	269	2,648	20	10	249	2,658	2,953	−2,705	−22,670 ↓
11 + 00	200	8	546	348	60	0	486	348	387	99	−22,571
12 + 00	560	0	1,407	15	65	0	1,342	15	16	1,326	−21,245
13 + 00	1,430	0	3,685	0	100	0	3,585	0	0	3,585	−17,660
14 + 00	3,580	0	9,278	0	120	0	9,158	0	0	9,158	−8,502
15 + 00	2,600	0	11,444	0	110	0	11,334	0	0	11,334	2,833

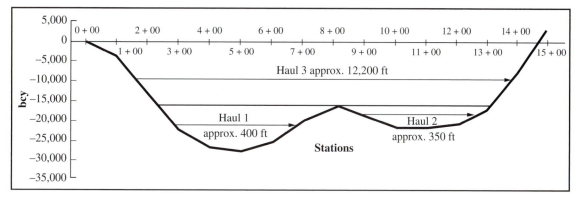

FIGURE 6.16   Mass diagram plotted from the data in Table 6.3.

Haul 1	11,459 bcy	4.0 stations	45,836 sta.-yd
Haul 2	5,590 bcy	3.5 stations	19,565 sta.-yd
Haul 3	17,080 bcy	12.5 stations	213,500 sta.-yd
	34,129 bcy		278,901 sta.-yd

If the individual station-yard values are summed and that value is divided by the total quantity moved, an average haul for the project is the result. In this case, the project average haul is 8.17 stations (278,901 sta.-yd/34,129 yd).

If all the hauls are of about the same length, the estimator can consolidate the production calculations by using an averaging process, as just described. In the case of the data presented in Table 6.3 and Fig. 6.16, it is clear that there are two very distinct haul situations on this project. There are two short-haul sections and one long-haul section. These two situations will most likely each require a different number and perhaps type of haul units. The short-haul situation will have shorter haul times and therefore will require fewer units to achieve continuous production. The long haul will require more units. Assuming the haul grades are about the same, the estimator will not calculate production for every individual part of the mass diagram. This is also driven by the result of the practical situation of not mobilizing and demobilizing various numbers of machines for small differences in hauls.

Considering the data in Table 6.3, the estimator would most likely develop two production scenarios. The first scenario would be for the short hauls of hauls 1 and 2.

Haul 1	11,459 bcy	4.0 stations	45,836 sta.-yd
Haul 2	5,590 bcy	3.5 stations	19,565 sta.-yd
	17,049 bcy		65,401 sta.-yd

Average Haul 1 and 2: $\dfrac{65,401 \text{ sta.-yd}}{17,049 \text{ bcy}} = 3.8$ stations

Haul 3	17,080 bcy	12.2 stations

The average haul that results from combining these two is 3.8 stations. The second scenario would be for the long-haul situation of haul 3.

There are other mass diagram applications that can be investigated. On some projects, it can prove economical to borrow material rather than undertake extremely long hauls. The same holds true for wasting material. The mass diagram provides the engineer the ability to analyze such situations.

PRICING EARTHWORK OPERATIONS

The cost of earthwork operations will vary with the kind of soil or rock encountered and the methods used to excavate, haul, and place the material in its final deposition. It is usually not too difficult to compute the volume of earth or rock to be moved, but estimating the cost of actually performing the work depends on both a careful study of the project plans and a diligent site investigation. The site work should seek to identify the characteristics of the subsurface soils and rock that will be encountered.

The earthwork quantities and average movement distances can be determined using the techniques described in the Earthwork Quantities and Mass Diagram sections of this chapter. Proper equipment and estimated production rates are determined by (1) selecting an appropriate type of machine (discussed in Chapter 11, Equipment Selection and Utilization) and (2) using machine performance data (as discussed in Chapter 10, Machine Power). The costs to be matched with the selected machines is derived as described in Chapter 12, Equipment Costs.

Spread Production

spread
A group of construction earthwork machines that work together to accomplish a specific construction task such as excavating, hauling, and compacting material.

To accomplish a task, machines usually work together and are supported by auxiliary machines. To accomplish a loading, hauling, and compacting task would involve an excavator (Fig. 6.17), several haul units, and auxiliary machines to distribute the material on the embankment and achieve compaction (Fig. 6.18). Such groups of equipment are referred to as an equipment **spread**. An excavator and a fleet of trucks can be thought of as a linked system, one link of which will control the spread production. If spreading and compacting of the hauled material is required, a two-link system is created. Because the systems are linked, the capabilities of the individual components of the spread must be compatible in terms of overall production (Fig. 6.19). The number of machines and specific types of machines in a spread will vary with the proposed task.

The production capacity of the total system is dictated by the lesser of the production capacities of individual systems. Our objective is to predict the spread production rate (*linked-system* production rate) and the cost per unit of production. In the case of the Fig. 6.16 data, the estimator would develop two production spreads. The first spread would be for the short 3.9-station haul and the second for the 12.2-station haul.

FIGURE 6.17 A hoe used to excavate and load haul trucks.

FIGURE 6.18 A roller used to compact the material on the fill.

FIGURE 6.19 Two-link earthwork system.

> Always ensure that a consistent set of units is used when estimating spread production. If the mass diagram quantities are expressed in bank cubic yards (bcy), excavator capacity, hauling capacity, and compaction should all be converted to bcy. The units used in the estimate are usually chosen to match those used in the owner's bid documents.

EXAMPLE 6.4

Develop the equipment spread for a short 4-station haul and for a long 12-station haul. A hoe excavator having the ability to excavate and load 280 bcy per hour will be used. Truck capacity is 12 bcy. Average truck travel speed is 10 mph for the short haul and 15 mph for the long haul. Dump time is 2 minutes.

$$\text{Time to load a 12 bcy truck: } \frac{12 \text{ bcy}}{280 \text{ bcy per hr}} \times 60 \text{ min/hr} = 2.6 \text{ min}$$

$$\text{Time to haul 400 ft: } \frac{400 \text{ ft}}{88 \frac{\text{ft}}{\text{mile}} \times \frac{\text{hr}}{\text{min}} \times 10 \text{ mph}} = 0.45 \text{ min}$$

$$\text{Time to haul 1,200 ft: } \frac{1{,}000 \text{ ft}}{88 \frac{\text{ft}}{\text{mile}} \times \frac{\text{hr}}{\text{min}} \times 15 \text{ mph}} = 0.91 \text{ min}$$

where 88 is the conversion factor to feet per minute

$$\frac{5{,}280 \text{ ft}}{\text{mile}} \times \frac{\text{hr}}{60 \text{ min}} = 88 \frac{\text{ft}}{\text{mile}} \times \frac{\text{hr}}{\text{min}}$$

Total cycle time:

	400 ft haul	**1,200 ft haul**
Load	2.60 min	2.60 min
Haul	0.45 min	0.91 min
Dump	2.00 min	2.00 min
Return	0.45 min	0.91 min
Total	5.50 min	6.42 min

$$\text{Number of trucks required 400-ft haul: } \frac{5.50 \text{ min total cycle time}}{2.6 \text{ min load time}} = 2.1 \text{ trucks}$$

$$\text{Number of trucks required 1,200-ft haul: } \frac{6.42 \text{ min total cycle time}}{2.6 \text{ min load time}} = 2.5 \text{ trucks}$$

We cannot have a fraction of a truck, therefore let us assume that 2 trucks will be used for the 400-ft haul and 3 for the 1,200-ft haul situation.

Production 400-ft haul (with only 2 trucks and a situation where 2.1 trucks are needed to keep the excavator fully utilized, the truck production capability will control the system production):

$$\frac{2 \text{ trucks} \times 12 \text{ bcy} \times 60 \text{ min per hour}}{5.5 \text{ min cycle time}} = 263 \text{ bcy per hr}$$

Production 1,200 ft haul [loader controls production because we are using more trucks (3) than the loader can support (2.5)]: 280 bcy per hr

Equipment spread 400-ft haul (production 263 bcy per hr):

Type of machines	Number of machines
Excavator—hoe	1
Trucks—12 bcy capacity	2
Dozer to spread the material on fill	1
Compactor	1
Water truck	1

Equipment spread 1,200-ft haul (production 280 cy per hr):

Type of machines	Number of machines
Excavator—hoe	1
Trucks—12 bcy capacity	3
Dozer to spread the material on fill	1
Compactor	1
Water truck	1

Time to Complete a Task

The next step in the estimating process is to calculate the working time required to complete a work task, Eq. [3.1].

$$\text{Time to complete a task} = \frac{\text{Task quantity}}{\substack{\text{Task production rate} \\ \text{(the minimum linked system production rate)}}}$$

EXAMPLE 6.5

Based on this information, calculate the time required to complete each task described in Ex. 6.4.

Task	Quantity	System production rate	Haul distance
Short haul	16,108 bcy	263 bcy per hr	4 stations
Long haul	17,080 bcy	280 bcy per hr	12 stations

$$\text{Short-haul time} = \frac{16,108 \text{ bcy}}{263 \text{ cy per hr}} = 61 \text{ hr}$$

$$\text{Long-haul time} = \frac{17,080 \text{ bcy}}{280 \text{ cy per hr}} = 61 \text{ hr}$$

Production Efficiency

In these calculations, it has been assumed that the equipment will be productive for 60 minutes during every hour of the workday. This is usually not the case and each situation must be analyzed very closely. Excavators that control a spread's production are typically kept in good operating condition and the best operators are assigned to these machines. In the case of the trucks, a breakdown or slow truck will affect production. In the case where 3 trucks are employed and only 2.5 are required, a slow truck would have minimal effect on production. However, in many cases it is realized that efficiency will not be 100% and a reduced number of working minutes per hour is used to reflect production efficiency. More realistically, we should assume something like 50 minutes of productive time per hour. In the production calculation, this efficiency assumption accounts for possible breakdown and other possible delays. Consider the two-truck situation in Examples 6.4 and 6.5. The revised production calculation would be:

Production 400-ft haul (trucks control production):

$$\frac{2 \text{ trucks} \times 12 \text{ bcy} \times 50 \text{ min per hour}}{5.5 \text{ min cycle time}} = 218 \text{ bcy per hr}$$

This would also change the time duration to complete the work.

$$\text{Short-haul time} = \frac{16{,}108 \text{ bcy}}{218 \text{ cy per hr}} = 74 \text{ hr}$$

Cost to Employ a Machine

The procedures to calculate machine ownership and operating (O&O) costs are explained in Chapter 12. Labor cost was examined in Chapter 5. When a machine is employed on a project the operator labor cost must be added to the machine O&O cost. Furthermore, when a spread of equipment is employed, there is often a requirement for additional personnel such as a spotter to guide the equipment when dumping.

Cost to Complete a Task

Contractors usually have their own standard formats for displaying estimated cost information in segregated and accumulated columns and rows. A typical format is presented in Table 6.4, which illustrates how to capture important cost information.

Estimating sheets similar to Table 6.4 are usually generated by special estimating programs or by using a spreadsheet program. The task quantity and production rate are calculated, and then entered into the program. The machine unit costs are extracted from the company's equipment database and then multiplied by the estimated number of use hours to generate total machine costs. Labor costs are generated in a similar manner; however as labor rates often vary from job to job, it is usually necessary to establish a labor rate database for each job.

TABLE 6.4 Estimate cost information for the short-haul task.

Task	16,108 bcy, short haul, 4 stations						
Production rate	263 bcy per hr		**Total hours**	61			
	No.	**Ownership**	**Operating**	**Labor**	**Other**	**Total**	**Unit cost ($ per bcy)**
Foreman w/ pickup	1	$ 610.00	$ 122.00	$1,830.00	$ 0	$ 2,562.00	$0.159
Hoe	1	7,320.00	2,745.00	1,159.00	0	11,224.00	0.697
Trucks	2	4,270.00	7,198.00	1,098.00	0	12,566.00	0.780
Dozer	1	2,745.00	1,403.00	1,098.00	0	5,246.00	0.326
Compactor	1	1,220.00	1,006.50	518.50	0	2,745.00	0.170
Water truck	1	732.00	366.00	518.50	805.40	2,421.90	0.150
Spotter	1	0	0	427.00	0	427.00	0.027
Total		$16,897.00	$12,840.50	$6,649.00	$805.40	$37,191.90	$2.309
Unit cost		$1.049	$0.797	$0.413	$0.050	$2.309	

TABLE 6.5 Estimate cost information for the long-haul task.

Task	17,080 bcy, long haul, 12 stations						
Production rate	280 bcy per hr		**Total hours**	61			
	No.	**Ownership**	**Operating**	**Labor**	**Other**	**Total**	**Unit cost ($ per bcy)**
Foreman w/ pickup	1	$ 610.00	$ 122.00	$1,830.00	$ 0	$ 2,562.00	$0.150
Hoe	1	7,320.00	2,745.00	1,159.00	0	11,224.00	0.657
Trucks	3	6,405.00	10,797.00	1,647.00	0	18,849.00	1.104
Dozer	1	2,745.00	1,403.00	1,098.00	0	5,246.00	0.307
Compactor	1	1,220.00	1,006.50	518.50	0	2,745.00	0.161
Water truck	1	732.00	366.00	518.50	854.00	2,470.50	0.144
Spotter	1	0	0	427.00	0	427.00	0.025
Total		$19,032.00	$16,439.50	$7,198.00	$854.00	$43,523.50	$2.548
Unit cost		$1.114	$0.963	$0.421	$0.050	$2.548	

To generate the Table 6.4 cost for the hoe, a machine ownership rate of $120 per hour, an operating rate of $45 per hour, and an operator labor rate of $19 per hour were used. The "Other" column is used for consumable materials. In this case, it has been assumed that it will be necessary to purchase the water that will be used for compaction moisture control. If there were permanent materials involved, the table would also have a material column.

Reviewing the Estimate

When we are reviewing an estimate, total cost is a necessary piece of information, but because experienced estimators usually have a feel for what certain types of work cost on a unit price basis, it is good practice to also display unit

FIGURE 6.20 Two transit-mix trucks delivering concrete to a building project. Concrete being lifted in buckets by a tower crane.

costs. It is easier to notice an unreasonable unit cost number and to use that as a starting point for checking an item in detail.

Comparing the task unit costs in Tables 6.4 and 6.5, there is a difference of $0.239 ($2.548 − $2.309). The unit cost is greater for the long-haul situation. This difference is in line with our expectations: longer haul distance means greater costs. In the case of the Table 6.5 estimate, the haul is much longer and, to maintain the production rate, a third truck had to be added to the spread. The variance in trucking cost per bcy is a positive $0.324 ($1.104 − $0.780), while all of the other costs have decreased in magnitude because of better utilization. The hoe is now fully utilized and its cost reduced $0.040; it does not experience idle time waiting for trucks.

CONCRETE WORK

Portland cement concrete is one of the most widely used structural materials. Its versatility, economy, adaptability, worldwide availability, and especially its low maintenance requirements make it an excellent building material. The term *concrete* is applicable for many products, but it is generally used with Portland cement concrete. It consists of Portland cement, water, and aggregate that have been mixed together, placed, consolidated, and allowed to solidify and harden. To a constructor, fresh concrete is *all important,* because it is the fresh concrete that must be mixed, transported, placed, consolidated, finished, and cured.

There are two types of concrete-mixing operations in use: (1) transit-mixed (Fig. 6.20) and (2) central-mixed (Fig. 6.21). Today, unless the project is in a

FIGURE 6.21 Central-mix concrete plant, drum mixer on the left, cement and fly ash silos overhead center.

remote location or is relatively large, the concrete is batched in a central batch plant and transported to the job site in transit-mix trucks, often referred to as *ready-mixed* concrete trucks. Concrete, a heterogeneous mixture of water and solid particles in a stiff condition, will normally contain a large quantity of voids when placed. The purpose of consolidation is to improve density, and engineers are interested in density because of its relation to strength. Consolidation also improves watertightness and wear resistance. The importance of proper consolidation cannot be overemphasized, as entrapped air can render the concrete totally unsatisfactory.

Along with placement and consolidation, proper curing of the concrete is extremely important. When Portland cement and water are mixed, chemical reaction of hydration takes place giving the material strength. Hydration is dependent on the availability of moisture. If complete hydration does not take place, the concrete will not reach its design strength. Curing encompasses all methods whereby the concrete is assured of adequate time, temperature, and *supply of water* for the cement to continue to hydrate. The time normally required is 3 days, and optimum temperatures are between 40°F and 80°F. As most concrete is batched with sufficient water for hydration, the only problem is to ensure that the concrete does not dry out. This can be accomplished by ponding with water (for slabs), covering with burlap or polyethylene sheets, or spraying with an approved curing compound. Curing is one of the least costly operations in the production of quality concrete and one that is all too frequently overlooked. If the concrete is not properly cured, it may not gain its proper strength, in which case it will have to be removed—a very costly operation.

Item no.	Description	Estimated quantity	Unit	Unit price	Estimated amount
	SECTION 00010				
	BIDDING SCHEDULE				
1.	Diversion and control of water	1	job	L.S.	$ _____
2.	Clear site and remove obstructions	1	job	L.S.	_____
3.	Construction water	1	job	L.S.	_____
4.	Excavation, foundation	132,600	cy	_____	_____
5.	Excavation, basin	517,000	cy	_____	_____
6.	Excavation, toe	18,590	cy	_____	_____
20.	Concrete, invert slab	73	cy	_____	_____
21.	Concrete, walls	123	cy	_____	_____
22.	Concrete, top slab	60	cy	_____	_____
23.	Concrete, ogee weir	1440	cy	_____	_____
24.	Reinforcing steel	39	ton (tn)	_____	_____

FIGURE 6.22 Partial bid schedule for a small project—note four concrete items by volume and one item for reinforcing steel by weight.

CONCRETE ESTIMATES

The quantity take-off for concrete work is made for two reasons: (1) to check work quantities in the bid schedule and (2) to establish work quantities that are not listed in the bid schedule as pay items, but are still required to complete the work. Typically, on a unit-price job, an owner pays for concrete work under only one or two bid items such as concrete by volume and reinforcement steel by weight (Fig. 6.22). Therefore, the quantity information needed to prepare a concrete estimate includes

- Concrete volumes
- Formwork contact surface area
- Ground contact area
- Cure areas
- Reinforcing steel weights
- Embedded items

Concrete Volume

The shapes of concrete features are given on the construction plans. The form of the concrete may be detailed in specific drawings, or there may be only a typical drawing and dimensions presented in a table either on the drawings or possibly in another location in the contract documents. When placing the concrete in

FIGURE 6.23 The excavation walls used to form the concrete.

constructed forms, the volume required will equal the volume of the formwork. However, when using the sides of an excavation as formwork (Fig. 6.23), an allowance for overexcavation must be added to the net volume.

Formwork

Form cost is a large portion of the total cost for concrete work. The thinner and/or more complicated the structure, the greater will be the number of square feet of forms per cubic yard of concrete, and, therefore, the form cost per cubic yard of concrete will be greater. The formwork take-off is concerned with *square feet* of concrete *contact area* (sfca) and should delineate the type of form to be used, for example, field-built timber, purchased form system, or steel forms (Fig. 6.24). The take-off should state the number of times a form will be reused. Form reuse is a function of shape, how many times the same shape will be formed, and the material used to create the form. In addition to the formwork that actually serves to shape the plastic concrete, it is necessary to determine the amount of blocking or shoring required to support elevated forms (Fig. 6.25). Forms are not shown in the construction plans; it is typically the responsibility of the contractor to design the required formwork. In some cases, it is possible to slipform the concrete and eliminate the requirement for constructed formwork (Fig. 6.26).

Form materials such as form oil, anchors, bolts, ties, clamps, and other form fasteners are additional cost items that must be included in the form cost. It may also be necessary to employ a crane or other equipment to lift and move constructed forms.

FIGURE 6.24   Steel column form.

FIGURE 6.25   Shoring system used to support concrete formwork.

FIGURE 6.26 Slipform paving an interstate highway.

Ground Contact Area

When constructing a slab on a grade, it is necessary to carefully prepare the excavation or embankment to the proper grade. It is very difficult and expensive to cut the grade to the exact elevation. If it happens that the grade is cut with a positive variance, the concrete slab will not have the contract-specified thickness and consequently the concrete work will be rejected by the owner. Therefore, it is always necessary to work the grade to a minus tolerance, as that will guarantee an acceptable slab thickness. In the case of thin slabs (less than one foot in thickness), the grade tolerance has a severe impact on the concrete waste factor. If the ground is prepared 0.1 ft below the prescribed grade, the resulting concrete waste factor for a 6-in. slab is 20%, and even for a 1-ft slab the waste factor would still be an unacceptable 10%. Therefore, the ground surface area must be properly prepared in order to minimize concrete waste, and this may require a significant effort.

Cure Area

The amount of exposed concrete area that will require either water curing or the application of curing compound must be calculated during the concrete take-off. The curing required for each type of concrete and the curing time duration is usually delineated in the project specifications.

Reinforcing Steel

Other than in the case of mass concrete, most concrete structures have reinforcement steel cast within the concrete. On many projects, reinforcement steel is paid for under a separate unit-price bid item, although on some jobs there is no separate item and its cost must be included within the concrete bid price. Usually, steel suppliers will independently make a take-off of required reinforcing and offer a material price to furnish steel in the prescribed lengths and shapes ready for installation. Prime contractors sometimes like to make their own take-off as a check. When this is done, the take-off should segregate bar sizes and differentiate straight and bent quantities. Some estimating references recommend that, as normal practice, an allowance of between 2% and 5% of calculated bar weight should be added to the total to account for waste and splices. Reinforcing steel bar weights are given in Table 6.6.

Embedded Items

There is frequently a requirement to embed nonstructural items in the cast-in-place concrete. These would include items such as anchor bolts; steel plates; corner protection angles; and sleeves for the passage of pipes, electrical conduits, and ductwork.

TABLE 6.6 Reinforcing steel data.

Bar size	Weight (pounds per linear ft)	Diameter (in.)	Cross-sectional area (sq in.)	Perimeter (in.)
#3	0.376	0.375	0.11	1.178
#4	0.668	0.500	020	1.571
#5	1.043	0.625	0.31	1.963
#6	1.502	0.750	0.44	2.356
#7	2.044	0.875	0.60	2.470
#8	2.670	1.000	0.79	3.142
#9	3.400	1.125	1.00	3.544
#10	4.303	1.270	1.27	3.990
#11	5.313	1.410	1.56	4.430
#14	7.650	1.693	2.25	5.320
#18	13.600	2.257	4.00	7.090

PRICING CONCRETE

Estimating the cost of concrete work is normally accomplished by applying historical production rates against the work task quantities. The historical production rates are based on specific crew (to include equipment) makeup by number and skill level. This is a different approach from that used to estimate earthwork cost, where productions are calculated based on specific project conditions such as haul distance, haul road grade, haul route rolling resistance, and specific machine type. Companies develop concrete production factors from their historical records of past jobs. If no corporate production data are available for a particular work task, estimating reference books [6] are available that present a broad range of production data for most construction tasks.

Crew Productivity

In Fig. 6.22, 123 cy of concrete are required by bid Item No. 21 for the construction of concrete walls. To determine the wall thickness and height, it is necessary to study the project drawing. That study revealed that the walls range in height from 6 ft to 7 ft and are 8 in. wide, and that number 6 reinforcing bars are used in the walls. Typical production data for the required wall construction tasks are shown in Table 6.7.

EXAMPLE 6.6

Estimate the time required to place the 123 cy of wall concrete using a concrete pump. Also consider using a crane and bucket. Using the Table 6.7 data:

Placing rate, using a pump (8-in. walls)	Placing rate, using a crane (8-in. walls)
$\dfrac{100 \text{ cy per day}}{8 \text{ hr per day}} = 12.5$ cy per hr	$\dfrac{80 \text{ cy per day}}{8 \text{ hr per day}} = 10.01$ cy per hr

Placing time, using a pump	Placing time, using a crane
$\dfrac{123 \text{ cy}}{12.5 \text{ cy per hr}} = 9.84$ hr	$\dfrac{123 \text{ cy}}{10 \text{ cy per hr}} = 12.3$ hr

Labor Cost

If a concrete pump is used to place the concrete, the labor crew would consist, based on the Table 6.7 data, of a labor foreman, five laborers, a cement mason, and an equipment operator. If the project is covered by the *Davis-Bacon Act,* the contractor will, as a minimum, have to pay the wage rates and fringes as stated in the Department of Labor wage decision. If the project is not covered by the *Davis-Bacon Act,* the contractor is free to offer wages at the lowest level that the market will bear. But low wage offers do not always attract qualified workers. The skill level of the workers hired for the project must, as a minimum, match that assumed by the task productivity used in developing the estimate. If the project will last more than 1 year, the contractor should consider wage inflation

TABLE 6.7   Concrete work task production data.

Work task	Productivity (normal 8-hr day)	Unit of measure	No.	Crew and equipment
Forms in place				
Job-built plywood wall forms, to 8 ft high, 1 use, below grade	300	sfca*	1 4 1	Carpenter foreman Carpenter Laborer
Job-built plywood wall forms, to 8 ft high, 2 use, below grade	365	sfca*	1 4 1	Carpenter foreman Carpenter Laborer
Reinforcing steel in place				
Walls, #3 to #7 bars	3	ton	4	Ironworkers
Walls, #8 to #18 bars	4	ton	4	Ironworkers
Placing concrete				
Walls, 8 in. thick, direct chute	90	cy	1 4 1 2	Labor foreman Laborer Cement mason Vibrators
Walls, 8 in. thick, pumped	100	cy	1 5 1 1 2 1	Labor foreman Laborer Cement mason Equipment operator Vibrators Concrete pump
Walls, 8 in. thick, crane and bucket	80	cy	1 5 1 1 1 2 1 1	Labor foreman Laborer Cement mason Equipment operator Assistant equipment operator Vibrators Concrete bucket Crane, 55 ton
Concrete curing				
Sprayed membrane curing compound	95	csf*	2	Concrete laborers

*sfca, square foot contact area; csf, hundred square feet.

in the out years. Table 6.8 presents a labor rate analysis framework. When establishing the rates that will be used in creating the estimate, the engineer should review the project schedule and consider when a particular work task will occur. If the task occurs toward the end of the project, the projected future rates should be used.

EXAMPLE 6.7

Estimate the labor and equipment cost component for placing the 123-cy of wall concrete using a concrete pump. Use the "estimate" labor rates of Table 6.8. The wall will be constructed in two equal parts so that only half the required formwork material will have to be purchased (two uses of the wall forms). From Example 6.6, the calculated placing time is 9.84 hr, therefore use 10 hr to calculate placement labor and equipment cost. The vibrators will cost $5.24 per hr and the combined O&O cost for the pump is $80 per hr.

Task	123 cy, place concrete, pump					
Production rate	12.5 cy per hr	**Total hours**	10			
	No.	**Labor rate**	**Labor**	**Equipment**	**Total**	**Unit cost ($ per cy)**
Foreman	1	$20.50	$ 205.00	$	$ 205.00	$ 1.666
Laborer	5	12.10	605.00		605.00	4.919
Cement mason	1	17.00	170.00		170.00	1.382
Equipment operator	1	16.00	160.00		160.00	1.301
Vibrator	2			104.80	104.80	0.852
Concrete pump	1			800.00	800.00	6.504
		Total	$1,140.00	$904.80	$2,044.80	$16.624

Concrete Estimate Summary

Some companies construct their estimates using *standard crews* similar to those shown in Table 6.7. Correspondingly, the company's production database is established based on performing work with the standard crews. Therefore, the individual members of the crew are often not delineated when producing estimate cost sheets such as that shown in Example 6.7. The "crew" labor cost per hour for the Example 6.7 concrete pumping task would be $114 per hour

TABLE 6.8 Labor rate analysis.

Worker classification	Davis-Bacon rate per hr	XYZ's current rate	Projected rate 1 yr out	Projected rate 2 yr out	Estimate rate
Carpenter foreman	$18.00	$22.00	$22.50	$23.00	$22.50
Labor foreman	17.00	20.00	20.50	21.00	20.50
Carpenter	16.00	14.00	14.50	15.50	16.00
Ironworker	15.00	13.00	13.50	14.50	15.00
Laborer	12.00	9.00	12.00	12.50	12.10
Cement mason	15.00	16.00	16.50	17.50	17.00
Equipment operator—crane	18.00	18.00	18.50	19.50	18.50
Equipment operator—pump	16.00	14.00	14.50	15.50	16.00
Equipment operator—assistant	12.50	14.00	14.50	15.50	14.00

($20.50 + (5 × $12.10) + $17.00 + $16.00), and the estimate sheet would be presented as shown here:

Task	123 cy, place concrete, pump					
Production rate	12.5 cy per hr	**Total hours**	10			
	No.	**Crew rate**	**Labor**	**Equipment**	**Total**	**Unit cost ($ per cy)**
Concrete crew, pump	1	$114.00	$1,140.00	$	$1,140.00	$ 9.268
Vibrator	2			104.80	104.80	0.852
Concrete pump	1			800.00	800.00	6.504
		Total	**$1,140.00**	**$904.80**	**$2,044.80**	**$16.624**

A complete estimate for the "concrete, walls," for the Fig. 6.22 proposal item 21, would be as shown in Table 6.9. It was determined by a study of the project drawing that the total length of wall is 765 ft and that the walls have an average height of 6.5 ft. This yields 9,945 sfca of formwork [(2 sides × 765 ft × 6.5 ft) + (2 ends × 0.66 ft × 6.5 ft)]. An average cost of $1.29 per sfca was used to calculate the cost of wall form materials. This value should come from the company's historical data, but an average cost can also be found in estimating data reference books. Wall-forming data are usually presented based on the number of times the forms will be reused and the average height of the wall. The $1.29 value is for forms that will be used twice and a wall 7 ft or less in height. Reuse of the forms reduces both the form material and labor cost. The cost reduction is dependent on the extra effort to carefully strip the forms, so that they can be reused, and the amount of forms that must be replaced after each reuse. The savings generated by the ability to use the forms twice can be as high as 40%. For this reason, it is important to carefully analyze the form use sequence. Form material is usually listed under a separate cost category (see the column "Other" in Table 6.9); this is because the forms are not a permanent part of the finished structure. The same is true for the curing compound that is also listed in the other classification. The material classification is reserved for those materials that actually become part of the finished facility, in this case the concrete.

Additional information concerning concrete estimating is presented in Chapter 7.

Risk

The percentage contribution of each cost category is denoted at the bottom of Table 6.9. Labor at 52% is the most significant cost component in this example. The cost category Other at 29% and the category Material at 17% are significant cost contributors. The material category (e.g., concrete cost) has two components: (1) quantity and (2) price. The accuracy of the quantity component is a function of the quality of the project plans and the estimator's attention to detail in computing the volume of concrete required. With a complete set of plans, the

TABLE 6.9 Concrete cost estimate.

Task	123 cy, wall concrete					
Forming	9945 sfca					
Production rate	45.6 sfca per hr	**Total hours**	218			

	Labor	Equipment	Material	Other	Total	Unit cost ($ / sfca)
Forming crew	$21,494.80	—	—		$21,494.80	$2.159
Form material		—	—	$12,840.66	$12,840.66	$1.290
Total	$21,494.80	—	—	$12,840.66	$34,335.46	$3.449
Unit cost	$2.159	—	—	$1.290	$3.449	

Placing concrete, pump	123 cy					
Production rate	12.5 cy per hr	**Total hours**	10			

	Labor	Equipment	Material	Other	Total	Cost ($ / cy)
Concrete crew	$1,140.00	$	$	$	$1,140.00	$9.268
Equipment	—	$904.80	—	—	$904.80	$7.356
Concrete	—	—	$7,564.50	—	$7.564.50	$61.500
Total	$1,140.00	$904.80	$7,564.50	—	$9,609.30	$78.124
Unit cost	$9.268	$7.356	$61.50	—	$78.124	

Cure concrete	510 sf					
Production rate	1,187 sf per hr	**Total hours**	1			

	Labor	Equipment	Material	Other	Total	Cost ($ / cy)
Labor	$24.20	$	$	$	$24.20	$0.047
Curing comp.	—	—	—	$17.95	$17.95	$0.035
Total	$24.20	—	—	$17.95	$42.15	$0.082
Unit cost	$0.047	—	—	$0.035	$0.082	

	Labor	Equipment	Material	Other	Total	Cost ($ / cy)
Item total	$22,659.00	$904.80	$7,564.50	$12,858.61	$43,986.91	$357.618
Unit cost ($ / cy)	$184.220	$7.356	$61.500	$104.542	$357.618	
Percent of total cost	52	2	17	29		

accuracy of the quantity take-off should not be an issue. Once the contract has been executed, the contractor should be able to lock in the price of material by means of a purchase order to the concrete supplier. Therefore, the 17% material cost component usually does not subject the contractor to a high degree of risk.

However, the 52% "labor" and the 29% "other" (forming material) components can represent significant risk. The forming material, like the concrete, has two cost components: (1) quantity and (2) price. It is important to realize the estimate was prepared based on the assumption that the forms would be used twice.

Labor risk is a different matter. Labor cost is very sensitive to the *forming production rate* used in creating the estimate. The estimate was constructed with a forming production rate of 45.6 sf per crew hour. If that forming production is not achieved, the cost of the work can increase significantly. Your attention is again directed to this relationship, as shown in Fig. 3.4 in Chapter 3. When form production falls to 40 sf per hr, the wall cost rises by $24.73 per cy; a 12% reduction in the forming production rate adds 7% to the total wall cost. The estimator must exercise judgment in selecting production rates. Even when using corporate historical data, it is necessary to carefully look at the proposed project and consider if the historical rates should be adjusted.

SUMMARY

The goal of estimating is to determine the value of specific construction products. Earthwork computations involve the calculation of earthwork volumes, the balancing of cuts and fills, and the planning of the most economical material hauls. The quantity take-off for concrete work is completed to check work quantities in the bid schedule and to establish work quantities that are not listed in the bid schedule as pay items. Typically, on a unit-price job, an owner uses only one or two bid items, such as concrete by volume and reinforcement steel by weight, to compensate the contractor for concrete work. Critical learning objectives that support engineer estimating include

■ Ability to calculate earthwork volume
■ Ability to construct an earthwork calculation sheet
■ Ability to construct an understanding of the mass diagram
■ Ability to use historical production data in constructing a concrete estimate
■ Ability to calculate crew cost
■ An understanding of the project risk that results from a high percentage of labor cost. Labor is the major risk a contractor has and it is within the manager's control to manage.

These objectives are the basis for the questions in this chapter.

REVIEW QUESTIONS

6.1 Using the average-end-area method, calculate the cut and fill volumes for stations 125 + 00 through 131 + 00.

Station (1)	End-area cut (sf) (2)	End-area fill (sf) (3)	Volume of cut (bcy) (4)	Volume of fill (ccy) (5)
125 + 00	0	785	—	—
126 + 00	652	0		
127 + 00	2,150	0		
128 + 00	3,210	0		
129 + 00	1,255	147		
130 + 00	95	780		
131 + 00	0	3,666		

6.2 Using the average-end-area method calculate the cut and fill volumes for stations 19 + 00 through 24 + 00.

Station (1)	End-area cut (sf) (2)	End-area fill (sf) (3)	Volume of cut (bcy) (4)	Volume of fill (ccy) (5)
19 + 00	326	0	—	—
20 + 00	157	0		
21 + 00	44	0		
21 + 50	0	0		
22 + 00	0	147		
23 + 00	0	165		
24 + 00	0	133		

6.3 Using the average-end-area method calculate the cut and fill volumes for stations 25 + 00 through 31 + 00.

Station (1)	End-area cut (sf) (2)	End-area fill (sf) (3)	Volume of cut (bcy) (4)	Volume of fill (ccy) (5)
25 + 00	0	3,525	—	—
26 + 00	355	985		
27 + 00	786	125		
28 + 00	2,515	55		
29 + 00	1,255	23		
29 + 25	620	0		
29 + 50	25	845		
30 + 00	0	3,655		
31 + 00	0	8,560		

6.4 Complete the earthwork calculation sheet here and plot the resulting mass diagram. Divide ccy by 0.9 to convert to bcy (column 10).

Station (1)	End-area cut (sf) (2)	End-area fill (sf) (3)	Vol. of cut (bcy) (4)	Vol. of fill (ccy) (5)	Stripping cut (bcy) (6)	Stripping fill (ccy) (7)	Total cut (bcy) (8)	Total fill (ccy) (9)	Adj. fill (bcy) (10)	Algebraic sum (bcy) (11)	Mass ordinate (12)
									0.90		
0 + 00	0	0									
1 + 00	0	90			0	24					
2 + 00	0	154			0	40					
3 + 00	147	0			15	19					
4 + 00	192	0			50	0					
4 + 50	205	0			57	0					
5 + 00	179	0			21	0					
6 + 00	121	0			43	0					
6 + 50	100	0			19	0					
7 + 00	52	10			8	0					
8 + 00	0	180			10	20					
9 + 00	0	231			0	69					
10 + 00	0	285			0	18					

6.5 Complete the earthwork calculation sheet here and plot the resulting mass diagram. Divide ccy by 0.9 to convert to bcy (column 10). Calculate the average haul (trapezoidal formula) for the balances on this project. Is this a waste or borrow project?

Station (1)	End-area cut (sf) (2)	End-area fill (sf) (3)	Vol. of cut (bcy) (4)	Vol. of fill (ccy) (5)	Strip-ping cut (bcy) (6)	Strip-ping fill (ccy) (7)	Total cut (bcy) (8)	Total fill (ccy) (9)	Adj. fill (bcy) (10)	Alge-braic sum (bcy) (11)	Mass ordi-nate (12)
									0.90		
10 + 00	0	0									
11 + 00	580	0			80	0					
12 + 00	2100	0			90	0					
13 + 00	4650	0			100	0					
14 + 00	6000	0			100	0					
15 + 00	3250	560			80	60					
16 + 00	1300	1620			80	80					
17 + 00	700	2450			80	85					
18 + 00	0	7800			0	100					
19 + 00	0	3620			0	90					
20 + 00	0	1980			0	80					
21 + 00	0	1310			0	80					
22 + 00	580	860			80	10					
23 + 00	1620	250			100	10					
24 + 00	3850	0			100	0					
25 + 00	2600	0			100	0					

6.6 Estimate the time required to place the 200 cy of wall concrete using a concrete pump; use the Table 6.7 data.

6.7 Estimate the time required to place the 45 cy of wall concrete by direct chute; use the Table 6.7 data.

6.8 Estimate the labor and equipment cost component for placing the 37.5 cy of wall concrete using a concrete pump. Use the "estimate" labor rates of Table 6.8. The wall will be constructed in two equal parts so that only half the required formwork material will have to be purchased (two uses of the wall forms). The vibrators will cost $5.24 per hr and the combined O&O cost for the pump is $80 per hr.

REFERENCES

1. *Construction Estimating & Bidding Theory Principles Process* (1999). Publication No. 3505, The Associated General Contractors of America, 333 John Carlyle Street, Suite 200, Alexandria, VA.

2. Oberlender, Garold D., and Steven M. Trost (2001). "Predicting Accuracy of Early Cost Estimates Based on Estimate Quality," *Journal of Construction Engineering and Management,* ASCE, 127(3), May–June.

3. Rignwald, Richard C. (1993). *Means Heavy Construction Handbook,* R. S. Means Company, Inc., Kingston, MA.

4. *RSMeans Building Construction Cost Data,* R. S. Means Company, Inc., Kingston, MA (published annually).

5. *RSMeans Heavy Construction Cost Data,* R. S. Means Company, Inc., Kingston, MA (published annually).

6. *Walker's Building Estimator's Reference Book,* Frank R. Walker Company, Lisle, IL (published regularly).

7. *Guide to Formwork for Concrete (ACI 347-04),* (2004). American Concrete Institute, Farmington Hills, MI, 7th edition.

8. *Formwork for Concrete (ACI SP-4),* (2005). American Concrete Institute, Farmington Hills, MI.

WEBSITE RESOURCES

1. www.usbr.gov/pmts/estimate/cost_trend.html Bureau of Reclamation, *Construction Cost Trends.* The Bureau of Reclamation's *Construction Cost Trends (CCT)* was developed to track construction relevant to the primary types of projects being constructed by the organization. All the various cost indexes consist of two elements, contractor labor and equipment costs, and contractor-supplied materials and equipment.

2. www.hcss.com Heavy Construction Systems Specialists Inc., (HCSS). HCSS specializes in construction estimating, bidding, and job cost software for contractors. HCSS, 6200 Savoy, Suite 1100, Houston, TX 77036.

3. www.uscost.com U.S. COST provides project control services and software.

4. www.trakware1.com TRAKWARE supplies excavation-estimating software for the construction industry. Trakware, Inc., 4425 Juan Tabo NE, Ste. 125, Albuquerque, NM 87111

5. www.agtek.com AGTEK Development Company, Inc., supplies construction computer products including earthwork quantity take-off and graphical grade-positioning systems to sitework, highway, paving, and underground contractors. AGTEK, 368 Earhart Way, Livermore, CA 94550.

C H A P T E R

Estimating Building Projects

The architects and engineers who design buildings use drawings and symbolic language to communicate their ideas to contractors. Calculating quantities of materials can be simplified if you understand a few principles of measurement. In the case of building estimates, uniformity and consistency are achieved by using the Construction Specification Institute (CSI) item-coding format. Estimates are organized to serve the needs of the contracting organization, with a critical need being identification and assessment of risk. The estimate serves to identify and quantify risk.

INTRODUCTION

A detailed and accurate take-off is the foundation of any estimate. The *units* of the take-off should reflect each work task. The architects and engineers who design commercial buildings communicate their ideas to constructors by means of the symbolic language of drawings and through written specifications. The drawings and their accompanying notes show and explain how the building is arranged and how it is constructed. Because of the amount and type of information that is shown on drawings, and because much of the information is repeated from job to job, symbols (Fig. 7.1) have been developed to help communicate this information. The lines, notes, and symbols are a shorthand language. This shorthand language is an indispensable tool for those who design and construct buildings. The drawings, or "plans," are part of the contract between the owner and contractor. To estimate the cost of the work, the contractor must understand the information that the plans communicate.

COMPUTER-AIDED DESIGN

The plans for many projects are developed using computer-aided drafting/drawing (CADD) tools. Developments in geometric modeling have advanced

212

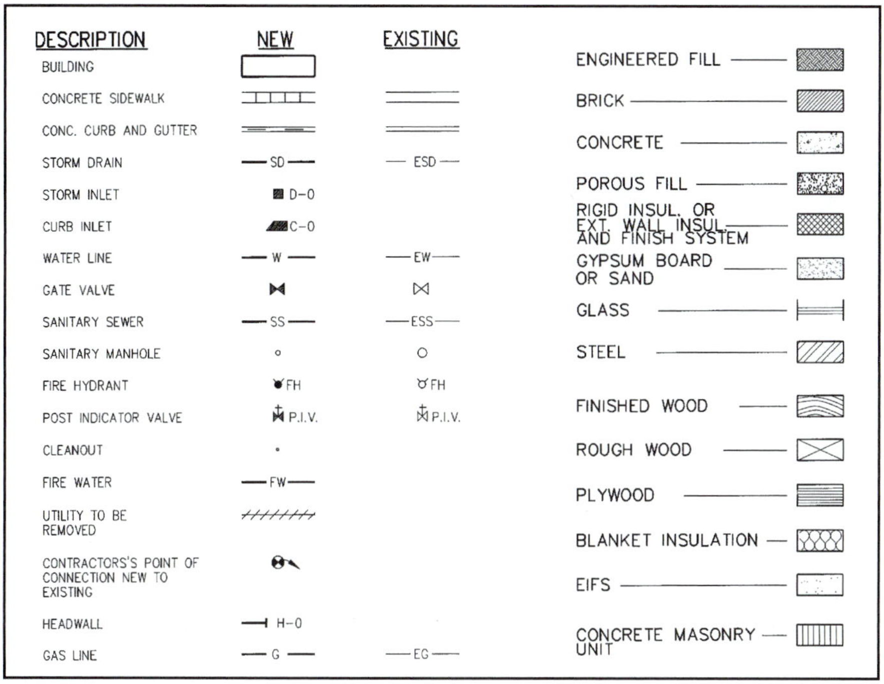

FIGURE 7.1 Some of the common symbols used in communicating construction information.

this technology so that now there is the ability to do computer-aided design (CAD). When plans are transmitted to the constructor in these formats, it is possible to extract construction quantities electronically. The estimator can automatically import quantities directly from CADD or CAD files and perform manual adjustments as necessary. Many of the computer-based estimating systems referenced at the end of this chapter have CAD integrator components as part of the software package.

UNDERSTANDING THE DRAWINGS

Building drawings are arranged according to broad user type groupings. These groupings may vary by architecture and engineering firm practice but generally there is a cover sheet with general information and a listing of the drawing followed by the boring log sheets. After that preliminary information, there will be the pavement, civil, architectural, structural, mechanical, plumbing, and electrical information.

General Information

The first page in a set of drawings is the title sheet. Depending on the preferences of the A/E, other information such as an index of drawings and a plot plan

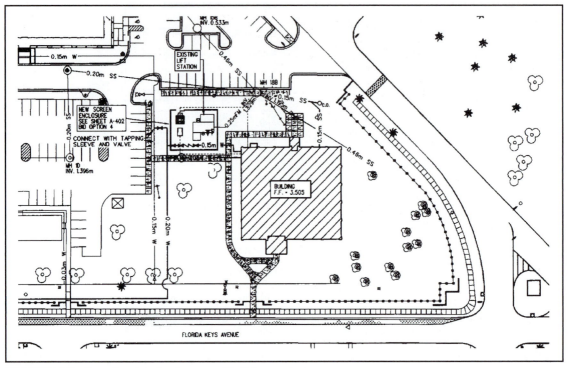

FIGURE 7.2   Part of a plot plan for a small building project.

may be included on the title sheet. Sometimes because of the size of the project, it takes more than one sheet to present all of this information. A plot plan, sometimes called a *site plan,* is drawn to scale and indicates the location of the proposed building on the site (Fig. 7.2.)

On the plot plan, dashed lines represent existing grades, while solid lines indicate finish grades. The elevation figures on the plot are based on a **benchmark**. A benchmark is a point whose elevation is used to reference all other elevations on the plans. Many times, the point's exact elevation in feet above sea level is used for this purpose, but it could also be a random elevation noted as a 100 ft datum. Existing buildings, roadways, and utilities are also displayed on the plot plan.

benchmark

Mark on some permanent object fixed to the ground from which land measurements and elevations are taken

Civil Plans

The civil plans contain sheets depicting the existing conditions, demolition work, landscaping, the erosion and sediment control requirements, and sheets showing the details of these items of work.

Architectural Plans

The architectural plans provide floor plans; elevations; sections; details; and finish, door, and window schedules.

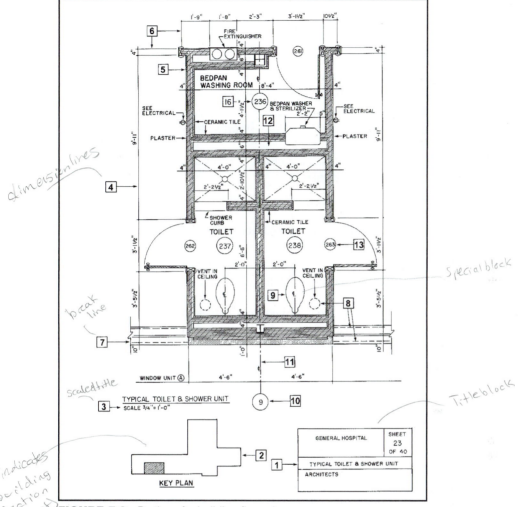

FIGURE 7.3 Portion of a building floor plan.
(Drawing reprinted courtesy of F. W. Dodge, McGraw-Hill Construction Information Group)

Floor Plans The floor plan represents a cut horizontally through a building showing the view looking downward. A separate floor plan drawing is made for each floor of the building. These plan sheets indicate the location and arrangement of **bearing walls**, **partitions**, doors, windows, and vertical transportation. Dimensions are provided on the floor plans for determining exact locations in the horizontal plane. These plans also provide information concerning materials used. A sample floor plan for only a portion of a building is shown in Fig. 7.3.

The Fig. 7.3 sample drawings of toilet and shower areas between two rooms has reference numbers inside square box symbols (□), these are not a part of the drawing but are used here to reference the discussion of information shown on the drawing.

bearing wall or partition
A wall supporting any vertical load other than its own weight.

1. The title block references the project and drawing.
2. A plan key is used if the whole building plan is not shown on one drawing. The shaded area indicates the shown section.
3. A scale reference is found on all drawings. There may be different scales used on the same drawing sheet.
4. Dimension lines show distances between points.

> Always use dimension values, as drawings may not be exactly to scale.

5. Main object lines indicate outlines of walls and structural objects, which the dimensions reference.
6. Extension lines extend beyond the main object lines for the purpose of easy dimensioning.
7. Broken lines indicate that the part continues but is not shown on the drawing.
8. Dashed lines indicate an object or area that cannot be seen from the vantage point of the cut in a plan, elevation, or section drawing. The objects are indicated to show their relationship to other parts of the building. In Fig. 7.3, the dashed lines indicate the exterior wall lines that are hidden from view because of the window frame and sill. The ceiling vent is above the cut level.
9. Equipment lines show fixtures and other equipment.
10. Column line identifiers are provided to aid in explicitly describing locations or points within the building; numbers are usually used for the left to right identifiers (usually increasing left to right) and letters for the bottom to top identifiers on the sheet (usually increasing top to bottom).
11. Centerlines are used to determine the location of equipment, fixtures, and structural elements such as columns. The symbol L superimposed on a C identifies the centerline (₵).
12. Pipe chases are void spaces between walls that are usually for the purpose of accommodating plumbing, HVAC ducts, and wiring.
13. Small circled numbers at door openings identify specific doors. Larger circled numbers in the center of a space identify the room or area. Using these numbers, additional information can be found on the door, window, and room schedules (see Fig. 7.4).

elevation

A drawing showing either the front, sides, or rear face of a building.

Elevations and Sections An **elevation** illustrates the vertical elements of a building, either exterior or interior. A section reveals the arrangement of materials and systems (Fig. 7.5). Section views enable a builder to both estimate and perform the work properly.

Details There are certain construction details that cannot be shown adequately by the floor plans or the elevations and sections. Therefore, there are special large-scale detail drawings found throughout a set of building plans. These spe-

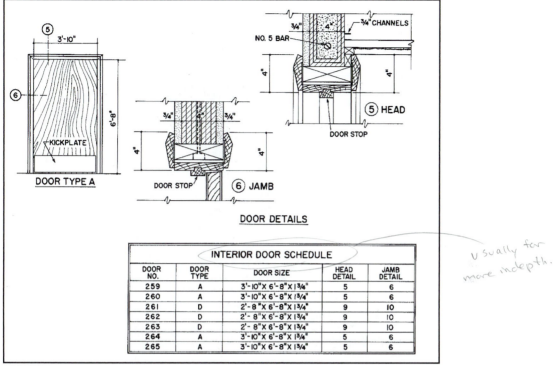

DOOR DETAILS

INTERIOR DOOR SCHEDULE				
DOOR NO.	DOOR TYPE	DOOR SIZE	HEAD DETAIL	JAMB DETAIL
259	A	3'-10"X 6'-8"X 1¾"	5	6
260	A	3'-10"X 6'-8"X 1¾"	5	6
261	D	2'-8"X 6'-8"X 1¾"	9	10
262	D	2'-8"X 6'-8"X 1¾"	9	10
263	D	2'-8"X 6'-8"X 1¾"	9	10
264	A	3'-10"X 6'-8"X 1¾"	5	6
265	A	3'-10"X 6'-8"X 1¾"	5	6

[handwritten: Usually for more indepth.]

FIGURE 7.4 Detail drawing of a door head and jamb, and a door schedule.
(Drawing reprinted courtesy of F. W. Dodge, McGraw-Hill Construction Information Group)

cial drawings explain construction and design features. Figure 7.4 is a special construction detail illustrating door details. The small circles enclosing numbers indicate the areas shown in details.

Schedules   On the bottom of Fig. 7.4 is a door schedule that lists door dimensions and other pertinent information. A set of plans will also contain schedules for windows and for each room detailing how the walls, ceiling, and floor will be finished.

Structural Plans

The structural drawings in a set of plans indicate the size and location of major framing pieces and give specific dimensions and details. These plans will depict the foundation and framing systems to include the roof.

Foundation   Footings and foundation walls transfer the building loads to the earth below. A second function of foundation walls is to keep moisture out of the underground parts of the building. Usually, a waterproofing compound is applied to the exterior surface of foundation walls. Figure 7.6 shows a footing and foundation wall drawing. On the drawing, the foundation walls are shown by two

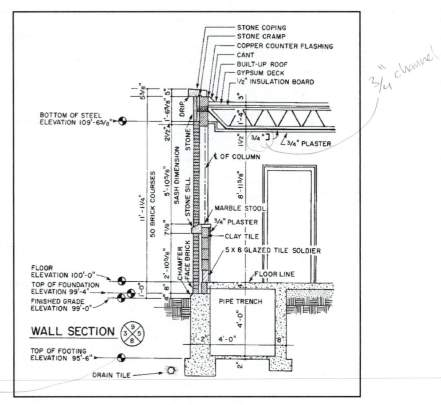

FIGURE 7.5 Section view of a building wall system.
(Drawing reprinted courtesy of F. W. Dodge, McGraw-Hill Construction Information Group)

solid lines (=). The other dashed lines (– – –) on either side of the solid lines are the footings on which the foundation wall is constructed. The circled letters and numbers are the column line identifiers (discussed in the floor plans section under item 10). They aid in identifying column locations.

Framing   Steel, concrete, and timber are widely used structural framing materials. The supplier prepares detailed drawings for steel framing work. These drawings are referred to as *shop drawings*. Shown in Fig. 7.7 is part of a steel roof-framing plan. The W12 × 26 terminology indicates a *wide flange beam*, nominally 12 in. deep, weighing 26 lb per linear foot. The P. ½" × 11½" indicates a steel plate attached to the beam to support masonry above (section S-S, Fig. 7.7). The OW JOIST 12H4 indicates an *open web steel joist* 12 in. deep, sometimes referred to as a bar joist. The last number, "4" in the 12H4, identifies the size of the top and bottom cords (section S-S, Fig. 7.7).

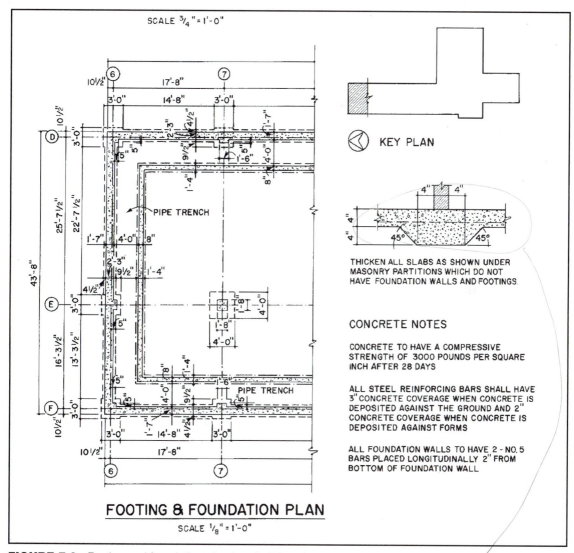

SCALE ¾" = 1'-0"

KEY PLAN

THICKEN ALL SLABS AS SHOWN UNDER
MASONRY PARTITIONS WHICH DO NOT
HAVE FOUNDATION WALLS AND FOOTINGS.

CONCRETE NOTES

CONCRETE TO HAVE A COMPRESSIVE
STRENGTH OF 3000 POUNDS PER SQUARE
INCH AFTER 28 DAYS

ALL STEEL REINFORCING BARS SHALL HAVE
3" CONCRETE COVERAGE WHEN CONCRETE IS
DEPOSITED AGAINST THE GROUND AND 2"
CONCRETE COVERAGE WHEN CONCRETE IS
DEPOSITED AGAINST FORMS

ALL FOUNDATION WALLS TO HAVE 2 - NO. 5
BARS PLACED LONGITUDINALLY 2" FROM
BOTTOM OF FOUNDATION WALL

PIPE TRENCH

PIPE TRENCH

FOOTING & FOUNDATION PLAN
SCALE ⅛" = 1'-0"

FIGURE 7.6 Footing and foundation plan for a building project.
(Drawing reprinted courtesy of F. W. Dodge, McGraw-Hill Construction Information Group)

*increased thickness
for extra shear resistance*

Mechanical, Plumbing, and Electrical

The mechanical, plumbing, and electrical drawings are used to locate pipes, fixtures, ducts, outlets, lights, and wiring runs. The detailed descriptions of these
and other equipment to be installed, and the work required, are described in the
written specifications (Fig. 7.8). A portion of the HVAC ductwork for a section
of building office space is shown in Fig. 7.9.

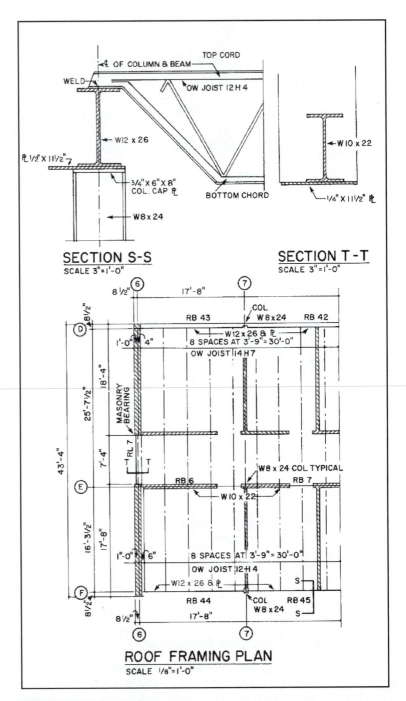

FIGURE 7.7 Roof framing plan.
(Drawing reprinted courtesy of F. W. Dodge, McGraw-Hill Construction Information Group)

PART 3 EXECUTION
3.1 GENERAL INSTALLATION REQUIREMENTS
Hubless cast-iron pipe shall not be installed under concrete floor slabs. Piping located in air plenums shall conform to NFPA 90A requirements. Unprotected plastic pipe shall not be installed in air plenum. Piping located in shafts that constitute air ducts or that enclose air ducts shall be noncombustible in accordance with NFPA 90A. Installation of plastic pipe where in compliance with NFPA may be installed in accordance with PPFA-01. The plumbing system shall be installed complete with necessary fixtures, fittings, traps, valves, and accessories. Water and drainage piping shall be extended 5 feet outside the building, unless otherwise indicated. A gate valve and drain shall be installed on the water service line inside the building approximately 6 inches above the floor from point of entry. Piping shall be connected to the exterior service lines or capped or plugged if the exterior service is not in place. Sewer and water pipes shall be laid in separate trenches, except when otherwise shown.

FIGURE 7.8 Execution portion of a plumbing specification.

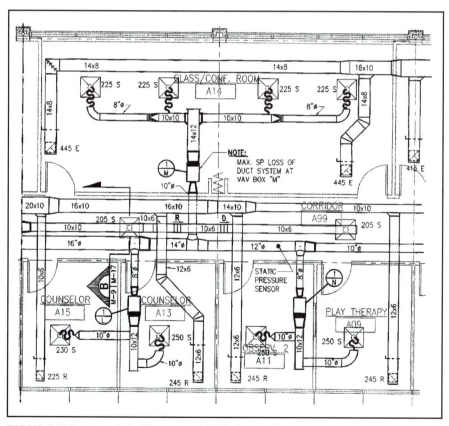

FIGURE 7.9 Layout of office space HVAC ductwork.

SPECIFICATIONS

In the case of building construction, the most commonly used specification organization and format is that of the Construction Specification Institute (CSI). The CSI MasterFormat 2004 divides the organization of the specifications into 50 divisions. The format callout numbers are six digits in length and arranged into three sets of paired numbers, one pair per level. The main six-digit number represents three levels of subordination, as the numbers did in previous editions of the MasterFormat. A comparison of the previous (1995) and the current MasterFormat is presented in Table 5.1. The basic building construction CSI divisions under the 2004 MasterFormat are shown in Table 7.1.

Organization of a CSI Technical Specification

The CSI prescribes a consistent organization of the information presented in a technical specification. Each subsection within a division that is used as part of the contract specifications will have three major sections: (1) general, (2) prod-

TABLE 7.1 CSI divisions.

Division name		Division name	
Procurement and Contacting Requirements Group			
Division 00	Procurement and Contracting Requirements		
Specifications Group			
General Requirements Subgroup			
Division 01	General Requirements		
Facility Construction Subgroup			
Division 02	Existing Conditions	Division 09	Finishes
Division 03	Concrete	Division 10	Specialties
Division 04	Masonry	Division 11	Equipment
Division 05	Metals	Division 12	Furnishings
Division 06	Wood, Plastic, and Composites	Division 13	Special Construction
Division 07	Thermal and Moisture Protection	Division 14	Conveying Systems
Division 08	Openings		
Facility Services Subgroup			
Division 21	Fire Suppression	Division 26	Electrical
Division 22	Plumbing	Division 27	Communications
Division 23	Heating, Ventilation, and Air Conditioning	Division 28	Electronic Safety and Security
Site and Infrastructure, Subgroup			
Division 31	Earthwork		

RESILIENT FLOORING

PART 1 GENERAL

1.1 REFERENCES

The publications listed below form a part of this specification to the extent refer-enced. The publications are referred to in the text by basic designation only.

AMERICAN SOCIETY FOR TESTING AND MATERIALS (ASTM)

ASTM D 4078   (1992; R 1996) Water Emulsion Floor Polish

ASTM E 648   (1997) Critical Radiant Flux of Floor-Covering
Systems Using a Radiant Heat Energy Source

FIGURE 7.10 General portion of a flooring specification.

RESILIENT FLOORING

PART 2 PRODUCTS

2.1 VINYL-COMPOSITION TILE TYPE VCT-1

Vinyl-composition tile shall conform to ASTM F 1066, Class 2 (through pattern tile), Composition 1, asbestos-free, and shall be 12 inches square and 1/8 inch thick. Tile shall have the color and pattern uniformly distributed throughout the thickness of the tile. Flooring in any one continuous area shall be from the same lot and shall have the same shade and pattern.

2.4 ADHESIVE

Adhesive for flooring and wall base shall be as recommended by the flooring manufacturer.

FIGURE 7.11 Parts of the products portion of a flooring specification.

ucts, and (3) execution. As the names imply, each addresses a particular part of the information the builder needs to properly perform the work.

The general section discusses the scope of work and general requirements under which the work must be executed (Fig. 7.10). The order of subsections for the general sections is (1) references; (2) scope; (3) related work; (4) system description; (5) quality assurance; (6) submittals; (7) delivery, storage, and han-dling; (8) project site conditions; (9) sequencing/scheduling; (10) alternatives/alterations; (11) allowances; (12) unit prices; and (13) warranty.

The products section details the materials that can be incorporated into the work (Fig. 7.11). The order of subsections for the products sections is (1) accept-able manufacturers, (2) materials, (3) equipment, (4) mixes, and (5) fabrication.

The execution section explains the workmanship, quality, and installation requirements (Fig. 7.8). The order of subsections for the execution sections is presented here with typical verbiage of a flooring specification.

1. *Inspection* The contractor shall examine and verify that site conditions are in agreement with the design package and shall report all conditions that will prevent a proper installation.

2. *Preparation* Flooring shall be in a smooth, true, level plane, except where indicated as sloped.

3. *Installation/Application/Erection* Wall base shall be installed with adhesive in accordance with the manufacturer's written instructions. Base joints shall be tight and base shall be even with adjacent resilient flooring. Voids along the top edge of base at masonry walls shall be filled with caulk.

4. *Field quality control, adjusting and cleaning* Immediately on completion of installation of tile in a room or an area, flooring and adjacent surfaces shall be cleaned to remove all surplus adhesive.

5. *Protection* From the time of laying until acceptance, flooring shall be protected from damage as recommended by the flooring manufacturer.

6. *Extra stock/Spare parts*

7. *Schedules* (this refers to reference schedules, see Fig. 7.4, not the time schedule for the work) The following painting schedules identify the surfaces to be painted and prescribe the paint to be used and the number of coats of paint to be applied.

The contractor is bound to produce the quality specified by the contract specifications, therefore it is necessary to carefully study information presented in the specifications before attempting to price the work.

MEASUREMENT

An estimator uses the project plans to calculate or count the quantities of material required to construct the work. The calculation-counting process is referred to as a *quantity survey* or *take-off*. Calculating quantities of materials can be simplified by understanding a few principles of measurement. Consider only the bottom footer (Fig. 7.12) of the foundation wall in Fig. 7.5. If it were necessary to compute the volume of concrete, one procedure would be to multiply the cross-sectional area by the total length of the wall centerline. But usually it is also necessary to compute the formwork contact area. The length of the inside and outside formwork will differ from the centerline length.

Perimeter

Three perimeters can be defined for the solid sections circumscribing a building (Fig. 7.12).

1. *Mean* The true measure of an object's length. It is the length of the line about which the object is symmetrical. It is the length of the object stretched out in a straight line.

2. *Inside perimeter*

3. *Outside perimeter*

The mean perimeter can be used to compute volumes; the inside and outside perimeters are useful in computing surface areas. If a wall and footing are sym-

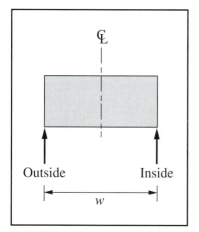

FIGURE 7.12 Measurement of a rectangular section.

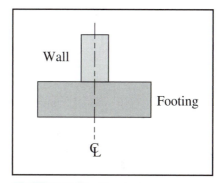

FIGURE 7.13 Two features symmetrical about the same centerline.

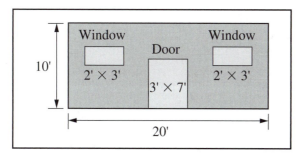

FIGURE 7.14 Computing wall surface area.

metrical about the same centerline, their mean perimeters will be the same (Fig. 7.13).

All descriptions in an estimate are assumed to describe a positive item, but it is necessary to enter negative items when developing a take-off. It is easier to calculate the total area of a wall and to then deduct the door and window space when computing wall surface area (Fig. 7.14). A deduction that is totally within a measured area is termed a *void.* A deduction that is not totally within a measured area is termed a *want,* and a deduction enclosed on three sides by a measured area is a *recess.* The empty spaces in the upper right and lower left of Fig. 7.15 are wants. The cavity in the lower center is a recess.

Because the dimensioning for the foundation drawing in Fig. 7.15 shows distance between exterior points, it is easy to calculate the building's outside perimeter (OP).

$$\text{Outside perimeter} = \sum \text{outside dimensions} \qquad \textbf{[7.1]}$$

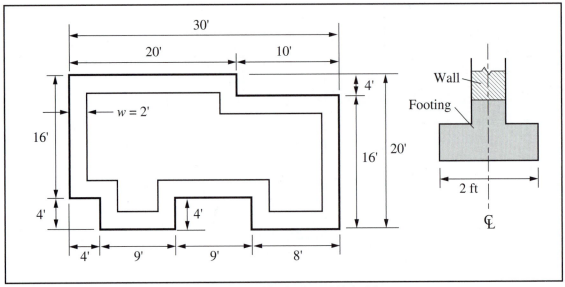

FIGURE 7.15 Foundation drawing for a building. The footing width is 2 ft.

EXAMPLE 7.1

Calculate the outside perimeter of the wall shown in Fig. 7.15.

$$\text{Outside perimeter} = 20 + 4 + 10 + 16 + 8 + 4 + 9 + 4 + 9 + 4 + 4 + 16 \Rightarrow 108 \text{ ft}$$

The total length of the building is 30 ft and in the OP calculation the length values used were 20, 10, 8, 9, 9, and 4 for a sum of 60, equaling twice the length. The width of the building is 20 ft and in the OP calculation the width values used were 4, 16, 4, and 16 for a sum of 40, equaling twice the width. The depth of the recess is 4 ft and in the OP calculation the recess values were 4 and 4 for a sum of 8, equaling twice the depth of the recess. From this example, a simple measurement rule can be stated.

$$\text{Outside perimeter} = 2 \times (\text{length} + \text{width} + \text{recess}) \qquad \textbf{[7.2]}$$

The outside perimeter can be calculated easily using Eq. [7.2] and the Fig. 7.15 information.

$$\text{Outside perimeter} = 2 \times (30 + 20 + 4) \Rightarrow 108 \text{ ft}$$

Relationship Between Perimeters

Examining Fig. 7.16 helps in gaining an understanding of the relationship between perimeters. The term "w" in the figure stands for the wall thickness. If

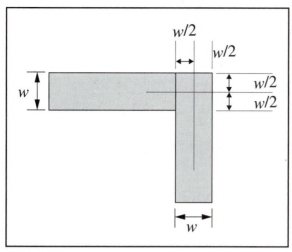

FIGURE 7.16 Measurement around corners.

the total length of the outside perimeter is known, the length of the mean perimeter can be calculated using Eq. [7.3].

$$\text{Mean perimeter} = \text{OP} - 4\left[2 \times \left(\frac{w}{2}\right)\right] \qquad [7.3]$$

where w = thickness of the section.

The equation states that at each corner it is necessary to twice deduct one-half the thickness. At a "want" or "recess," the deductions are canceled out by an addition at the next corner. Equation [7.4] is a general formula for calculating perimeter distance once one perimeter is known.

$$\text{Perimeter}_1 = \text{Perimeter}_2 + 4 \times \left[2 \times \left(\frac{\pm \text{ horizontal distance}}{\text{between } P_1 \text{ and } P_2}\right)\right] \quad [7.4]$$

A positive sign is employed when the perimeter distance is increasing, moving from an inside perimeter to an outside perimiteter. A negative sign is used when the perimeter is decreasing, moving from the outside to the inside of the space.

EXAMPLE 7.2

The thickness of the foundation wall shown in Fig. 7.15 is 2 ft. Calculate the volume of concrete required to build this foundation wall if its cross-sectional area is 4 sf. If the wall is 2 ft in height, what is the required formwork in sfca?

$$\text{Outside perimeter} = 2 \times (30 + 20 + 4) \Rightarrow 108 \text{ ft}$$

$$\text{Mean perimeter} = 108 - 4\left[2 \times \left(\frac{2}{2}\right)\right] \Rightarrow 100 \text{ ft}$$

$$\text{Volume concrete} = 100 \text{ ft} \times 4 \text{ sf} \Rightarrow 400 \text{ cf or } 14.8 \text{ cy}$$

Inside perimeter = OP + 4 × [2 × (−2)]

Inside perimeter = 108 + 4 × [−4] ⇒ 92 ft

Outside forms = 108 × 2 ⇒ 216 sfca

Inside forms = 92 × 2 ⇒ 184 sfca

Total forms required = 216 + 184 ⇒ 400 sfca

CONCRETE CONSTRUCTION (CSI DIVISION 3)

Concrete is a universal construction material because of its unique properties of strength, durability, and workability. Estimates are organized to serve the needs of the contracting organization, with a critical need being identification and assessment of risk. Therefore, as discussed in Chapter 6, columns in the estimate segregate project risk. The typical column headings are labor, equipment, material, other, and subcontracts. It is also necessary to organize the estimate for uniformity and consistency. In the case of building estimates, uniformity and consistency are achieved by using the Construction Specification Institute (CSI) item-coding format. Concrete forms and accessories are in the 03 10 00 subsection, concrete reinforcement is in subsection 03 20 00, cast-in-place concrete is in subsection 03 30 00, precast concrete is in subsection 03 40 00, and cast decks and underlayment are in subsection 03 50 00.

Placement of concrete is in units of volume, while forming is in units of form contact area. The basic concrete operations are

1. Preparation for placement of the wet concrete:
   - Prepare surface—fine grading and compaction of material below footings and on-grade slabs
   - Fabricate forms—job constructed or manufactured, quantity required to satisfy the placement schedule, number of reuses
   - Erect forms—type of form (wall, column, beam, or slab), height above working surface
   - Embedded items—angle iron ledgers, channel door frames, ladder rungs and miscellaneous fabrications may be furnished and installed under CSI Division 05500 Metal Fabrications
2. Reinforcing steel:
   - Fabrication—prefabricated or job fabricated
   - Placement—hand placed in the forms, or cages assembled and lifted into place by machine
3. Placing concrete—size of individual placement:
   - Delivery to the site—from ready mix plant, onsite plant, haul distance affects job site handling time

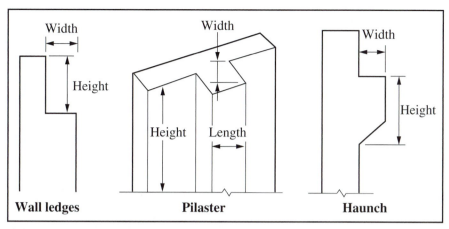

FIGURE 7.17 Wall variations.

- Transport to point of placement—chute, bucket, pump
- Placement—wall, column, beam, or slab
- Consolidation
4. Finishing—refer to the specifications, rubbing, point and patch
5. Curing and protection of concrete—exposed surfaces, ponding, compound, hot and cold weather protection

The concrete material cost will be influenced by the specified strength requirement. Strength requirements, which are usually expressed in pounds per square inch (psi), will be found in the project specifications. The concrete quantity take-off should separately quantify concrete volumes by mix design (strength), admixtures, and color and/or any other special features.

When performing the quantity take-off, the estimator keeps informational notes. These notes identify the specific sections and details found in the plans that support the recorded measurements—*All reinforcing is to be fabricated in accordance with CRSI standards; Proposed forming system does not match the rebar spacing.*

Formwork

Building layouts are normally repeated from bay to bay. Such repetition allows a formwork cost savings and dictates the quantity of formwork material. The formwork material must be able to withstand repeated usage if the sequence of work is to be maintained. The architecture floor plans usually provide schedules specifying finish quality. Smoother finishes require higher quality form-facing materials. The structural plans will provide information concerning expansion joints, placement joints or strips, reinforcing steel configuration, and penetrations.

Walls The cost of wall forming depends on wall height, number and size of openings, intersections, **pilasters** (Fig. 7.17), haunches, and blockouts. If there

pilaster
Rectangular pier attached to a wall for the purpose of strengthening the wall; also a decorative column attached to a wall.

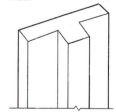

are openings in the wall, their areas are usually not deducted from the form quantity. However, there would be a deduct in figuring concrete volume. Good construction practice is to make the form continuous over the opening so as not to destroy the forming material for future reuse. This practice also minimizes cutting form material. It is necessary to calculate the additional blockout forms that are placed around the opening. If ledges are present, it is necessary to add forming material for ledge forms. Ledges are formed by adding a box to the inside of a wall form. This means that the contact area is created twice.

In the case of many features, it is easiest to measure through the feature and then add back the extra forming material. Consider the pilaster wall shown in Fig. 7.17. When calculating the amount of form material, measure the length and height of the wall, and then add the required side material for the pilasters. The extra material for the pilaster form is calculated by the formula

$$\text{Extra material per pilaster} = 2 \times w \times h \qquad [7.5]$$

where w = pilaster projection width
h = pilaster projection height

It is sometimes required to fasten a *chamfer* strip to the inside of the forms where the concrete would make a right angle corner. Chamfering keeps the concrete edges from easily chipping.

When a wall is being constructed in segments, it will be necessary to calculate the extra forming material required to form the bulkheads at the end of each segment. The quantity of bulkhead is often kept as a separate quantity because these forms require extra drilling work for the protruding reinforcing bars.

Columns   The form quantity for a column is the perimeter multiplied by the column height. For circular columns, the formulas are:

$$\text{Volume, circular column} = \pi \times r^2 \times h \qquad [7.6]$$

where r = radius of the column
h = column height

$$\text{Surface area, circular column} = \pi \times d \times h \qquad [7.7]$$

where d = diameter of the column
h = column height

beam
A horizontal structural member that carries a load.

Beams   The formwork for a beam consists of the **beam** bottom and sides. The quantity of forming material for one side of an internal beam is the depth of the beam below the slab, times the length ($D \times L$). The forming for the inside of an exterior beam conforms with an interior beam (depth to the bottom of the slab), but the outside form extends to the top of the slab. The amount of beam bottom forming material that is not actually used when the beam crosses a column is not usually deducted from the total. When two beams intersect at a column, the quantities of the deepest beam should be calculated based on the beam running

FIGURE 7.18 Floor slab formwork.

through the column. The shallow beam should be figured as running only to the edge of the deeper beam. Beam quantities should be accumulated by height of required shoring. Accumulation increments of 5 ft are common because that matches shoring tower heights.

Slabs   A slab-on-grade rests directly on a prepared ground surface. Under the slab, it is common to find a stone base and a vapor barrier. The slab may have reinforcing steel bars, but typically reinforcing is wire mesh. Edge forms may constitute the only formwork required, but many times even these are not needed as previously constructed walls serve as the edge forms.

Elevated flat slabs require a bottom form supported by shoring (Fig 7.18). The outside edges of the beam forms provide the perimeter forms of the slab.

Reinforcing

The quantity of reinforcing steel is calculated by summing bar lengths according to size and then converting (Table 6.6) the length into weight, either pounds (lb), hundred weight (CWT), or tons (tn). Bar laps must be accounted for when calculating reinforcing steel quantities. Lap lengths are usually specified. The cost of reinforcing steel is affected by

1.   Bar grade, tensile strength, specified
2.   Whether they are deformed, have ridges of indentations on their surface, or if they can be smooth

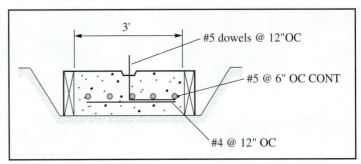

FIGURE 7.19   Placement of reinforcing steel in a concrete footing.

3. Extent to which they must be fabricated, cut, and bent

4. Whether they can be plain or must be galvanized or epoxy coated

on center

Method of indicating spacing of framing members by stating the distance from center of one to the center of the next.

The designation #5 dowels @ 12"OC (upper right Fig. 7.19) means that the #5 "L"-(90° angle) shaped dowels are placed every 12 in. on centers along the length of the e footing. A dowel is a bar that will penetrate into the next part of the constructed system. In this case the next system is probably a wall. The dowel will provide reinforcing steel continuity between the wall and the footing. The designation #5 @ 6" OC CONT, means that #5 bars are spaced 6 in. apart on centers across the footing width and placed lengthwise continuously. The #4 @ 12" OC means that #4 bars are placed perpendicularly across the longitudinal #5 bars and that they have a 12-in. **on-center** spacing. The #4 transverse steel serves to hold the longitudinal steel in the proper position.

EXAMPLE 7.3

Calculate the reinforcing steel required for the footing shown in Fig. 7.19. The footing is 36 ft in length. The dowel has a height of 2 ft. The longitudinal bars require 6 in. of cover at their ends and a transverse bar is placed at the end of the longitudinal bars.

Length of longitudinal bars: 36 ft − 0.5 ft − 0.5 ft = 35 ft

Number of longitudinal bars: 5

Total length of longitudinal bars: 35 ft × 5 = 175 ft of #5 bars

Length of transverse bars: 2 ft

Number of transverse bars: 35 + 1 = 36

Total length of transverse bars: 36 × 2 ft = 72 ft of #4 bars

Length of dowel bars: 1 ft + 2 ft = 3 ft

Number of dowel bars: 35 + 1 = 36

Total length of dowel bars: 36 × 3 ft = 108 ft of #5 bars

Weight of #4 bars: 72 ft total length × 0.668 lb per lf = 48.1 lb

Weight of #5 bars: 283 ft total length (175 ft + 108 ft) × 1.043 lb per lf
                     = 295.2 lb

Total weight reinforcing steel: 48.1 lb + 295.2 lb = 343.3 lb

Curing

To control the rate of its hydration, the concrete must be cured. Proper curing produces concrete with excellent long-performing service. Curing serves two purposes:

- It retains moisture in the slab so that the concrete continues to gain strength.
- It delays drying shrinkage until the concrete is strong enough to resist shrinkage cracking.

Curing methods may be specified in the project's specifications or selection of a method may be at the contractor's discretion. Common concrete curing methods are

- Leave the forms in place.
- Water cure—the concrete is flooded, ponded, or mist sprayed.
- Water-retaining methods—use coverings such as sand, canvas, burlap, or straw that are kept continuously wet.
- Waterproof paper or plastic film seal—these are applied as soon as the concrete is hard enough to resist surface damage.
- Chemical membranes—the chemical application should be made as soon as the concrete is finished.

The surface areas to receive a curing treatment should be totaled based on types of treatment. Many of the calculations should have been performed during the formwork take-off. Historical production rates are used to calculate the labor cost to accomplish the curing. Material cost and application rates for a curing compound must be obtained from the manufacturers.

Pricing Concrete

The concrete work items can be priced as was described in Chapter 6 or historical unit costs can be applied to the quantities. Most construction companies develop historical unit costs for the types of work they perform. These costs are developed from actual experience on previous projects. Therefore, once the formwork quantity (by type, slab, column, beam) is known, the work can be priced. If the company does not have experience with a particular work item, average pricing information can be found in published reference books [11, 13]. The reference books will state the assumed production rate; the labor, equipment, material, and other prices; and crew composition and pay rates (see Table 7.2), equipment rates, and the overhead and profit percentages used in their cost calculations.

TABLE 7.2 Concrete crew composition and pay rate data from a reference book.

Concrete crew	
1 Carpenter foreman	$31.15/hr
4 Carpenters	29.15/hr
1 Labor	22.85/hr

EXAMPLE 7.4

Calculate the bare cost (without overhead and profit) to place 75 cy of 3000 psi concrete. The concrete is to be placed in 8-in.-thick wall forms by use of a concrete pump truck. These data were found in an estimating cost data publication.

	Production rate	Material cost/cy	Labor cost/cy	Equipment cost/cy	Total cost/cy
Walls, 8" thick pumped	12.5 cy/hr		$15.80	$6.40	$22.20

The publication did not provide a price for concrete but a local supplier has quoted a price of $75 per cy delivered.

	Production rate	Material cost/cy	Labor cost/cy	Equipment cost/cy	Total cost/cy
Walls, 8" thick pumped	12.5 cy/hr	$75.00	$15.80	$6.40	$97.20
75 cy	6 hr	$5,625	$1,185	$480	$7,290

Therefore, the total bare cost is $7,290 (this is just to provide the concrete and to place it in the forms). It will take 6 hr to complete the placement.

MASONRY CONSTRUCTION (CSI DIVISION 4)

Brick, subsection 04 21 clay unit masonry, and concrete masonry units, subsection 04 22, are the two most common types of masonry, but Division 4—Masonry also includes glass block (subsection 04 23) and stone (subsection 04 43). Many varieties of texture and appearance can exist under the classification of brick or concrete masonry and the estimator should always understand local custom and terms used to describe masonry units. The term *concrete masonry units* (CMU) refers to concrete block or concrete brick. Concrete block is usually a relatively large unit with hollow cores. An accurate estimate of masonry cost requires knowledge of various factors, including the type of structural element, thickness of the section, kind of mortar, style of mortar joint, class of workmanship, and skill of the workers. There are several categories of masonry walls:

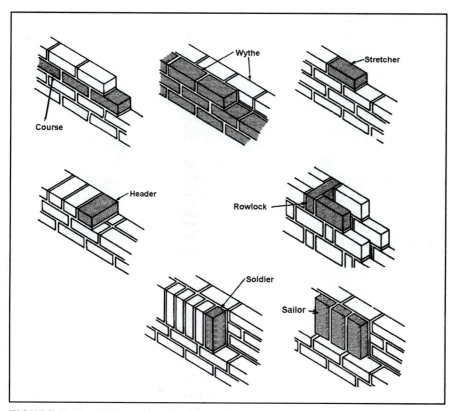

FIGURE 7.20 Brick positions in a wall to form patterns.

1. Conventional bonded walls
2. Drainage-type walls, including masonry veneer, cavity type, and masonry-bonded hollow walls; the success of this type wall depends on the care taken to provide continuous means for water to escape
3. Barrier-type walls—metal tied and reinforced masonry

Position of the Brick

Most brick is laid as *stretchers* so the long, narrow face dimension is horizontal. The terms applied to brick surface positions are (1) stretcher, (2) header, (3) rowlock, (4) rowlock stretcher, (5) soldier, and (6) sailor (Fig. 7.20).

Bond

The term *bond*, in reference to masonry, can have three different meanings. *Structural bond* refers to the method by which individual masonry units are interlocked or tied together to cause the entire assembly to act as a single structural unit. The adhesion of mortar to the masonry units is termed *mortar bond*.

FIGURE 7.21 Concrete block wall with a running bond lay pattern.

The term used to describe how the masonry units are presented in the face of a structure is *pattern bond*. The pattern formed by the units and the mortar joints on the face of a wall make up the pattern bond. The pattern may be dictated for structural reasons or for purely decorative design. Three of the basic structural bonds are

- *Running bond.* The simplest of the basic pattern bonds, the running bond consists of all stretchers (Fig. 7.21). Because there are no headers in this bond, metal ties are ordinarily used. Running bond is used largely in cavity wall construction.

- *Common* or *American bond.* This is a variation of the running bond with a **course** of full-length headers at regular intervals.

- *English bond.* This is made up of alternating courses of stretchers and headers.

course

A continuous row of stone or brick of uniform height.

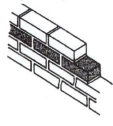

Joints

The mortar between masonry units must be struck to form the joint and the specifications may require a specific type of joint. A *flush* joint (sometimes referred to as a rough cut) is created when the mason simply strikes off the excess mortar with the edge of a trowel. *Concave* and *V-shaped* joints are formed using a

FIGURE 7.22 Climbing scaffolding for blocking a building wall.

steel-jointing tool. They are effective in resisting rain penetration and are recommended for use in areas subjected to heavy rains.

Factors That Impact Masonry Cost

Many factors influence the labor cost of placing building masonry. There are no set quantities of masonry units that a mason will lay per hour, though some unions have rules that *limit* the number a mason can place per day. Solid brick walls are used less frequently than in the past. Today, most brick is a facing backed up with concrete blocks.

Production will vary with the required grade of workmanship, the style of bond in which the units are laid, the width and kind of mortar joint, and the type of building—straight walls or walls with multiple door and window openings.

Common bond is the most economical method of placing face masonry units. Other styles, such as English, require more labor because of the various patterns and the additional cutting required.

The cost of cleaning face brick varies with the kind of brick. In the case of smooth face brick, it may be necessary only to wash the wall, but other masonry cleaning requirements might include (1) brush hand cleaning, (2) high-pressure water, or (3) chemical cleaning.

The height of the wall will affect production by adding **scaffolding** (Fig. 7.22) and lifting equipment requirements (Fig. 7.23).

scaffolding
A temporary structure, designed and built by the contractor, to provide elevated working space.

FIGURE 7.23 Using a forklift to handle brick.

Masonry Quantities

The quantity take-off steps for a masonry estimate are (1) determine the surface area of masonry and (2) convert the area to the number of masonry units based on the type of unit and the required placement pattern. All openings should be deducted in full regardless of their size. From the elevation view plan sheets (Fig. 7.24), it is usually easy to identify masonry surfaces; but to calculate the area of those surfaces, it is generally necessary to also consult the floor plans, which give longitudinal dimensions, and the section drawing (Fig. 7.25), which provides height information.

The section drawings will be referenced on the elevation views. The two dark left-pointing arrows at the top of Fig. 7.24 let the viewer know that TYP. 1 and 2, located on sheet A8-1 should be consulted. The A7-1 indicates the sheet of this elevation view. TYP stands for typical. TYP. 1 will show a view through the window area and TYP. 2 will show a view of the solid wall. The typicals in this particular case do not by themselves provide sufficient information to calculate the vertical dimension of the brick. It is necessary also to consult several special details.

Above the window in WALL SECTION @ WINDOW view 1, Fig. 7.25 special detail 1 on sheet A9-1 (Fig. 7.26) is referenced. From studying the elevation drawing and the two section views in Fig. 7.25, it is clear that the majority of the brick is laid in a running bond pattern. Because there are no headers in a running bond pattern, metal ties are ordinarily used to tie the brick exterior wall to the inner wall of the structure.

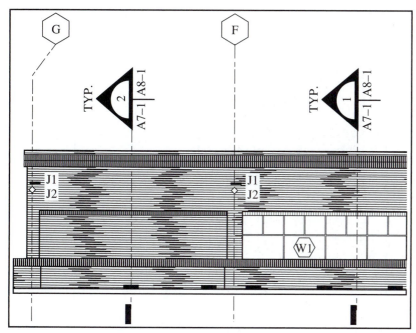

FIGURE 7.24 Elevation view of a building having facing brick.

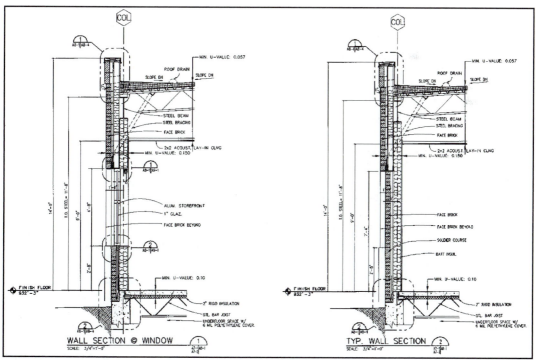

FIGURE 7.25 Section views of the wall shown in Fig. 7.24.

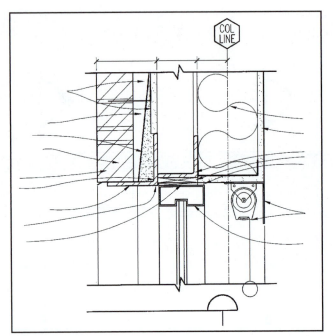

FIGURE 7.26   Detail of window head, referenced from left drawing Fig. 7.25.

The plans show that there is some variation in the brick pattern (Fig. 7.24). At the level directly under the window, there is a soldier course and at the level directly above the window there is a rowlock course (also see Fig. 7.25). It should be clear that before starting a masonry take-off, the estimator must carefully study many different drawings in the plan set.

Brick walls are usually designated on the plans as 4 in. or 4½ in., or 8 in. or 9 in. The estimator need not consider this variation in thickness because a 4-in. or 4½-in. wall is one brick thick, and an 8-in. or 9-in. wall is two bricks thick.

After the surface area of the wall is obtained, it is necessary to determine the number of bricks required per square foot. The nominal dimensions for a standard nonmodular brick is 8 in. × 2¼ in. × 3¾ in. The nominal dimensions for a standard modular brick are 8 in. × 2⅔ in. × 4 in. Sizes will vary somewhat depending on the brick's location in the kiln during burning, the brick in the center of the kiln usually being more thoroughly burned, but the nominal dimensions yield satisfactory estimating results.

The thickness of the mortar joint must be added to the brick dimensions. Mortar thickness will vary, but ⅜ in. is the normal thickness. Therefore, when a standard nonmodular brick is laid as a stretcher and mortar thickness is included, the dimensions are 8⅜ in. long by 2⅝ in. in height, and it has a surface area of 21.984 sq in. (8⅜ in. × 2⅝ in.). To obtain the number of bricks required per square foot of wall, divide 144 sq in./sf by 21.984 sq in. For a 4-in. (one brick

FIGURE 7.27 Pallet of bricks stacked at the job site.

thick) wall, 6.55 bricks (144 sq in./sf ÷ 21.984 sq in.) are required per sf of wall surface.

The waste factor for brick wall construction should not exceed 1½% to 2% unless irregular or poor-quality bricks are used. However, bricks and CMUs are ordered by the pallet (Fig. 7.27) so the order must be in pallet multiples.

The nominal dimensions for standard modular concrete blocks are 7⅝ in. × 7⅝ in. × 15⅝ in. When laid up with ⅜-in. mortar joints, the unit is 8 in. high and 16 in. long, requiring 112.5 units per 100 sf, not including an allowance for waste.

EXAMPLE 7.5

The outside perimeter of a brick wall is 30 ft. Calculate the bricks required to build this wall if its height is 12 ft and the bricks will all be laid as soldiers. Standard modular bricks have been specified for the project. Assume a 1.5% waste factor.

Area of wall = 30 × 12 ⇒ 360 sq ft

Area of one soldier brick = $(2\frac{2}{3} + \frac{3}{8}$ in.$) \times (8 + \frac{3}{8}$ in.$) \Rightarrow 25.471$ sq in

Bricks per sf = $\left(\dfrac{144 \text{ sq in./ft}}{25.474 \text{ sq in.}}\right) \Rightarrow 5.653$

Number of bricks without waste = 360 sq ft × 5.653 bricks per sf ⇒ 2,035 bricks

Number of bricks with waste = 2,035 bricks × 1.015 ⇒ 2,066 bricks

TABLE 7.3 Typical masonry crew composition.

Masonry crew
1 Bricklayer (or block) foreman
3 Bricklayers
3 Bricklayer helpers

Pricing Masonry

Many construction companies use historical unit costs to price masonry work. Again, if the company does not have experience with a particular type of masonry work, average pricing information can be found in published reference books [11, 13]. Table 7.3 presents information on the composition of masonry crews.

EXAMPLE 7.6

Calculate the bare cost (without overhead and profit) to place 360 sf of concrete block. The blocks are being used to construct an exterior wall. Do not include the cost of scaffolding. The wall is to be 8 in. thick. The following data were found in an estimating cost data publication.

	Production rate	Material cost/sf	Labor cost/sf	Equipment cost/sf	Total cost/sf
CMU walls, 8" thick, exterior	45.6 sf/hr	$2.39	$2.95		$5.34

	Production rate	Material cost/sf	Labor cost/sf	Total cost/sf
CMU walls, 8" thick, exterior	45.6 sf/hr	$2.39	$2.95	$5.34
360 sf	7.9 hr	$860.40	$1,062.00	$1,922.40

Therefore, with a standard crew, the total bare cost is $1,922.40 and it will take 8 hr to complete the placement.

METALS (CSI DIVISION 5)

CSI Division 5 includes structural steel framing, subsection 05 12 (Fig. 7.28); metal joists, subsection 05 21 (Fig. 7.29); steel decking, subsection 05 31; structural metal stud framing, subsection 05 41 (Fig. 7.30), and formed metal fabrications, subsection 05 58. Fabricated steel can be purchased by the general contractor and erected by company forces. It is more common, however, to subcontract both the fabrication and erection.

A structural steel estimate involves a large amount of detail work. The quantities of structural members are listed from the plans and weights computed. The cost of shop fabrication (purchase cost) is determined, to which is added freight

FIGURE 7.28 Steel frame building under construction.

FIGURE 7.29 Steel framing with metal joist used to support second-story floor.

FIGURE 7.30 Structural metal stud framing used in an office building.

cost; and finally, the cost of erecting the steel on the job is ascertained. Material prices depend on where material is purchased. Mill prices apply to large orders. Warehouse prices apply to small orders from a local supplier. Prices may be quoted as of *time of delivery;* therefore, the estimator should be very cognizant of escalation clauses.

Steel Shapes and Nomenclature Information

When estimating the quantity of structural steel required for any job, each class of work (column bases, columns, girders, beams, trusses) should be estimated separately because each involves different labor operations in fabrication and erection. The steel member structural shapes are shown in Figs. 7.31 and 7.32, and a nomenclature explanation is provided beside each shape in the figures.

Sequence of Take-Off

When performing a steel take-off, it is best to maintain sequence of execution. One strategy is to work by structural types, columns, beams, and then flooring,

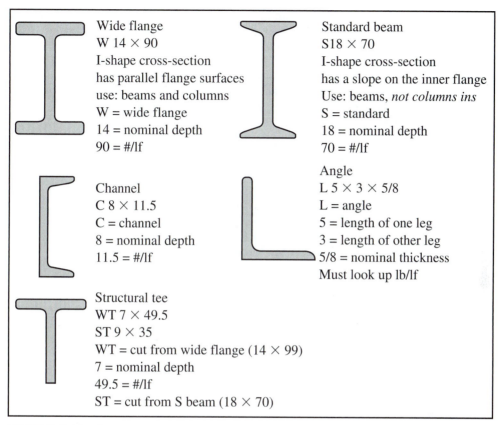

Wide flange
W 14 × 90
I-shape cross-section
has parallel flange surfaces
use: beams and columns
W = wide flange
14 = nominal depth
90 = #/lf

Standard beam
S18 × 70
I-shape cross-section
has a slope on the inner flange
Use: beams, *not columns ins*
S = standard
18 = nominal depth
70 = #/lf

Channel
C 8 × 11.5
C = channel
8 = nominal depth
11.5 = #/lf

Angle
L 5 × 3 × 5/8
L = angle
5 = length of one leg
3 = length of other leg
5/8 = nominal thickness
Must look up lb/lf

Structural tee
WT 7 × 49.5
ST 9 × 35
WT = cut from wide flange (14 × 99)
7 = nominal depth
49.5 = #/lf
ST = cut from S beam (18 × 70)

FIGURE 7.31 Common steel member structural shapes.

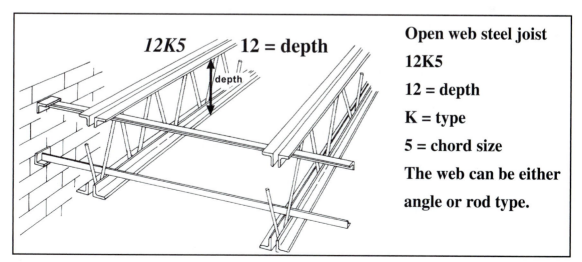

12K5 **12 = depth**

depth

Open web steel joist

12K5

12 = depth

K = type

5 = chord size

The web can be either

angle or rod type.

FIGURE 7.32 Steel joist.

and to work across the structure by column line identifiers. It is also good practice to work up the structure one floor at a time. This is illustrated in Example 7.7.

EXAMPLE 7.7

Calculate the total weight of steel (tons) required to frame one floor of the building shown in Fig. 7.33. The 10J3 steel joists (4.83 lb/ft) are over all four bays, but for clarity are shown only once.

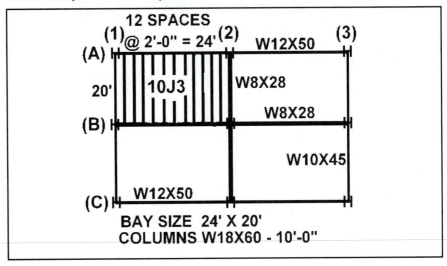

FIGURE 7.33   Steel arrangement for a small building.

Steel take-off			1st floor		
Type of section or member	Structural shape	Number	Length	Weight/ft	Total pounds
Columns	W18 × 60	9	10 ft	60 #/ft	5,400
Beams	W12 × 50	4	24 ft	50 #/ft	4,800
	W10 × 45	4	20 ft	45 #/ft	3,600
	W8 × 28	2	24 ft	28 #/ft	1,344
	W8 × 28	2	20 ft	28 #/ft	1,120
Joist	10J3	(4 3 11)	20 ft	4.83 #/ft	4,250
				Total weight lb	20,514

$$\text{Total weight} = \frac{20{,}514 \text{ lb}}{2{,}000 \text{ lb/ton}} \Rightarrow 10.26 \text{ tons}$$

Pricing Steel

The contractor typically receives a material price from a fabricator for the cost of the steel members required in the project. On a simple job, it may be possible to buy standard members from a warehouse, but for most projects, the estimator will submit a set of project plans to a manufacturer or fabricator so that they can prepare their own quantity take-off. The manufacturer or fabricator will then supply a material quotation. Steel joists are usually sold directly from the manufacturer to the contractor (or subcontractor).

Few contractors have the trained personnel and equipment (cranes or hoists) to erect structural steel. Therefore, most general contractors rely on experienced specialty subcontractors to handle this task. Based on historical project data, contractors relate labor hours to tons of steel by building type. Such information can be found in estimating reference books. An example would be: industrial buildings, one story, beam and girder construction 12.90 tons per 8-hr day with a specific crew. The labor cost to erect the steel can be calculated by applying the project labor rates to the labor crew and multiplying by the crew time. Crew time is calculated by dividing the project tonnage by the crew production, in tons per hour or per day.

EXAMPLE 7.8

The total weight of the framing steel for a building is 38.7 tons. A steel erection crew can complete 12.9 tons in an 8-hr day. The daily cost for the crew is $2,673.60. Calculate the number of days needed to erect the steel and the erection cost on a per ton basis.

$$\text{Time required} = \frac{38.7 \text{ tons}}{12.9 \text{ tons/day}} \Rightarrow 3 \text{ days}$$

$$\text{Cost} = 3 \text{ days} \times \$2,673.60 \Rightarrow \$8,020.80$$

$$\text{or } \frac{\$8,020.80}{38.7 \text{ ton}} = \$207.26 \text{ per ton}$$

WOOD, PLASTIC, AND COMPOSITES (CSI DIVISION 6)

Lumber is a commodity sensitive to supply and demand. Therefore, it is important to always check on the latest market prices when estimating a project. To accurately perform a carpentry take-off, the estimator should have an understanding of timber design and construction procedures. While it may not be clear from the plans, joists and **studs** should be doubled in many locations. Wood construction can be divided into several specific types of work:

■ Heavy framing and planking to include glued and laminated pieces
■ Rough carpentry—framing, sheathing, roofing, and door and window installation
■ Finish carpentry
  – Interior—flooring, stairs, trim
  – Exterior—siding, porches, corners, doors, and windows
■ Custom woodwork—millwork, factory built, cabinets

studs
Vertical framing members in a wall normally spaced at 16" or 24" on center.

Lumber

The dimensions used to designate lumber sizes refer to the cross-sectional dimensions of the board, dimensioned lumber, or timber, as shown in Fig. 7.34. Boards are thin pieces 1 in. in thickness. **Dimensioned lumber** refers to the intermediate-sized pieces used for rough carpentry work. Timbers are the large-sized pieces used for heavy framing. Boards and dimensioned pieces are usually dressed (finished) on all four sides, the designation being S4S, surfaced four sides. The numbers used to designate lumber size refer to the **nominal size**. The actual, or **dressed lumber**, dimensions will be less as can be seen by the data in Fig. 7.34. A 2 by 4 is actually 1½ in. thick by 3½ in. in width after surfacing. Framing lumber can usually be purchased only in lengths that are multiples of 2 ft. (i.e., 8, 10, 12, 14 ft). The one common exception is precut studs, which are $92\frac{5}{8}$ in (7 ft $8\frac{5}{8}$ in) in length.

Board Measure   The common system used to measure lumber is known as **board measure**. Many types of lumber are sawed in random widths and lengths. This is particularly the case for valuable hardwoods. To provide a unit of measure for such lumber, the board foot (BF) was developed. It is a volume measure

dimensioned lumber
Framing lumber that is of nominal thickness.

nominal size
Size of lumber before dressing, rather than its actual or finished size.

dressed lumber
Lumber machined and smoothed at the mill, to usually ½ in. less than nominal (rough) size.

board measure
System of lumber measurement. The unit is 1 BF, which is 1 ft square by approximately 1 in. thick.

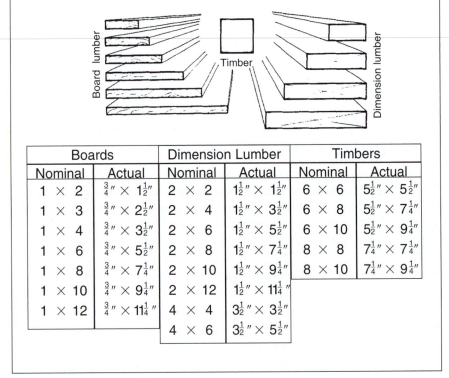

Boards		Dimension Lumber		Timbers	
Nominal	Actual	Nominal	Actual	Nominal	Actual
1 × 2	$\frac{3}{4}"\times 1\frac{1}{2}"$	2 × 2	$1\frac{1}{2}"\times 1\frac{1}{2}"$	6 × 6	$5\frac{1}{2}"\times 5\frac{1}{2}"$
1 × 3	$\frac{3}{4}"\times 2\frac{1}{2}"$	2 × 4	$1\frac{1}{2}"\times 3\frac{1}{2}"$	6 × 8	$5\frac{1}{2}"\times 7\frac{1}{4}"$
1 × 4	$\frac{3}{4}"\times 3\frac{1}{2}"$	2 × 6	$1\frac{1}{2}"\times 5\frac{1}{2}"$	6 × 10	$5\frac{1}{2}"\times 9\frac{1}{4}"$
1 × 6	$\frac{3}{4}"\times 5\frac{1}{2}"$	2 × 8	$1\frac{1}{2}"\times 7\frac{1}{4}"$	8 × 8	$7\frac{1}{4}"\times 7\frac{1}{4}"$
1 × 8	$\frac{3}{4}"\times 7\frac{1}{4}"$	2 × 10	$1\frac{1}{2}"\times 9\frac{1}{4}"$	8 × 10	$7\frac{1}{4}"\times 9\frac{1}{4}"$
1 × 10	$\frac{3}{4}"\times 9\frac{1}{4}"$	2 × 12	$1\frac{1}{2}"\times 11\frac{1}{4}"$		
1 × 12	$\frac{3}{4}"\times 11\frac{1}{4}"$	4 × 4	$3\frac{1}{2}"\times 3\frac{1}{2}"$		
		4 × 6	$3\frac{1}{2}"\times 5\frac{1}{2}"$		

FIGURE 7.34   Types and sizes of lumber.

and is the common quantity unit for lumber. A board foot is 1 in. $\times$ 1 ft $\times$ 1 ft or 144 cubic inches. The formula used to calculate BF is given by Eq. [7.8]. Nominal material dimensions are used in the equation.

$$BF = \text{length (ft)} \times \frac{\text{dressed size (in.} \times \text{in.)}}{12 \text{ in./ft}} \qquad [7.8]$$

EXAMPLE 7.9

A building requires 256 studs for wall framing. A stud is a precut piece of lumber 2 $\times$ 4 by 7.5 ft in length. How many board feet of studs should be ordered?

$$BF = 256 \times 7.5 \text{ ft} \times \frac{2 \times 4}{12} \Rightarrow 1{,}280$$

Framing Take-Off

The term *timber frame* refers to a variety of different construction methods used to create a load-bearing structure from timber (Fig. 7.35). The method used in most timber-frame buildings today is **platform framing**. This consists of a number of modular panel-like components. A platform-framed structure is assembled one story level at a time. Although the components are made from relatively lightweight softwood timber, when they are fastened together they form a very strong, rigid structure. The principal parts of a platform-framed structure are identified in Fig. 7.36. A second type of wood-frame construction is the **balloon framing** system. With balloon framing, the studs are continuous from the foundation sill to the top wall plate. Floor structures (one, two, or more) are hung from the studs (Fig. 7.36). A second timber construction method is "post and beam." Post-and-beam structures are characterized by the concentration of structural loads on fewer and larger sized pieces of timber.

platform framing
Framing in which each story is built upon the other.

balloon framing
The building-frame construction in which each stud is one piece from the foundation to the roof of a two-story structure.

FIGURE 7.35 Wood-frame construction for an apartment complex.

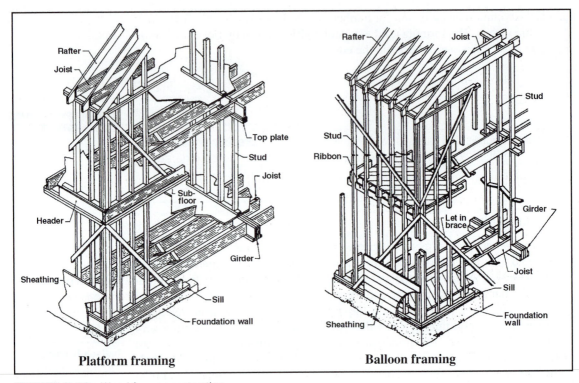

Platform framing   **Balloon framing**

FIGURE 7.36   Wood-frame construction.

Joist   To obtain the number of joists required for any floor, take the length of the floor, divide by the distance the joists are spaced, and add 1 to allow for the extra joist required at the end of the span (Fig. 7.37).

Sill Plates   The length of sill plate lumber is equal to the perimeter of the structure.

Studs   To obtain the number of studs, take the length of each wall or partition and divide by the stud spacing, and to that number add:

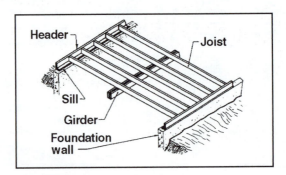

FIGURE 7.37   Joists floor framing.

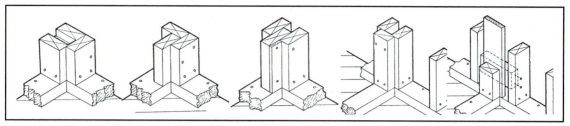

FIGURE 7.38 Examples of corner stud assemblies.

- Two studs per wall opening—doors and windows
- Two studs per corner—interior and exterior (see Fig. 7.38)
- One stud per wall lead

Studs are 7.5 ft in length for a structure having an 8-ft ceiling height. This is because there will be a sole and two top plates to complete the framing system.

Wall Plates Use the length of wall multiplied by the number of plates, normally two on top and one sole.

Headers The length of header (end of joist) lumber is twice the length of the structure, because there is a header at each end of the joists.

Rafters A rafter take-off is similar to that for joists. To obtain the number of rafters required, double the number found from taking the length of the structure divided by the rafter spacing and the one added to allow for the extra rafter required at the end of the span (Fig. 7.39). Treat a hip roof just like a gable roof as the total jack rafter requirement will equal the regular rafter requirement, but add two hip rafters. When figuring the length of a rafter, it is necessary to account for the **pitch** (slope) of the roof. Separate rafters by size.

Sheathing Take-off the area of sheathing and convert to the number of panels based on the size of panels that will be used. Some sheathing and siding is

pitch
Slope of a roof usually expressed as a ratio.

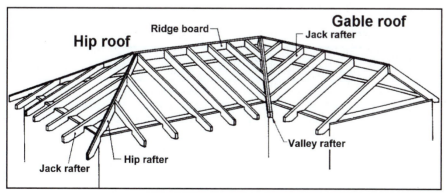

FIGURE 7.39 Roofing system for a timber-framed structure.

lapped, so account for the lap when calculating gross amount of material. Sheathing waste is about 5%.

EXAMPLE 7.10

Take-off the number of studs required for the house shown in Fig. 7.40. The studs are to be spaced 16 in. on centers.

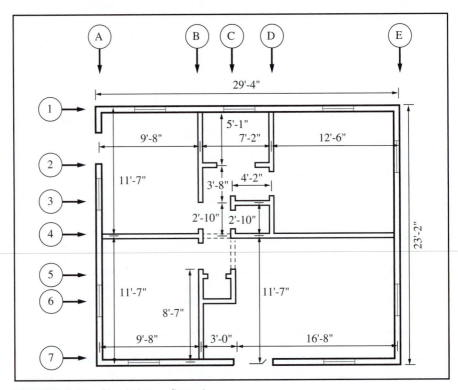

FIGURE 7.40 Simple house floor plan.

The length of the wall is

Column line	Length	Column line	Length
1	29 ft-4 in.	A	23 ft-2 in.
2	7 ft-2 in.	B	23 ft-2 in.
3	4 ft-2 in.	C	14 ft-5 in. [(2 ft-10 in.) + (11 ft-7 in.)]
4	29 ft-4 in.	D	11 ft-7 in.
5	3 ft-0 in.	E	23 ft-2 in.
6	3 ft-0 in.		
7	29 ft-4 in.		
Total	105 ft-4 in.		95 ft-6 in.

Total length of wall = (105 ft-4 in.) + (95 ft-6 in.)
= 200 ft-10 in. or 2,410 in.

$$\text{Number of studs} = \frac{2{,}410 \text{ in.}}{16 \text{ in.}} \Rightarrow 151$$

Add 2 studs per wall opening or corner:

Column line	Number of openings	Number of corners	Column line	Number of openings	Number of corners
1	3	2	A	3	2
2	1	2	B	2	2
3	0	2	C	2	1
4	1	2	D	1	2
5	1	2	E	2	2
6	0	2			
7	3	2			
Total	9	14		10	9

Total number of wall openings or corners: 9 + 14 + 10 + 9 = 42

Total number of studs to add: 42 × 2 = 84

Total number of studs: 151 + 84 = 235

THERMAL AND MOISTURE PROTECTION (CSI DIVISION 7)

There are many types of roof coverings to include asphalt shingles, wood shingles or shakes, and built-up. Roofing comes under CSI Division 7, Thermal and Moisture Protection. **Asphalt shingles**, part of subsection 07 22, are manufactured with a thermally activated asphaltic sealant that bonds the shingles together once applied to the roof and exposed to sufficient heat from sunlight. A **built-up bituminous roof**, subsection 07 51, is made of alternating layers of fiberglass or asphalt-saturated felts (Fig. 7.41). They are usually covered with small aggregate. Built-up roofs are common in commercial construction. There are many other roof types such as sheet metal, subsection 07 61, and tile, subsection 07 32.

In preparing a detailed man-hour estimate for roofing, it is necessary to include allowances for unloading, hoisting, and storing materials at the construction site. Factors that affect roofing cost are

asphalt shingles
Composition roof shingles made from asphalt-impregnated felt covered with mineral granules.

built-up bituminous roof
Roofing for low-slope roofs composed of several layers of felt and hot asphalt or coal tar, usually covered with small aggregate.

- Size of the roof (roof area)—The first step in estimating roof cost is to calculate the size of the roof (roof area). Obviously, this is the most important factor affecting quantity of materials and labor.
- Slope of the roof—It is important to note the slope of the roof. The steeper the roof the more expensive the installation. A rise of more than 5 in. per foot may require that workers be tied off.

FIGURE 7.41   Sealing the insulation layer of a built-up roof for a commercial building in Arizona. Yes, it's a cowboy hard hat.

■ Number of penetrations in the roof—Either a hole has to be cut into the membrane or the penetration must be covered.

■ Type of flashing system—Flashing may be plastic, copper, asphalt, rubber, or other material. Usually it is taken off in linear feet.

■ New or retrofit roof

■ Geographic location of project/weather conditions

square
In roofing, 100 sf of roofing.

Many roofing and siding products are bought by the **square**. One square is equal to an area of 100 sq. ft.

OPENINGS (CSI DIVISION 8)

bid
The offer of a bidder, submitted in the prescribed manner, to furnish all labor, equipment, and materials to complete the specified work.

Door and windows are included in CSI Division 8, Openings. There are separate subsections for metal doors, 08 13; wood doors, 08 14; composite doors, 08 16; and different window frame types (metal, wood, and plastic). Some of the items in this division (CSI Division 8) are furnished by material suppliers and installed by the general contractor, but it is also possible that some items may be **bid** by subcontractors as "in-place" items. Care must be exercised that all costs are covered, but the estimator must also be careful that they are counted only once as several different installation trades may be involved in completing the work. Hollow metal frames could be installed by the general contractor, or by subcontractors involved in masonry or metal stud and **drywall construction**. Metal windows could be supplied by one company, installed by the general contractor, and glazed (glass glazing is subsection 08 81) by a subcontractor. In the estimate, it is important to include labor cost under the proper trade.

drywall construction
Interior wall covering other than plaster, usually referred to as "gypsum board" or "wallboard."

TABLE 7.4 Door checklist.

Metal doors	Wood doors	Glass
Type of metal	Hollow or solid core	Type of glass
Metal gauge	Fire rating	Thickness
Door core material	Flush or raised panel	Size—measure to the next
Fire rating	Finish	even inch, list width first
Finish		and then height
		Type of frame

Door and Window Schedules

Doors and windows can look alike but still have significant differences. Table 7.4 is a door check list.

Information concerning these requirements will be found in the specifications or on the door schedule (Fig. 7.42). The door hardware cost will add considerably to the price of the work. The hardware requirements will be delineated in the door schedule, note the "HDW SET" column in Fig. 7.42. Therefore, the estimator should become very familiar with the schedules the A/E includes in the plans. Door schedules normally list the door openings by "mark" numbers or letters (e.g., #1, #2, #3 or A, B, C). After the mark designation there will be the door size, material, frame, fire rating, etc., and remarks. The "mark" designation refers to the particular type and has nothing to do with the total quantity required.

EXAMPLE 7.11

From the door schedule shown in Fig. 7.42, identify the fire rating for door 113A. Door 101 is listed as Type A. What is a Type A door? What kind of door is door 114B?

Door 113A must have a fire rating of ¾ hr.

Door 101 is flush solid core wood.

Door 114B is an overhead door.

Pricing Door or Window Work

The labor time required to install a door or window unit is dependent on the quality of workmanship required, the type of tools and equipment that will be used, and whether the work involves only a few units or multiple units in a production line-type operation. Estimating tables in reference texts can be consulted if a company does not have a historical record of labor time for this type of work. When using such tables, it is important to carefully select the proper classification of work. As an example, one reference states that a single carpenter can fit and hang by hand one wood door per hour, but reading further it is noted that if there is a large number of doors to hang (production line job), then a carpenter should only require 0.6 hr per door.

DOOR SCHEDULE

DOOR NO.	WIDTH	HEIGHT	THICK	TYPE	FRAME	FIRE LABEL (HRS)	WEATHER STRIP	HDW. SET	JAMB NO.	HEAD NO.	THRESHOLD	REMARKS
101	914	2134	44	A	STL-1	—	—	1	J1	H1	—	—
102	PR. 914	2134	44	C	STL-4	—	X	100	J7, J7A	H2A/A5-3	—	—
103	914	2134	44	B	ALUM.	—	X	101	3/A9-3	2/A9-3	—	SEE ⟨W⟩A9-1
104	914	2134	44	A	STL-1	—	—	2	J1	H1	—	—
105	914	2134	44	C	STL-2	3/4	—	4	J3	H3	—	—
106	914	2134	44	A	STL-1	—	—	2	J1	H1	—	—
107	914	2134	44	A	STL-1	—	—	1	J1	H1	—	—
108	914	2134	44	A	STL-1	—	—	3	J1	H1	—	—
109	914	2134	44	C	STL-2	—	—	5	J2	H2	—	—
110	914	2134	44	A	STL-1	—	—	1	J1	H1	—	—
111	914	2134	44	A	STL-1	—	—	1	J1	H1	—	—
112	914	2134	44	C	STL-2	—	—	5	J2	H2	—	—
113	914	2134	44	C	STL-2	—	X	102	J5, J5A	H5	—	—
113A	914	2134	44	C	STL-2	3/4	—	4	J3	H3	—	—
114	1219	2134	44	C	STL-2	—	X	104	J4	H4	—	—
114A	PR. 914	2134	44	C	STL-2	—	X	103	J6, J6A	H6	—	—
114B	3658	4267	—	D	—	—	X	—	H3/A5-3 SM	H3/A5-3 SM	—	OVERHEAD DOOR
114C	3658	4267	—	D	—	—	X	—	J3/A5-3	H3/A5-3	—	OVERHEAD DOOR
116	914	2134	44	C	STL-2	—	—	5	Jt	H1	—	—
117	914	2134	44	C	STL-3	—	X	105	J4	H1/A5-3	—	—
118	914	2134	44	E	—	—	—	—	—	—	—	VAULT
119	914	2134	44	C	STL-2	—	—	1	Jt	H1	—	—
121	PR. 914	2134	44	D	STL-2	—	X	100	J4	H4	—	—

FLUSH SOLID CORE WOOD

TYPE A

DOOR TYPE

SCALE : 1:50

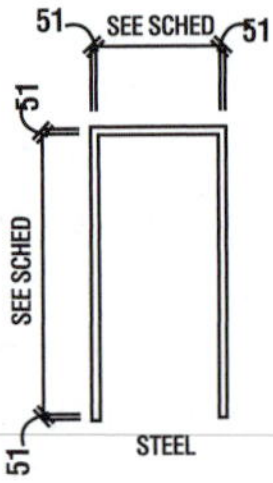

TYPE STL–

FRAME TYPE

SCALE : 1:50

FIGURE 7.42 Door schedule for a small project. This project was dimensioned in metric units so width, height, and thickness are given in mm.

In the case of glass, it is important to note whether glazing is from the inside or outside of the building. If from the outside, ladders, scaffolding, or temporary structures may be necessary to do the work. As the height above ground increases, additional labor and equipment are needed for distributing materials.

FINISHES (CSI DIVISION 9)

Before beginning an estimate of the cost to accomplish finish work (CSI Division 9), the estimator should carefully review the entire set of project plans for additional work that may apply to and be related to the finish work and the necessary coordination with other work. Structural plans are examined to determine the type of exterior walls and other specifics bearing on finish work. The architectural plans and details provide wall, ceiling, and floor dimension information.

Finish Schedules

A room finish schedule should be included in the plans (Figs. 7.43, 7.44, and 7.45). This schedule should delineate by room number and name: (1) floor material, (2) base material, (3) wall material (each wall), and (4) ceiling materials and ceiling height. Additionally, there will be notes with special instructions. If the plans do not contain a finish schedule, the estimator should create one.

Finish Work Quantity Take-Off

Measure the actual surfaces as accurately as possible from the plans. On typical commercial work, the labor cost for finish work is often much higher than the material cost. Therefore, the quantity survey should reflect labor requirements. The labor productivity when painting a flat surface is different compared to painting a small-diameter pipe, so in the first case, square footage of surface area is a good measure of work, but in the second case, linear foot measure is more appropriate. Surfaces are grouped based on type of surface, finish, application method, and expected labor production rates. Application method and labor production are directly related, so the estimator must make assumptions about how the work will be accomplished. Some general measurement rules are

1. No object is considered to be less than one linear foot.
2. Small openings in a continuous surface are not deducted (it is just as difficult to work around the opening).

Most types of flooring are measured on a square foot basis; however, sheet flooring is measured by the square yard. **Baseboard** is measured by the linear foot. Drywall (gypsum board or wallboard) is measured in square feet, but it may be calculated by the number of sheets. With drywall, it is also necessary to know the linear feet of joints that must be taped (Fig. 7.46). The estimator must consult the specifications concerning the type of board that is required, such as water resistant or low-sag. There may be thickness requirements stated either on the drawings or in the specifications. In the case of painting, it is necessary to consult the specifications to determine the number of coats required. Additionally, with painting it is necessary to ascertain the amount of protection and masking that will be required.

baseboard
Finish board covering the interior wall where the wall and floor meet.

Estimating Finishes

Labor production rates for finish work can be found in estimating reference manuals. Estimating data publications such as the *RSMeans Building Construction*

| ROOM NO. | ROOM NAME | FLOOR | | BASE | | WALLS | | | | | | | | CEILING | | | REMARKS |
| | | MAT. | COLOR | MAT. | COLOR | NORTH | | EAST | | SOUTH | | WEST | | | | | |
						MAT.	COLOR	MAT.	COLOR	MAT.	COLOR	MAT.	COLOR	MAT.	COLOR	HEIGHT	
A01	SOCIAL WORK	CP1	CPT-1	RB	RB-1	GWP	PT-1	GWP	PT-1	GWP	PT-1	GWP	AL	ACT*	ACT-1	9'-0"	* NOTE 1
A02	COUNSELOR	CP1	CPT-1	RB	RB-1	GWP	PT-1	GWP	PT-1	GWP	PT-1	GWP	PT-1	ACT*	ACT-1	9'-0"	* NOTE 1
A03	SOCIAL WORK	CP1	CPT-1	RB	RB-1	GWP	PT-1	GWP	PT-1	GWP	PT-1	AL		ACT*	ACT-1	9'-0"	* NOTE 1
A04	COUNSELOR	CP1	CPT-1	RB	RB-1	GWP	PT-1	GWP	PT-1	GWP	PT-1	GWP	PT-1	ACT*	ACT-1	9'-0"	* NOTE 1
A05	SOCIAL WORK	CP1	CPT-1	RB	RB-1	GWP	PT-1	GWP	PT-1	AL		GWP	PT-1	ACT*	ACT-1	9'-0"	* NOTE 1
A06	COUNSELOR	CP1	CPT-1	RB	RB-1	GWP	PT-1	GWP	PT-1	GWP	PT-1	GWP	PT-1	ACT*	ACT-1	9'-0"	* NOTE 1
A07	OBSERV. 1	CP1	CPT-1	RB	RB-1	GWP	PT-1	GWP	PT-1	AL		GWP	PT-1	ACT*	ACT-1	9'-0"	* NOTE 1
A09	PLAY THERAPY	CP1	CPT-1	RB	RB-1	GWP	PT-1	GWP	PT-1	AL		GWP	PT-1	ACT*	ACT-1	9'-0"	* NOTE 1
A10	GROUP THER. 1	CP1	CPT-1	RB	RB-1	GWP	PT-1	GWP	PT-1	GWP	PT-1	GWP	PT-2	ACT*	ACT-1	9'-0"	* NOTE 1
A11	OBSERV. 2	CP1	CPT-1	RB	RB-1	GWP	PT-1	GWP	PT-1	AL		GWP	PT-1	ACT*	ACT-1	9'-0"	* NOTE 1, ** NOTE 5
A12	GROUP THER. 2	CP1	CPT-1	RB	RB-1	GWP	PT-1	GWP	PT-2	GWP	PT-1	GWP	PT-1	ACT*	ACT-1	9'-0"	* NOTE 1
A13	COUNSELOR	CP1	CPT-1	RB	RB-1	GWP	PT-1	GWP	PT-1	AL		GWP	PT-1	ACT*	ACT-1	9'-0"	* NOTE 1, ** NOTE 5
A14	CLASS/CONF. ROOM	CP1	CPT-1	RB	RB-1	GWP	PT-1	GWP	PT-3	GWP	PT-1	GWP	PT-3	ACT*	ACT-1	9'-0"	* NOTE 1
A15	COUNSELOR	CP1	CPT-1	RB	RB-1	GWP	PT-1	GWP	PT-1	AL		GWP	PT-1	ACT*	ACT-1	9'-0"	* NOTE 1, ** NOTE 5
A99	CORRIDOR	CP1	CPT-1	RB	RB-1	GWP	PT-1	GWP	PT-1	GWP	PT-1	GWP	PT-1	ACT*	ACT-1	9'-0"	
A99E	VESTIBULE	CP1	CPT-1	RB	RB-1	GWP	PT-1	AL		GWP	PT-1	AL		ACT*	ACT-1	9'-0"	* NOTE 1

FIGURE 7.43 Finish schedule for a small project.

Finish schedule legend			
ACT	2"×"2 Acoustical Ceiling Tile	EFS	Exterior Wall Insulation and Finish System
AL	Aluminum Storefront	EXP	Exposed
BRK	Exposed Brick Front	GWP	Gypsum Wall Board—Painted
CMU	Concrete Masonry Unit	NONE	None
CON	Concrete	RB	Rubber Base
CPT	Carpet	VCT	Vinyl Composition Tile
CS	Concrete Sealed	VWC	Vinyl Wall Covering
CT	Ceramic Tile		

FIGURE 7.44 Abbreviation legend for the Fig. 7.43 finish schedule.

Color design legend

Floor:

CPT-1   Mannington Carpets
Pattern: The Cambrian Collection
     "Caldera"
Color # Villarrica (VIRI)

CPT-2   Mannington Carpets
Pattern: Nepenthe II
Border Carpet—Broadloom
Color # Forest Shade (FOSH)

CT-1   American Olean
Color # A50 Bone
Size 2"×2"

CT-2   American Olean
Color # R22 Woodland

VCT-1   Armstrong
Pattern: Imperial Texture
Color # 51911 Classic White

VCT-2   Armstrong
Pattern: Imperial Texture
Color # 51901 Taupe

VCT-3   Armstrong
Pattern: Imperial Texture
Color # 51915 Charcoal

CS-1   Concrete Sealer
Natural Grey

Base:

RB-1   Roppe Rubber
Color # 40 Fawn
Size 4" Cove Base

CT-3   American Olean
Color # S13 Classic Bone
Size 4"×4"

Walls:

VWC-1   Tower Wallcovering
Pattern: Corfu
Color # T2-CF-02 Bisque

PT-1   Devoe Raynolds
Color # 2H42G Ice Beige
Latex Semigloss

PT-2   Devoe Raynolds
Color # 1M501D Chocola Tint
Latex Semigloss

PT-3   Devoe Raynolds
Color # 1U57B Deacon
Latex Semigloss

Ceilings:

ACT-1   Armstrong World Industries
Pattern: Minaboard # 770,
Natural fissured
Size 2"×2" Lay-in with white
suspension grid

ACT-2   Armstrong World Industries
Pattern: Scored beaded cirrus #624
Size 2"×2" Tegular edge lay-in with
white suspension grid

PT-1   Devoe Raynolds
Color # 2H42G Ice Beige

FIGURE 7.45 Color legend for the Fig. 7.43 finish schedule.

FIGURE 7.46 Finishing a drywall joint.

Cost Data [11] book provide insight into the factors that affect costs. In the case of floors, these could include underlayment or subflooring, stringent leveling, staining of concrete, and, for wood floors, sanding and finishing requirements.

Flooring Estimate   There are many types of flooring. In the case of tile flooring, it is necessary to estimate both the time and materials required to prepare and prime the receiving surface, and the time to actually lay the flooring. The productivity of placing tile flooring varies with the size and shape of the room or space. Carpeting is manufactured in rolls of standard width, usually 12 feet. Because of this, the estimator must carefully determine how the carpet will be placed in the spaces. Irregular-shaped areas or unusually sized rooms can result in high waste factors.

Painting Estimate   Factors to consider when estimating painting include (1) covering capacity, (2) thickness of coat, (3) roughness of surface, and (4) absorption of material—prime, second, or third coat. Do not overlook the cost of protection, scaffolding, and other equipment cost. Labor costs vary exten-

FIGURE 7.47 Workman using a manlift while installing a ceiling.

sively with the grade of workmanship specified. The production rates and required quality of paint to cover a fixed area will vary with the coat. In the case of interior walls constructed of hard-finished wallboard, these average coverage rates illustrate the point: primer or sealer 575–625 sf/gal, second coat 500–550 sf/gal, and third coat 575–625 sf/gal. Manufacturer's data should always be consulted.

Ceiling Estimate Acoustical tile ceiling treatments are used to absorb and deaden sound in buildings. They may be either cemented directly to dry plaster, concrete, or gypsum board, or installed into a suspended grid system. Tiles come factory cut and prefinished. Labor production rates can be found in estimating reference manuals, but always remember to adjust rates based on working conditions (Fig. 7.47).

PLUMBING (CSI DIVISION 22) AND HEATING, VENTILATION, AND AIR-CONDITIONING (CSI DIVISION 23)

Building piping, plumbing fixtures, and heating/ventilation/air-conditioning equipment are all found under CSI Divisions 22 and 23. Piping may be related

FIGURE 7.48 Piping for a commercial building's HVAC system.

to plumbing; fire suppression (Division 21); or heating, ventilation, and air-conditioning systems (HVAC) (Fig. 7.48). It is necessary to review the structural and architectural plans and specifications to note requirements that impact the cost of mechanical items. Mounting requirements, penetration requirements, and depths of excavation should be checked. Specialty subcontractors perform most mechanical work but the general contractor may be responsible for furnishing support. It is important to have a clear understanding of which contractor is performing each work item. A general contractor can arrive at a budget estimate for plumbing installations using a percentage of total project cost. The range is between 3% and 12%, with stores and warehouses at the lower end and apartments and motels at the high end. Commercial buildings may have a slightly higher upper range.

Plumbing is a system of piping, apparatus, and fixtures for water distribution, subsection 22 11, and waste disposal, subsection 22 13 sanitary sewerage, within a building. Pipes and fittings for plumbing systems are classified into four basic groups: (1) cast-iron soil pipe, (2) galvanized-steel/iron pipe, (3) copper tubing, and (4) plastic pipe. When preparing a detailed plumbing estimate, it will be necessary to calculate the length of runs for each pipe size, as piping is priced by the linear foot. The quantity and type of joints and fittings must be determined. Fittings can add another 25% to 50% to material costs. Suspended pipe hangers are usually spaced every 10 feet; however, for cast iron pipe, the hanger spacing is 5 feet, and for plastic pipe three hangers are needed every 10 feet. The number of each type of fixture must also be determined. Cost for many plumbing items reflects three separate costs: (1) the fixture cost, (2) the rough-in cost, and (3) final setting of the fixture.

FIGURE 7.49 Ductwork for a building's HVAC system.

Ductwork, subsection 23 31 (Fig. 7.49), for HVAC systems is taken-off by linear footage for rectangular ducts. The linear number is converted to pounds as cost is usually computed based on the poundage required. It is necessary to itemize fittings and diffusers, registers, and intakes. Additionally, it is often necessary to have a crane available to set rooftop HVAC units.

ELECTRICAL (CSI DIVISION 26)

Power distribution, branch power, and branch lighting are parts of CSI Division 26, Electrical. As in the case of mechanical work, specialty subcontractors perform most electrical work, but the general contractor may be responsible for furnishing support or coordinating the electrical and mechanical subcontractors. Electrical work can include construction of electrical distribution lines, outdoor lighting, and interior electrical work. On average, electrical work is about 8% to 11% of office building construction cost.

Before beginning a quantity take-off, the estimator should study the

- Plot plan—for obstruction to underground cable work
- Floor plans—wall construction types (wood, block, concrete, metal stud)
- Ceiling plans
- Roof framing plan—connections to rooftop HVAC units

The take-off itself will involve counting fixtures, counting devices, and measuring conduit and wire. On the electrical drawings, there should be a fixture schedule, an electrical panel schedule (Fig. 7.50), and a legend that describes the

FIGURE 7.50 Panel boxes in a commercial building.

different symbols used. The number of fixtures shown on the schedule should be checked against a room-by-room count. Smaller conduit and wire is detailed only on the plans, whereas larger runs for feeders are found on the riser diagram. The riser diagram must be superimposed on the plans to measure run distance.

Building electrical work is a two-step process. First, the electrician must perform rough-in work, once the building is framed but before the walls and ceiling are closed in. This work includes installing service mains, switches, panels, conduits, and outlet boxes and pulling cable through conduit and splicing in electrical boxes. After the walls and ceilings are finished, the electrician returns to perform the electrical finish work of installing and connecting receptacles, switches, light fixtures, controls, and appliances.

SUMMARY

The drawings and their accompanying notes show and explain how the building is laid out, how it is constructed, what materials must be used, and how systems operate. To estimate the cost of doing the work, the contractor must understand the information that the plans communicate. Calculating quantities of materials can be simplified by understanding a few principles of measurement. Estimates are organized to serve the needs of the contracting organization, with a critical need being identification and assessment of risk.

A detailed and accurate take-off is the foundation of any estimate. The *units* of the take-off should reflect work task. Placement of concrete is in units of volume, while forming is in units (sf) of form contact area. An accurate estimate of masonry cost requires knowledge of various factors, including the type of structural element, thickness of the section, type of mortar, style of mortar joint, class of workmanship, and skill of the workers. A structural steel estimate involves a large amount of detail work. The quantities of structural members are listed from the plans and weights are computed; the cost of shop fabrication (purchase cost) is determined, to which is added freight cost; and finally, the cost of erecting the steel on the job is determined.

To perform a carpentry take-off accurately, the estimator should have an understanding of timber design and construction procedures. There are many types of roof coverings including shingles, built-up, roll, and tile. In preparing a detailed man-hour estimate for roofing, it is necessary to include allowances for unloading, hoisting, and storing materials at the construction site. Some of the items in CSI Division 8, for example, door and windows, are furnished by material suppliers and installed by the general contractor, but it is also possible that some items may be bid by subcontractors as "in-place" items.

Before beginning an estimate of the cost to accomplish finish work, the estimator should review the entire set of project plans for additional work that may apply and related work and coordination with other work. A room finish schedule should be included in the plans. Building piping and plumbing fixtures are included in CSI Division 22 and heating/ventilation/air-conditioning equipment are all found under CSI Division 23. Power distribution, branch power, and branch lighting are parts of CSI Division 26, Electrical. As in the case of mechanical work, specialty subcontractors perform most electrical work, but the general contractor may be responsible for furnishing support or coordinating the electrical and mechanical subcontractors. Critical learning objectives that support building estimating include

- Ability to understand the information that the plans communicate
- Ability to understand how a set of plans is organized
- Ability to calculate an unknown perimeter when the distance to a known perimeter is given
- Ability to calculate concrete and reinforcing steel quantities
- Ability to calculate masonry quantities

- Ability to take-off steel quantities
- Ability to take-off lumber framing quantities
- Ability to read door and window schedules
- Ability to read finish schedules

These objectives are the basis for the review questions in this chapter.

REVIEW QUESTIONS

7.1 Calculate the outside perimeter of the wall shown in the figure.

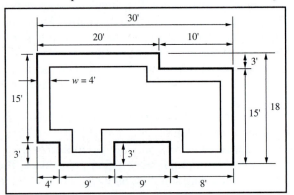

7.2 The figure shows the outside dimensions of a building perimeter wall. (a) Calculate the outside perimeter of the wall. (b) What is the inside perimeter of the footing? (c) If the footing must be formed, what is the required contact surface area of formwork in sf?

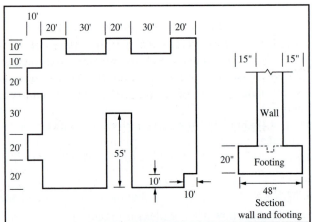

7.3 The width of the foundation wall shown in Question 7.1 is 4 ft. Calculate the volume of concrete required to build this foundation wall if its cross-sectional area is 8 sf. If the wall is 2 ft in height, what is the required formwork in sfca?

7.4 Calculate the longitudinal #6 reinforcing steel in the wall (not the footing) shown in the figure. The wall is 60 ft long and 6 in. of cover is required at each end. The bars come in 20-ft lengths and will have to be lapped 6 in. How many tons of #6 steel should be ordered?

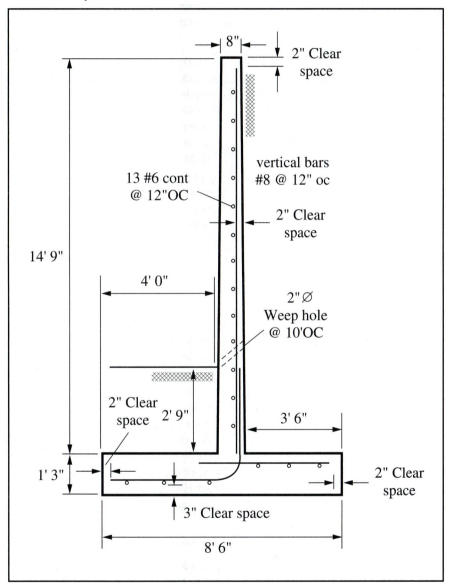

7.5 The outside perimeter of a brick wall is 53 ft. Calculate the bricks required to build this wall if its height is 8 ft and the bricks will all be laid in a running bond. Standard nonmodular bricks have been specified for the project. Assume a 1% waste factor.

7.6 The house shown in the plan will have a hip roof. Calculate in BF the total quantity of 2 × 8 rafters required to construct the roof if a 19 in. spacing is used. Because of the pitch, the rafters are 18 ft in length.

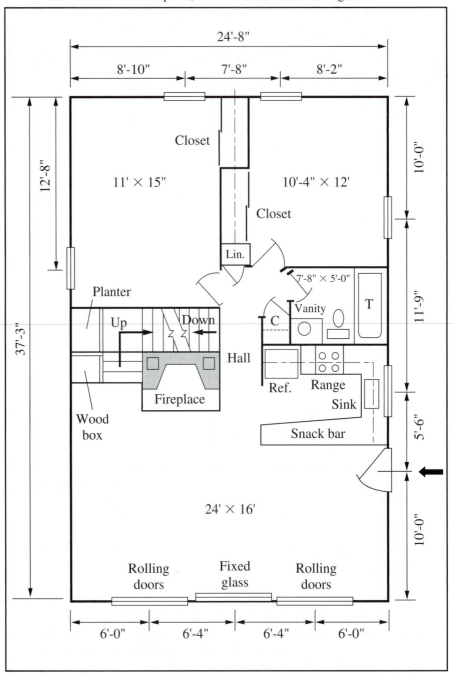

7.7 The exterior studs for the house in Question 7.6 will have 2×6 exterior studs with a 16-in. spacing. Calculate the number of studs required to construct the exterior walls.

7.8 Perform a steel take-off of the framing plan show here. The columns are W 21×82, and 10 ft in height.

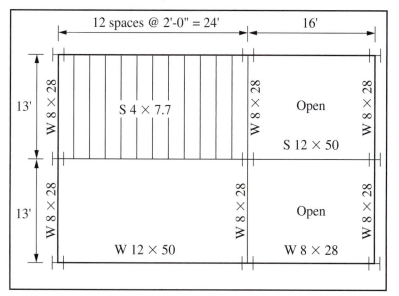

REFERENCES

1. *Concrete Construction Formwork,* 7th ed. (2005). ACI International, Farmington Hills, MI.

2. *Concrete Masonry Handbook,* 5th ed. (1991). Portland Cement Association, Skokie, IL.

3. *Construction Estimating & Bidding Theory Principles Process* (1999). Publication No. 3505, The Associated General Contractors of America, 333 John Carlyle Street, Suite 200, Alexandria, VA.

4. *Estimating Guide,* Volume 2, *Rates and Tables* 21st ed. (2004). Painting and Decorating Contractors of America, Fairfax, VA.

5. Foley, J. D., A. van Dam, S. K. Feiner, and J. F. Hughes (1990). *Computer Graphics: Principles and Practice,* 2nd ed., Addison-Wesley Publ. Co., Reading, MA.

6. Hurd, M.K. (1995). *Formwork for Concrete,* 7th ed. (2005). American Concrete Institute Special Publication No. 4, ACI International, Farmington Hills, MI.

7. Mäntylä, M. (1988). *An Introduction to Solid Modeling,* Computer Science Press, Inc., Rockville, MD.

8. *Manual of Standard Practice,* 27th ed. (2001). CRSI, Schaumburg, IL.

9. *MasterFormat 2004* (2004). The Construction Specifications Institute, 99 Canal Center Plaza, Suite 300, Alexandria, VA 22314.

10. Oberlender, Garold D., and Steven M. Trost (2001). "Predicting Accuracy of Early Cost Estimates Based on Estimate Quality," *Journal of Construction Engineering and Management,* ASCE, 127(3), May–June.

11. *RSMeans Building Construction Cost Data,* R. S. Means Company, Inc., Kingston, MA (published annually).

12. *Standard Estimating Practice,* 6th ed. The American Society of Professional Estimators, Wheaton, MD.

13. *Walker's Building Estimator's Reference Book,* 28th ed. (2006). Frank R. Walker Company, Lisle, IL.

WEBSITE RESOURCES

1. www.csinet.org   Construction Specifications Institute (CSI). CSI is an individual membership technical society. The institute provides technical information and professional conferences to enhance communication among all disciplines of nonresidential building design and construction, and to meet the industry's need for a common system of organizing and presenting construction information. The Construction Specifications Institute, 99 Canal Center Plaza, Suite 300, Alexandria, VA.

2. www.crsi.org/index.html   Concrete Reinforcing Steel Institute (CRSI). CRSI supports research and engineering for the safe and proper use of materials in reinforced concrete construction. CRSI, 933 North Plum Grove Road, Schaumburg, IL.

3. www.aci-int.org   American Concrete Institute (ACI). ACI provides knowledge and information for the best uses of concrete. ACI International, P.O. Box 9094, Farmington Hills, MI.

4. www.hcss.com Heavy Construction Systems Specialists, Inc. (HCSS). HCSS specializes in construction estimating, bidding, and job cost software for contractors. HCSS, 6200 Savoy, Suite 1100, Houston, TX.

5. www.uscost.com   U.S. COST provides project control services and software.

6. www.asphaltroofing.org   The Asphalt Roofing Manufacturers Association (ARMA). ARMA is the North American trade association that represents the majority of the asphalt roofing industry's manufacturing companies and their raw material suppliers. Asphalt Roofing Manufacturers Association, 4041 Power Mill Road, Suite 404, Calverton, MD.

7. www.pdca.com   Painting and Decorating Contractors of America (PDCA). PDCA develops industry-specific publications and resources for painting companies. Painting and Decorating Contractors of America, 3913 Old Lee Highway, Second Floor, Fairfax, VA.

8. www.smacna.org   Sheet Metal and Air Conditioning Contractors' National Association (SMACNA). SMACNA standards and manuals address all facets of the sheet metal industry, from duct construction and installation to air pollution control. Sheet Metal and Air Conditioning Contractors' National Association, 4201 Lafayette Center Drive, Chantilly, VA.

9. www.aisc.org   American Institute of Steel Construction (AISC). AISC is a nonprofit association representing and serving the structural steel industry in the United States. Its purpose is to expand the use of fabricated structural steel through research and development, education, technical assistance, standardization, and quality control.

10. www.aitc-glulam.org   The American Institute of Timber Construction (AITC). AITC is the national technical trade association of the structural glued laminated timber industry.

11. www.ashrae.org   American Society of Heating, Refrigerating and Air-Conditioning Engineers, Inc. (ASHRAE). ASHRAE is organized for the purpose of advancing the arts and sciences of heating, ventilation, air-conditioning, and refrigeration through research, standards writing, continuing education, and publications. American Society of Heating, Refrigerating and Air-Conditioning Engineers, Inc., 1791 Tullie Circle, N.E., Atlanta, GA.

12. www.wwpa.org/woodinfo.htm   Wood Information on the Internet. A linking site of forest products associations, other lumber, and timber informational sites.

13. www.assoc-spec-con.org   Associated Specialty Contractors (ASC). ASC is an "umbrella organization" of nine national associations of construction specialty contractors (Ceilings and Interior Systems Construction Association, Finishing Contractors' Association, Mechanical Contractors' Association of America, National Electrical Contractors' Association, National Insulation Association, National Roofing Contractors of America, Painting and Decorating Contractors' of America, Plumbing-Heating-Cooling Contractors' Association, and the Sheet Metal and Air Conditioning Contractors' National Association). Associated Specialty Contractors, Inc., 3 Bethesda Metro Center, Suite 1100, Bethesda, MD.

8

Construction Contract Administration

Everyone involved in the construction process must understand contracts—the sections of the contract itself, such as the agreement and the specifications and other required contract documents such as bonds and insurance—and the processes involved in contract administration. Some contracts are more complicated than others, and no two projects are the exact same size, duration, or type of design; but all contracts require similar administrative processes. For instance, a complicated contract might require a complicated monthly pay estimate system, with as many as 10 or 11 copies of all documents, while a simple contract might have a more basic monthly pay estimate system; but they both require a monthly pay estimate and processes of developing and approving the monthly pay estimate.

DESCRIPTION OF A CONTRACT

There are many federal, state, and local laws that affect construction contracts, and the successful contractor must be well advised in the legal requirements of construction contracting. Contracts are the vehicles used for the procurement of everything in the construction business, both goods and services. The makeup of a contract, whether the owner is a public agency or a private corporation, is essentially the same. The form of the documents may change, but the elements of the contracts are the same.

Definition

In its simplest, most essential form, a construction contract is defined as an agreement between two parties that is enforceable by law. It requires a "meeting of the minds" and there must be a service and consideration. One party must agree to perform work (service) for the other party and receive payment (consideration) for the work. The contract must be enforceable by law, which for all practical purposes means the contract must be for a service that is legal. Although there is no actual requirement that a contract be written, realistically it must be. Oral contracts are essentially impossible to enforce in construction because of the lack of evidence regarding the scope of the agreement. Generally, written contracts eliminate problems by removing any doubt about the agreed-on terms.

For a contract to be valid and enforceable by law, it must meet certain criteria. These criteria are

■ There must be mutual agreement or a meeting of the minds.

■ There must be an offer. An offer can normally be withdrawn up until the time it is accepted, except the bid documents used in public works generally state that the bid may not be withdrawn once submitted.

■ The offer must be accepted. The acceptance completes the meeting of the minds. In low-bid construction, the acceptance is the award of the contract to the low bidder.

■ There must be consideration for the service performed—payment.

■ The subject matter of the contract must be lawful. A contract to commit a crime is not legal and not enforceable.

■ The contracting parties must have the legal capacity to enter a contract. A contract with a minor is not lawful. Contracts are signed by representatives of both the owner and contractor who have the legal authority to sign for their organizations.

The purpose of the contract should be to produce a safe, quality construction project on schedule and within budget.

ESSENTIAL CONTRACT DOCUMENTS

There are many documents that make up a construction contract. The four *essential* documents are:

■ The Agreement
■ The General Conditions
■ The Special Conditions
■ The Drawings and Specifications

Agreement

The agreement is the document that represents and reflects the legal contract between two parties. This can be between the owner and the contractor or between the owner and the designer or between the general contractor (GC) and the subcontractors or between the contractors and material suppliers. The purpose of the agreement is to record in written form those items agreed to by the two parties. The most common agreement used in the construction industry may be those published by the American Institute of Architects (AIA) and is commonly referred to as AIA Document A101. The full title of the document is AIA Document A101-2007 Standard Form of Agreement Between Owner and Contractor—Stipulated Sum (see Appendix C). It is simply a letter that constitutes legal evidence that a contract exists, and forms the basis for its enforcement. The agreement must contain

- Date of the agreement
- Names and addresses of the contracting parties
- Description of the scope of work
- Time limitations
- Contract considerations
- Payment conditions
- Reference to other documents
- Signatures

General Conditions

A document called General Conditions is an essential part of the contract. It defines the responsibilities of the parties involved in the contract—the owner and the general contractor. It describes the guidelines that will be used in the administration of the contract. It is often referred to as *boilerplate,* implying that the same documents are standard to all contracts. Contractors must know exactly what is contained in the boilerplate.

Various standard forms of General Conditions have been developed by different organizations. These forms are familiar to all parties concerned, and the wording not only is clearly understood, but also has been tested in the courts.

AIA Document A201

The most common General Conditions. May be purchased from the American Institute of Architects for use in contracts.

The most common general conditions for use by the owner and general contractor are those published by the AIA, and are commonly referred to as **AIA Document A201**. The full title of the document is AIA Document A201 General Conditions of the Contract for Construction (Appendix D). Other standard general conditions forms are available for use in other types of contracts, such as the contract between a general contractor and subcontractor (AIA Document A401), or the contract between the owner and architect (AIA Document B141). A listing of other common AIA documents is shown in Appendix E and a complete listing can be found on the AIA website at www.aia.org. There are also many other agencies that produce standard forms of General Provisions, including the General Services Agency (GSA), U.S. Army Corps of Engineers (CE), and state departments of transportation (DOTs).

Standard documents, such as AIA Document A201, are intended to be used in their entirety, and not in a "cut and paste" format. The most recent version of AIA Document A201 was published in 2007. Changes to the *General Conditions* are made by writing the change into the *Supplementary Conditions,* not by making changes to the general conditions themselves. The AIA Document A201 contains the following 14 articles:

- General Conditions
- The Owner
- The Contractor
- Administration of the Contract
- Subcontractors
- Construction by Owner or by Separate Contract
- Changes in the Work
- Time
- Payments and Completion
- Protection of Persons and Property
- Insurance and Bonds
- Uncovering and Correction of Work
- Miscellaneous Provisions
- Termination or Suspension of the Contract

Most experts feel that it is not wise to use nonstandard general conditions, mostly because all of the published standard general conditions have been tested in court, and their legal interpretations are well known. However, if the contract does not incorporate standard general conditions, such as those written by the AIA or the Associated General Contractors of America (AGC), and instead the designer or the owner writes the document, as a minimum, the nonstandard general conditions must include

- A definition of the contract documents that lists and gives a brief description of the documents that form the contract
- Document precedence, which clearly states which document will have precedence over the other in case of discrepancy (Fig. 8.1)
- Duties and responsibilities of the owner and contractor while the construction is in progress
- A definition of how a portion of the work can be awarded to a subcontractor, and the working relationships between subcontractors
- The rights of the owner to self-perform construction, or to award work to separate contractors
- The time for completion of the work
- The mode and the frequency of payment, or the stages of work that determine when the contractor will be paid. This section also contains the rights of both parties with respect to **retainage**, and the definition of completion.
- The requirements and penal value of bonds and insurance. This section includes bid bonds as well as payment and performance bonds.

retainage
A project owner, in order to have protection that the contractor will complete the work, often holds back (retains) a portion of what a contractor earns until the end of the project. Ten percent is a common retainage percentage value.

GENERAL CONDITIONS TO THE CONSTRUCTION CONTRACT
ARTICLE 1   GENERAL PROVISIONS
1.1 DEFINITIONS, CORRELATION AND INTENT

    1.1.7   Contract Document Order of Precedence. The Drawings, Specifications, and other Contract Documents will govern the Work. The Contract Documents are intended to be complementary and cooperative and to describe and provide for a complete Project. Anything in the Specifications and not on the Drawings, or on the Drawings and not in the Specifications shall be as though shown or mentioned in both.

    1.1.7.1   If there is a conflict between Contract Documents, the document highest in precedence shall control. The precedence for the Contract Documents shall be:

        1.1.7.1   Permits from other agencies required by law

        1.1.7.2   Change Orders, Supplemental Agreements, and approved revisions to Contract Specifications and Contract Drawings.

        1.1.7.3   Contract Specifications

        1.1.7.4   Contract Drawings—Material/Equipment Schedules

        1.1.7.5   Contract Drawings—Detailed Plans

        1.1.7.6   Contract Drawings—General Plans

        1.1.7.7   Standard Plans, i.e., standard structural details, devices, or instructions referred to on the Plans or Specifications by title or number.

        1.1.7.8   Reference Specifications, i.e., Building Codes, Test References, etc.

FIGURE 8.1   General Conditions clause addressing precedence.

■ The conditions that constitute a change of work must be clearly defined, and the steps to be taken when a change order is required must also be clear.

■ The system for dispute resolution, such as arbitration or mediation, must be defined. Also, the rights of each party to terminate the contract must be detailed.

■ There are always some miscellaneous provisions such as the governing laws, delegation of work, requirements for inspections and tests, approvals during the work, and statutory limits.

Supplementary Conditions

The supplementary conditions are sometimes known as special provisions or special conditions. The purpose of the supplementary conditions is to provide an extension of the general provisions of the contract to fit the specific project at hand. They serve as amendments or augmentation to the general conditions. Items included in the supplementary conditions are entirely subject to the discretion of the parties to the contract, and may include topics such as

- The number of copies of contract documents to be received by the contractor
- Survey information to be provided by the owner
- Materials provided by the owner
- Changes in insurance requirements
- Phasing requirements
- Site visits
- Start date of the construction
- Requirements for security and temporary facilities
- Procedures for submittal and processing of shop drawings
- Cost and schedule reporting requirements
- Traffic control and street cleaning requirements
- Responsibilities for testing of materials
- Actions to be taken in the event of discovery of artifacts or items of historical value

Drawings

Drawings are the means by which the designer conveys the physical, quantitative, and visual description of the project to the contractor. The drawings are a two-dimensional representation of the physical structure that meets the objectives of the owner. They are also known as plans or blueprints.

The contract drawings are organized into sections. The architectural drawings show the layout of the project. They are numbered sequentially beginning with page A-1 (mechanical drawings begin with M-1, electrical drawings begin with E-1, structural drawings begin with S-1, etc.). For a building project, the architectural drawings are the core drawings of the contract. They show the floor plans, exterior elevations, interior elevations, details, windows, doors, and finish schedules. The structural drawings show the load-carrying systems. They show how the structural members will support the building and transmit the loads to the foundation and the ground. The mechanical drawings show the plumbing; heating, venting, and air-conditioning (HVAC); and fire protection. The electrical drawings show the various electrical and communications systems in the building. Site drawings show the relation of the site to adjoining property, the drainage plan, and systems such as sanitary sewer and utilities.

Specifications

Specifications may also be known as Technical Provisions. They are written instruments to be used in conjunction with the drawings, so together the drawings and the specifications fully describe and define the requirements of the contract, to include the quality that is to be achieved. They supplement the drawings and provide information that cannot be shown in graphic form, or information that is too lengthy to be placed within the drawings. They guide bidders in the preparation of cost proposals as well as field execution of the work. They also guide the contractor through the processes of ordering materials and construction and installation of the facility. Specifications provide information regarding

- The quality of materials
- The quality of workmanship
- Erection and installation methods
- Test and inspection requirements and methods

Specifications have restricted application, usually to a specific item or work operation. The designer or specification writer is able therefore to assign responsibility for each provision of the specifications to the desired specific party. Specifications must satisfy these basic criteria:

- Technical accuracy and adequacy
- Definite and clear stipulations
- Fair and equitable requirements
- A format that is easy to use during bidding and construction
- Legal enforceability

There are several general types of specifications. Performance specifications state how the finished product must perform without dictating how the contractor is to do the work. An example of performance specifications would be a requirement for concrete that has a 28-day strength of 3,500 psi. It is the contractor's responsibility to provide concrete that meets the requirement.

Design specifications, also known as materials and workmanship specifications or prescriptive specifications, state how work is to be performed. For example, a typical wall specification requires that the studs be placed 16 in. on center, and may even specify the nailing pattern. The contractor is not free to make changes to the spacing of the studs. By using a design specification, the designer assumes liability for the performance of the product. Looked at another way, the designer warrants the performance of the product.

Open specifications allow any product that will meet the requirement. Closed specifications allow only products of a certain type. Proprietary specifications allow only one product, without any allowance for substitutions. Equal specifications normally specify one product, but allow the substitution of others that are "equal."

In case of a conflict between specifications and drawings, the specifications have precedence. The other option is to require that the architect resolve any conflicts between plans and specifications.

Addenda

bid opening
The opening and tabulation of bids at the prescribed time and place.

Any change to the bid documents after they are released for bidding but before bids are actually received requires the issuance of an addendum. This formal document changes the original bid documents and becomes a part of the bid package. At the time of **bid opening**, in their bid documents, bidders must acknowledge all addenda. Addenda may be issued to change the bid opening date, to modify the original design, to delete or add items, or to correct errors. Addenda should not be issued close to the bid opening unless the bid date is also extended accordingly.

Documents Included by Reference

Contracts may make reference to other documents, which become part of the contract by reference. Most common among documents of this nature are laws, regulations, codes, covenants, and ordinances. Federal laws such as the Davis-Bacon Act (prevailing wage) and the Miller Act (**payment** and **performance bonds**) are usually included by reference within the General Conditions. Building codes may include national or regional codes, such as the **Uniform Building Code (UBC)**, and local building codes that are particular to a specific city or area. There may be occasions during construction when it is necessary to refer to local codes and ordinances and it is wise to have copies of the building codes available on the project.

CONTRACTUAL RELATIONSHIPS

Owners

On most construction projects, there is a contract between the owner and the general contractor (see Fig. 8.2). There are also subcontracts that are between the general contractor and the specialty contractors. It is important to understand that there is no contractual relationship between the owner and the subcontractors. In

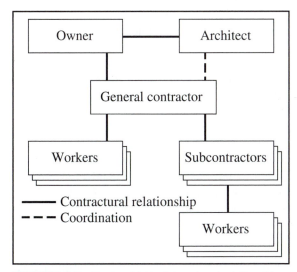

FIGURE 8.2 Design-bid-build contract relationships between owner, architect, general contractor, and subcontractors.

payment bonds
Guarantee that the contractor will pay the subcontractors and material suppliers.

performance bonds
Guarantee that the contractor will perform the work in accordance with the plans and specifications.

Uniform Building Code (UBC)
First enacted by the International Conference of Building Officials in 1927. The Code is dedicated to the development of better building construction and greater safety to the public by uniformity of building laws.

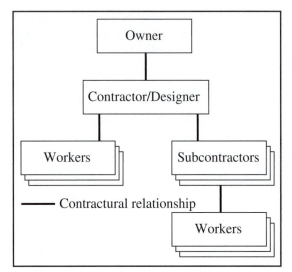

FIGURE 8.3 Design-build project procurement contract relationships.

the design-bid-build relationship, the owner also has another contract with the architect, but there is not a contractual relationship between the contractor and the architect. The architect has many active functions on the job, such as the processing and approval of shop drawings, approval of progress schedules, approval of periodic pay estimates, and acceptance of the work; but there is no contract between the contractor and the architect. They have a working relationship based on their separate contracts with the owner. For a design-build contract, the contract structure is different (Fig. 8.3).

The owner may be represented on the job site by the architect, by a construction manager (CM), in-house staff, or a consultant team. The manner in which the owner is represented has no impact on the owner's responsibilities—they do not change. Under the provisions of AIA A201, the owner has the responsibility to provide a safe workplace, respond to the contractor's questions, secure adequate funding, and pay the contractor promptly. The normal period of payment is monthly. The owner can impact the project cost either upward or downward by developing a reputation for slow or prompt payment. In today's world of electronic information transfer, there is no reason owners cannot pay contractors within three days, and require that the general contractor also pay the subcontractors and material suppliers promptly. But most do not and they increase the overall price of construction by failing to pay promptly.

Designers

The most obvious relationship is the contractual relationship between the owner and designer. The designers have no contract with the contractors, but they have several functions that involve the contractors and they must work together. The designers review the shop drawings. In recent years, architects have moved away from approving shop drawings to protect themselves from unnecessary liability. They review and approve monthly pay requests, and have the potential to increase the owner's costs by delaying the process. Delays to the process will result in higher-priced change orders and a higher incidence of claims. Designers must provide comments and recommendations to the owner in case of claims and disputes. The contractor and the designer have no contractual relationship, but they must work together on several matters that are important to the success of the project, and ultimately to each other's financial success.

General Contractors

The general contractor has a direct contractual relationship with the owner, subcontractors, and suppliers. The general contractor has no contractual relationship with the designer, second-tier subcontractors, materials suppliers of the subcontractors, or the construction manager, if the project has one. However, the general contractor interacts with all these organizations, and they depend on each other for their mutual success.

The biggest complaint that subcontractors have is that general contractors hold their money too long. The general contractor incurs an obligation to pay the subcontractors promptly when paid by the owner. Inflation and the cost of borrowing money will take all the profit out of a project for a subcontractor who is not paid promptly by the general contractor. This has been such a problem that some states have enacted legislation to require contractors to pay their subcontractors promptly.

Subcontractors

The subcontractors, or specialty contractors, have a contract with the general contractor, and have no contractual relationship with any other party on the site. All their communications must be routed through the GC. There is usually only one monthly pay estimate on a project, and it is submitted by the GC to the owner. Each subcontractor's progress, and hence payment, is contained within that pay estimate. Subcontractors are totally dependent on the GC for timely processing of their pay estimate, and timely payment once the owner has paid the GC. Change orders often have a larger impact on a subcontractor than they have on the GC, simply because one change order can represent a higher percentage of a subcontractor's total earnings on the project. Subcontractors often must rely on the GC to conduct the change order negotiations on their behalf.

Material Suppliers

Examples of materials suppliers are electrical wholesalers, lumberyards, concrete batch plants, plumbing supply stores, and other businesses that sell materials to the contractors. Their contract is between themselves and the contractor or sub-subcontractor with whom they are dealing. Suppliers may even be required to provide warranties to the owner, or even to provide technical training to the owner's staff, but their contract is only with the contractor or subcontractor. Receiving payment for their materials has sometimes been problematical. They can attach a lien to private property until they are paid, and the Miller Act protects them on public contracts.

Construction Managers

During the 1980s and 1990s, the practice of hiring a construction manager became very popular. The idea was to bring the construction manager onto the project team in time to help select the architect, and then to work with the architect during the design, and then to manage the construction process. The goal is to get construction input early in the design process. There are now many large general contractors who refer to their companies as construction managers. They bid projects, and then subcontract 100% of the work, performing none of the work themselves. This form of construction management is nothing more than a variation of the general contractor role.

THE BID AND AWARD PROCESS

Government projects must be advertised and are normally awarded to the lowest responsive bidder (see Fig. 8.4). Private projects do not need to be advertised and are usually negotiated, with only a limited number of selected contractors

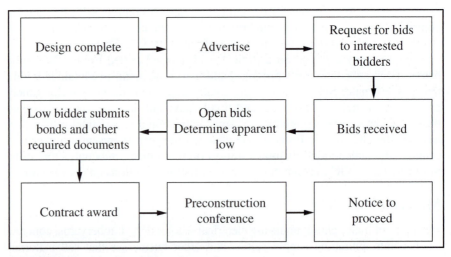

FIGURE 8.4 Advertise, bid, and award.

involved in the negotiation. Advertisement is at the discretion of the private owner.

Advertising of government projects is required to ensure that all contractors who might have an interest in the project are offered the same opportunity. Federal projects are advertised in the *Commerce Business Daily,* which is printed by the U.S. Government Superintendent of Documents every working day in both an electronic and a printed version. The *Dodge Report* is another widely distributed commercial publication used exclusively for advertising construction projects. Publications such as *Engineering News-Record,* which is published weekly, also are widely used for advertising major projects. Local newspapers are the other source of construction project advertisements. Contractors who bid government projects review these publications every day.

Construction project advertisements contain sufficient technical information to enable contractors to make a preliminary decision regarding their interest in bidding. As a minimum, the project name, owner, location, required start and completion dates, a short description of the project scope, estimated cost range, and the bid opening location and date are included. The advertisement contains the contact name and telephone number for those desiring to obtain copies of the Instructions to Bidders.

Bid documents can now be obtained in more than one format. For many years, interested bidders had to go to the office of the owner or the designer to obtain the bid package, which consisted of the invitation to bid, specifications, drawings, and bid documents. Usually there is a fee involved in obtaining these documents. Many government agencies such as the U.S. Army Corps of Engineers and the Bureau of Reclamation now make the entire packet of bid documents available by CD-ROM or over the Internet. There is added convenience for the bidder in being able to obtain the documents in this manner because it is quicker and easier. There is cost saving for the agency because fewer copies of the documents must be printed and shipped. Potential bidders can download the bid documents as they are needed. To see an example of bid documents available by Internet, students should go to tsn.wes.army.mil/. This is the site for the military Tri-Service Solicitation Network. The site contains links to government agencies. The site does not issue bid documents, but provides links to the issuing government agencies.

Bidding

Bids must be submitted on time, at the location specified, on the correct forms, with acknowledgment of all addenda, and with a valid **bid bond** (if required). Otherwise, the bids are considered nonresponsive and are returned unopened. Most bid-opening rooms have a large clock clearly visible to all in attendance. When the time for accepting bids is reached, an announcement is made that the time for accepting bids has passed and any bids turned in late will be returned unopened. In the case of government projects, the person responsible for opening bids then opens the bids publicly and reads the bid amounts to those assembled.

bid bond
A bond that assures that the contractor will accept the contract when awarded and will supply the required performance and payment bonds.

The low apparent bidder may be announced at the bid opening, but the low bidder cannot be determined until the bid is checked, and bonds and certificates of insurance are examined. Many agencies accept bids by mail, and it is possible for a bid to be mailed on time, but not be received at the time of bid opening. Bids that are submitted by mail are accepted without regard to the date and time received, provided they meet the requirements for date and time of mailing. It is also necessary to check the bids for mistakes and completeness of the bid documents. After the agency is certain that all bids have been received and they have examined all bids, the low bidder is announced.

Bids are sometimes advertised with a **base bid** and alternates, which can be **additive or deductive**. This is done only in cases when the adequacy of the project funding is questionable, and the agency wants to maximize the probability of receiving an awardable bid by accepting enough alternates to bring the total price below the project programmed amount. Government projects are approved with a *programmed amount* plus *contingency funds*. It is very difficult to obtain approval to award projects above the programmed amount.

Alternates may consist of features to be added to the base bid, at an additional cost for each additive, or they may be deductive with a cost reduction for each alternate accepted. Alternates are considered sequentially until the low bidder has provided everything that can be included in the project and the bid still remains below the programmed amount. Alternates must be accepted in numerical order, without skipping, and the total bid for each bidder must be considered with the addition of each alternate. In the case of additive alternates, the alternates are added in sequence until accepting the next alternate would result in the bid being too high. In the case of deductive alternates, the alternates are accepted until a sufficient amount has been subtracted from the base bid to bring the total bid below the project budget. Consider Example 8.1 involving five deductive alternates.

base bid
Used for projects that may be in danger of exceeding the capital budget. Provides a mechanism for the owner to accept bids on a portion of the facility, and not construct some of the features that would cause the budget to be exceeded.

additive or deductive
Amounts that are added to or subtracted from the base bid on a project to ensure the bid is within the capital budget.

EXAMPLE 8.1

A school district has determined that there is a need to add a new gymnasium to an existing high school. The programmed amount for the project, which is the amount of funds available, based on the enabling legislation, is $12.5 million. The architect's estimate is $12.9 million, but the school district desires to retain as many of the features in the gymnasium as possible. These deductive alternates and their estimated values have been identified:

1.	Bleachers for the basketball court	$240,000
2.	Racquetball courts	$125,000
3.	Additional parking	$55,000
4.	Replace brick exterior with metal siding	$98,000
5.	Replace air-conditioning with ventilation	$135,000

The table shows the five bids that were received, and the evaluation of the bids to determine the low bidder.

Spreadsheet of base bids and five deductive alternates
New high school gymnasium

Bidders	A	B	C	D	E
Base bid	$12,879,384	$13,045,900	$13,243,765	$12,999,001	$13,684,700
Alternate 1	−224,318	−237,982	−264,280	−274,460	−312,238
Cumulative	12,655,066	12,807,918	12,979,485	12,724,541	13,372,462
Alternate 2	−97,210	−114,000	−107,340	−145,600	−99,000
Cumulative	12,557,856	12,693,918	12,872,145	12,578,941	13,273,462
Alternate 3	−43,500	−62,300	−65,000	−68,500	−73,600
Cumulative	12,514,356	12,631,618	12,807,145	12,510,441	13,199,862
Alternate 4	−97,500	−89,450	−99,650	−92,450	−89,650
Cumulative	**12,416,856**	**12,542,168**	**12,707,495**	**12,417,991**	**13,110,212**
	Low bidder				
Alternate 5	Does not need to be considered				

Using the total of the base bid and all alternates, the low bidder is bidder A, with base bid and alternates 1 through 4 subtracted from the contract. That means that the alternate 5 deduction, replace air-conditioning with ventilation, was not accepted. It remains part of the contract. The contract will be awarded with air-conditioning, but without bleachers, racquetball courts, additional parking, and with a metal siding exterior.

The contracting officer could have decided to advertise the project with a base bid and then add the alternates to the base bid until the next one could not be included without exceeding the project budget. In that case, the base bid would be reduced in scope, and then the alternates would be added to the scope until adding the next alternate would cause the total bid to exceed the project budget. Both approaches to the problem (additive or deductive alternates) are perfectly acceptable and legal. Construction contractors are used to dealing with both methods. Example 8.2 shows a base bid with additive alternates.

EXAMPLE 8.2

A new research laboratory building is to be constructed on a university campus. The programmed amount is $5.6 million. The architect's estimate is $5.8 million. The contract will not be awardable above $5.6 million. The decision is made to advertise the building with a base bid and four additive alternates. The four additive alternates and their estimated values are

1. Additional installed testing equipment $74,000
2. Upgraded administrative area $88,000
3. Additional climate control areas $48,000
4. Landscaping $76,000

The table shows the bids that were received and the evaluation of the bids to determine the low bidder.

Spreadsheet of base bids and four additive alternates
New university research laboratory

Bidders	A	B	C	D	E
Base bid	$5,440,200	$5,195,165	$5,380,760	$5,582,685	$5,234,678
Alternate 1	56,480	69,485	71,900	46,425	68,945
Cumulative	5,496,680	5,264,650	5,452,660	5,629,110	5,303,623
Alternate 2	82,450	79,675	56,900	82,000	80,450
Cumulative	5,579,130	5,344,325	5,509,560	5,711,110	5,384,073
Alternate 3	40,800	43,580	44,687	49,784	42,136
Cumulative	5,619,930	5,387,905	5,554,247	5,760,894	5,426,209
Alternate 4	74,690	79,300	72,345	70,985	76,890
Cumulative	**5,694,620**	**5,467,205**	**5,626,592**	**5,831,879**	**5,503,099**
		Low bidder			

The low bidder is bidder B at the total bid price of $5,467,205 for the base bid and all four alternates. The total bid is below the programmed amount, and will be awarded.

Owners do not all use the same method for determining the low bidder when there are alternates, but the method must be stated in the bid documents and adhered to in awarding the contract. Some owners compare the total of the base bid plus all alternates that are accepted. The low bidder would have the low total of the base bid plus alternates. Other owners determine the low bidder based only on the base bid, and then accept as many of the alternates as funding allows. The method of determining the low bidder will affect bid strategy. In the case of the low bidder determined by the base bid only, a contractor may be tempted to bid the base project abnormally low and increase the amount of the alternates to offset the low base bid. However, that tactic could be extremely costly if only the base bid, and none of the alternates, is awarded. In every case of base bid and alternates, bidders must read the contract to determine the procedure for selecting the low bidder and determine their bid strategy wisely.

Bid Depository

bid depository
An organization used by owners, general contractors, subcontractors, and materials suppliers to help ensure fair bidding.

A **bid depository** provides a service to general contractors and subcontractors by receiving subcontractor bids and distributing them to general contractors. There are many types of bid depositories. Historically, bid depositories have been created by the trade associations in an effort to protect member companies from unscrupulous general contractors. The main objective of any bid depository is the prevention of bid shopping, which can occur after award of the contract when a general contractor requests that a specialty contractor lower a bid. The request may or may not involve the disclosure of the bids already submitted to the general contractor by competitor specialty contractors. The practice is not ethical, but it does happen.

The bid depository process begins by developing a bid form that represents a consensus by the general contractors and subcontractors who are bidding the project about which items should be provided by which trade. The bid form is designed to eliminate any areas of disagreement or overlap. Subcontractors then submit sealed bids to the bid depository, marked for specific general contractors. Bid envelopes are usually coded by trade in some manner such as color coding. Subcontractors are free to bid as they normally would, by bidding to some or all contractors at whatever price they select, without obligation to bid the same price to all contractors. General contractors receive their bids several hours before bid opening, instead of minutes before bid opening, as is so often the practice when a bid depository is not used. This allows time for the general contractors to verify bids and bonding capacity, determine if the subcontractor is properly licensed, and discuss any issues that may deviate from the standard bid.

The bid depository does not become involved in the bid and is not responsible for the actions of the bidders. Its function is simply to level the playing field for all specialty contractors who choose to take advantage of the opportunity. Nothing in the functioning of the bid depository prevents either the general or specialty contractors from accepting bids outside the bid depository. Bid depositories are nonprofit corporations, but they do have daily operating expenses. They derive their income from a fee charged to only the successful subcontractors. The service is free to all others. They are open to all licensed subcontractors in their state or region, in any of the craft areas served by the depository. The general contractor is not legally bound to award to the low bidder and remains responsible for determining who is responsive and responsible.

Award

Government agencies that use the design-bid-build contracting method award the contract to the low responsive, responsible bidder. During the time between the bid opening and bid award, the contractor must provide the required payment and performance bonds. The contract is awarded when the general contractor and the owner sign the contract documents. Owners mail two copies of the contract to the general contractor for signature and return. The owner then signs both copies, keeps one for the owner's contract file, and returns the other signed copy to the general contractor. At that point, the general contractor will begin to finalize contracts with all subcontractors. A general rule is that the subcontracts should contain the same contract requirements as the contract between the owner and the general contractor, including but not limited to insurance, bonds, completion date, **liquidated damages**, and claims requirements that can be punitive to the general contractor.

Notice to Proceed

The Notice to Proceed is the document that establishes the contractor's right to have access to the owner's property. It is issued after the contract between the owner and the general contractor is executed. It establishes the official

liquidated damages
Daily amount, specified by the contract, paid by the contractor to the owner to compensate the owner for late completion. Liquidated damages are not penalties. They represent compensation for expenses and lost income.

start date of the contract, and thereby the official date all work on the project must be complete, because contracts normally require that the work be complete in some specified number of days after Notice to Proceed. Change orders may alter the required completion date by adding days for the accomplishment of the additional work. Most contracts establish a maximum number of days after issuance of the Notice to Proceed by which the contractor must begin physical work on the site. The owner and contractor should use the preconstruction conference to discuss and agree on the contractor's plan for beginning work. Open and efficient communications on such matters are essential.

STANDARD CONTRACT DOCUMENTS

Government

All government agencies have standard contract forms, and many use the AIA forms with amendments. Just as the AIA Document A201 is used in many private contracts, government agencies have their own standard General Provisions. Their General Provisions enjoy the same advantages as A201. The requirements are known and understood by all. The General Provisions have been court tested, so their meaning is well established. Owners should be discouraged from using locally written or "homegrown" contract forms because the parties cannot be as sure of how the courts would interpret those documents.

Agencies also have standard specifications or technical provisions. The agencies are intimately familiar with such standard provisions because they use them all the time. The agencies vary from local school districts to state departments of transportation and state general services offices. At the federal level, there are many agencies that contract for construction. The most common are the U.S. Army Corps of Engineers, the Navy Facilities Engineering Command, and the General Services Administration. In between, there are many highway administrations, water and sewer districts, sports authorities, and rail and ports authorities. All have their own standard contract provisions and forms, or use other standard documents. Contractors doing business with a government agency for the first time must quickly and thoroughly familiarize themselves with the standard forms used by that agency.

American Institute of Architects

The American Institute of Architects is the professional voice of the architecture profession. AIA publishes many standard contract forms, all of which they sell to their members. Their General Provisions, AIA Document A201, for example, is designed for a contract between the owner and general contractor, but the AIA also provides standard forms for design contracts, construction management contracts, and others as shown in Appendix F.

Construction Owners Association of America

The Construction Owners Association of America (COAA), like the AIA, provides several standard contract forms to its members. As an organization of owners, they feel that the forms provided by other construction industry interest groups are biased in favor of the groups that produce them. To provide a balanced alternative, COAA has produced a group of interrelated contracts for use by owners who want an alternative.

TYPES OF CONSTRUCTION CONTRACTS

There are many types of construction contracts, and the type of contract selected depends on the kind of work being performed and the conditions under which it is being performed. A summary of common contract types is presented in Table 8.1.

Lump Sum

Lump sum contracts are typically used for buildings. The quantities of the materials required can be calculated with sufficient accuracy during the bidding process to allow contractors to submit a single lump sum price for the work. The quantity of items such as drywall, door frames, bathroom sinks, electrical conduit and wiring, and roof tiles can be calculated accurately from the plans. When a clear definition of the quantities and quality of work required is provided by the contract documents, submitting a single lump sum bid is fair to the contractors.

Unit Price

Unit-price contracts are used for work where it is not possible to calculate the exact quantity of materials that will be required. Unit-price contracts are

TABLE 8.1 Common types of contracts and their normal application.

Contract type	Application	Contract type	Application
Lump sum	Structures such as buildings that permit the bidders to estimate the quantities and costs accurately	Unit price	Projects such as roads that do not permit accurate estimation of materials quantities
Cost plus	Any type of facility that has conditions making it impossible for either the owner or the contractor to compute costs accurately; an example would be work in remote areas	A + B	Contracts that provide a financial incentive to contractors with the expertise to complete the project quickly, by rewarding early completion

commonly used for heavy/highway work. The designer may calculate that 1,000,000 cy of earth needs to be moved, but the owner and contractors know that after the work has been completed, the contractor will not have moved *exactly* 1,000,000 cy. The exact quantity will vary. The fair solution to the problem is for the owner to tell the contractor what the estimated quantity is, and then to pay the contractor for the exact measured amount after the work is complete.

Contractors submit a price for each item on a unit-price contract. Unit prices are multiplied by the engineer's estimated quantities and totaled. The low bidder is the bidder with the low total of all items. Items whose actual quantity varies from the estimated quantity by more than 15% or 20%, either above or below the estimated quantity, are sometimes subject to renegotiation of the unit price. When the actual quantity is low, the contractor may request a renegotiated higher price because the anticipated amount of earned overhead has been reduced. When the actual quantity is high, the owner may request a renegotiated unit price for the opposite reason. The goal of unit-price contracts is fairness to both parties.

Cost Plus

Cost plus (cost reimbursable) contracts are used in situations that make it difficult or impossible for either the owner or the contractor to predict their costs during the negotiation, bid, and award process. Factors that may make the calculation of costs impossible include unpredictable and extreme weather conditions such as would be encountered in the Antarctic, unknown transportation requirements to remote locations, combat or war, or contracts where the amount of effort that will be required depends on another contractor's work.

Cost plus contracts take many forms, the most common being cost plus a fixed fee and cost plus a percentage. Most owners prefer cost plus fixed fee because then the amount of profit the contractor will earn cannot increase, thereby removing any incentive for the contractor to be anything less than thrifty, or to produce poor-quality work. Cost plus percent contracts may be fair in situations that are very difficult, or when the time to complete the work is not known with any certainty, but some incentives to maintain productivity are needed. A profit that can be earned in six months may not be attractive if there is a possibility that the work may require a year or two.

Incentive Contracts

incentive contracts
Contracts with a bonus for early completion, A + B contracts that reward contractors who accomplish the work most efficiently, or contracts that offer a higher pay based on quality.

The most common types of **incentive contracts** are those that offer a bonus for early completion. Contracts that require a penalty for late completion are required by law to also offer a bonus for early finish. Penalty for late finish and liquidated damages are not the same thing and should not be confused. A penalty is just that—a penalty. Penalties may be in any amount that the contractor is willing to accept. Liquidated damages are designed to pay the cost an owner experiences if the construction continues past a preset completion date.

"A + B" contracts are a relatively new innovation developed by the departments of transportation to reduce highway construction time and minimize the inconvenience to the traveling public. Many highway projects involve renovation, repair, improvement, or widening. Very few are true new construction projects. The purpose of the A + B contract is to provide a way to compensate the contractor for the expertise in sequencing and staging to complete the work quickly and efficiently. "A + B" means cost plus time. The Department of Transportation will assign a value to each day the highway work is being performed and the contractors will then bid a price plus a number of days. The low bidder is determined as the low total of cost plus time. Therefore, it is very possible to select a low bidder who bids a higher price, but fewer days, than the other contractors.

For "A + B" contracts, a value is placed on each day of construction, and that amount becomes the B component of the bid. For example, the estimated cost to the public in additional delays and inconvenience resulting from closing a highway or a bridge might be estimated at $10,000 per day. Contractors are then required to bid an amount in dollars, the A component of the bid, plus a number of days, the B component. The low bidder is determined by converting the days to dollars, and adding that cost to the A component. It is possible to bid a construction cost amount higher than another bidder and still be the low bidder if a contractor knows how to accomplish the project in fewer total days.

Consider this example using $10,000 per day for the B component:

Bidder 1: $5,000,000 and 200 days = $7,000,000
Bidder 2: $5,350,000 and 160 days = $6,950,000

Bidder 2 is the low bidder in spite of a higher bid price.

In northern California, about a mile east of the Oakland-San Francisco Bay Bridge, the eastbound I-580 connector bridge over I-880 collapsed after the intense heat from a gas tanker fire caused the steel bridge to weaken. The California Department of Transportation (Caltrans) used an incentive contract to repair the 165-foot long bridge. The contract included an incentive of $200,000 per day for re-opening the bridge early and a penalty of $200,000 per day for exceeding the 50 day deadline. C.C. Meyers Construction Company completed the repairs in just 18 days, 32 days early and collected the contract's maximum $5 million bonus [16].

The Kentucky Transportation Cabinet let an I-275 project using an A + B + C contract. The "A" part was the sum of the unit prices to perform the actual work items. The "B" part was the number of contract days the bid stated were required to complete the work. The price per contract day was established at $25,000. The contract additionally called for penalties of $25,000 per day for every day the contract used beyond the contract time. The "C" part was a project warranty. The contractor had to provide a five-year warranty for the work but for every year beyond that up to a 10-year maximum, the bidder was given a $500,000 bidding credit [10].

The reason these contracts are gaining in acceptance is that they offer the owner an opportunity to buy and pay for the contractor's knowledge. Contractors are forced to focus on time. Those who can get in and get out the soonest will be the most successful. Contractors who know how to organize and work more efficiently will win the bid, and the owner will have the facility completed sooner.

Negotiated Contracts

Just as the name implies, negotiated contracts are awarded through a negotiation process rather than through a bid process. In its purest form, an owner will ask a contractor for a project price. A negotiation process will ensue, culminating in an agreement to build the project at an agreed on, negotiated price. This process works especially well when private owners are negotiating the expenditure of their own money and are working with contractors with whom they have a previous working relationship. It does not work as easily when the government is trying to do the same thing with the public's money.

The process called competitive negotiation involves the taking of bids, after which the final price is then negotiated. There are normally two or more contractors involved in the negotiation process. Negotiated contracts may be fast-tracked, a system that allows work to begin before all design work is complete. In that case, the drawings needed first, such as site drawings and foundation design, will be completed and given to the contractor.

Job Order Contracts

job order contracts
Contracts used mainly for maintenance work. Contractors submit a bid in the form of a decimal multiplier for the owner-published book of standard rates for work. The contractor then performs work against a series of work orders. This process is designed to avoid a design contract for each work order.

Job order contracts are used by large organizations such as universities and industries that own large facility complexes. These facilities require repair and maintenance on a continuing basis. There are two basic methods for accomplishing all the on-going painting, electrical repairs, clean up, and minor projects that must be done. One method is to hire an in-house crew to do the work. The other way is to hire a contractor. The purpose of the job order contract is to be able to hire the contractor and to eliminate the need to hire a designer to prepare a design for each small item of work. The need for a designer separate from the contractor is not only too expensive, it is also too slow.

For small construction projects under $10,000, there is little time or need to go to a designer to design the work, draw up plans and specifications, and advertise the work for bid. The job order contract (JOC) avoids that process and accomplishes the work more quickly. The contractor assumes responsibility for completing the design work, if any design is required, and for completing the construction task, all on the same work order.

Three documents are provided to the contractors to bid the project, standard specifications, a book of owner cost estimates, and general conditions. Usually contractors are required to bid by submitting two coefficients, one for normal work hours and one for overtime work. To bid a project, contractors are given a number of theoretical projects. They then review the specifications, determine the owner's cost from the book of owner cost estimates, calculate their own bid price, and divide the owner cost by their company cost. The ratio will produce a "coefficient" such as 0.967 for estimates below the owner's estimate, or a number such as 1.0224 for estimates a little above the owner's estimate. The bid process is low bid, with the coefficient replacing a dollar cost. The contract is awarded to the contractor submitting the lowest coefficient.

The contractor who is awarded the contract will receive work orders or "call orders" for each required task. Every effort is made to define tasks that can be accomplished for less than $10,000. JOC is not appropriate for large tasks. JOC provides a mechanism for the owner to obtain quality work with the least possible loss of time. Most owners award more than one JOC at a facility, and then award more call orders to the contractors doing the best work.

JOC is a competitive bid process with specifications and cost data provided by the owner. Design is the responsibility of the contractor. The process provides a continuing incentive for quality work, while reducing time and cost for the owner.

There are other forms of contracts, each with its own specialized purpose. For instance, guaranteed maximum price contracts are sometimes used for negotiated work that is started before the design is complete, making it difficult or impossible for the contractor to estimate the price accurately. This type of contract is also sometimes used when one contractor defaults on a partially complete contract, and a second contractor finishes the project.

BONDS (GUARANTEES)

An early legend, that some say is fact, says that in the late 1800s in Baltimore County, Maryland, the county began requiring the county sheriff to deposit a sum of money with the county treasurer as a guarantee that he would faithfully perform the requirements of his job. Sheriffs who had substantial funds would withdraw the money from the bank, and deliver it to the county treasurer, who deposited it back into the bank in the county's account. Sheriffs who were not wealthy would borrow the money from their parents and in-laws, and then deposit it in the county's account.

As the tale goes, an enterprising banker decided it would be easier to write a certificate to the county saying that if the sheriff did not faithfully perform, the bank would give the county the funds. The certificate amounted to a "preapproved loan" to the sheriff. Again, for sheriffs who did not qualify for the "preapproved loan," the bank might require their parents or in-laws to cosign (indemnify) for the certificate. If the sheriff failed in his duties, and the county took the money, then the bank would collect the money back from the sheriff and his friends and family who might have guaranteed his loan. As you can see, this process is a banking (credit) obligation. The early form of the certificate was called a bond.

That is exactly how bonds operate today, except that in 1890, the Supreme Court decided that bonds could not be issued by bankers, and bond companies

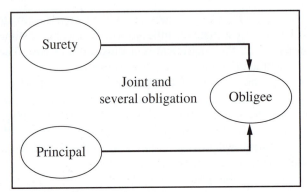

FIGURE 8.5 Relationship between the surety, principal, and obligee.

today are typically insurance companies. It was a strange decision because bonds are more like bank credit than insurance. As in any bank loan, the bank expects to have its money repaid. Bonds are a guaranteed loan waiting to happen. Unlike an insurance policy, no losses are expected, and if there are losses, they are expected to be repaid. The mechanism for repayment is called an indemnity agreement. Contractors sign indemnity agreements agreeing to reimburse the bond company if the bond company must use its funds to complete a construction project (in the case of a performance bond). There are provisions made for the stockholders and their spouses to individually indemnify the bond company (surety), just as the less-wealthy sheriffs had their parents and in-laws do in the late 1800s.

In any bond there are three parties—the principal (contractor), the surety (bonding company), and the obligee (owner) (see Fig. 8.5). The surety and the principal make a joint and several guarantee to the owner, that is, both parties are obligated to the owner.

Bid Bonds

Bid bonds have two purposes: (1) to guarantee the contractor will enter into a contract if determined to be the lowest responsible bidder and (2) to guarantee the contractor will provide the required payment and performance bonds and insurance policies. When the performance and payment bonds have been submitted, the contractor is released from the bid bond obligations. Most bid bonds are what we call spread bid bonds: they require only forfeiture of the difference between the low bid and the next bidder's price if the two guarantees are not met. There are also penal sum bonds. A penal sum bond is stated in the form of actual damages up to a limiting amount.

Performance Bonds

Performance bonds guarantee the performance of the contract requirements at the stated bid price. In effect, the surety is saying it guarantees the performance of the contractor, or it will complete the project as described in the plans and specifications. The surety is in the position of being asked to guarantee the contractor's performance. Therefore, the contractor must demonstrate an ability to perform before the surety is willing to issue payment and performance bonds. The surety will visit the contractor's home office and job sites, and will contact the owners of recently completed contracts. In addition, contractors can be expected to provide to the surety the information and material outlined here:

- Résumés of key personnel in the firm
- List of work successfully completed
- Names and addresses of owners, designers, suppliers, subcontractors, general contractors, and others in the industry with whom the contractor has done business
- Organization chart and responsibilities of key individuals in the company
- Business continuity plan
- Explanation of how the project fits with the company's other work
- Strategic plan of the company
- Risk management and safety programs
- Safety performance
- List of business relationships with accountants, attorneys, consultants, and insurance companies
- Financial information such as balance sheets, statement of earnings, owner's equity, schedule of indirect costs, general and administrative expenses, and a cash flow statement
- Progress schedules of current work, list of completed contracts, and a list of current contracts

Contractors must develop a long-standing working relationship with their surety, and they must learn to present their company in its most favorable light when applying for payment and performance bonds. They must work consistently with a surety who will underwrite projects with specific constraints of project type, location, and size.

If a contractor defaults on performance of the contract, the surety has three basic choices:

- Buy back the bond. This amounts to giving the owner a check for the amount of the penal value of the bonds.
- Replace the contractor. Negotiate or advertise for bids for the purpose of retaining another contractor to finish the work.
- Finance the contractor. The bonding company runs the risk of spending more than the value of the bond, but this is still a common option because the contractor is familiar with the project.

Today most sureties try to replace the contractor.

TABLE 8.2 Summary of bonds.

Bond type	Purpose
Bid	Ensures the low bidder will accept the contract if offered and will submit the required payment and performance bonds
Performance	Guarantees the contractor will complete the work in accordance with the plans and specifications
Payment	Guarantees the contractor will pay the subcontractors and suppliers

Payment Bonds

The Miller Act of 1935 provides that on federal projects over $200,000, the contractor must furnish payment and performance bonds in a satisfactory amount. Most states have passed similar legislation commonly referred to as Little Miller Acts. Government agencies require bonds because contractors cannot place a lien on government-owned property. The Miller Act and state-level Little Miller Acts provide the mechanism to guarantee that the subcontractors, labor, and suppliers will be paid. Table 8.2 presents a summary of the types of construction bonds.

Bonding Limits

There are many factors that control how much bonding capacity the bonding company is willing to provide to a contractor. Bonds cost about ¾% to 4% of the contract amount, so the failure rate must be kept very low. Therefore, bonding companies evaluate contractors very carefully. The cost is a function of the risk assumed by the surety. There is not enough premium in bonds to cover the potential exposure. Sureties have to be correct in their evaluation of contractor performance and financial capability. As a rule of thumb, contractors can be bonded up to about 10 times their working capital (working capital = current assets minus current liabilities) or 4 times their net worth. No more than half of a contractor's bonding capacity should be used on any one project. These rules are further discussed in Chapter 9.

CONTRACTOR INSURANCE

Risk management is a vital component of any successful construction project. In the AGC *Guide to Insurance,* the risk management process is explained as having five steps: (1) risk identification, (2) risk analysis, (3) selection of the appropriate treatment technique, (4) implementation of the selected technique, and (5) measurement of the results. The project owner, as the party ultimately responsible for the construction work, is seeking to enhance control over project safety and risk. Requiring contractors to have certain types of insurance is one way owners manage their risk. Contractors also seek to control risk, and insurance is a method to protect against certain events.

Worker's Compensation Insurance

Until about 1900, depending on the state, workers injured on the job had to pay for their own medical costs, and there was no provision for any insurance to cover lost wages. To sue the employer, the worker had to prove negligence on the part of the employer. Usually the negligence was at least in part attributable to the worker, making it very difficult for the worker to prevail. Today all 50 states have passed workers' compensation legislation that requires employers to insure workers for injuries on a no-fault basis. Workers' compensation insurance awards to a worker can be denied for willful misconduct, drug use, and intoxication.

Workers' compensation protects workers who are injured on the job by covering medical and hospitalization expenses, plus a percentage (usually 75%) of the hourly wage during the time the employee is out of work. Serious injuries, such as loss of a limb or death, are compensated by a lump sum for a predetermined monetary amount. Workers' compensation also protects the contractor by limiting the workers' remedy from the employer to the amount that is covered by the workers' compensation law. Workers who are covered by Workers' Compensation Insurance cannot sue their employer for additional compensation, but some subcontractor employees, who cannot sue their employer because workers' compensation is their sole remedy from their employer, have sued the general contractor under the safe-place-to-work doctrine, which says it is the general contractor's responsibility to provide a safe place to work. These suits are known as "employee over-claims." Unlike other kinds of insurance, contractors cannot name other additional insureds on their workers' compensation policy.

Workers' compensation is a very expensive required benefit that contractors must maintain for their workers. Premiums are based on payroll. The rates are set according to trade or craft by the rating bureaus retained by the state. In addition, each company's *experience modification ratio* (EMR) impacts the cost of workers' compensation. Companies with a higher accident rate history pay more for their worker's compensation insurance than companies with a better safety record. Companies with a high EMR have higher overhead, and therefore are less competitive at bid time. Many general contractors will not use subcontractors with an EMR above 1.0. Likewise, many owners will not use GCs with an EMR greater than 1.0. This is especially true in private sector industrial construction.

COMMERCIAL GENERAL LIABILITY INSURANCE

General contractors should require subcontractors to name the general contractor as an additional insured on the subcontractor's commercial general liability insurance policy. This should be done on a primary (the subcontractor's policy is the primary policy) and noncontributory (the general contractor's policy will not contribute to the claim) basis to protect the general contractor from sub-contractor employee over-claims. This situation occurs when an employee of a subcontractor is injured on the job and collects workers' compensation from the subcontractor but sues the general contractor for additional compensation.

There are four coverages provided in commercial general liability insurance policies.

- *Premises / operations liability insurance* Operations liability insurance provides protection from risks associated with on-going construction operations. Most liability claims result from on-going construction operations. For example, when a member of the public is hit by a falling tool, or the painter makes a mistake and someone's car is covered by over-spray, premises/operations liability is the coverage that protects the contractor. All contractors carry premises/operations liability insurance.

- *Completed operations and product liability insurance* Completed operations and product liability insurance protects the contractor from liability resulting from construction defects after a building has been completed. If the ceiling in a building falls on a crowded room several years after the building is complete, the contractor is protected through completed operations and product liability insurance. This insurance begins where premises/operations insurance ends.

- *Contractor's protective insurance* This type of policy provides coverage for the general contractor for occurrences caused by their subcontractors. Almost all of the time when an accident occurs that is caused by a subcon-tractor, the injured party will sue not only the subcontractor but also the general contractor. This is the coverage part that provides insurance for the general contractor.

- *Contractual liability insurance* This policy covers liabilities that the contractor accepts by contract. Contractual liability results from the indemnification or hold harmless clauses in the construction contract; assessing the contractor's contractual risk requires an examination of the contract. There are many forms of indemnification clauses in construction contracts, with varying degrees of liability transferred to the contractor. Most of these clauses begin, "The contractor shall indemnify and save harmless. . . ."

Builder's Risk

Builder's risk insurance is an "all risk" policy that protects the owner, the general contractor, the subcontractors, and the material suppliers while the facility is under construction, and is normally purchased by the owner, but sometimes by the general contractor. Floods and earthquakes are normally excluded from coverage, so if the facility is in a flood- or earthquake-prone area, that coverage needs to be purchased separately. Builder's risk covers risks such as fire, theft, and wind, that damage the work during the construction period.

Equipment Insurance

Equipment insurance covers theft of or damage to construction equipment.

Umbrella Insurance

An umbrella policy is one that would be purchased by a contractor to add to the limits of coverage of the Commercial General Liability Policy, the automobile policy, and the employer's liability section of the worker's compensation policy. An umbrella policy comes into use if the contractor suffers a loss that is greater than the limits of the contractor's other policies.

Wrap-Up Insurance (Owner-Controlled Insurance Programs—OCIP)

Owners sometimes purchase wrap-up insurance to save money on the contract price. In theory, the contractors have no need for general liability and worker's compensation insurance because the owner covers all parties for those coverages under the terms of a wrap-up policy. Contractors are able to eliminate the need to buy insurance for the project so there is a cost saving to the owner. Wrap-up projects also have extensive safety requirements, which help control losses.

A summary of construction insurance is presented in Table 8.3.

Certificates of Insurance

Most contracts require that a contractor provide a certificate of insurance to the owner to show that the contractor has the required insurance. The certificate will also show, where required, that the owner is an additional insured on the contractor's policy, for that project. It is important to note that it is not adequate to just receive a certificate of insurance. The actual endorsement naming the owner as additional insured, which changes the policy, is needed to have a third party named as an additional insured. The GC must ensure subcontractors have certificates of insurance or the GC can be liable for the subcontractors' losses.

Subrogation

Subrogation is the right of the insurance company to recover money from the responsible third party who is at fault in an accident. Subrogation is best described as the right to stand in the shoes of another. The insurance company stands in the shoes of its insured to recover damages from the insurance company representing the third party who caused the accident. When your car gets hit in the rear, your insurance company will pay to repair the damages, but will

TABLE 8.3 Types of construction insurance.

Type of insurance	Coverage
Worker's compensation	Pays claimant in case of injury, disability, or death of employees resulting from work on the job.
General liability	Protects the owners and the contractors from the financial consequences of various risks, such as hazardous operations, or accidents during construction and after work is completed. The insurance pays for a variety of benefits, including legal defense expenses, injuries to people, and damage to property.
Builder's risk	Pays for damages and losses to a project that occur while it is being built.
Excess liability	An umbrella policy that pays for losses that exceed primary policy limits, such as general liability, automobile liability, and employers' liability on workers' compensation.
Pollution liability	Pays for environmental losses associated with accidental chemical spills and the leakage or disbursement of dangerous vapors.
Design professional liability	Pays for architects' and engineers' professional liability for errors and omissions. This coverage is usually purchased by the architectural and engineering firms but could be included under wrap-up insurance for a design-build project.
Design-build errors and omissions	For companies working in the design-build arena, provides coverage for contractor errors.
Railroad protective	Liability insurance coverage for railroads, purchased by those who conduct operations (construction) on or adjacent to railroad property.
Longshoremen/maritime	Liability insurance similar to workers' compensation that provides coverage for workers, including construction workers, on the water (working on barges) or those working over water.
Automobile liability	Pays for damage caused by the policyholder's vehicles. Also pays medical costs of persons injured in or by the vehicles. This insurance is typically not included in wrap-up insurance because vehicles are operated outside the confines of the project.
Tools and equipment	Pays when a contractor's tools, equipment, field offices, or other property are destroyed, damaged, or stolen. This insurance is not included in wrap-up insurance because these items are considered mobile and therefore difficult to manage. In addition, the premium costs for these policies are not material and would be difficult to isolate from bids.

recover its expense from the insurance holder of the driver who was at fault. That is the right of subrogation. Another example would be a building that was damaged by a truck running into it. The builder's risk insurance would pay for the damage and would subrogate against the truck owner.

Under the terms of paragraph 11.4.7 of AIA Document A201, General Conditions, "the owner and the contractor waive all rights against (1) each other, any of their subcontractors, sub-subcontractors, agents and employees, each of the other, and (2) the Architect, Architect's consultants, separate contractors described in Article 6, if any." To create a valid working relationship on a project, the parties cannot have their insurance companies have the ability to recover from members of the contraction team for damage that occurs on the site because of the fault of one of the team members. Each party is required to carry commercial general liability insurance. Subcontractors are required to name the GC as an additional insured to make sure the subcontractor provides the GC with subcontractor's liability insurance covering the suboperations on the project. As an additional insured of the subcontractor, the GC would send any claims resulting from the subcontractor's negligence to the subcontractor's policy.

ADMINISTERING A CONSTRUCTION CONTRACT

Preconstruction Conference

Nearly all construction projects begin with a *preconstruction conference.* This is a formal meeting, with an agenda distributed by the owner before the meeting, and minutes of the meeting published by the owner and distributed to all attendees following the meeting. Attendees include the owner's representative, the designer, the general contractor and as many subcontractors as possible, municipal authorities, utility companies, emergency organizations such as fire and police, and others who may be impacted by the work. The preconstruction conference is a business meeting, and is not open to the public. Its main purposes are to open the lines of communication between owner and contractor to establish agreements on the rules of doing business, and to address any technical questions the general contractors and subcontractors may have.

The meeting is used to review important policies of the owner, such as minority subcontracting objectives, restrictions on the hours of the day the contractor may work, critical dates, or any other matters of importance to any of the parties. The safety of the public and traffic control are always discussed.

Underground utilities and their location are always a major concern whenever excavation is required. The requirements to locate underground gas, water, sewer, communications cables, and any other underground utilities require considerable discussion. Some communities have very accurate information regarding underground facilities; others do not. The correctness of the information provided on the contract drawing should be verified to prevent change orders and claims.

The preconstruction conference is used to discuss the technical details of the project and address the contractor's questions. For example, some specifications call out a particular model of a specific manufacturer for a piece of installed equipment, but the specification will allow an "or equal" product. The contractor may need to know what products the designer will accept as "equal" substitutes. Questions that are not answered in the meeting must be answered in writing as soon as possible but not later than a specific agreed-on date.

The one topic that is probably the most important to all contractors is the processing of pay estimates. At the preconstruction conference, the monthly progress cutoff date and the date of submission for the pay estimates are selected, or the requirements already in the specifications are noted. The most important issue to the contractor is the number of days the owner will require before making payment to the general contractor, and the maximum number of days the general contractor has to pay the subcontractors. There are many efforts to shorten the time between completion of the work and receipt of payment. For example, the State of Arizona requires that owners pay the general contractor within 10 days of receipt of a monthly pay estimate, and that general contractors pay subcontractors within 10 days after receipt of their funds. If for any reason the general contractor cannot or will not pay the subcontractors, the general contractor is obliged to return the progress payment to the owner. The Arizona law applies to both public and private owners.

Subcontracts

A subcontractor is a construction firm that performs work in a specialty area such as electrical or mechanical construction. They contract directly with the general contractor and usually do not have any contractual relationship with the owner. The general contractor remains responsible to the owner for the quality of the subcontractor's performance; however, the general contractor must ensure the contract between the GC and subcontractors indicates that the GC is not responsible for the subcontractors' negligence. If approval of the subcontractor by the owner is required, the approval must be obtained before the subcontract is signed, even though disapproval of subcontractors is a rare event. Some projects, such as housing, may have several subcontractors on the site, while others, such as highways, have fewer subcontractors.

The practice of subcontracting enables contractors to maintain a stable workforce because the general contractors have the workforce of the subcontractors available when they are needed, without the requirement to maintain a huge workforce of their own at all times. Subcontractors are generally able to provide a better-trained and properly equipped workforce in their area of specialization.

Some subcontracts are small enough to allow the use of an informal contract, such as a letter of proposal and an acceptance, but it is preferable that the general contractor use a formal written document for all subcontracts. There are standard subcontract forms available from organizations such as the Associated General Contractors of America, the American Institute of Architects, or the

American Subcontractors Association (ASA), or contractors may develop their own forms. Preparation of a subcontract is an area where contractors should seek the advice and assistance of legal counsel. Use of a standard subcontract form helps the subcontractors because they will gain experience with the general contractor and know what contract terms to expect when they submit a subsequent subcontract proposal. In addition, change orders require that the subcontracts be changed to reflect the requirements of the change. Standard contract forms contain clear change order language that both the general contractor and the subcontractor understand.

Submittals

Submittals contain information concerning products to be incorporated in a construction project. The owner or designer must approve the submittals before products are ordered or used in the work. They consist of drawings, diagrams, catalog cuts, data sheets, and other data prepared by the contractor, subcontractor, manufacturer, or supplier to illustrate what items will be installed in the facility or how specific portions of the work will be performed. Submittals provide a method by which the owner verifies that the materials being ordered and subsequently installed meet the project specifications. They provide the design professional the opportunity to confirm or reject the proposed product, and to ensure that the proposed product will function to its intended use. Submittals serve to confirm that the proposed product complies with the intent of the contract documents. They provide data so that the designer can ensure the dimensional information is correct, and they develop a standard of quality assurance ensuring the proper methods and materials will be employed. As a general rule, everything that will be installed in the facility requires a submittal. Submittals are normally processed as shown in Fig. 8.6.

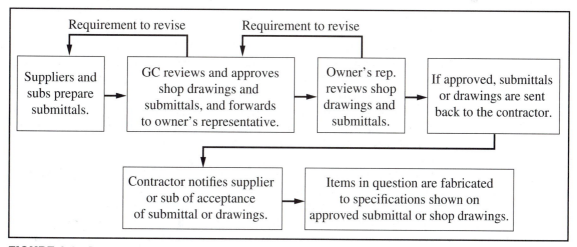

FIGURE 8.6 Processing submittals and shop drawings.

The required routing of formal communications such as submittals is very important because inefficient routing can cause delays. For instance, submittals must be sent up the chain from the materials supplier to the specialty contractor to the general contractor to the designer for review and approval. If the designer rejects the submittal and requires a resubmittal, the submittal must make its way back down the chain to the originator, and then back up the chain again through the general contractor to the designer. This can be a time-consuming process, and it has the potential to delay the overall completion of the project by delaying the material procurement process. The preconstruction conference is the opportunity for the contractors to work out an efficient use of technology to track the submittal process.

It is becoming common for the owner, designer, general contractor, and subcontractor to agree that the contractor originating the submittal may route it directly to the designer for review, with an information copy to the general contractor. The designer then returns an information copy to the general contractor, and the action copy directly to the subcontractor who initiated the submittal. This system has two advantages:

■ It is more efficient (saves time).
■ It enables the general contractor to become actively involved only when required.

There are some disadvantages to such a system. The most important disadvantage may be that the general contractor may not be informed of all possible concerns in a timely manner.

Request for Information

Requests for information (RFIs) are a specific type of written communication used to document questions stemming from the contract documents and to record the resulting answers. The subject matter addressed by RFIs in many cases relates to design intent, affects the project schedule and budget, and may or may not be urgent. RFIs serve to document the question and to require the other party to write out and document the answer. The party requesting the information should always ensure the designer issues the appropriate pricing documentation (change order) if the response to the RFI results in extra work or cost change. Most general contractors have a standard form they use for all RFIs (Fig. 8.7), and they require their subcontractors to use the same form. General contractors also maintain a log of RFIs, sorted by urgency, to track the RFIs and the responses.

The practice of requesting answers to questions has grown or evolved from the need to maintain written records in the field. It is not addressed in the contract documents. AIA Document A201 addresses communications in paragraph 4.2.4, but does not mention RFIs. The standard subcontractor agreement, A401, does not mention communications between the general contractor and subcontractors.

FIGURE 8.7 Example of one contractor's RFI form.

Pay Estimates

Contractors are normally paid monthly; however, it is necessary to read the contract to understand the details. One of the most important aspects of any owner's reputation with contractors is the time required to process a payment after submission of the monthly pay estimate by the contractor. Slow payment or **delayed payment** leads to higher bids. Fast payment leads to lower bids.

The actual process of submitting the monthly pay estimate involves estimating the percentage of completion of each job activity, calculating the amount due for each of those activities, and subtracting the retainage from the total. A portion of the monthly pay estimate belongs to the general contractor and a portion belongs to the subcontractors. Therefore, the obvious second question relates to the time taken by the general contractor to pay the subcontractors after receipt of payment from the owner. The same formula applies—general contractors who

delayed payment
Sometimes, because of credit or cash flow problems, owners fail to pay contractors for services according to the schedule set by the contract. To justify it, they may claim the contractor's work is defective or substandard.

pay their subcontractors quickly receive lower bids than those who hold subcontractors' money.

The payment clauses in subcontracts can be different from the payment clauses in general contracts. In subcontracts, payment clauses fall into one of two categories, either "pay when paid" or "pay if paid." Subcontractors prefer a "pay when paid" clause because such a contract clause creates a contractual obligation for the general contractor to pay the subcontractor. "Pay if paid" clauses do not. Subcontractors should read the payment provisions carefully.

Change Orders

Change orders are part of every construction contract because each project is a new prototype. No two are exactly alike, and owners do not fund the design process at a high enough level to produce a "perfect" design. Even houses in a large development, that are very similar, have differences caused by the terrain, existing utilities, or other factors such as subsurface conditions. Change orders are accepted as a normal part of the contract administration process.

Change orders do not change the scope of work; in fact, the requirements of the change order must be within the original scope of work. Changes that are outside the scope of work require a supplemental agreement, and are legally quite different from change orders. Change orders change the details or conditions of the work, and they are used to add "extra" work or to delete work. The contract does not allow the contractor the option of refusing change order work (they are not change *options*); but it does entitle the contractor to additional time and compensation for work that significantly modifies the original requirement and involves changes to the existing work, as well as additional work. Change orders must be written and must be issued by a party with the proper authority to issue the change; however, the parties involved can waive that requirement by their actions. Owners normally provide a letter to the contractor to explain who has authority to issue and approve changes, and the exact monetary limits on each person's authority. The designer, for example, usually has no authority to obligate the owner, and cannot approve change orders. Many construction companies use field orders for minor no-cost changes that have no impact on the contract provisions. Field orders change the work but do not modify the contract; they simply provide a record of the change.

There is an important difference between *extra* work and *additional* work. Extra work is added by change order, and was not included in the original requirements of the contract, but was within the original scope of work. A simple example of *extra* work would be the addition of a new door opening that is not in the original plans, but is added by change order. The contractor is entitled to additional compensation. An example of *additional* work could occur in a project to add a fire sprinkler system to an existing building. If the contractor had to remove a ceiling to gain access to the work area, and then restore it to its original condition, that is additional work that should be included in the original estimate of the work. There would be no additional compensation for this work.

Change orders are subject to negotiation, and the contractor is rightfully interested in gaining as much compensation as possible for the extra work. Change order work is more expensive than the original contract work because it tends to be out-of-sequence work and there is additional overhead to track and implement the change. Most owners feel that a contractor is entitled to fair compensation for the change order work. The contractor's problem is to convince the owner of what is fair and reasonable. Some contracts are written with prenegotiated markup and profit for change order work in place as part of the original contract. Large change orders on most government projects are subject to audit. The contractor agrees to open the company's books to the auditors to determine the actual costs associated with the change order work.

Change orders can be very simple or they can become very complicated. A change order to add a new door opening is very simple. A change order to protect a project from the effects of the weather can be very difficult.

Claims

Claims are the unfortunate result of change orders that cannot be resolved or job conditions that the contractor considers to be a materially changed condition but the owner does not agree. Often claims result from a disagreement on the meaning of the terms of the contract. Example causes of claims include late progress payments, delays caused by the owner, or rejection by the designer of "or equal" substitutes. Claims often result from poor communications and a claim will be compounded by poor communication.

Contracts require that a contractor notify the owner of claims within a specified number of days and then proceed with the work. The contractor is required to rely on the claims process for fair recovery. The contractor cannot refuse to proceed with work that is within the original scope of work, but it is sometimes difficult to determine whether work is within the original scope. In the case of federal contracts, the contractor is required to file a claim with the contracting officer, and then to pursue the claim in the judicial system if the contracting officer's decision is not favorable. Contractors must read the contract to determine what process must be followed to file a claim. In all cases, public or private, the administrative remedies specified must be exhausted before any judicial remedy can be sought.

Successful settlement of a claim depends on adequate documentation. Contractors have learned to maintain high-quality daily records of the events in the field. Good records include meeting minutes, records of telephone conversations, project photographs taken regularly, and other records that would recreate an accurate picture of the events that caused the claim.

Dispute Resolution

The predominant method of dispute resolution in construction is litigation, even though everyone in the industry recognizes that resorting to the courts is not the best way to resolve disputes. Just the opposite is true. There has been a trend over the last decade to find ways to resolve disputes without litigation. The results are mixed, with some methods more successful than others. Disputes are unresolved claims, unresolved disagreements, or unresolved change orders that somehow resulted in a court case. They always involve money. The key word is *unresolved*. Disputes are not caused by illegal actions. They are caused by unresolved disagreements about compensation for what the contractor sees as just cause for payment, and what the owner sees as an unnecessary expense that should have been included in the original contract price.

What should a contractor do when a situation appears to be developing into a dispute? The first response to that question is to keep detailed and accurate records. A more appropriate response is to maintain or improve communications between the contractor and the owner. Construction is, by the nature of the contract, an "arm's length" business characterized by adversarial relationships and broken communications. Contractors need to communicate. When there are problems, they need to be communicated immediately and clearly. Bad news will not improve with time. Open and honest business practices that enable good communications with the owner may help avert a situation that could otherwise lead to an unresolved dispute. In a potential dispute situation, the best advice that can be given to the contractor is to resolve the dispute as rapidly as possible, and let it disappear into the history of the project. Project management by an owner based on a philosophy that says "do it my way and sue me when the project is over if you don't like it" is not good business, but it can happen.

Contractors are trying harder to resolve disputes when they occur. They are trying to resolve them in the field and not in the courts. Many alternative dispute resolution methods are being tried, all with the objective of resolving disputes before they advance to the litigation stage.

Alternative Dispute Resolution

Common alternative dispute resolution methods are

- Partnering
- Mediation
- Arbitration
- Mini trials
- Project neutral

Partnering has gained favor as a philosophy of construction management based on trust. The basis of partnering is trust. Partnering meetings are held before the project starts in an effort to develop a level of trust between the owner and contractors. Common goals and objectives are identified. A charter, in which all parties resolve to operate in an atmosphere of trust, is written and signed. A dispute resolution framework is put in place to require problems to be solved in the field within a very limited number of days, maybe only 3 days, or elevated to the next level of management. Monthly partnering meetings are held to evaluate the partnering process. All meetings are chaired by an outside facilitator to create a "level playing field" for the contract partners. The move toward partnering appears to be very successful.

Mediation is a system of selecting a neutral third party to assist the parties to reach a mutually acceptable agreement. The mediator serves as a "go-between" to improve communications and does not impose a settlement on the parties. There is nothing binding about a mediation process. It is undertaken by mutual agreement of the parties involved, in an attempt to resolve the disagreement before it becomes a dispute. Often it is the last step before litigation and represents the efforts of both parties to avoid litigation.

Arbitration is slightly closer to litigation in philosophy than mediation because the arbitrator imposes a solution on the parties to the dispute, and the parties agree to accept the arbitrator's solution. Arbitration is not court and is not binding except by prior agreement between the parties to accept the findings of the arbitrator or when court directed. The arbitrator is selected by mutual agreement. The parties to the dispute can make the arbitration process binding by mutual agreement because both sides want to avoid the expense of a lawsuit. There are no arbitration records as there would be court transcripts, so the parties maintain a higher degree of privacy. The process is less costly than a lawsuit. The arbitrators are usually experts in the kind of construction issue in dispute, providing a more informed decision than might be obtained in court where the judge cannot possibly be an expert in every area of litigation presented. Most judges are not as knowledgeable about construction matters as arbitrators.

Mini trials allow each side to present its position in the dispute, and then a judge renders a decision. They are judicial proceedings. They have not proven to be very successful because the opposing sides have a tendency to escalate the proceedings into a normal hearing and leave the "mini" description behind. The level of effort that goes into them tends to grow, and as a result they fail to accomplish their intended purpose.

Project neutral is a new approach to dispute resolution that is being used on very large, highly technical projects. The project neutral is a person (or team) that is hired before the construction contract is awarded, and that participates in the project throughout the construction period. The project neutral is technically knowledgeable and extremely well informed in matters of construction and the contract requirements. As the name implies, the project neutral does not represent the position of any of the project parties and does not advocate positions taken by any party. The project neutral is analogous to an official in an athletic contest providing expert on-the-spot decisions when they are needed. Typically, the project neutral visits the project often enough to remain well informed of the status of work, and of the beginnings of any disputes or potential claims. The function of the project neutral is to reach a solution to disputes. The finality of the decisions made by the project neutral depends on the provisions of the contract.

Liquidated Damages and Substantial Completion

Liquidated damages are paid by the contractor to the owner if such amounts are required by the contract, and if the contractor exceeds the allowed contract time period. A contract may require liquidated damages of a few hundred dollars per day, or it may require liquidated damages of many thousands of dollars per day. The amount is determined before the contract is advertised, and it is based on an estimate of actual costs the owner or the public would incur by the late completion of the work.

To avoid the imposition of liquidated damages, contractors must actively manage the project schedule. The project manager must have a realistic schedule to complete the work. In addition, change order work must always include additional time for the added work as part of the negotiation process.

Most contracts require liquidated damages for failure to reach "substantial completion" by the specified time. Contractors must know the requirement of each contract and the definition of substantial completion. The general rule (not a legal rule) is that the work must be done. The facility must be ready for the owner to move into and use. There can still be a "punch list" of items the contractor needs to correct to reach final completion, and those items can be completed while the owner is occupying the facility. Punch list items can be anything from touching up paint to completing the landscaping. Punch list items cannot be substantial items of work such as installing the windows and doors or completing the air-conditioning system. If those items remain, the facility may not be "substantially complete."

Final Inspection

When the contractor feels the facility is substantially complete and the owner agrees that the facility is ready for acceptance, as defined in the contract documents, the "final inspection" will be scheduled. Contractors should treat the final inspection as a formal contract requirement. It is the owner's last formal opportunity to walk through the facility to identify remaining work items and to formally accept ownership of the facility. The final inspection is used to generate the final punch list. Owners should prepare for the final inspection by working proactively with the contractor as the work progresses, and staying knowledgeable of the status of the facility. This should be a "one bite out of the apple" event, meaning the final inspection is the one last opportunity for the owner to identify remaining work items. Those items are written onto the punch list. The owner and the contractor agree on the punch list items. At that point, the contractor should be able to proceed with the knowledge that the punch list items are, in fact, the last remaining items of work on the facility. When the punch list items are done, the contract requirements are fulfilled. The contractor should be required to return to the site only for possible corrective work covered in the warranty.

SUMMARY

A construction contract is an agreement between two parties that is enforceable by law. It requires a "meeting of the minds" and there must be a service and consideration. The purpose of the contract should be to produce quality construction safely, on time, and within budget. The essential contract documents are the Agreement, the General Conditions, the Special Conditions, and the Drawings. There are many organizations such as the AIA and government agencies that produce standard contract documents.

Unit-price contracts are commonly used on heavy/highway projects. Lump sum contracts are commonly used for buildings. Cost plus contracts are used to share the risk of unknown factors between the owner and the contractor. Incentive contracts such as A + B contracts provide a mechanism for the owner to obtain and pay for the benefit of the contractor's knowledge.

The employment of labor and the procurement of materials and subcontracts require that contractors be aware of and comply with multiple federal, state, and local laws. Government projects are advertised, the bids are opened publicly, and award is made to the low responsive, responsible bidder. Private projects are usually negotiated. They do not need to be advertised. Government contracts require bonds because contractors cannot place a lien on government property. Bid bonds guarantee that the contractor will enter into the construction contract and provide the required payment and performance bonds. Performance bonds guarantee that the contractor will perform the work in accordance with the contract terms. Payment bonds guarantee that subcontractors and material suppliers will receive payment.

Bids must be submitted on time, at the location specified, on the correct forms, with receipt of all addenda acknowledged, and with a valid bid bond. Nonresponsive bids are returned unopened. The low bidder is required to provide evidence of insurance before the contract is awarded. General contractors require subcontractors to submit evidence of insurance before their construction operations begin. Payment and performance bonds and insurance are ways for the owner to manage risk.

The preconstruction conference is a formal meeting of the owner, the general contractor, and key subcontractors to review important policies of the owner and address the questions of the contractors. Submittals contain information concerning products to be incorporated in a construction project that must be approved by the owner or designer before they are used. They allow the designer the opportunity to confirm or reject the proposed product, and ensure that the proposed product will function to its designed use and the owner's intended use. Contractors are normally paid monthly. The speed with which owners pay contractors has a great impact on the price contractors are willing to bid.

Change orders change the details or conditions of work. They are used to add "extra" work or to delete work. Change orders must be written. The contractor is obligated to perform the work but has the right to negotiate a fair price and time extension.

Claims are the unfortunate result of change orders that cannot be resolved, or changed job conditions. Claims can lead to judicial action, resulting from the inability of the contractor and owner to reach an agreement. Alternative dispute methods include partnering, mediation, arbitration, mini trials, or a project neutral. When there is the possibility that a situation will develop into a dispute, a contractor's first response should be to keep detailed and accurate records. A very appropriate response that can often solve the problem is to improve communications with the owner.

Liquidated damages are charged to the contractor if the contract period exceeds the allowed contract time. Liquidated damages are not a penalty; they compensate the owner for costs incurred after the work should be complete.

Worker's compensation is a very expensive required employee benefit that protects workers who are injured on the job by insuring medical and hospital expenses and a percentage (usually 75%) of their hourly wage. Commercial general liability insurance has four coverage parts: (1) premises/operations liability, (2) completed operations and product liability, (3) contractor's protective insurance, and (4) contractual liability. Builder's risk insurance protects against loss to the project while it is under construction. Contractors are normally required to provide certificates of insurance as proof that they have purchased the insurance that is required by the contract. Subrogation is the right of the insurance company to recover damage payments from a negligent third party.

REVIEW QUESTIONS

8.1 What are the essential elements of a construction contract? What are the four essential documents?

8.2 Describe the purpose of the Agreement.

8.3 What are the necessary elements of the general conditions? Why are general conditions used in their entirety, without any deletions or additions?

8.4 Explain why the supplementary or special conditions take precedence over the general provisions.

8.5 Explain the difference between performance, design, open, and proprietary specifications. Provide an example of an "or equal" specification.

8.6 What is the contractual relationship between the GC and the A/E? Between the owner and the subcontractors? Between the contractor and the subcontractors? What role does a CM play?

8.7 What are the major differences between lump sum and unit-price contracts? When is it appropriate to use each? When is it appropriate to use a cost plus percentage or fixed fee contract?

8.8 Owners have a preference for design-build contracts. Explain the basis for that preference.

8.9 What is the purpose of the Miller Act of 1935?

8.10 Name and define the purpose of the three types of bonds. Who do bonds protect? The principal and surety are said to make a joint and several guarantee to the owner. Define the meaning of "joint and several."

8.11 Should subcontractors be required to provide bonds?

8.12 What does worker's compensation insurance cover? Who is protected? Conduct an Internet search to determine the specific benefits in different states across the nation for commercial construction.

8.13 What protection is offered by contractor's protective liability insurance? Contractual liability insurance?

8.14 Explain subrogation.

8.15 Is it necessary for construction contracts to be in writing? Explain why or why not.

8.16 Make a list of topics that would be discussed at a typical preconstruction conference. Use a highway project as an example. Use a project found on tsn.wes.army.mil as an example.

8.17 Why do contractors use subcontractors? Does the subcontracting process make construction more or less expensive? Does the practice improve or hurt overall project quality?

8.18 Explain the purpose of shop drawings. Use a mechanical system component such as a thermostat to illustrate.

8.19 What is the purpose of requests for information? Why has the process developed into a formal system? Why do contractors use standard forms?

8.20 What is retainage? Do you agree or disagree with the practice of retainage?

8.21 Contractors submit pay estimates monthly. Some contractors "front load" their bids, and hence their pay estimates. What is front loading? Is it an ethical or an unethical practice?

8.22 Why do construction projects have change orders?

8.23 What causes claims? What must contractors do to support a claim?

8.24 Discuss the current trends in dispute resolution. Why is there growing interest in dispute resolution systems?

8.25 Explain the concept of liquidated damages. Why are they not categorized as a "penalty"?

8.26 Which contractor in the following A + B contract is the low bidder? The daily estimated cost is $24,000.

Contractor A: $2,000,000 and 340 days

Contractor B: $2,340,000 and 317 days

Contractor C: $1,990,600 and 352 days

REFERENCES

1. Bockrath, Joseph T. (1995). *Contracts and the Legal Environment for Engineers and Architects,* 5th ed. McGraw-Hill, New York.

2. Clough, Richard H., and Glenn A. Sears (1994). *Construction Contracting,* McGraw-Hill, New York.

3. Collier, Keith (2001). *Construction Contracts,* 3rd ed. Prentice Hall, Upper Saddle River, NJ.

4. Collier, Keith (1994). *Managing Construction—The Contractual Viewpoint,* Delmar Publishers, Inc., Albany, NY.

5. Coombs, William E., and William J. Palmer (1989). *Construction Accounting and Financial Management,* McGraw-Hill, New York.

6. Davis, Steven D., and Ron Prichard (2000). *Risk Management and Insurance Bonding for the Construction Industry,* The Associated General Contractors of America, 333 John Carlyle Street, Suite 200, Alexandria, VA.

7. *Guide to Construction Insurance,* 2nd ed. (1992). Publication #1131, The Associated General Contractors of America, 333 John Carlyle Street, Suite 200, Alexandria, VA.

8. Haltenhoff, C. Edward (1999). *The CM Contracting System: Fundamentals and Practice,* 3rd ed. Prentice Hall, Upper Saddle River, NJ.

9. Hinze, Jimmie (2001). *Construction Contracts,* 2nd ed. McGraw-Hill, New York.

10. Onnen, Jim (2002). "Kentucky's I-275 Reconstruction Project: Guaranteed to Last," *Better Roads,* October.

11. Palmer, William J., James M. Maloney, and John L. Heffron, III (1996). *Construction Insurance, Bonding, and Risk Management,* McGraw-Hill, New York.

12. Russell, Jeffrey S. (2001). *Surety Bonds for Construction Contracts,* ASCE Press, Reston, VA.

13. *Owner Controlled Insurance Programs,* NCHRP Synthesis 308 (2002). Transportation Research Board of the National Academies, Washington, DC.

14. *Quality in the Constructed Project: A Guide for Owners, Designers, and Constructors,* 2d ed. (2000). ASCE, Reston, VA.

15. Welch, John W., James Morelewicz, Andrew J. Ruck, and Stephen J. Trecker (1992). *Contract Surety,* Vols. I and II, Insurance Institute of America, Malvern, PA.

16. *ENR,* June 4, 2007, p. 12.

WEBSITE RESOURCES

1. www.fmozeleski.com   This is an example of a commercial Internet site. The Mozeleski Group, Inc., offers construction management support, and their site offers a comprehensive guide to construction contract management.

2. www.cmaanet.org   Homepage of the Construction Management Association of America. Supports construction managers by providing a directory and certification information. Also offers standard contracts and publications.

3. www.coaa.org   Provides information for construction owners who build as a secondary objective of their business.

4. www.formsresource.com   Sells forms for construction contractors, insurance claims, and real estate.

5. www.cityofla.org/BCA   Los Angeles Bureau of Contract Administration. The Bureau of Contract Administration is part of the Department of Public Works and protects the city's interests in construction matters. Many cities have similar sites.

6. www.acq.osd.mil/ar/bestprac   Best Practices for Contract Administration. Office of Federal Procurement Policy (OFPP), October 1994. This guide was edited April 15, 1998, to reflect the practices of the Department of Defense, especially with regard to audit practice.

7. www.surety.org   Provides information relating to insurance industry issues such as fidelity, surety, statistical services, and others.

8. www.nasbp.org   Home page of the National Association of Surety Bond Producers.

9. tsn.wes.army.mil   Tri-Service Solicitation Network, which provides links to government agencies advertising solicitations through the Internet.

Construction Accounting

Two important accounting documents that are used to monitor a company's performance are the balance sheet and the income statement. The balance sheet can be likened to a snapshot of financial health at a particular moment, and the income statement is like a moving picture over a period that at the end of the period becomes the snapshot. The analysis of a balance sheet can identify potential liquidity and leverage challenges. A company's income statement shows all of the money a company earned (revenues) and all of the money a company spent (costs and expenses) during a specific period, usually the fiscal year. The income statement provides a map of how a company is changing and the pace of that change. It is essential for management to know how to analyze different elements of these important documents. These documents serve as the basis for important decisions affecting company capability and for decisions made by outside parties such as banks and sureties.

accounting
The practice of recording, classifying, reporting, and interpreting the financial data of an organization.

cost accounting
The phase of accounting that deals with collecting and controlling the cost of producing a product or service.

INTRODUCTION

There are three types of **accounting**: (1) generally accepted accounting principles (GAAP accounting) for financial statements (lots and lots of rules), (2) management accounting (no rules) for use in running the firm, and (3) tax accounting. This chapter is going to discuss only the first two and leave the tax aspect to those accounting professionals who are constantly trying to understand all the tax laws that politicians can dream up and constantly change.

Accounting is the process of recording, classifying, reporting, and interpreting the financial data of an organization. One of the main purposes of accounting is to aid decision makers in choosing among alternative courses of action—management accounting. Effective management of a construction company requires proper **cost accounting**, and the availability of current and accurate cost information. For information to be useful, there must be consistent use of accounting principles from one period to another (e.g., a method of recording and reporting information). Accounting documentation that has a consistent base provides information needed for making business decisions that enable management to guide a company on a profitable course.

Financial statements are the primary means of communicating useful financial information and are the result of simplifying, condensing, and aggregating transactions. No one financial statement provides sufficient information by itself and no one item or part of each statement can summarize the information. The most important accounting document used to monitor a company's performance (earnings) is the **income statement**. The income statement shows the sources of a company's production costs, other expenses, and income. Another important document is the **balance sheet**; it shows the company's overall financial health and potential for future growth.

INCOME STATEMENT

A company's income statement is a record of its earnings or losses for a given period. This can be thought of as a moving picture of company performance. It shows all of the **revenue** a company earned and all of its expenditures (costs and expenses) during a specific period, usually the fiscal year. It accounts for the effects of some basic accounting principles such as depreciation.

The income statement provides a record of a company's performance across time. Most important, the income statement tells if the business is profitable. It provides a map (see Fig. 9.1) of how a company is changing and the pace of that change. It is essential for management to know how to analyze different elements of this important document.

Operating income (revenue) does not include interest earned, nor does it include income generated outside the normal activities of the company, such as income on investments. Operating income is particularly important because it is a measure of profitability based on a company's operations. In other words, it assesses whether or not the core operations of a company are profitable. It ignores income or losses outside a company's normal domain. Earnings before taxes are the sum of operating and nonoperating income. Net earnings or net income (loss) after taxes is the proverbial bottom line. It is the amount of profit a company makes after all of its income and all of its expenses.

Retained Earnings

Retained earnings are the amount of equity from profits that a company has accumulated through all of its transactions since the beginning of the firm. If the company continually makes substantial profits, it is a stable company, or if it is like the Low Bid Corporation (see Fig. 9.1, "Income (loss) from operations"), there are major problems and the viability of the company is threatened. In addition to gross profits that are minimal, it appears that Low Bid had a problem with growth of its general and administrative expenses for the year 2001. These could include an unnecessary fancy office, a company airplane, or even unnecessary estimating cost caused by attempting to bid everything that comes along.

income statement
A financial statement showing revenues earned by a business, the expenses incurred in earning the revenues, and the resulting net income or net loss.

balance sheet
A financial report showing the assets, liabilities, and owner's equity of a company on a specific date.

revenue
An inflow of assets, not necessarily cash, in exchange of goods and services sold.

LOW BID CONSTRUCTION CORPORATION
STATEMENT OF INCOME AND RETAINED EARNINGS

	For the Years Ended December 31,		
	2001	**2000**	**1999**
Revenue (Notes 10 and 16)	$174,063,148	$163,573,258	$210,002,272
Cost of revenue (Note 10)	169,404,787	158,935,103	200,070,826
Gross profit	4,658,361	4,638,155	9,931,446
General and administrative expenses (Notes 1 and 10)	8,062,623	6,777,840	6,671,035
Income (loss) from operations	(3,404,262)	(2,139,685)	3,260,411
Other income (expense):			
Interest income	319,797	646,480	668,928
Interest expense	(485,937)	(250,996)	(209,872)
Other income (expense)	377,840	(115,246)	211,119
	211,700	280,238	670,175
Income (loss) before income taxes	(3,192,562)	(1,859,447)	3,930,586
Income tax benefit (expense)	668,631	284,861	(1,590,480)
Net income (loss) (Note 17)	$ (2,523,931)	$ (1,574,586)	$ 2,340,106
BEGINNING RETAINED EARNINGS	$3,090,214	$4,664,800	$ 4,664,800
ENDING RETAINED EARNINGS	$566,283	$3,090,214	

FIGURE 9.1   Sample income statement.

Notes

See Fig. 9.1.

10. Related Party Transactions:

Management believes that the fair value of the following transactions reflect current amounts that the Company could have consummated transactions with other third parties.

> *Revenue:*
> *During the years ended December 31, 2001 and 2000, the Company provided construction materials to various related parties in the amounts of $108,112 and $26,556, respectively. Included in accounts receivable at December 31, 2001 and 2000 are amounts due from related parties in the amounts of $27,337 and $15,132, respectively.*
>
> *Professional Services:*
> *During the years ended December 31, 2001, 2000, and 1999, a related party rendered professional services to the Company in the amounts of $14,573, $23,342, and $7,944, respectively. During the years ended December 31, 2001,*

2000, and 1999, the Company paid $30,000, $30,000, and $5,000, respectively, to outside members of the board of directors.

Subcontractor/Supplier:

Various related parties provided materials and equipment used in the Company's construction business during the years ended December 31, 2001, 2000, and 1999, in the amounts of $4,114,319, $535,694, and $65,441, respectively. Included in accounts payable at December 31, 2001 and 2000, are amounts due to related parties, in the amounts of $1,046,908 and $154,861, respectively, related to supplies.

Royalties:

During the years ended December 31, 2001, 2000, and 1999, the Company paid a related party mining royalties in the amounts of $390,144, $328,310, and $182,061, respectively. Included in accrued liabilities December 31, 2001 and 2000, are amounts due to related parties, in the amounts of $30,464 and $49,983, respectively, related to royalties.

Commitments:

The Company leased office space in the state on a month-to-month basis, at a rental rate of $840 per month, from a related party of the Company. The lease terms also required the Company to pay common maintenance, taxes, insurance, and other costs. Rental expense under the lease for the years ended December 31, 2001, 2000, and 1999, amount to $9,240, $10,080, and $10,080, respectively.

16. Significant Customers:

For the years ended December 31, 2001, 2000, and 1999, the Company recognized a significant portion of its revenue from four Customers (shown as an approximate percentage of total revenue):

For the years ended December 31,		
	2001	**2000**	**1999**
A	21.9%	17.5%	26.2%
B	12.5%	16.3%	28.7%
C	9.6%	23.0%	17.2%
D	14.7%	6.1%	5.8%

At December 31, 2001 and 2000, amounts due from the aforementioned Customers included in restricted cash and accounts receivables are as follows:

For the years ended December 31,	
	2001	**2000**
A	$3,809,567	$2,968,786
B	$6,255,403	$1,855,666
C	$350,962	$1,124,196
D	$2,218,976	$762,181

17. Stock Option Plan:

In November 1994, the Company adopted a Stock Option Plan providing for the granting of both qualified incentive stock options and nonqualified stock options. The Company has reserved 1,200,000 shares of its common stock for issuance under the plan. Granting of the options is at the discretion of the board of directors and may be awarded to employees and consultants. Consultants may receive only nonqualified stock options. The maximum term of the stock options is 10 years and may be exercised as follows: 33.3% after one year of continuous service, 66.6% after two years of continuous service, and 100% after three years of continuous service. The exercise price of each option is equal to the market price of the Company's common stock on the date of grant.

BALANCE SHEET

A company's balance sheet (see Figs. 9.2 and 9.3) describes its financial position at a specific date, a single point in time. This is like a snapshot picture of the company's financial condition. Because it is a single-point-in-time representation, it is sometimes referred to as a position statement. It is prepared at least once per year, but it also may be presented quarterly, semiannually, or even monthly. The balance sheet provides information on what the company owns (its **assets**), what it owes (its **liabilities**), and the value of the business to its owners (**equity**). The name, balance sheet, is derived from the fact that these accounts must always be in balance. Assets must always equal the sum of liabilities and **owner's equity**.

assets
A property or economic resource owned by a company.

liability
A debt owed.

equity
A right, claim, or interest in property.

owner's equity
The equity of the owner or owners of a business in the assets of that business.

Assets

Assets are economic resources that are expected to produce economic benefits for their owners. The company's assets are presented first on a balance sheet (Fig. 9.2). The assets of a construction company are usually money (cash), receivables, equipment and plant, property, and materials used to construct projects.

Compared to other types of businesses, construction companies typically have less in the way of physical assets. A manufacturing business will have a large investment in buildings and machinery, and may even at times carry significant inventory of raw material with which to produce the product or an inventory of completed products that have not been sold. These are physical assets that a lender can rely on as collateral against a loan to the company.

By comparison, a construction company conducts its business on property owned by someone else and only needs to actually possess a small office for corporate management. A large heavy construction company might have a fleet of equipment, which represents substantial asset holdings, but many times such equipment is leased or rented so it does not belong to the construction company. The one exception would be a construction company that produces concrete and asphalt. Many of these companies own their own quarries from which they get

CONSOLIDATED BALANCE SHEET
(Thousands of dollars)

December 31,	2001	2000
ASSETS		
Current assets		
Cash	$ 11,330	$ 10,025
Short-term investments, at cost, which approximates market	8,620	14,738
Accounts receivable including retentions $44,943 and $65,064	208,007	182,319
Refundable federal income taxes	—	11,225
Costs and earnings in excess of billings on uncompleted contracts	192,727	164,232
Equity in joint ventures	96,655	82,311
Real property held for development and sale at the lower of cost or market	45,163	33,000
Other	11,462	7,790
Total current assets	573,964	505,640
Investments and other assets		
Equity in affiliated enterprises	9,815	16,349
Marketable securities, at cost, market $9,667 and $10,741	259	259
Other investments, at cost, which approximates market	41,318	15,896
Notes receivable and other assets	18,024	8,959
Total investments and other assets	69,416	41,463
Property and equipment, at cost		
Land	2,596	2,566
Buildings, ways, and wharves	164,322	148,900
Construction and other equipment	223,144	212,186
Total property and equipment	390,062	363,652
Less accumulated depreciation	(140,169)	(123,702)
Property and equipment—net	249,893	239,950
Total assets	$893,273	$787,053

The accompanying notes are an integral part of the financial statements.

FIGURE 9.2 Sample balance sheet, assets part.

the raw aggregate for their product. Quarries with their aggregate reserves represent significant asset holdings. Therefore, because the value of actual physical assets possessed by a construction company is usually minimal, lenders must carefully look to the continuing operational performance of the company when evaluating the risk of a loan.

December 31,	2001	2000
LIABILITIES AND STOCKHOLDERS' EQUITY		
Current liabilities		
Short-term and current portion of long-term debt	$ 72,742	$ 29,350
Accounts payable and accrued expenses	228,115	200,761
Billings in excess of costs and earnings on uncompleted contracts	66,982	70,846
Advances from clients	19,687	32,076
Income taxes		
Currently payable	11,462	3,476
Deferred	6,047	46,866
Dividends payable	3,236	2,672
Total current liabilities	408,271	386,047
Non-current liabilities		
Deferred income taxes	43,666	20,611
Deferred income and compensation	15,513	15,602
Accrued workmen's compensation	27,973	23,189
Long-term debt	103,611	76,063
Total non-current liabilities	190,763	135,465
Commitments		
Stockholders' equity		
Common stock, par value $3.33⅓, authorized 20,000,000 shares,		
issued 10,616,391 shares	35,388	35,388
Capital in excess of par value	74,056	73,273
Retained earnings	196,764	170,421
	306,208	279,082
Less cost of treasury stock, 802,878 and 899,744 shares	(11,969)	(13,541)
Total stockholders' equity	294,239	265,541
Total liabilities and stockholders' equity	$893,273	$787,053

FIGURE 9.3 Sample balance sheet, liabilities and stockholders' equity part.

> In reading a balance sheet, it is important to remember that property and equipment are recorded at cost and not the price at which an asset may be sold or the cost to replace the asset.

Current Assets

Current assets are those assets that turn themselves into cash within 1 year.

- *Cash* is the most basic current asset.

- *Cash equivalents* are not cash but can be converted into cash so easily that they are considered equal to cash. Cash equivalents are generally highly liquid, short-term investments such as U.S. government securities, commercial paper, and money market funds. Short term is generally defined as converting or maturing into cash within 90 days if it is to be considered cash.

- *Accounts receivable* represent money that clients (project owners) owe to the firm for services rendered or for goods sold (e.g., producers of asphalt and concrete) and include retentions receivable.

- *Inventory,* in the case of a construction firm, is the material that is kept on hand for use in constructing projects.

Long-Term Assets

Many assets are economic resources that cannot be readily converted into cash. These long-term, or "other," assets are expected to provide benefit for future operations.

- *Fixed assets* are tangible assets. Generally, fixed assets refer to items such as equipment, vehicles, plant, buildings, and property. On the balance sheet, these are valued at their cost. Depreciation is subtracted from all classes except land.

Liabilities

Liabilities are obligations a company owes to outside parties. They represent rights of others to money or services of the company. Examples include bank loans, debts to suppliers, debts to employees, debts to subcontractors, and both deferred taxes and taxes currently payable to government entities. On the balance sheet, liabilities are generally broken down into current liabilities and long-term liabilities.

- *Current liabilities* are those obligations that are to be paid within the year, such as accounts payable, interest on long-term debts, and taxes payable. The most pervasive item in the current liability section of the balance sheet is accounts payable. Note that any "**overbilling**" is recognized as a current liability (see Fig. 9.3 "Billings in excess of costs and earnings on uncompleted contracts").

- *Accounts payable* are debts owed to subcontractors and to suppliers for the purchase of goods and services on an open account.

> Because current liabilities are usually paid with current assets, it is important to examine the degree to which current assets exceed current liabilities. This difference is called working capital.

overbilling
Placing a higher value on work performed early in the project and lower values on work to be completed near the end of the project so that the contractor does not experience an out of pocket cash flow—have to finance the project.

EXAMPLE 9.1

Review the balance sheet asset and liability information contained in Figs. 9.2 and 9.3. Does this company appear to be in a good position in regard to meeting current obligations?

	Dec. 31, 2001	Dec. 31, 2000
Total current assets	$573,964	$505,640
Total current liabilities	408,271	386,047
Working capital	165,693	119,593

The company's current assets are greater than its current liabilities. It should be able to meet current obligations.

■ *Long-term debt* is a liability of a period greater than 1 year. It usually refers to loans made to the company. These debts are often paid in installments. These could include loans to purchase vehicles and equipment and mortgage payments for buildings and land. If this is the case, the portion of the principal to be paid in the current year is considered a current liability.

Owner's Equity

Owner's equity or the net worth of the company is the excess of assets over liabilities. It represents what the owners have invested in the business plus accumulated profits, which are retained in the company, from ongoing operations. Losses will reduce owner's equity. The formulation of the report's balance is shown by Eq. [9.1]:

$$\text{Equity (net worth)} = \text{Total assets} - \text{Total liabilities} \qquad \textbf{[9.1]}$$

Considering the balance sheet asset and liability information contained in the Figs. 9.2 and 9.3, it is possible to calculate the net worth of the company without knowing exactly the types of equity the company owners possess (see Table 9.1).

At the bottom of the balance sheet in the "stockholders' equity" section (Fig. 9.3), the composition of the equity is presented. Retained earnings is that part of equity resulting from earnings in excess of losses.

TABLE 9.1 Net worth calculations.

	Dec. 31, 2001	Dec. 31, 2000
Total assets	$893,273	$787,053
Total current liabilities	408,271	386,047
Total noncurrent liabilities	190,763	135,465
Total liabilities	599,034	521,512
Net worth	294,239	265,541

Liquidity and Working Capital

Management must study the balance sheet to measure a firm's liquidity, financial flexibility, profit generation ability, and debt payment ability. Financial flexibility is the ability to take effective actions to alter the amounts and timing of cash flow in response to unexpected needs and opportunities.

Bonding companies will carefully study the company's balance sheet before bonding a project. The picture that the balance sheet reveals concerning the company's stability will, therefore, affect the surety's decision concerning the bonding limit (volume of work) of the construction company. Without the ability to secure a bond, the amount of work that the company can undertake is very limited.

The analysis of a balance sheet can identify potential debt payment problems. These may signify the company's inability to meet financial obligations. Liquidity is the quality that an asset has of being readily convertible into cash, and cash is necessary to keep the business operating—to pay the bills. Working capital (Eq. [9.2]) is simply the amount that current assets exceed current liabilities (see Example 9.1). The relative amount of working capital is an indication of short-term financial strength.

$$\text{Working capital} = \text{Current assets} - \text{Current liabilities} \qquad [9.2]$$

Working capital is the cash that flows into, through, and out of a construction company. It flows through the company (a circulatory system) as operations are conducted. This flow of working capital through the company is a lifeblood cycle and without the "blood" the company dies. The flow of cash includes (see Fig. 9.4):

1. Clients (project owners) input *contract revenues.*

FIGURE 9.4 Flow of funds through a construction company.

2. A major portion of that revenue stream flows out of the company as direct costs to pay *direct project costs:* subcontractors, material suppliers, the contractor's labor force, and other costs.

3. Part of the remainder is used for job *overhead costs* (this cost is allocated to and absorbed by all jobs).

4. That which is left is classified as gross profit (operating profit).

5. Nonjob overhead, general and administrative expenses (G&A) are then deducted.

6. Finally a small portion makes it through as *income* to the company. But this is not the portion that the company gets to keep because taxes must be paid on income.

7. However, if there is even one bad project (total cost is greater than the contract amount), profits from profitable projects are used to complete the work (pay direct cost).

If the company is operating profitably, it should generate a cash surplus. If it does not generate surpluses, it will eventually run out of cash and expire. From the data presented in the balance sheet for Low Bid Construction Corporation (see Fig. 9.5), it is clear that the contractor developed a problem with liquidity and working capital in 2001 (see Table 9.2). The Low Bid Corporation was forced to use up much of its past profits (retained earnings).

Recognizing its problem, Low Bid has made a provision in the assets part of the 2001 balance sheet to generate needed working capital. See the "Assets held for sale (Note 1)" row in the Assets section of Low Bid's balance sheet (see Fig. 9.5). A note that accompanies the balance sheet states "in order to improve working capital, the Company executed a definite agreement to sell certain assets. Accordingly, $3,213,484 of assets consisting of inventories and equipment has been classified as assets held for sale on the balance sheet."

Current Ratio

current ratio
The relation of a company's current assets to its current liabilities.

While the magnitude of a firm's working capital, as calculated using Eq. [9.2], provides information about ability to prosecute the firm's workload, it is also an indication of the ability to pay short-term obligations. There are other financial ratios designed to measure a company's ability to meet these obligations. Ratios provide a means to identify changes in position that are not as apparent when only reviewing the magnitude of values. The **current ratio**, Eq. [9.3], is a tool to identify changes in a company's ability to meet short-term obligations.

TABLE 9.2   Liquidity analysis for Low Bid Construction Corp.

	Dec. 31, 2001	Dec. 31, 2000
Current assets	$35,072,542	$33,543,119
Current liabilities	36,267,328	28,596,945
Working capital	−1,194,786	4,946,174

LOW BID CONSTRUCTION CORPORATION
CONSOLIDATED BALANCE SHEETS

	December 31, 2001	December 31, 2000
Assets (Note 9):		
Current Assets:		
Cash and cash equivalents (Notes 1 and 2)	$ 2,228,506	$ 1,822,598
Restricted cash (Notes 1, 2, and 16)	2,401,548	1,783,005
Accounts receivable, net (Notes 1, 3, 10, and 16)	21,377,904	14,297,564
Prepaid expenses and other	404,780	749,708
Inventory (Note 1)	3,365,750	4,288,235
Income tax receivable (Notes 1 and 11)	—	774,000
Costs and estimated earnings in excess of billings on uncompleted contracts (Notes 1 and 4)	5,294,054	9,828,009
Total Current Assets	35,072,542	33,543,119
Property and equipment, net (Notes 1, 5, 8, and 12)	15,267,791	18,111,506
Assets held for sale (Note 1)	3,213,484	—
Deferred tax asset (Notes 1 and 11)	1,957,923	873,441
Refundable deposits	55,110	176,565
Goodwill, net (Note 1)	—	1,500,733
Mineral rights and pit development, net	533,608	1,180,666
Claims receivable (Notes 1, 4, and 14)	5,968,026	—
Other assets	80,558	—
Total Assets	$ 62,149,042	$ 55,386,030
Liabilities and Stockholders' Equity:		
Current Liabilities:		
Accounts payable (Notes 6 and 10)	$ 27,025,984	$ 17,606,113
Accrued liabilities (Notes 7 and 10)	1,811,998	2,289,698
Notes payable (Note 8)	1,685,634	1,604,399
Obligations under capital leases (Note 12)	1,118,055	1,041,921
Billings in excess of costs and estimated earnings on uncompleted contracts (Notes 1 and 4)	4,625,657	6,054,814
Total Current Liabilities	36,267,328	28,596,945
Deferred tax liability (Notes 1 and 11)	2,718,734	2,272,700
Notes payable, less current portion (Notes 8 and 9)	9,484,479	7,674,608
Obligations under capital leases, less current portion (Note 12)	2,964,195	3,603,540
Total Liabilities	51,434,736	42,147,793
Commitments and contingencies (Notes 9, 10, 12 and 14)		
Stockholders' Equity:		
Preferred stock — $.001 par value; 1,000,000 shares authorized, none issued and outstanding (Note 13)	—	—
Common stock — $.001 par value; 15,000,000 shares authorized 3,559,438 issued and outstanding (Notes 13 and 17)	3,601	3,601
Additional paid-in capital	10,943,569	10,943,569
Capital adjustments	(799,147)	(799,147)
Retained earnings	566,283	3,090,214
Total Stockholders' Equity	10,714,306	13,238,237
Total Liabilities and Stockholders' Equity	$ 62,149,042	$ 55,386,030

FIGURE 9.5 Balance sheet, Low Bid Construction Corp. (the notes that accompanied the balance sheet are not shown here).

$$\text{Current ratio} = \frac{\text{Current assets}}{\text{Current liabilities}} \qquad [9.3]$$

Financial analysis is the process of tracking changes in a company's financial health. Ratios are one method of analysis that often causes one to look deeper into a company's financial statements. As a rule of thumb, the current ratio for a construction company should be above 1.3:1. Example 9.2 illustrates the power of the current ratio to give an indication of trouble.

EXAMPLE 9.2

Review the working capital situation for the Fast Buck Construction Co. A summary of Fast Buck's balance sheet data is presented in the table here.

Fast Buck Construction Co. balance sheet data (thousands).

Year	1995	1994	1993	1992	1991
Current assets	$ 8,354	$7,539	$3,587	$1,649	$819
Current liabilities	10,559	6,514	3,127	1,089	467
Working capital	−2,205	1,025	460	560	352
Current ratio	0.79	1.15	1.14	1.51	1.75

Looking at the change in working capital across the years 1991 to 1994, it would appear the company has the liquidity to meet its obligations. The amount of working capital actually grows from $352,000 to $1,025,000 between 1991 and 1994. The amount of working capital is growing, but the current ratio alerts management to the fact that the growth is not keeping pace with the growth in current liabilities. The drop in the current ratio value in 1993 provides a warning of this situation.

When the adequacy of working capital is studied, the composition of the company's current assets should be considered. A company with a high proportion of cash to accounts receivable is in a better position to meet its current obligations. The company with a high proportion of accounts receivable must turn those into cash before it can pay the bills. Using the Low Bid balance sheet (Fig. 9.5), the liquidity deterioration is reflected by the cash to accounts receivable proportional change (Table 9.3).

Working capital *increases* when a company *makes a profit on a job,* sells equipment or assets, or borrows money on a long-term note. Contractors typically resist selling equipment or engaging in long-term borrowing.

TABLE 9.3   Cash to accounts receivable analysis for Low Bid Construction Corp.

	Dec. 31, 2001	Dec. 31, 2000
Current assets—cash	$ 2,228,506	$ 1,822,598
Current assets—accounts receivable	21,377,904	14,297,564
Proportion cash to accounts receivable	1:9.6	1:7.8

Working capital *decreases* when the company *loses money on a job,* buys equipment, or pays off long-term debt. Contractors are typically eager to buy equipment and pay off debt.

CONSTRUCTION CONTRACT REVENUE RECOGNITION

In construction accounting, everything follows accounting practices used in other industries except for revenue recognition. In most circumstances, revenue is recognized at the point of sale, because most of the uncertainties related to the earning process are removed and the exchange price is known. The exception to the general rule of recognition at point of sale is caused by the long-term nature of construction projects. The accounting measurements associated with long-term construction projects are a bit more complicated because events and amounts must be estimated for a period of years, and even good estimates are modified because of owner-desired change orders or unexpected conditions [4 and 8].

There are several different methods a construction company can use to identify revenue including the cash, straight accrual, completed contract, and percentage-of-completion methods, but for financial statements, the only recognized method is percentage of completion.

Cash Method of Revenue Recognition

With the cash method, a contractor records revenue when payments for services are received (actual receipts to date) and records costs as bills are paid (costs paid to date). There is never an accurate picture of the company's financial condition or how individual projects are progressing (at a profit or loss). Considering the project data in Table 9.4 and using the cash method of revenue recognition, the revenue to date for the project would be

Payments received to date	$630,000
Costs paid to date	$400,000
Revenue to date	**$230,000**

TABLE 9.4 OK Project financial data (sample data).

Project financial data	Amount
Contract amount	$1,000,000
Original estimated cost	900,000
Billed to date	700,000
Payments received to date	630,000
Costs incurred to date	450,000
Forecasted cost to complete	400,000
Costs paid to date	400,000

Straight Accrual Method of Revenue Recognition

The straight accrual method considers cost as costs to date (even if not yet paid) and revenue as what has been billed to date. Considering the project data in Table 9.4 and using the straight accrual method of revenue recognition, the revenue to date for the project would be

Billed to date	$700,000
Costs incurred to date	$450,000
Revenue to date	**$250,000**

Completed-Contract Method of Revenue Recognition

The completed-contract method should almost never be used and then only when

- the contractor has predominantly very short-term contracts,
- the conditions for using the percentage-of-completion method cannot be met, or
- there are inherent hazards in the contract beyond normal, recurring business risks.

Considering the project data in Table 9.4 and using the completed-contract method of revenue recognition, the revenue to date for the project would be zero at this time, because the project has not been completed.

Percentage-of-Completion Method of Revenue Recognition

The percentage-of-completion method should be used when estimates of progress toward completion, revenues, and costs are reasonably dependable, and all these conditions exist:

- The contract clearly specifies the enforceable rights regarding services to be provided and received by the parties, the consideration to be exchanged, and the manner and terms of settlement.
- The owner (buyer) can be expected to satisfy all obligations under the contract.
- The contractor can be expected to perform contractual obligations.

When using the percentage-of-completion method to calculate revenue, it is first necessary to calculate the percentage of work that has been completed (Eq. [9.4]):

$$\text{Percentage completion} = \frac{\text{Cost to date}}{\text{Cost to date} + \text{cost to complete}} \times 100\% \quad \textbf{[9.4]}$$

Considering the project data in Table 9.4, the percentage of completion for the project would be

$$\frac{\$450,000}{\$450,000 + \$400,000} \times 100\% = 52.9\%$$

The revenue to date for the project would be

$$\text{Revenue to date} = \text{Percentage of completion} \times \text{Contract price} \quad \textbf{[9.5]}$$

Therefore, revenue is 52.9% $\times$ $1,000,000 = $529,000

$$\text{Gross profit} = \text{Revenue} - \text{Cost} \quad \textbf{[9.6]}$$

Revenue to date	$529,000
Cost incurred to date	$450,000
Gross profit to date	**$79,000**

Overbillling/Underbilling

When the OK Project was originally bid, the anticipated profit was $100,000 (Table 9.4, contract amount minus original estimated cost). As the work has progressed, the cost to complete the work has been continually revised and, at the point in time that the data represents, the new estimated final cost is $850,000 (Table 9.4, cost incurred to date plus forecasted cost to complete, $450,000 + $400,000). The construction company has billed the owner for $700,000 and has actual cost to date of $450,000; this leaves the contractor with $250,000 in cash. For this discussion, the fact that the contractor has not received all $700,000, because the owner is holding 10% retainage, will be ignored. Part of that $250,000 is true gross profit (what the contractor has really earned, $79,000) and part is overbilling, $171,000. The way to calculate overbilling is by subtracting the revenue to date (percentage-of-completion method) from the billed to date amount. This reveals the excess of the billing over the revenue. Such an undertaking demonstrates whose money is whose (e.g., what belongs to the contractor and what belongs to the owner).

Billed to date	$700,000
Revenue to date (percentage-of-completion method)	$529,000
Overbilling or job borrow	**$171,000**

Most contractors are not in the banking business, in the sense that they do not have the funds to finance construction of the project for the owner. Therefore, contractors allocate more profit to those items of work that will be completed early in the project. In the case of a lump sum project, this is done when the schedule of values is negotiated with the owner or general contractor. The total amount of revenue for the total work is the same; it is just that more of the revenue is received on the early items of work. The formula for calculating overbilling is given by Eq. [9.7]:

$$\text{Overbilling} = \text{Billed to date} - \text{Revenue} \quad \textbf{[9.7]}$$

Early in a project, a contractor can be in an underbilling situation where actual earned revenue is greater than the amount billed to the owner (Eq. [9.8]):

$$\text{Underbilling} = \text{Revenue} - \text{Billed to date} \qquad \textbf{[9.8]}$$

Assuming that overbillings have been paid, they indicate that the contractor is "borrowing" money from the owner by billing ahead of the revenue the contractor has actually earned. While overbillings are preferable to significant underbillings, they can pose difficulties. For example, if the cash borrowed from a project is used for other purposes, it may not be available for payment of job costs when they are incurred later.

Managers should look for underbilling late in a project. Significant underbillings *always* indicate problems and the possibility of unapproved change orders that may not get approved or paid. Underbillings indicate that the contractor is allowing the project owner to "borrow" money, in the sense that the contractor has performed work (and incurred expenses) without yet billing for it. Underbillings reduce cash flow for the individual contract as well as other contracts and the general operations of the company. An underbilling can be characterized as an unbilled receivable.

CONTRACT STATUS REPORT

A contract status report (work in progress schedule) summarizes the financial position of each project that is not completely finished. It presents a comparison of work accomplished (revenue earned) against costs and a projection of the final financial outcome of each project. Table 9.5 presents the base data required to develop a contract status report. In preparing a contract status report, it is the project manager who is usually responsible for evaluating the accuracy of the estimated cost to complete a project. The contract price is the sum of the original contract amount and all approved change orders.

The first step in developing a complete contract status report is to calculate a new or current (at the present point in time) estimate of the total project cost. Summing the cost to date of each project with its re-estimated cost to complete does this (col. 2, Table 9.6). Once that is done, it is possible to use Eq. [9.4] (cost to date as a ratio of projected total cost) to calculate the percentage complete of each project (col. 5, Table 9.6). Additionally, once the current (new) estimate of

TABLE 9.5 It Will Cost You a Million Construction Co. contract financial data.

Contract	Contract price ($)	Cost to date ($)	Total amount billed to date ($)	Estimated cost to complete ($)
School 1	1,000,000	450,000	700,000	400,000
School 2	1,000,000	450,000	400,000	400,000
Store 1	1,000,000	450,000	600,000	500,000
Store 2	1,000,000	450,000	450,000	460,000
Hospital	1,000,000	475,000	525,000	575,000
	5,000,000	2,275,000	2,700,000	2,335,000

the total project cost is calculated, it can be subtracted from the project contract price to obtain a current estimated profit/loss at completion (col. 3, Table 9.6). By multiplying the estimated profit at completion by the percentage complete, profit to date is obtained (col. 6, Table 9.6). Note in Table 9.6 that, in the case of the hospital project, a loss is projected at project completion. An expected loss must be recognized in the period when the loss is identified—the total expected loss.

Table 9.7 presents the complete status report with revenue and over/underbillings recognized. Project managers are expected to analyze and explain the progress of their projects. If the project manager fails to forecast a cost overrun, the report will show an underbilling. This can be particularly troublesome if the cause of the cost overrun cannot be identified.

FINANCIAL STATEMENT ANALYSIS

In and of themselves, very few of the individual figures in a financial statement are exceptionally significant. It is their relationship to other quantities, or the magnitude and direction of change with time (since a previous statement), that is important as an indicator of the company's condition. Analysis of financial statements is largely a matter of studying relationships, changes, and trends. The three commonly used analysis techniques are (1) dollar or percentage change, (2) component percentages, and (3) ratios.

Financial statement/report analysis requires that relationships and changes between items and groups of items be described. A **comparative statement** is an analysis report that shows changes in statement items by presenting item amounts for two or more successive accounting periods placed side by side in a columns format on a single statement (see Fig. 9.3). As an aid to controlling operations, a comparative income statement is usually more valuable than a comparative balance sheet.

comparative statement
A financial statement with data for two or more successive accounting periods placed in columns side by side to illustrate changes.

Dollar and Percentage Change

The dollar magnitude (amount) of change from one reporting period to another is significant, but expressing the change in percentage terms adds perspective.

TABLE 9.6 It Will Cost You a Million Construction Co. contract status.

Contract	Contract price ($)	Current estimated total cost ($)	Estimated profit at completion	Cost to date ($)	Percentage complete	Total profit (loss) recognized to date ($)	Estimated cost to complete ($)
	(1)	(2)	(3)	(4)	(5)	(6)	(11)
School 1	1,000,000	850,000	150,000	450,000	53	79,500	400,000
School 2	1,000,000	850,000	150,000	450,000	53	79,500	400,000
Store 1	1,000,000	950,000	50,000	450,000	47	23,500	500,000
Store 2	1,000,000	910,000	90,000	450,000	49	44,100	460,000
Hospital	1,000,000	1,050,000	−50,000	475,000	45	−50,000	575,000
	5,000,000	4,610,000		2,275,000			2,335,000

TABLE 9.7 Contract status report for It Will Cost You a Million Construction Co.

Contract description	Contract price	Current re-est. total cost	Estimated gross profit at completion	Cost to date	Percentage complete	Total gross profit (loss) recog. to date
Column	(1)	(2)	(3)	(4)	(5)	(6)
Formula		4 + 11	1 − 2		4 ÷ 2	7 − 4 or 5 × 3
School 1	$1,000,000	$ 850,000	$150,000	$ 450,000	53%	$79,412
School 2	$1,000,000	$ 850,000	$150,000	$ 450,000	53%	$79,412
Store 1	$1,000,000	$ 950,000	$ 50,000	$ 450,000	47%	$23,684
Store 2	$1,000,000	$ 910,000	$ 90,000	$ 450,000	49%	$44,505
Hospital	$1,000,000	$1,050,000	−$ 50,000	$ 475,000	45%	−$22,619
	$5,000,000			$2,275,000		

	Amount earned to date (revenue)	Total amount billed to date	Over-billings	Under-billings	Re-est. cost to complete	Contract balance
Column	(7)	(8)	(9)	(10)	(11)	(12)
Formula	5 × 1 or 4 + 6		8 − 7	7 − 8		1 − 8
School 1	$529,412	$ 700,000	$170,588		$ 400,000	$300,000
School 2	$529,412	$ 400,000		$129,412	$ 400,000	$600,000
Store 1	$473,684	$ 600,000	$126,316		$ 500,000	$400,000
Store 2	$494,505	$ 475,000		$ 19,505	$ 460,000	$525,000
Hospital	$452,381	$ 525,000	$ 72,619		$ 575,000	$475,000
		$2,700,000			$2,335,000	

The dollar magnitude of change is the difference between the amount of a comparison statement (year) and a base statement (year). The percentage change is computed by dividing the amount of the change between the statements by the magnitude for the base statement. Table 9.8 presents the dollar magnitude and percentage change of Low Bid Construction Corporation's revenue and cost of revenue for the 3-year period 1999 through 2001.

Though the company cannot be happy about the decrease in revenue shown by the comparison, one possible cause can be eliminated. Because the percentage changes are approximately the same (e.g., revenue decreases 22.1% and likewise the cost of revenue decreases 20.6%), it seems reasonable to assume that an increase in cost was not the cause of the revenue decrease experienced in 2000.

Component Percentages

Component percentages indicate the relative size of each item included within a total. Each item on a balance sheet can be expressed as a percentage of total current assets, total assets, total current liabilities, or total liabilities. Such a display

TABLE 9.8 Analysis of Low Bid Construction Corporation's financial situation.

Year	2001	2000	1999
Revenue	$174,063,148	$163,573,258	$210,002,272
Cost of revenue	169,404,787	158,935,103	200,070,826
Dollar change in revenue	$10,489,890	−$46,429,014	
Percentage change in revenue	6.4%	−22.1%	
Dollar change cost of revenue	$10,469,684	−$41,135,723	
Percentage change cost of revenue	6.6%	−20.6%	

TABLE 9.9 Component percentage analysis of current assets (thousands).

Current assets	1998		1997		1996	
Cash	$67,054	15.7%	$53,215	13.1%	$48,310	10.5%
Accounts receivable	$175,513	41.0%	$187,311	46.3%	$222,341	48.1%
Unbilled receivables	$74,552	17.4%	$74,514	18.4%	$101,564	22.1%
Total current assets	$427,722		$405,014		$459,249	

of total assets illustrates the relative importance of current and noncurrent assets as well as the relative amount of financing obtained from current creditors, by long-term borrowing, and from the owners.

Table 9.9 presents a component percentage of current assets analysis. The analysis shows that the company is increasing its cash position and decreasing accounts receivable and unbilled receivables (underbillings).

Ratios

A ratio is the relationship expression of two quantities (items) expressed as the quotient of one divided by the other. To compute a meaningful ratio, there must be a significant relationship between the two items considered. The current ratio (discussed before, Eq. [9.3]) summarizes the relationship between current assets and current liabilities. Besides the current ratio, some other important financial ratios for construction companies are

$$\text{Quick ratio} = \frac{\text{Cash} + \text{receivables}}{\text{Current liabilities}} \qquad [9.9]$$

The quick ratio measures the short-term liquidity of a company.

$$\text{Debt to worth} = \frac{\text{Total liabilities}}{\text{Total shareholders' equity}} \qquad [9.10]$$

The debt ratio measures the firm's leverage. The margin of protection to creditors is higher when the ratio is lower because the owners are contributing the bulk of funds for the business. Properly leveraged contractors have debt to worth ratios of 2:1 or less.

$$\text{Receivables to payables} = \frac{\text{Receivables}}{\text{Payables}} \qquad \text{[9.11]}$$

The suggested range for the receivables-to-payables ratio varies with the type of construction activity undertaken by the firm.

Based on the type of industry being studied, there are rules of thumb for evaluating financial ratios. In the case of construction companies,

■ Current ratio should be above 1.3:1.

■ Quick ratio should be above 1.1:1.

■ Debt to worth should be below 2.0:1.

■ Receivables to payables (commercial general contractors) should be around 1.5:1.

■ Receivables to payables (trade/heavy contractors) should be around 2.0:1.

■ Receivables to payables (labor-intensive trade contractors) should be around 3.0:1.

One further rule of thumb concerns a construction company's capacity to do work. A firm's backlog of work should not be more than approximately 10 times the company's working capital amount. This is something that a surety will consider before granting a bond for new work.

SUMMARY

Financial statements are the primary means of communicating useful financial information and are the result of simplifying, condensing, and aggregating transactions. No one financial statement provides sufficient information by itself and no one item or part of each statement can summarize the information. The two most common accounting documents that are used to monitor a company's performance are the balance sheet and the income statement. The balance sheet shows, at a particular point in time, the company's overall financial strength and potential for future growth. The income statement shows the sources of a company's income, production costs, and other expenses. It shows the company's income for the period.

The analysis of a balance sheet can identify potential liquidity problems. These may signify the company's inability to meet financial obligations. A company's income statement shows all of the money a company earned (revenues) and all of the company's expenses during a specific period, usually the fiscal year.

The percentage-of-completion method for calculating revenue should be used when estimates of progress toward completion, revenues, and costs are reasonably dependable. When using the percentage-of-completion method to calculate revenue, it is first necessary to calculate the percentage of work that has been completed.

A contract status report summarizes the financial position of each project that is not completely finished. It presents a comparison of work accomplished against cost and a projection of the final financial outcome of each project.

REVIEW QUESTIONS

9.1 Based on the project data presented in the table, calculate for each of the projects the revenue that would be recognized using (a) cash and (b) straight accrual methods. In addition, calculate (c) the revenue using the percentage-of-completion method and (d) the gross profit to date using the percentage-of-completion method.

Project financial data	Candy store project	Toy store project	Ice rink project
Contract amount	$5,000,000	$5,000,000	$5,000,000
Original estimated cost	4,700,000	4,800,000	4,850,000
Billed to date	1,700,000	1,700,000	1,700,000
Payments received to date	1,530,000	1,530,000	1,530,000
Cost incurred to date	1,450,000	1,350,000	1,550,000
Forecasted cost to complete	3,500,000	3,400,000	3,100,000
Costs paid to date	1,400,000	1,400,000	1,400,000

9.2 Based on the project data presented in the table, calculate for each of the projects: the revenue that would be recognized using (a) cash and (b) straight accrual methods; additionally calculate (c) the revenue using the percentage-of-completion method and (d) the gross profit to date.

Project financial data	Jail project	Office project	Hotel project
Contract amount	$15,000,000	$15,000,000	$15,000,000
Original estimated cost	14,400,000	14,600,000	14,800,000
Billed to date	10,700,000	10,700,000	10,700,000
Payments received to date	10,900,000	10,530,000	10,630,000
Cost incurred to date	11,450,000	10,350,000	10,550,000
Forecasted cost to complete	3,000,000	4,400,000	4,100,000
Costs paid to date	9,400,000	9,300,000	9,600,000

9.3 What is the amount of over/underbilling for each of the Question 9.1 projects?

9.4 What is the amount of over/underbilling for each of the Question 9.2 projects?

9.5 Using the corporate balance sheet data shown on pages 338 and 339, calculate the company's net worth, working capital, and current ratio for both 1996 and 1995.

9.6 Using the corporate balance sheet data shown on pages 338 and 339, calculate the company's net worth, working capital, and current ratio for both 1998 and 1997.

9.7 Locate several construction company corporate balance sheets on the Web and calculate the company's net worth, working capital, and current ratio.

CONSOLIDATED BALANCE SHEETS
(In thousands except per share data)

November 30,	1996	1995
ASSETS		
Current assets		
Cash and cash equivalents	$ 48,310	$ 30,035
Accounts receivable, including retentions of $28,348 and $14,513	222,341	41,327
Unbilled receivables	101,564	5,033
Refundable income taxes	10,806	—
Current portion of notes receivable	3,000	3,576
Investments in and advances to construction joint ventures	24,538	1,846
Deferred income taxes	31,291	514
Other	17,399	3,390
Total current assets	459,249	85,721
Investments and other assets		
Securities available for sale, at fair value	30,494	—
Investments in mining ventures	56,210	—
Land held for sale or lease	—	8,266
Assets held for sale	18,853	—
Cost in excess of net assets acquired, net of accumulated amortization		
of $2,177 and $1,013	140,677	15,777
Long-term notes receivable, net of current portion	5,087	7,935
Deferred income taxes	31,555	—
Other	12,178	886
Total investments and other assets	295,054	32,864
Property and equipment, at cost		
Construction equipment	179,483	99,608
Land and improvements	7,110	28,374
Buildings and improvements	25,062	1,941
Equipment and fixtures	30,388	576
Total property and equipment	242,043	130,499
Less accumulated depreciation	(156,709)	(63,783)
Property and equipment, net	85,334	66,716
Total assets	$839,637	$185,301

LIABILITIES AND STOCKHOLDERS' EQUITY

	1996	1995
Current liabilities		
Accounts payable	$ 68,926	$ 10,195
Subcontracts payable, including retentions of $27,006 and $9,778	75,036	16,658
Billings in excess of cost and estimated earnings on uncompleted contracts	49,626	4,789
Advances from customers	11,280	—
Estimated costs to complete long-term contracts	100,832	—
Accrued salaries, wages, and benefits	49,136	8,718
Income taxes payable	8,255	501
Other accrued liabilities	37,513	1,327
Total current liabilities	400,604	42,188
Noncurrent liabilities		
Postretirement benefit obligation	53,433	—
Accrued workers' compensation	26,061	—
Pension and deferred compensation liabilities	20,563	—
Environmental remediation obligations	8,972	—
Long-term debt	—	5,042
Deferred income taxes	—	9,120
Total noncurrent liabilities	109,029	14,162

CONSOLIDATED BALANCE SHEETS
(In thousands except per share data)

November 30,	1998	1997
ASSETS		
Current assets		
Cash and cash equivalents	$ 67,054	$ 53,215
Accounts receivable, including retentions of $18,627 and $26,970	175,513	187,311
Unbilled receivables	74,552	74,514
Refundable income taxes	780	14,331
Investments in and advances to construction joint ventures	70,855	29,270
Deferred income taxes	26,489	30,173
Other	12,479	16,200
Total current assets	427,722	405,014
Investments and other assets		
Securities available for sale, at fair value	45,985	39,314
Investments in mining ventures	67,967	57,439
Assets held for sale	14,169	13,301
Cost in excess of net assets acquired, net of accumulated amortization		
of $9,330 and $5,755	112,994	136,150
Deferred income taxes	30,965	31,183
Other	8,077	7,594
Total investments and other assets	280,157	284,981
Property and equipment, at cost		
Construction equipment	179,337	172,154
Land and improvements	6,993	6,993
Buildings and improvements	6,341	6,276
Equipment and fixtures	63,534	53,483
Total property and equipment	256,205	238,906
Less accumulated depreciation	(175,933)	(158,657)
Property and equipment, net	80,272	80,249
Total assets	$788,151	$770,244

The accompanying notes are an integral part of the financial statements.

LIABILITIES AND STOCKHOLDERS' EQUITY

	1998	1997
Current liabilities		
Accounts payable	$ 56,388	$ 53,448
Subcontracts payable, including retentions of $22,843 and $20,266	59,857	42,513
Billings in excess of cost and estimated earnings on uncompleted contracts	40,959	34,163
Estimated costs to complete long-term contracts	49,228	73,103
Accrued salaries, wages, and benefits	58,939	52,618
Income taxes payable	1,535	1,371
Other accrued liabilities	36,118	42,679
Total current liabilities	303,024	299,895
Noncurrent liabilities		
Postretirement benefit obligation	53,456	53,689
Accrued workers' compensation	39,625	34,088
Pension and deferred compensation liabilities	16,390	15,334
Environmental remediation obligations	4,753	6,107
Total noncurrent liabilities	114,224	109,218

9.8 Locate a construction company's corporate balance sheets for several years on the Web and calculate the company's current ratio for each of the years. Did the company maintain a good position?

REFERENCES

1. Bersch, Dennis W. (1973). "Do You Manage or Just Watch?" *Constructor,* The Associated General Contractors of America, June.

2. Bersch, Dennis W. (1973). "Conclusion of a Series, Do You Manage or Just Watch?" *Constructor,* The Associated General Contractors of America, Washington, DC, September.

3. *Construction Cost Control* (1982). ASCE Manuals and Reports of Engineering Practice No. 65, ASCE, New York.

4. Delaney, Patrick R., Barry J. Epstein, Ralph Nach, and Susan Weiss Budak (2001). "Special Accounting and Reporting Issues," *Wiley GAAP 2002.* John Wiley & Sons, Inc., New York.

5. Korman, Richard, Tony Illia, and Stephen H. Daniels (2002). "Don't Fade Away," *ENR,* July 15, pp. 24–26.

6. Larson, Kermit D., John Wild, and Barbara Chiappetta (2001). *Fundamental Accounting Principles,* 16th ed., McGraw-Hill/Irwin, Burr Ridge, IL.

7. Meigs, Robert F. (2001). *Financial and Managerial Accounting: The Basis for Business Decisions,* 12th ed., McGraw-Hill Education Group, New York.

8. Schleifer, Thomas C. (1995). "Measuring Financial Risk." *Constructor,* The Associated General Contractors of America, February.

9. Trottman, Melanie (2002). "Halliburton Faces an SEC Probe of Its Accounting," *The Wall Street Journal,* May 30.

WEBSITE RESOURCES

1. www.centex.com   Centex Corporation. The corporate website includes current financial data in the investors section.

2. www.fluor.com   Fluor Corporation. The corporate website includes current financial data in the investors section.

3. www.graniteconstruction.com   Granite Construction Inc. The corporate website includes current financial data in the investor relations section.

4. www.planware.org   Invest-Tech Limited, 27 Ardmeen Park, Blackrock, Co. Dublin, Ireland. Interesting site containing white papers on financial matters.

5. www.cfma.org   The Construction Financial Management Association (CFMA). 29 Emmons Drive, Suite F-50, Princeton, NJ 08540. 609-452-8000 (phone); 609-452-0474 (fax). CFMA is a resource of information and insight into the unique discipline of construction financial management.

Construction Practice

10

Machine Power

The constructor must select the proper equipment to relocate and/or process materials economically. The analysis procedure for matching the best possible machines to the project task requires consideration of a machine's mechanical capabilities. The engineer must first calculate the power required to propel the machine and its load. This power requirement is established by two factors: (1) rolling resistance and (2) grade resistance. Equipment manufacturers publish performance charts for individual machine models. These charts enable the equipment planner to analyze a machine's ability to perform under a given set of job and loading conditions.

KNOWLEDGE FOR ESTIMATING

On heavy construction projects, the majority of the work consists of handling and processing large quantities of bulk materials. The constructor must select the proper equipment to relocate and/or process these materials economically. The decision process for matching the best possible *machine* to the project task requires that the estimator take into account both the properties of the material to be handled and the mechanical capabilities of the machine.

When the estimator considers a construction material-handling problem, there are three crucial *material* considerations: (1) total *quantity* of material, (2) rate at which it must be moved, and (3) *size* of the individual material pieces. The quantity of material to be handled and the time constraints resulting from the project contract specifications or from expected weather conditions influence the selection of machines as to type, size, and number to be employed. Larger units generally have lower unit-production cost, but there is a trade-off in higher mobilization and fixed costs. The size of the individual material pieces will affect the machine size alternatives that can be considered. A loader used in

FIGURE 10.1 A loader must be able to handle the largest rock sizes of the project.

a quarry to move blasted rock (shot rock) must be capable of handling the largest rock sizes produced (Fig. 10.1).

Payload

The payload capacity for construction excavation and hauling equipment can be expressed *volumetrically* or *gravimetrically*. Volumetric capacity can be stated as struck or heaped volume, and either volume can be expressed in terms of loose cubic yard (lcy), bank cubic yard (bcy), or compacted cubic yard (ccy).

- *Loose cubic yard*—one cubic yard of material after it has been disturbed, usually by the excavation process
- *Bank cubic yard*—one cubic yard of material as it lies in the natural or undisturbed state
- *Compacted cubic yard*—one cubic yard of material after compaction

The payload capacity of excavation buckets and hauling unit is often stated by the manufacturer in terms of the volume of loose material, assuming that the material is heaped in some specified angle of repose. A gravimetric capacity represents the safe operational weight that the axles or structural frame of the machine is designed to handle.

From an economic standpoint, overloading a truck or any other haul unit to improve production looks attractive and overloading by 20% might increase the haulage rate by 15%, allowing for slight increases in time to load and haul. The cost per ton hauled should show a corresponding decrease, since direct labor costs will not change and fuel costs will increase only slightly. This apparently favorable situation is only temporary, for the advantage is being bought at the cost of premature aging of the truck and a corresponding increased replacement capital expense.

Machine Performance

Cycle time and payload determine a machine's production rate, and machine travel speed directly affects cycle time. "Why does a loaded machine travel at only 12 mph when its top speed is 33 mph?" To answer the travel speed question, it is necessary to examine three power questions:

1. Required power
2. Available power
3. Usable power

REQUIRED POWER

rolling resistance
The resistance of a level surface to constant-velocity motion across it.

Power required is the power needed to overcome resisting forces and cause machine motion. The magnitude of resisting forces establishes this power requirement. The forces resisting the movement of mobile equipment are (1) **rolling resistance** and (2) **grade resistance**. Therefore, the power required is the power necessary to overcome the *total resistance* to machine movement, which is the sum of rolling and grade resistance.

$$\text{Total resistance (TR)} = \text{Rolling resistance (RR)} + \text{Grade resistance (GR)} \qquad \textbf{[10.1]}$$

grade resistance
The force-opposing movement of a machine up a frictionless slope.

Rolling Resistance

Rolling resistance is the resistance of a level surface to constant-velocity motion across it. This is sometimes referred to as *wheel resistance* or *track resistance.* Rolling resistance results from friction of the driving mechanism, tire flexing, and the force required to shear through or ride over the supporting surface (Fig. 10.2).

This resistance varies considerably with the type and condition of the surface over which a machine moves. Soft earth offers a higher resistance than hard-surfaced roads such as those constructed of concrete or asphalt. For machines that move on rubber tires, the rolling resistance varies with the size of, pressure on, and tread design of the tires. For equipment that moves on crawler tracks, such as tractors, the resistance varies primarily with the type and condition of the road surface.

A narrow-tread, high-pressure tire gives lower rolling resistance than a broad-tread, low-pressure tire on a hard-surfaced road. This is the result of the small area of contact between the tire and the road surface. If the road surface is soft earth such a tire sinks into the earth; a broad-tread, low-pressure tire will

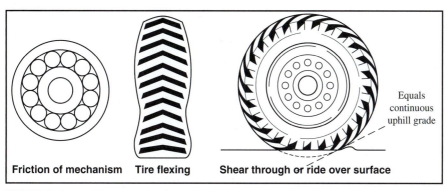

Friction of mechanism   Tire flexing   Shear through or ride over surface

Equals continuous uphill grade

FIGURE 10.2 Mechanisms of rolling resistance.

offer a lower rolling resistance than a narrow-tread, high-pressure tire when soft ground conditions are encountered. The reason for this condition is that the narrow tires sink farther into the earth than the broad tires and thus are always having to climb out of a deeper rut, which is equivalent to climbing a steeper grade.

The rolling resistance of an earthen-haul road probably will not remain constant under varying weather conditions or with the varying types of soil that exist along the road. If the earth is stable, highly compacted, and well maintained by a grader, and if the moisture content is kept near optimum, it is possible to provide a surface with a rolling resistance about as low as that of concrete or asphalt. Moisture can be added, but following an extended period of rain, it may be difficult to remove the excess moisture and the haul road will become soft and rutted; when this happens, rolling resistance will increase. Providing good surface drainage will speed the removal of the water and should enable the road to be reconditioned quickly. For major earthwork projects, it is economical to provide graders, water trucks, and even rollers to keep the haul road in good condition.

> The maintenance of haul roads is one of the best financial investments an earthmoving contractor can make. The cost of having a grader to maintain the haul road is repaid in increased production.

A tire sinks into the soil until the product of bearing area (A) and bearing capacity (P) is sufficient to sustain the load (F): $F = PA$. Then, after that equilibrium point is reached, the tire is always attempting to climb out of the resulting rut. The rolling resistance will increase about 30 lb/ton for each inch of penetration. Total rolling resistance is a function of the riding gear characteristics (independent of speed), the total weight of the vehicle, and torque. It is usually expressed as pounds of resistance per ton of vehicle weight, or as an equivalent grade resistance. Consider a loaded truck that has a gross weight equal to 20 tons and is moving over a level road having a rolling resistance of 100 lb/ton, the tractive effort required to keep the truck moving at a uniform speed will be 2,000 lb (20 tons × 100 lb/ton).

The estimation of off-road rolling resistance is based largely on empirical information, which may include experience with similar soils. Rarely are rolling resistance values based on actual tests. Much of the actual test data available comes from aircraft tire performance research at the U.S. Army Waterways Experiment Station. Although it is impossible to give completely accurate values for the rolling resistances for all types of haul roads and wheels, the values given in Table 10.1 are reasonable estimates.

If desired, one can determine the rolling resistance of a haul road by towing a truck or other vehicle whose gross weight is known along a level section of the haul road at a uniform speed. The tow cable must be equipped with a dynamometer or some other device that will enable determination of the average tension in the cable. This tension is the total rolling resistance of the gross weight for the truck. The rolling resistance in pounds per gross ton will be

$$R = \frac{P}{W} \qquad \text{[10.2]}$$

where R = rolling resistance in pounds per ton
          P = total tension in tow cable in pounds
          W = gross weight of truck in tons

When tire penetration is known, an approximate rolling resistance value for a wheeled vehicle can be calculated using Eq. [10.3]:

$$RR = [40 + (30 \times TP)] \times GVW \qquad \text{[10.3]}$$

where RR = rolling resistance in pounds per ton
           TP = tire penetration in inches
          GVW = gross vehicle weight in tons

Grade Resistance

The force-opposing movement of a machine up a frictionless slope is known as grade resistance. It acts against the total weight of the machine, whether track type or wheel type. When a machine moves up a slope (see Fig. 10.3), the power required to keep it moving increases approximately in proportion to the slope of the road. If a machine moves down a sloping road, the power required to keep it moving is reduced in proportion to the slope of the road. This is known as **grade assistance**.

grade assistance
The effect of gravitational force in aiding movement of a vehicle down a slope.

The most common method of expressing a slope is by gradient in percent. A 1% slope is one where the surface rises or drops 1 foot vertically in a horizontal distance of 100 feet. If the slope is 5%, the surface rises or drops 5 feet per 100 feet of horizontal distance. If the surface rises, the slope is defined as plus, whereas if it drops, the slope is defined as minus. This is a physical property not affected by the type of machine or the condition of the road, but in

TABLE 10.1 Representative rolling resistances for various types of wheels and surfaces.

| Type of surface | Steel tires, plain bearings | | Crawler type track and wheel | | Rubber tires, antifriction bearings | | | |
| | | | | | High pressure | | Low pressure | |
	lb/ton	kg/m ton	lb/ton	kg/m ton	lb/ton	kg/m ton	lb/ton	kg/m ton
Smooth concrete	40	20	55	27	35	18	45	23
Good asphalt	50–70	25–35	60–70	30–35	40–64	20–33	50–60	25–30
Earth, compacted and maintained	60–100	30–50	60–80	30–40	40–70	20–35	50–70	25–35
Earth, poorly maintained	100–150	50–75	80–110	40–55	100–140	50–70	70–100	35–50
Earth, rutted, muddy, no maintenance	200–250	100–125	140–180	70–90	180–220	90–110	150–200	75–100
Loose sand and gravel	280–320	140–160	160–200	80–100	260–290	130–145	220–260	110–130
Earth, very muddy, rutted, soft	350–400	175–200	200–240	100–120	300–400	150–200	280–340	140–170

Note: Rolling resistance is in pounds per ton or kilograms per metric ton of gross vehicle weight.
Source: Peurifoy and Schexnayder, 2002.

FIGURE 10.3 Scraper moving up an adverse slope.

respect to analyzing forces, its effect is dependent upon the machine's direction of travel.

For slopes of less than 10%, the effect of grade is to increase, for a plus slope, or decrease, for a minus slope, the required tractive effort by 20 lb per gross ton of machine weight for each 1% of grade. This can be derived from elementary mechanics by calculating the required driving force.

From Fig. 10.4, the following relationships can be developed:

$$F = W \sin \alpha \qquad \textbf{[10.4]}$$
$$N = W \cos \alpha \qquad \textbf{[10.5]}$$

For angles less than 10°, $\sin \alpha \approx \tan \alpha$ (the small-angle assumption); with that substitution:

$$F = W \tan \alpha \qquad \textbf{[10.6]}$$

But

$$\tan \alpha = \frac{V}{H} = \frac{G\%}{100}$$

where $G\%$ is the gradient. Hence,

$$F = W \times \frac{G\%}{100} \qquad \textbf{[10.7]}$$

If we substitute $W = 2{,}000$ lb/ton, the formula reduces to

$$F = 20 \text{ lb/ton} \times G\% \qquad \textbf{[10.8]}$$

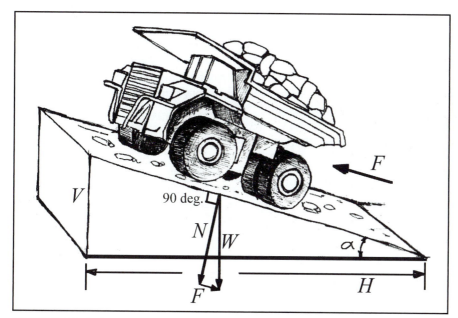

FIGURE 10.4 Frictionless slope-force relationships.

This formula is valid for a *G* up to about 10%, that is, the small-angle assumption (sin α ≈ tan α).

Total Resistance

Total resistance equals rolling resistance plus grade resistance or rolling resistance minus grade assistance (Eq. [10.1]). It can also be expressed as an *effective grade.*

Using the relationship expressed in Eq. [10.8], a rolling resistance may be equated to an equivalent gradient:

$$\frac{\text{Rolling resistance expressed in lb/ton}}{20 \text{ lb/ton}} = G\% \qquad [10.9]$$

Table 10.2 gives values for the effect of slope, expressed in pounds per gross ton or kilograms per metric ton (m ton) of weight of the vehicle.

By combining the rolling resistance, expressed as an equivalent grade, and the grade resistance, expressed as a gradient in percent, one can express the total resistance as an effective grade. The three terms, *power required, total resistance,* and *effective grade,* all denote the same thing. Power required is expressed in pounds. Total resistance is expressed in pounds or pounds per ton of machine weight, and effective grade is expressed in percent.

EXAMPLE 10.1

The haul road from the borrow pit to the fill has an adverse grade of 4%. Wheel-type hauling units will be used on the job, and it is expected that the haul-road rolling resistance will be 100 lb per ton. What will be the effective grade for the haul? Will the units experience the same effective grade for the return trip?

Using Eq. [10.9], we obtain

$$\text{Equivalent grade (RR)} = \frac{100 \text{ lb/ton rolling resistance}}{20 \text{ lb/ton}} = 5\%$$

$$\text{Effective grade (TR}_\text{haul}) = 5\% \text{ RR} + 4\% \text{ GR} = 9\%$$

$$\text{Effective grade (TR}_\text{return}) = 5\% \text{ RR} - 4\% \text{ GR} = 1\%$$

where   RR = rolling resistance

GR = grade resistance

Note that the effective grade is not the same for the two cases. During the haul, the unit must overcome the uphill grade; on the return the unit is aided by the downhill grade.

EXAMPLE 10.2

If the haul unit in Example 10.1 has a gross vehicle weight (GVW) of 47 tons and an empty vehicle weight (EVW) of 22 tons, what is the total resistance experienced during the haul and during the return?

Total resistance$_\text{haul}$: 47 tons $\times$ 9% $\times$ 20 lb/tn = 8,460 lb

Total resistance$_\text{return}$: 22 tons $\times$ 1% $\times$ 20 lb/tn = 440 lb

It is clear that good job management would plan the work so that the haul (loaded) would be down the slope and the return (empty) up the slope.

> Efficiency is achieved by thoughtful planning of haul routes.

Haul Routes   During the life of a project, the haul-route grades (and, therefore, grade resistance) may remain constant. One example of this is trucking aggregate from a rail-yard off-load point to the concrete batch plant. In most cases, however, the haul-route grades change as the work progresses. On a linear highway project, the tops of the hills are excavated and hauled into the valleys. Early in the project, the grades are steep and reflect the existing natural ground. Over the life of the project, the grades begin to assume the design profile. Therefore, the engineer must study the project's mass diagram to determine the direction that the material has to be moved. Then the natural ground and the design pro-

TABLE 10.2 The effect of grade on the tractive effort of vehicles.

Slope (%)	lb/ton*	kg/m ton*	Slope (%)	lb/ton*	kg/m ton*
1	20.0	10.0	12	238.4	119.2
2	40.0	20.0	13	257.8	128.9
3	60.0	30.0	14	277.4	138.7
4	80.0	40.0	15	296.6	148.3
5	100.0	50.0	20	392.3	196.1
6	119.8	59.9	25	485.2	242.6
7	139.8	69.9	30	574.7	287.3
8	159.2	79.6	35	660.6	330.3
9	179.2	89.6	40	742.8	371.4
10	199.0	99.5	45	820.8	410.4
11	218.0	109.0	50	894.4	447.2

*Ton or metric ton of gross vehicle weight.

files depicted on the plans must be checked to determine the grades that the equipment will encounter during haul and return cycles.

On a mass diagram graph, the horizontal dimension represents the stations of a project and the vertical dimension (the mass ordinate) represents the cumulative sum of excavation and embankment from some point of beginning on the project profile. Positive mass ordinate values are plotted above the zero datum line and negative values below. Figure 10.5 shows a mass diagram with the profile view of the same project plotted above the diagram itself.

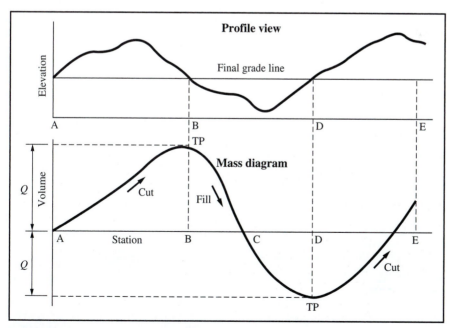

FIGURE 10.5 Mass diagram and the project profile view.

A mass diagram is a running total of the quantity of material that is surplus or deficient along the project profile. An excavation operation produces an *ascending* mass diagram slope; the excavation quantity exceeds the embankment quantity requirements. Conversely, if the operation is a fill situation, there is a deficiency of material and a *descending* slope is produced; the embankment requirements exceed the excavation quantity being generated.

Site-work projects are usually not linear in extent; therefore the use of a mass diagram is not appropriate. The engineer in that case must look at the cut-and-fill areas, lay out probable haul routes, and then check the natural and finish grade contours to determine the haul-route grades.

This process of laying out haul routes is critical to machine productivity. If a route can be found that results in less grade resistance, machine travel speed can be increased and production will likewise increase. In planning a project, a constructor should always check several haul-route options before deciding on a final construction plan.

Hauling efficiency is achieved by careful planning of haul routes.

Equipment selection is affected by travel distance because of the time factor distance introduces into the production cycle. All other factors being equal, increased travel distances will favor the use of high-speed large-capacity machines. The difference between the self-loading scraper (see Fig. 10.6) and a push-loaded scraper (see Fig. 10.7) can be used as an illustration. The self-loading scraper will load, haul, and spread without any assisting equipment, but

FIGURE 10.6 Self-loading scraper.

the extra weight of the loading mechanism reduces the unit's maximum travel speed and load capacity. A scraper, which requires a push tractor to help it load, does not have to expend power to haul a loading mechanism with it on every cycle. It will be more efficient in long-haul situations as it does not have to expend fuel transporting extra machine weight.

AVAILABLE POWER

There are two factors that determine available power: (1) horsepower and (2) speed. Horsepower is the time rate of doing work and is a constant value for any given machine. Since horsepower is a machine-specific constant, available pounds of pull or push will change as machine speed is varied. Fast travel results in low machine pulling ability; whereas slow travel enables the machine to exert a higher pulling or pushing force.

Horsepower Rating

Manufacturers rate machine horsepower as either gross or flywheel (sometimes listed as net horsepower). Gross horsepower is the actual power generated by the engine prior to load losses for auxiliary systems, such as the alternator, air conditioner compressors, and the water pump. Flywheel horsepower (fwhp) can be considered as *usable* horsepower. It is the power available to operate a machine—power the driveline—after deducting for power losses in the engine. This horsepower is sometimes listed as brake horsepower (bhp). Prior

FIGURE 10.7 Push-loaded scraper.

to electronic bench testing, horsepower was quantified as the amount of resistance against a flywheel brake. Although the method is no longer used, the term remains in the industry.

Flywheel power	209 kW (280 hp)
Rated power SAE (net)	198 kW (265 hp)

The Society of Automotive Engineers (SAE) standardized engine-rating procedure (J1349) measures horsepower at the flywheel, using an engine dynamometer. The engine is tested with all accessories installed, including a full exhaust system, all pumps, the alternator, the starter, and emissions controls. So today, equipment manufacturers measure torque on a dynamometer and then calculate horsepower by converting the radial force of torque into work units of horsepower.

The power output from the engine (fwhp) becomes the power input to the transmission system. This system consists of the drive shaft, a transmission, planetary gears, drive axles, and drive wheels.

When analyzing a piece of equipment, we are interested in the usable force developed at the point of contact between the tire and the ground (**rimpull**) for a wheel machine. In the case of a track machine, the force in question is that available at the drawbar (**drawbar pull**). The difference in the name is a matter of convention; both rimpull and drawbar pull are measured in the same units, pounds pull.

rimpull
The tractive force between the rubber tires of a machine's driving wheels and the surface on which they travel.

drawbar pull
The available pull that a crawler tractor can exert on towed load.

coefficient of traction
The factor that determines the maximum possible tractive force between the wheels or tracks of a machine and the surface on which it is traveling.

Rimpull

Rimpull is a term that is used to designate the tractive force between the tires of a machine's driving wheels and the surface on which they travel. If the **coefficient of traction** is high enough to eliminate tire slippage, the maximum rimpull is a function of the power of the engine and the gear ratios between the engine and the driving wheels. If the driving wheels slip on the haul surface, the maximum effective rimpull will be equal to the total pressure between the tires and the surface multiplied by the coefficient of traction. Rimpull is expressed in pounds.

EXAMPLE 10.3

A rubber-tired tractor is operating in dry clay, which has a coefficient of traction of 0.60. The total weight of the tractor is 108,000 lb, of which 50% is on the drive wheels. What is the maximum possible rimpull prior to slippage of the tires?

$$108,000 \text{ lb} \times 0.50 \times 0.60 = 32,400 \text{ lb}$$

If the rimpull of a vehicle is not known, it can be determined from the equation

$$\text{Rimpull} = \frac{375 \times \text{hp} \times \text{efficiency}}{\text{speed (mph)}} \text{ (lb)} \qquad \textbf{[10.10]}$$

TABLE 10.3 Maximum rimpull values for a 140-hp tractor.

Gear	Speed (mph)	Rimpull (lb)
First	3.3	13,523
Second	7.1	6,285
Third	12.9	3,542
Fourth	21.5	2,076
Fifth	33.9	1,316

The efficiency of most tractors and trucks will range from 0.80 to 0.85. For a rubber-tired tractor with a 140-hp engine and a maximum speed of 3.3 mph in first gear, the rimpull will be

$$\text{Rimpull} = \frac{375 \times 140 \times 0.85}{3.3} = 13,523 \text{ lb}$$

The maximum rimpull in all gear ranges for this tractor will be as shown in Table 10.3.

In computing the pull that a tractor can exert on a towed load, it is necessary to deduct from the rimpull of the tractor the force required to overcome the total resistance—the combination of rolling and grade. Consider a tractor that weighs 12.4 tons and whose maximum rimpull in the first gear is 13,523 lb. If it is operated on a haul road with a positive slope of 2% and a rolling resistance of 100 lb/ton, the full 13,523 lb of rimpull will not be available for towing, as a portion will be required for overcoming the total resistance arising from the haul conditions. Given the stated conditions, the pull available for towing a load will be

Maximum rimpull = 13,523 lb

Force required to overcome
  grade resistance,   12.4 ton × (20 lb/ton × 2%) = 496 lb

Force required to overcome
  rolling resistance,   12.4 ton × 100 lb/ton = 1,240 lb

Total resistance, 496 lb + 1,240 lb = 1,736 lb

Power available for towing a load (13,523 lb − 1,736 lb) = 11,787 lb

Drawbar Pull

The towing force a crawler tractor can exert on a load is referred to as *drawbar pull*. Drawbar pull is typically expressed in pounds. To determine the drawbar pull available for towing a load, it is necessary to subtract from the total pulling force available at the engine the force required to overcome the total resistance imposed by the haul conditions. If a crawler tractor tows a load up a slope, its drawbar pull will be reduced by 20 lb for each ton of weight of the tractor for each 1% slope.

The performance of crawler tractors, as reported in the specifications supplied by the manufacturer, is usually based on the Nebraska tests. In testing a tractor to determine its maximum drawbar pull at each of the available speeds, the haul road is calculated to have a rolling resistance of 110 lb/ton. If a tractor is used on a haul road whose rolling resistance is higher or lower than 110 lb/ton, the drawbar pull will be reduced or increased, respectively, by an amount equal to the weight of the tractor in tons multiplied by the variation of the haul road from 110 lb/ton.

EXAMPLE 10.4

A tractor whose weight is 15 tons has a drawbar pull of 5,685 lb in the sixth gear when operated on a level ground having a rolling resistance of 110 lb/ton. If the tractor is operated on a level ground having a rolling resistance of 180 lb/ton, the drawbar pull will be reduced by what amount?

$$15 \text{ tons} \times (180 \text{ lb/ton} - 110 \text{ lb/ton}) = 1,050 \text{ lb}$$

Thus, the effective drawbar pull will be 5,685 − 1,050 = 4,635 lb.

The drawbar pull of a crawler tractor will vary indirectly with the speed of each gear. It is highest in the first gear and lowest in the top gear. The specifications supplied by the manufacturer should give the maximum speed and drawbar pull for each of the gears.

USABLE POWER

Usable power depends on project conditions: primarily haul-road surface condition and type, altitude, and temperature. Underfoot conditions determine how much of the available power can be transferred at the wheel-surface (or track-surface) interface to propel the machine. As altitude increases, the air becomes less dense. Above 3,000 feet, the decrease in air density may cause a reduction in horsepower output of some engines. Manufacturers provide charts detailing appropriate altitude power reductions. Temperature will also affect engine output.

Coefficient of Traction

The total energy of an engine in any machine designed primarily for pulling a load can be converted into tractive effort only if sufficient traction can be developed between the driving wheels (or tracks) and the travel surface. If there is insufficient traction, the full power of the engine will not be available to do work, as the wheels or tracks will slip on the surface.

The coefficient of traction can be defined as the factor by which the total weight on the drive wheels or tracks should be multiplied to determine the maximum possible tractive force between the wheels or tracks and the surface just before slipping will occur.

$$\text{Usable force} = \frac{\text{Coefficient}}{\text{of traction}} \times \frac{\text{Weight on powered}}{\text{running gear}} \qquad \textbf{[10.11]}$$

The power that can be developed to do work is often limited by traction. The factors controlling usable horsepower are the weight on the powered running gear (drive wheels for wheel type, total weight for track type), the characteristics of the running gear, and the characteristics of the travel surface.

The coefficient of traction between rubber tires and travel surfaces will vary with the type of tread on the tires and with the travel surface material and condition of that material. For crawler tracks, it will vary with the design of the grouser and the travel surface. A grouser is the ridge or cleat across a track shoe of a track-type machine. These cleats run perpendicular to the long direction of the track and improve its grip on the ground. Because of these variations in tires and tracks, exact coefficient of traction values cannot be given.

Table 10.4 gives approximate values for the coefficient of traction between rubber tires or crawler tracks and road surfaces that are sufficiently accurate for most estimating purposes.

EXAMPLE 10.5

Assume that the rubber-tired tractor has a total weight of 18,000 lb on its driving wheels. The maximum rimpull in low gear is 9,000 lb. If the tractor is operating in wet sand, with a coefficient of traction of 0.30, what is the maximum possible rimpull prior to slippage of the tires?

$$0.30 \times 18,000 \text{ lb} = 5,400 \text{ lb}$$

Regardless of the horsepower of the engine, because of wheel slippage not more than 5,400 lb of force (power) is available to do work. If the same tractor is operating on dry clay, with a coefficient of traction of 0.60, what is the maximum possible rimpull prior to slippage of the wheels?

$$0.60 \times 18,000 \text{ lb} = 10,800 \text{ lb}$$

For this surface, the engine will not be able to cause the tires to slip. Thus, the full 9,000 lb of rimpull is available to do work.

EXAMPLE 10.6

A large off-highway truck is used on a road project. When the project initially begins, the truck will experience high rolling and grade resistance at one work

TABLE 10.4 Coefficients of traction for various road surfaces.

Surface	Rubber tires	Crawler tracks
Dry, rough concrete	0.80–1.00	0.45
Dry, clay loam	0.50–0.70	0.90
Wet, clay loam	0.40–0.50	0.70
Wet sand and gravel	0.30–0.40	0.35
Loose, dry sand	0.20–0.30	0.30
Dry snow	0.20	0.15–0.35
Ice	0.10	0.10–0.25

area. The rimpull required to maneuver in this work area is 42,000 lb. In the fully loaded condition, 67% of the total vehicle weight is on the drive wheels. The fully loaded vehicle weight is 149,000 lb. What minimum value of coefficient of traction between the truck wheels and the traveling surface is needed to maintain maximum possible travel speed?

Weight of the drive wheels $= 0.67 \times 149{,}000 \text{ lb} = 99{,}830 \text{ lb}$

Minimum required coefficient of traction $= \dfrac{42{,}000 \text{ lb}}{99{,}830 \text{ lb}} \Rightarrow 0.42$

Altitude Effect on Usable Power

The Society of Automotive Engineers standard J1349, *Engine Power Test Code—Spark Ignition and Compression Ignition—Net Power Rating Standard,* specifies a basis for a net engine power rating. The standard conditions for the SAE rating are a temperature of 60°F (15.5°C) and sea-level barometric pressure of 29.92 in. mercury (Hg) [103.3 kilopascals (kPa)].

> Altitude will affect machine performance.

The important point here is that the ratings are based on a specific barometric pressure. For naturally aspirated engines, operation at altitudes above sea level will cause a significant decrease in available engine power as the barometric pressure decreases. A decrease in barometric pressure causes a corresponding decrease in air density, and to operate at peak efficiency, the engine must have the proper amount of air. A reduction in air density affects the combustion fuel-to-air ratio in the engine's pistons. For specific machine applications, the manufacturer's performance data should be consulted.

The effect of the loss in power due to altitude can be lessened by the installation of a turbocharger or a supercharger. These are mechanical forced-induction systems that compress the air flowing into the combustion chamber of the engine, thus permitting sea-level performance at higher altitudes. The fundamental difference between a turbocharger and a supercharger is the unit's source of power. With a turbocharger, the exhaust stream powers a turbine, which in turn spins the compressor. The power source for a supercharger is a belt connected directly to the engine. If equipment is to be used at high altitudes for long periods of time, the increased performance will probably pay for installing one of these two devices.

performance charts
Graphical representations of the power and corresponding speed the engine and transmission of a machine can deliver.

PERFORMANCE CHARTS

Equipment manufacturers publish **performance charts** for individual machine models. These charts enable the equipment estimator/planner to analyze a machine's ability to perform under a given set of project-imposed load conditions. The performance chart is a graphical representation of the power and cor-

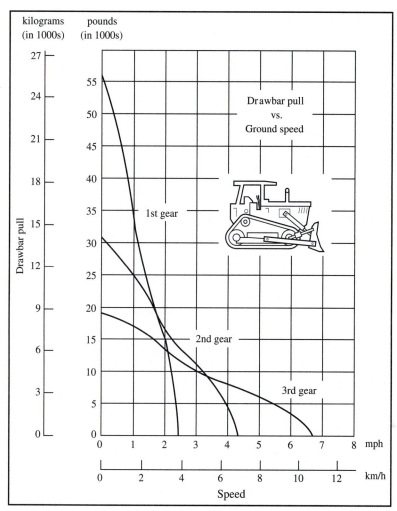

FIGURE 10.8 Drawbar pull performance chart.
(Reprinted courtesy Caterpillar Inc.)

responding speed the engine and transmission can deliver. The load condition is stated as either rimpull or drawbar pull. It should be noted that the drawbar pull/rimpull-speed relationship is inverse since vehicle speed increases pull decreases.

Drawbar Pull Performance Charts

In the case of the track machine whose drawbar pull performance chart is shown in Fig. 10.8, the available power ranges from 0 to 56,000 lb (the vertical scale) and the speed ranges from 0 to 6.7 mph (the horizontal scale).

Assuming the power required for a certain application is 25,000 lb, this machine would travel efficiently at a speed of approximately 1.4 mph in first gear. This is found by first locating the 25,000-lb mark on the left vertical scale and then moving horizontally across the chart. At the intersect point of this horizontal projection and the gear curve, project a vertical line downward to the speed scale at the bottom of the chart. The horizontal projection at 25,000 lb in this case also intersects the second gear curve. If the tractor were operated in second gear, it could only obtain a speed of 1 mph.

Rimpull Performance Charts

Each manufacturer has a slightly different graphical layout for presenting performance chart information. However, the procedures for reading a performance chart are basically the same. The steps described here are based on the chart shown in Fig. 10.9.

Required Power/Total Resistance The graphical arrangement of a rimpull chart enables a determination of machine speed using total resistance expressed either in terms of force (rimpull) or percent effective grade. The procedures to determine machine speed from a rimpull performance chart are

1. Ensure that the proposed machine has the same engine, gear ratios, and tire size as those identified for the machine on the chart. If the gear ratios or rolling radius of a machine is changed, the performance curve will shift along both the rimpull and speed axes.

2. Estimate the rimpull (power) required—total resistance (rolling resistance plus grade resistance)—based on the probable job conditions.

3. Locate the power requirement value on the left vertical scale and project a line horizontally to the right intersecting a gear curve. The point of intersection of the projected horizontal line with a gear curve defines the operating relationship between horsepower and speed.

4. From the point at which this horizontal line intersects the gear curve, draw a line vertically to the bottom x axis, which indicates the speed in mph and km/h. Sometimes the horizontal line from the power requirement (rimpull) will intersect the gear range curve at two different points. In such a case, the speed can be interpreted in two ways.

A guide in determining the appropriate speed is

■ If the required rimpull is *less* than that required on the previous stretch of haul, use the higher gear and speed.

■ If the required rimpull is *greater* than that required on the previous stretch of haul, use the lower gear and speed.

Effective Grade—Total Resistance Assuming that the total resistance has been expressed as an effective grade, the procedures to determine speed are

1. Ensure that the proposed machine has the same engine, gear ratios, and tire size as those identified for the machine on the chart.

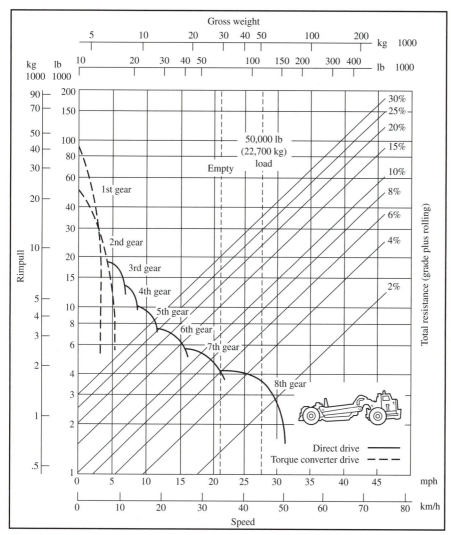

FIGURE 10.9 Rimpull performance chart.
(Reprinted courtesy Caterpillar Inc.)

2. Determine the machine weight both when the machine is empty and loaded. The empty weight is the operating weight and should include coolants, lubricants, full fuel tanks, and an operator. Loaded weight depends on the density of the carried material and the load size (volume). These two weights, empty and loaded, are often referred to as the *empty vehicle weight* (NVW) and the *gross vehicle weight* (GVW), respectively. The NVW (empty weight) is usually marked on the chart. Likewise the GVW (gross weight), based on the gravimetric capacity of the machine, will also usually be indicated on the chart. Note that these horizontal weight data are presented as a log scale.

3. Based on the probable job conditions, calculate a total resistance (the sum of rolling plus grade resistance both expressed as percent grade). For the performance chart shown in Fig. 10.9, the total resistance values are those shown as the vertical scale on the right side of the chart. Caution: the percent grade values are tick marked as a vertical scale, but the actual total percent resistance lines run diagonally, falling from right to left. The intersection of a vertical projection from vehicle weight and a diagonal total resistance line establishes the conditions under which the machine will operate and, correspondingly, the power requirement.

4. Project a line horizontally (do not proceed down the total resistance diagonal) from the intersection point of the vertical vehicle-weight projection and the appropriate total-resistance diagonal. The point of intersection of this horizontal projection with a gear curve defines the operating relationship between horsepower and speed.

5. From the point at which the horizontal line intersects the gear range curve, project a line vertically to the bottom *x* axis, which indicates the machine speed in mph and km/h. This is the vehicle speed for the assumed job conditions.

Performance charts are established assuming machine operation under standard conditions. When the machine is utilized under a differing set of conditions, the rimpull force and speed must be appropriately adjusted. Operation at higher altitudes will require a percentage derating in rimpull that is approximately equal to the percentage loss in flywheel horsepower.

Retarder Performance Charts

When operating on steep downgrades, a machine's speed may have to be limited for safety reasons. A retarder is a dynamic speed control device. By the use of an oil-filled chamber between the torque converter and the transmission, machine speed is retarded. The retarder will not stop the machine; rather it provides speed control for long downhill hauls, reducing wear on the service brake. Figure 10.10 presents a **retarder chart** for a wheel-tractor scraper.

retarder chart
A graph that identifies the speed that can be maintained when a machine is descending a grade having a slope such that the magnitude of the grade resistance is greater than the rolling resistance.

The retarder performance chart (see Fig. 10.10) identifies the speed that can be maintained when a machine is descending a grade having a slope such that the magnitude of the grade resistance is greater than the rolling resistance. This retarder-controlled speed is steady state, and use of the service brake will not be necessary to prevent acceleration.

A retarder performance chart is read in a manner similar to that already described, remembering that the total resistance (*effective grade*) values are actually negative numbers. As with the rimpull chart, the horizontal line can intersect more than one gear. In a particular gear, the vertical portion of the retarder curve indicates maximum retarder effort and resulting speed. If haul conditions dictate, the operator will shift into a lower gear and a lower speed would be applicable. Many times the decision as to which speed to select is answered by

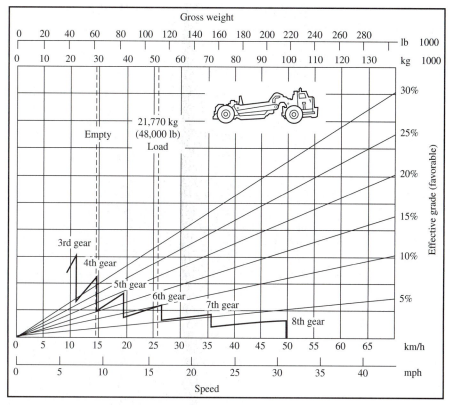

FIGURE 10.10 Retarder performance chart.
(Reprinted courtesy Caterpillar Inc.)

the question: "How much effort will be expended in haul-route maintenance?" Route smoothness is often the controlling factor affecting higher operating speeds. Route smoothness also impacts wear and tear on the equipment.

EXAMPLE 10.7

A contractor proposes to use scrapers on an embankment job. The performance characteristics of the machines are shown in Figs. 10.9 and 10.10. The scrapers have a rated capacity of 14 cy struck. Operating weight empty is 69,000 lb. The weight distribution of the scraper when loaded is 53% on the drive wheels.

The contractor believes that the average scraper load will be 15.2 bcy. The haul from the excavation area is a uniform adverse gradient of 5.0% with a rolling resistance of 60 lb/ton. The material to be excavated and transported is a common earth with a bank unit weight of 3,200 lb/bcy.

a. Calculate the maximum travel speeds that can be expected.

Machine weight:

Empty operating weight	69,000 lb
Payload weight, 15.2 bcy × 3,200 lb per bcy	48,640 lb
Total loaded weight =	117,640 lb

Haul conditions:

	Loaded (haul) (%)	Empty (return) (%)
Grade resistance	5.0	−5.0
Rolling resistance 60/20	3.0	3.0
Total resistance	8.0	−2.0

Loaded speed (haul): Using Fig. 10.9, enter the upper horizontal "Gross Weight" scale at 117,640 lb, the total loaded weight, and project a vertical down to the intersection with the 8.0% total resistance diagonal. From that intersection point, project a horizontal line to intersect with the gear curves. This horizontal line intersects fifth gear. By projecting a vertical line from the fifth-gear intersect point to the lower horizontal scale, the scraper speed is determined. The loaded speed will be approximately 11 mph.

Empty speed (return): Because the total resistance for the return is negative, use Fig. 10.10, the retarder chart, to determine the return speed. Enter the upper horizontal scale of Fig. 10.10 at 69,000 lb, the empty operating weight. Because this is a commonly used weight, it is marked on the chart as a dashed line. The intersection with the 2.0% effective resistance line defines the point from which to construct the horizontal line. The horizontal line intersects eighth gear, and the corresponding speed is 31 mph.

b. Calculate the one-way travel time for the haul and for the return.

$$\text{Travel time per segment (min)} = \frac{\text{Segment distance (ft)}}{88 \times \text{travel speed (mph)}} \quad \textbf{[10.12]}$$

Time to haul 2,800 ft:

$$\frac{2,800 \text{ ft}}{88 \times 11 \text{ mph}} = 2.9 \text{ min}$$

Time to return 2,800 ft:

$$\frac{2,800 \text{ ft}}{88 \times 31 \text{ mph}} = 1.0 \text{ min}$$

c. If the job is at elevation 12,500 ft, what will be the operating speeds when consideration of the nonstandard pressure is included in the analysis? Operating at an altitude of 12,500 ft, the manufacturer has reported that these scrapers, which have turbocharger engines, can deliver 82% of rated flywheel horsepower.

At an altitude of 12,500 ft, the rimpull that is necessary to overcome a total resistance of 8% must be adjusted for the altitude derating:

Altitude adjusted effective resistance: $\dfrac{8.0}{0.82} = 9.8\%$

Now proceed as in part (a) by locating the intersection of the total loaded weight line and the altitude adjusted total resistance line. Projecting a horizontal line from the 9.8% altitude-adjusted effective resistance diagonal yields an intersection with the fourth-gear curve, and by projecting a vertical line at that point, a speed of approximately 8 mph is determined.

In this problem, the nonstandard altitude would not affect the empty (return) speed because the machine does not require rimpull force for downslope motion. The retarder controls the machine's downhill momentum. Therefore, the empty (return) speed is still 31 mph at 12,500-ft altitude.

SUMMARY

The payload of hauling equipment can be expressed either *volumetrically* or *gravimetrically*. The power required is the power necessary to overcome the *total resistance* to machine movement, which is the sum of rolling and grade resistance. Rolling resistance is the resistance of a level surface to constant-velocity motion across it. The force-opposing movement of a machine up a frictionless slope is known as *grade resistance.*

The *coefficient of traction* is the factor by which the total weight on the drive wheels or tracks should be multiplied to determine the maximum possible tractive force between the wheels or tracks and the surface just before slipping will occur. A performance chart is a graphical representation of power and corresponding speed the engine and transmission can deliver. Critical learning objectives include

- An ability to calculate vehicle weight
- An ability to determine rolling resistance based on anticipated haul road conditions
- An ability to calculate grade resistance
- An ability to use performance charts to determine machine speed

These objectives are the basis for the questions in this chapter.

REVIEW QUESTIONS

10.1 A four-wheel tractor whose operating weight is 63,000 lb is pulled up a road whose slope is +4% at a uniform speed. If the average tension in the towing cable is 6,680 lb, what is the rolling resistance of the road? (132 lb/ton)

10.2 A wheel-type tractor-pulled scraper having a combined weight (tractor and scraper) of 172,000 lb is push-loaded down a 6% slope by a crawler tractor whose weight is 106,000 lb. What is the equivalent gain in loading force for the tractor and scraper resulting from loading the scraper down slope instead of up slope?

10.3 Consider a wheel-type tractor-pulled scraper whose gross weight is 118,700 lb. What is the equivalent gain, in horsepower (hp), resulting from operating this vehicle down a 4% slope instead of up the same slope at a speed of 12 mph? One horsepower equals 33,000 ft-lb of work per minute. (303.9 hp)

10.4 An articulated truck with a 260-hp engine has a maximum speed of 4.65 mph in first gear. Determine the maximum rimpull of the truck in each of the indicated gears if the efficiency is 85%.

Gear	Speed (mph)
First	4.0
Second	6.6
Third	11.5
Fourth	19.0
Fifth	32.2

10.5 If the truck in Question 10.4 weighs 46.7 tons and is operated over a haul road whose slope is +3% with rolling resistance of 80 lb per ton, determine the maximum external pull by the tractor in each of five gears. (14,181 lb; 6,019 lb; 669 lb; none; none)

10.6 If the truck in Questions 10.4 and 10.5 is operated down a 4% slope whose rolling resistance is 80 lb per ton, determine the maximum external pull by the truck in each of the five gears.

10.7 A truck tractor unit, operating in its second gear range and at its full rated rpm, is observed to maintain a steady speed of 1.5 mph, when operating under the conditions described here. Ambient air temperature is 60°F, altitude is sea level. The truck is climbing a uniform 6.5% slope with a rolling resistance of 65 lb per ton, and it is towing a pneumatic-tired trailer loaded with fill material. The two-axle truck tractor unit has a total weight of 60,000 lb, 55% of which is distributed to the power axle. The loaded trailer has a weight of 75,000 lb.

   a. The manufacturer, for the environmental conditions just described, rates the tractive effort of the new truck at 60 rimpull horsepower. What percentage of this rated rimpull does the truck actually develop? In performing your calculations, it is acceptable to assume that the component of weight normal to the traveling surface is equal to the weight itself (i.e., cos 0 taken as 1.00, where 0° is the angle of total resistance). The "20 lb of rimpull required per ton of weight per % of slope" approximation is acceptable. You may also disregard the power required to overcome wind resistance and to provide acceleration. Assume traction is not a limiting factor. (87.7%)

   b. What is the value of the coefficient of traction if the drive wheels of the truck are at the point of incipient slippage for the conditions just described? (0.4)

10.8 A truck tractor unit, operating in its fourth gear range and at its full-rated rpm, is observed to maintain a steady speed of 7.5 mph, when operating under the conditions described here. Ambient air temperature is 60°F. Altitude is sea level. The tractor is climbing a uniform 5.5% slope with a rolling resistance of 55 lb per ton, and it is towing a pneumatic-tired trailer loaded with fill material. The single-axle tractor has an operating weight of 66,000 lb. The loaded trailer has a weight of 48,000 lb. The weight distribution for the combined tractor trailer unit is 55% to the truck drive axle and 45% to the trailer rear axle.

    a. The manufacturer, for the environmental conditions just described, rates the tractive effort of the new tractor at 330 rimpull horsepower. What percentage of this rated rimpull horsepower does the tractor actually develop? In performing your calculations, it is acceptable to assume that the component of weight normal to the traveling surface is equal to the weight itself (i.e., cos 0 taken as 1.00, where 0° is the angle of total resistance). The "20 lb of rimpull required per ton of weight per % of slope" approximation is acceptable. You may also disregard the power required to overcome wind resistance and to provide acceleration. Assume traction is not a limiting factor.

    b. What is the value of the coefficient of traction if the drive wheels of the tractor are at the point of incipient slippage for the conditions just described?

10.9 A wheel-type tractor-scraper is operating on a level grade. Assume no power derating is required for equipment condition, altitude, temperature, etc. Use equipment data from Fig. 10.9.

    a. Disregarding traction limitations, what is the maximum value of rolling resistance (in lb per ton) over which the fully loaded unit can maintain a speed of 20 mph?

    b. What minimum value of coefficient of traction between the tractor wheels and the traveling surface is needed to satisfy the requirements of part (a)? For the fully loaded condition, 67% of the weight is distributed to the drive axle. Operating weight of the empty scraper is 70,000 lb.

10.10 A wheeled tractor with high-pressure tires and weighing 81,000 lb is pulled up a 5% slope at a uniform speed. If the tension in the tow cable is 15,200 lb, what is the rolling resistance of the road? What type of surface would this be?

10.11 A wheel-type tractor-scraper is operating on a 4% adverse grade. Assume that no power derating is required for equipment condition, altitude, and temperature. Use equipment data from Fig. 10.9. Disregarding traction limitations, what is the maximum value of rolling resistance (in lb per ton) over which the empty unit can maintain a speed of 14 mph?

REFERENCES

1. *Caterpillar Performance Handbook*, Caterpillar Inc., Peoria, IL. Published annually. (www.cat.com)

2. Peurifoy, Robert L., Clifford J. Schexnayder, Aviad Shapira (2006). *Construction Planning, Equipment, and Methods*, McGraw-Hill Companies, New York.

3. Schexnayder, Cliff, Sandra L. Weber, and Brentwood T. Brooks (1999). "Effect of Truck Payload Weight on Production," *Journal of Construction Engineering and Management*, ASCE, 125(1), Jan./Feb., pp. 1–7.

4. Peurifoy, Robert L., and Clifford J. Schexnayder (2002). *Construction Planning, Equipment, and Methods*, McGraw-Hill Companies, New York.

WEBSITE RESOURCES

1. www.aem.org   The Association of Equipment Manufacturers (AEM) is the international trade and business development resource for companies that manufacture equipment for the construction, mining, forestry, and utility industries.

2. www.cat.com   Caterpillar is the world's largest manufacturer of construction and mining equipment. In the Products, Equipment section of the website can be found the specifications for Caterpillar-manufactured scrapers.

3. www.constructionequipment.com   *Construction Equipment* magazine online.

4. www.equipmentworld.com   Equipmentworld.com is a news and e-commerce website for construction industry and equipment news.

5. www.hitachiconstruction.com   Hitachi Construction Machinery Co., Ltd., manufactures construction and transportation equipment.

6. www.terex.com   Terex Corporation is a diversified manufacturer of construction equipment. The specifications for their scrapers can be found in the Construction equipment section of the website.

C H A P T E R

11

Equipment Selection and Utilization

Hydraulic power is the key to the utility of many excavators. Hydraulic hoe-type excavators are used primarily to excavate below the natural surface of the ground on which the machine rests. The loader is a versatile piece of equipment designed to excavate at or above wheel/track level. Unlike a hoe, a loader must maneuver and travel with the load to position the bucket to dump. Cranes are used to lift material vertically on the job site. It is necessary to know the lifting capacity and working range of a crane selected to perform a given service.

HYDRAULIC EXCAVATORS

Hydraulic excavators can be either crawler- or rubber-tire-carrier-mounted. Either of these mounts can accommodate a variety of operating attachments. With the options in types, attachments, and sizes of machines, there are differences in appropriate applications and therefore variations in economical advantages. This chapter takes into account the important operating features and calls attention to cost consequences of specific machine applications.

One of the most significant operating features common to all these machines is hydraulic power. The hydraulic control of machine components provides

- Faster cycle times
- Positive control of attachments
- Precise control of attachments
- High overall efficiency
- Smoothness and ease of operation

If an excavator is considered as an independent unit (a one-link system), its production rate can be estimated using these steps.

Step 1. Obtain the heaped bucket load volume from the manufacturer's data sheet. This would be a loose volume (lcy) value.

369

fill factor

A numerical value used to adjust rated heaped excavator bucket capacity based on the type of material being handled and the type of excavator.

Step 2. Apply a bucket **fill factor** based on the type of machine and the class of material being excavated.

Step 3. Estimate a peak cycle time. This is a function of machine type and job conditions, including angle of swing, depth (height) of cut, and in the case of loaders, travel distance.

Step 4. Apply an efficiency factor.

Step 5. Conform the production units to the desired volume or weight (lcy to bcy or tons).

Step 6. Calculate the production rate.

The basic production formula is: Material carried per load × cycles per hour. In the case of excavators, this formula can be refined and written as

$$\text{Production} = \frac{3600 \text{ sec} \times Q \times F \times (\text{AS:D})}{t}$$
$$\times \frac{E}{60\text{-min-hr}} \times \frac{1}{\text{volume correction}} \qquad [\textbf{11.1}]$$

where
Q = heaped bucket capacity (lcy)
F = bucket fill factor
$AS{:}D$ = angle of swing and depth (height) of cut correction
t = cycle time in seconds
E = efficiency (min per hour)
volume correction for loose volume to bank volume,
$\dfrac{1}{1 + \text{swell factor}}$; for loose volume to tons, $\dfrac{\text{loose unit weight, lb}}{2{,}000 \text{ lb/ton}}$

A flow chart of this production analysis process is shown in Fig. 11.1.

Table 11.1 gives representative swell values for different classes of earth. These values will vary with the extent of loosening and compaction. If more accurate values are desired for a specific project, tests should be made on several samples of the earth taken from different depths and different locations within the proposed cut. The test can be made by weighing a given volume of undisturbed, loose, and compacted earth.

EXCAVATOR ACCIDENTS

A 26-year-old construction worker was killed while working in an 8-ft-deep trench, trying to remove a concrete sewer casing. Because it was impossible for the excavator operator to see the bottom of the trench where the casing was located, the victim was standing inside a trench box, giving hand signals to the operator above him. While pulling off the encasement, the bucket teeth slipped off the edge of the concrete and the excavator arm and bucket swung toward the victim, crushing him against the side of the trench box (Iowa NIOSH Fatality Assessment and Control Evaluation Investigation 96IA0).

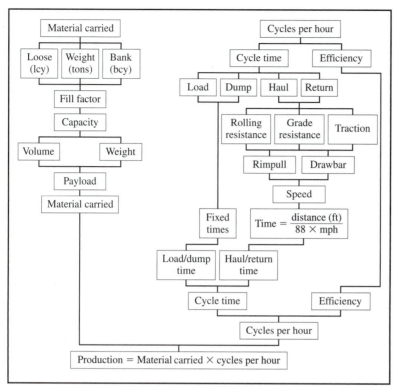

FIGURE 11.1 Excavator production analysis process.

TABLE 11.1 Representative properties of earth and rock.

| Material | Bank weight | | Loose weight | | Percent swell | Swell factor* |
	lb/cy	kg³/m	lb/cy	kg/m³		
Clay, dry	2,700	1,600	2,000	1,185	35	0.74
Clay, wet	3,000	1,780	2,200	1,305	35	0.74
Earth, dry	2,800	1,660	2,240	1,325	25	0.80
Earth, wet	3,200	1,895	2,580	1,528	25	0.80
Earth and gravel	3,200	1,895	2,600	1,575	20	0.83
Gravel, dry	2,800	1,660	2,490	1,475	12	0.89
Gravel, wet	3,400	2,020	2,980	1,765	14	0.88
Limestone	4,400	2,610	2,750	1,630	60	0.63
Rock, well blasted	4,200	2,490	2,640	1,565	60	0.63
Sand, dry	2,600	1,542	2,260	1,340	15	0.87
Sand, wet	2,700	1,600	2,360	1,400	15	0.87
Shale	3,500	2,075	2,480	1,470	40	0.71

Source: Peurifoy, Schexnayder, and Shapira, 2006.

*The swell factor is equal to the loose weight divided by the bank weight per unit volume.

Bureau of Labor Statistics (BLS) data identified 346 deaths associated with excavators or backhoe loaders during the period 1992 to 2000. A review of the data by the National Institute for Occupational Safety and Health (NIOSH) identified two common causes of injury:

- Being struck by the moving machine, swinging booms, or other machine components
- Being struck by quick-disconnect excavator buckets that unexpectedly detach from the stick [1]

Other leading causes of fatalities are rollovers, electrocutions, and machines sliding into trenches after cave-ins.

Managers have a responsibility to ensure that excavator operators and those working around such equipment always follow safe practices. Safe practice dictates that

- *Operators* keep machine attachments at a safe distance from workers at all times.
- *Workers* are trained regarding safe practices when working in close proximity to heavy equipment.
- *Supervisors* consider alternative working methods that eliminate the need to place workers in close proximity to heavy equipment.

BACKHOE-LOADERS

The success of the backhoe-loader (see Fig. 11.2) comes from the multiple tasks it can perform. This is not a high-production machine for any one task, but it provides flexibility to accomplish a variety of work tasks. The small machines (70 to 85 hp) have digging depths of approximately 15 feet. These are commonly used in residential construction applications for digging footers, trenching utility lines, general grading, and load and carry jobs. The larger machines (above 85 hp) have digging depths greater than 16 feet. They are used on larger industrial and commercial jobs requiring deeper trenches and handling heavier loads.

Loader functions A loader bucket attached to the front of the tractor enables this machine to excavate above wheel level.

Backhoe functions A hoe attached to the rear of the tractor enables it to excavate below wheel level. The hoe bucket can be replaced with a breaker or hammer, turning the unit into a demolition machine.

Production

The backhoe-loader is an excellent excavator for digging loosely packed moist clay or sandy clay. It is average over a broad range of soil conditions from sand

FIGURE 11.2 Backhoe-loader excavating a trench, front bucket flat on the ground for stability.

to hard clay. It is not really suitable for continuous high-impact diggings, such as with hard clay or caliche. Because of its four-wheel-drive capability, the backhoe-loader can work in unstable ground conditions.

HOES

GENERAL INFORMATION

Hoes are used primarily to excavate below the natural surface of the ground on which the machine rests (see Fig. 11.3). A hoe is sometimes referred to by other names, such as backhoe or back shovel. Hoes are adept at excavating trenches and pits for basements, and the smaller machines can handle general grading work. Because of their positive bucket control, they are superior to draglines for close-range work and loading into haul units.

Wheel-mounted hydraulic hoes (see Fig. 11.4) are available with buckets up to 1½ cy. Maximum digging depth for the larger machines is about 25 feet. With all four outriggers down, the large machines can handle 10,000-lb loads at a 20-foot radius. These are not production excavation machines. They are designed for mobility and general-purpose work.

Basic Parts and Operation of a Hoe

The basic parts and operating ranges of a hoe are illustrated in Fig. 11.5. Table 11.2 gives representative dimensions and clearances for hydraulic crawler-mounted hoes. Buckets are available in varying widths to suit the job requirements.

FIGURE 11.3 Crawler-mounted hydraulic hoe.
(Source: Kokosing Fru-Con.)

FIGURE 11.4 Wheel-mounted hydraulic hoe.

Penetration force into the material being excavated is achieved by the stick and bucket cylinders. Maximum crowd force is developed when the stick cylinder operates perpendicular to the stick. The ability to break material loose is best at the bottom of the arc because of the geometry of the boom, stick, and bucket and the fact that at that point, the hydraulic cylinders exert the maximum force drawing the stick in and curling the bucket.

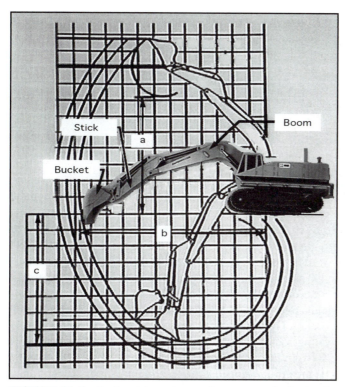

FIGURE 11.5 Basic parts and operating ranges of a hydraulic hoe: (a) dumping height; (b) digging reach; (c) maximum digging depth.

TABLE 11.2 Representative dimensions, loading clearance, and lifting capacity of hydraulic crawler hoes.

Size bucket (cy)	Maximum Stick length (ft)	Maximum reach at ground level (ft)	Maximum digging depth (ft)	Maximim loading height (ft)	Lifting capacity at 15 ft			
					Short stick		Long stick	
					Front (lb)	Side (lb)	Front (lb)	Side (lb)
⅜	5–7	19–22	12–15	14–16	2,900	2,600	2,900	2,600
¾	6–9	24–27	16–18	17–19	7,100	5,300	7,200	5,300
1	5–13	26–33	16–23	17–25	12,800	9,000	9,300	9,200
1½	6–13	27–35	17–21	18–23	17,100	10,100	17,700	11,100
2	7–14	29–38	18–27	19–24	21,400	14,500	21,600	14,200
2½	7–16	32–40	20–29	20–26	32,600	21,400	31,500	24,400
3	10–11	38–42	25–30	24–26	32,900*	24,600*	30,700*	26,200*
3½	8–12	36–39	23–27	21–22	33,200*	21,900*	32,400*	22,000*
4	11	44	29	27	47,900*	33,500*		
5	8–15	40–46	26–32	25–26	34,100†	27,500†	31,600†	27,600†

Source: Peurifoy, Schexnayder, and Shapira, 2006.
*Lifting capacity at 20 ft.
†Lifting capacity at 25 ft.

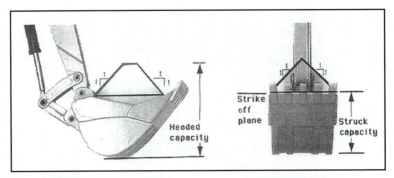

FIGURE 11.6 Hydraulic hoe bucket-capacity-rating dimensions.

BUCKET RATING FOR HYDRAULIC HOES

Hoe buckets are rated by Power Crane and Shovel Association (PCSA) and Society of Automotive Engineers (SAE) standards using a 1:1 angle of repose for evaluating heaped capacity (see Fig. 11.6). Buckets should be selected based on the properties of the material being excavated. By matching bucket width and bucket tip radius to the resistance of the material, full advantage can be taken of the hoe's potential. For easily excavated materials, wide buckets should be used. When excavating rocky material or blasted rock, a narrow bucket with a short tip radius is best. In the case of utility trenching work, the width of the required trench may be the critical consideration. Fill factors for excavator buckets adjust rated bucket heaped capacity based on the type of material being handled and the type of excavator. Fill factors for hoe buckets are presented in Table 11.3.

SELECTING A HOE

In the selection of a hoe for use on a project, these factors must be considered:

1. Maximum excavation depth required
2. Maximum working radius required for digging and dumping
3. Maximum dumping height required
4. Hoisting capability required (where applicable, e.g., handling pipe)

TABLE 11.3 Fill factors for hydraulic hoe buckets.

Material	Fill factor* (%)
Moist loam/sandy clay	100–110
Sand and gravel	95–110
Rock—poorly blasted	40–50
Rock—well blasted	60–75
Hard, tough clay	80–90

Source: Reprinted courtesy of Caterpillar Inc.
*Percentage of heaped bucket capacity.

Rated Hoist Load

In storm drain and utility work, the hoe can perform the trench excavation and handle the pipe, eliminating a second machine. Manufacturers provide machine-lifting capacities (rated hoist load) based on reach from the center of gravity of the bucket load to either the front (Fig. 11.7) or side of the track rails. Some typical data are provided in Table 11.2.

Rated hoist load is typically established based on these guidelines:

1. Rated hoist load shall not exceed 75% of the tipping load.
2. Rated hoist load shall not exceed 87% of the excavator's hydraulic capacity.
3. Rated hoist load shall not exceed the machine's structural capabilities.

HOE PRODUCTION

There are four elements in the production cycle of a hoe:

1. Load bucket
2. Swing with load
3. Dump load
4. Return swing

It should be noted that a hoe does not travel during the digging and loading cycle. Travel is limited to moving away from the face of the excavation as the work progresses.

FIGURE 11.7 Crawler-mounted hydraulic hoe lifting capacity.

TABLE 11.4 Excavation cycle times for hydraulic crawler hoes under average conditions.*

Bucket size (cy)	Load bucket (sec)	Swing loaded (sec)	Dump bucket (sec)	Swing empty (sec)	Total cycle (sec)
<1	5	4	2	3	14
1–1½	6	4	2	3	15
2–2½	6	4	3	4	17
3	7	5	4	4	20
3½	7	6	4	5	22
4	7	6	4	5	22
5	7	7	4	6	24

Source: Peurifoy, Shexnayder, and Shapira, 2006.
*Depth of cut 40%–60% of maximum digging depth; swinging angle 30–60°; loading haul units on the same level as the excavator.

The optimum depth of cut for a hoe will depend on the type of material being excavated and bucket size and type. As a rule, the optimum depth of cut for a hoe is in the range of 30%–60% of the machine's maximum digging depth. Table 11.4 presents cycle times for hydraulic track hoes based on bucket size and average conditions. No one has developed tables to relate average hoe cycle time to variations in depth of cut and horizontal swing. Therefore, when using Table 11.4, consideration must be given to those two factors when deciding on a load bucket time and the two swing times.

The basic production formula for a hoe used as an excavator is

$$\text{Hoe (excavator) production} = \frac{3600 \text{ sec} \times Q \times F}{t} \times \frac{E}{60\text{-min-hr}} \times \frac{1}{\text{volume correction}} \quad \text{[11.2]}$$

where
Q = heaped bucket capacity (lcy)
F = bucket fill factor for hoe buckets
t = cycle time in seconds
E = efficiency (min per hour)
volume correction for loose volume to bank volume, $\frac{1}{1 + \text{swell factor}}$; for loose volume to tons, $\frac{\text{loose unit weight, lb}}{2{,}000 \text{ lb/ton}}$

EXAMPLE 11.1

A crawler hoe having a 3½-cy bucket is being considered for use on a project to excavate very hard clay from a borrow pit. The clay will be loaded into trucks having a loading height of 9 ft 9 in. Soil-boring information indicates that below

8 ft, the material changes to an unacceptable silt. What is the estimated production of the hoe in cubic yards bank measure, if the efficiency factor is equal to a 50-min-hour?

Step 1. Size of bucket, 3½ cy

Step 2. Bucket fill factor (Table 11.3), hard clay 80–90%, use average 85%

Step 3. Typical cycle element times

Optimum depth of cut is 30–60% of maximum digging depth. From Table 11.2 for a 3½-cy-size hoe maximum digging depth is 23 to 27 ft. Depth of excavation, 8 ft:

$$\frac{8 \text{ ft}}{23 \text{ ft}} \times 100 = 34\% \geq 30\% \text{ okay}$$

$$\frac{8 \text{ ft}}{27 \text{ ft}} \times 100 = 30\% \geq 30\% \text{ okay}$$

Therefore, under average conditions and for a 3½-cy-size hoe, cycle times from Table 11.4 would be

1. Load bucket   7 sec   very hard clay

2. Swing with load   6 sec   load trucks

3. Dump load   4 sec   load trucks

4. Return swing   5 sec

Cycle time   22 sec

Step 4. Efficiency factor, 50-min-hr

Step 5. Class of material, hard clay, swell 35% (Table 11.1)

Step 6. Probable production:

$$\frac{3600 \text{ sec/hr} \times 3\frac{1}{2} \text{ cy} \times 0.85}{22 \text{ sec/cycle}} \times \frac{50 \text{ min}}{60 \text{ min}} \times \frac{1}{(1 + 0.35)} = 300 \text{ bcy/hr}$$

Check maximum loading height to ensure the hoe can service the trucks, from Table 11.2, 21–22 ft.

9 ft 9 in. < 21 okay

Hoe cycle times are usually of greater duration than those of other excavating machines. Part of the reason for this increase in cycle time is that after making the cut, the hoe bucket must be raised above the ground level to load a haul unit or to get above a spoil pile. If the trucks can be spotted on the floor of the pit, the bucket will be above the truck when the cut is completed. Then it would not be necessary to raise the bucket any higher before swinging and dumping the load. Every movement of the bucket equals increased cycle time. The spotting of haul units below the level of the hoe will increase production. A study by the first author found a 12.6% total cycle-time savings between loading at the same level and working the hoe from a bench above the haul units [9].

In trenching operations, the volume of material moved is not usually the question. The primary concern is to match the hoe's ability to excavate linear feet of trench per unit of time matched with the pipe-laying production.

LOADERS

GENERAL INFORMATION

Loaders are used extensively in construction work to handle and transport bulk material such as earth and rock; to load trucks; to excavate earth; and to charge aggregate bins at asphalt and concrete plants. The loader is a versatile piece of equipment designed to excavate at or above wheel/track level (see Fig. 11.8). The hydraulic-activated lifting system exerts maximum breakout force with an upward motion of the bucket. It does not require other equipment to level, smooth, or clean up the area in which it has been working.

Types and Sizes

Classified on the basis of running gear, there are two types of loaders: (1) the crawler-tractor-mounted type (see Fig. 11.9) and (2) the wheel-tractor-mounted type. They can be further grouped by the capacities of their buckets or the weights that the buckets can lift. Wheel loaders can be steered by the rear wheels or they can articulate. To increase stability during load lifting, the tracks of crawler loaders are usually longer and wider than those found on tractors.

FIGURE 11.8 Wheel-tractor loader.

FIGURE 11.9 Crawler-tractor type loader.

Operating Specifications

Tables 11.5 and 11.6 present the operating specifications across the ranges of commonly available wheel loaders and track loaders.

LOADER BUCKETS/ATTACHMENTS

The loader's hydraulic system provides the control necessary for operating a range of bucket types or attachments.

Buckets

The most common buckets are the one-piece conventional type, *general purpose;* the hinged-jaw, *multipurpose;* and the heavy-duty, *rock bucket.* The selected bucket is attached to the tractor by a push frame and lift arms.

 *General purpose.* The general purpose (one-piece) bucket is made of heavy-duty, all-welded steel. These buckets are usually equipped with replaceable teeth that bolt onto the cutting edge, but they also come with a straight lip (edge) and no teeth.

 *Multipurpose.* The segmented (two-piece) hinged-jawed bucket is made of heavy-duty, all-welded steel. It has bolted replaceable cutting edges. Bolt-on-type replaceable teeth are provided for excavation of medium-type materials. The two-piece bucket provides capabilities not available with a single-piece bucket. It enables the loader to also be used as a dozer and to grab material.

TABLE 11.5 Representative specifications for wheel loaders.

Size, heaped bucket capacity (cy)	Bucket dump clear- ance (ft)	Static tipping load, at full turn (lb)	Maximum forward speed				Maximum reverse speed				Raise/ dump/ lower cycle (sec)
			First (mph)	Second (mph)	Third (mph)	Fourth (mph)	First (mph)	Second (mph)	Third (mph)	Fourth (mph)	
1.25	8.4	9,600	4.1	7.7	13.9	21	4.1	7.7	13.9	—	9.8
2.00	8.7	12,700	4.2	8.1	15.4	—	4.2	8.3	15.5	—	10.7
2.25	9.0	13,000	4.1	7.5	13.3	21	4.4	8.1	14.3	23	11.3
3.00	9.3	17,000	5.0	9.0	15.7	26	5.6	10.0	17.4	29	11.6
3.75	9.3	21,000	4.6	8.3	14.4	24	5.0	9.0	15.8	26	11.8
4.00	9.6	25,000	4.3	7.7	13.3	21	4.9	8.6	14.9	24	11.6
4.75	9.7	27,000	4.4	7.8	13.6	23	5.0	8.9	15.4	26	11.5
5.50	10.7	37,000	4.0	7.1	12.4	21	4.6	8.1	14.2	24	12.7
7.00	10.4	50,000	4.0	7.1	12.7	22	4.6	8.2	14.5	25	16.9
14.00	13.6	98,000	4.3	7.6	13.0	—	4.7	8.3	14.2	—	18.5
23.00	19.1	222,000	4.3	7.9	13.8	—	4.8	8.7	15.2	—	20.1

Source: Peurifoy, Schexnayder, and Shapira, 2006.

TABLE 11.6 Representative specifications for track loaders.

Size, heaped bucket capacity (cy)	Bucket dump clearance (ft)	Static tipping load (lb)	Maximum forward speed (mph)	Maximum reverse speed (mph)	Raise/dump/ lower cycle (sec)
1.00	8.5	10,500	6.5	6.9	11.8
1.30	8.5	12,700	6.5	6.9	11.8
1.50	8.6	17,000	5.9*	5.9*	11.0
2.00	9.5	19,000	6.4*	6.4*	11.9
2.60	10.2	26,000	6.0*	6.0*	9.8
3.75	10.9	36,000	6.4*	6.4*	11.4

Source: Peurifoy, Schexnayder, and Shapira, 2006.
*Hydrostatic drive.

 Rock. The rock bucket is of one-piece, heavy-duty construction, having a protruding V-shaped cutting edge. This protruding edge can be used for prying up and loosening shot rock.

Other. Other buckets and attachments available include side dump buckets for use in cramped quarters; demolition buckets; plow blades for snow removal; brush rakes for clearing applications; heavy-duty sweeper brooms; and front booms, designed for lifting and moving sling loads.

Fill Factors for Loaders

The heaped capacity of a loader bucket is based on SAE standards. That standard specifies a 2:1 angle of repose for the material above the struck load. This repose angle (2:1) is different from that specified by SAE and PCSA for shovel and hoe buckets (1:1). The fill factor correction for a loader bucket (see Table 11.7) adjusts heaped capacity based on the type of material being handled and the type of loader, wheel, or track. Mainly because of the relationship between traction and developed breakout force, the bucket fill factors for the two types of loaders are different.

Operating Loads

Once the bucket volumetric load is determined, a check must be made of payload weight. Unlike a hoe, to position the bucket to dump, a loader must maneuver and travel with the load. A hoe simply swings about its center pin and does not require travel movement when moving the bucket from loading to dump position. SAE has established operating-load-weight limits for loaders. A wheel loader is limited to an operating load, by weight, that is less than 50% of rated full-turn static tipping load considering the combined weight of the bucket and the load, measured from the center of gravity of the extended bucket at its maximum reach, with standard counterweights and nonballasted tires. In the case of track loaders, the operating load is limited to less than 35% of static tipping load. The term *operating capacity* is sometimes used interchangeably for operating load. Most buckets are sized based on 3,000 lb per lcy material.

TABLE 11.7 Bucket fill factors for wheel and track loaders.

Material		Wheel loader fill factor (%)	Track loader fill factor (%)
Loose material			
Mixed moist aggregates		95–100	95–100
Uniform aggregates:	up to ⅛ in.	95–100	95–110
	⅛–⅜ in.	90–95	90–110
	½–¾ in.	85–90	90–110
	1 in. and over	85–90	90–110
Blasted rock			
Well blasted		80–95	80–95
Average		75–90	75–90
Poor		60–75	60–75
Other			
Rock dirt mixtures		100–120	100–120
Moist loam		100–110	100–120
Soil		80–100	80–100
Cemented materials		85–95	85–100

Source: Reprinted courtesy of Caterpillar Inc.

PRODUCTION RATES FOR WHEEL LOADERS

Two critical factors to be considered in choosing a loader are (1) the type of material and (2) the volume of material to be handled. Wheel loaders are excellent machines for soft to medium-hard material. However, wheel loader production rates decrease rapidly when used in medium to hard material. Another factor to consider is the height that the material must be lifted. To be of value in loading trucks, the loader must be able to reach over the side of the truck's dump bed. A wheel loader attains its highest production rate when working on a flat smooth surface with enough space to maneuver. In poor underfoot conditions or when there is a lack of space to maneuver efficiently, other equipment may be more effective.

Wheel loaders work in repetitive cycles, constantly reversing direction, loading, turning, and dumping. The production rate for a wheel loader will depend on the

1. Fixed time required to load the bucket, maneuver with four reversals of direction, and dump the load
2. Time required to travel from the loading to the dumping position
3. Time required to return to the loading position
4. The actual volume of material hauled each trip

Table 11.8 gives fixed cycle times for both wheel and track loaders. Figure 11.10 illustrates a typical loading situation. Because wheel loaders are more maneuverable and can travel faster on smooth haul surfaces, their production rates should be higher than those of track units under favorable conditions requiring longer maneuver distances.

When travel distance is more than minimal, it will be necessary to add a travel time to the fixed cycle time. For travel distances of less than 100 feet, a wheel loader should be able to travel with a loaded bucket at about 80% of its maximum speed in low gear and return empty at about 60% of its maximum speed in second gear. In the case of distances over 100 feet, return travel should be at about 80% of its maximum speed in second gear. If the haul surface is not well maintained or is rough, these speeds should be reduced accordingly.

TABLE 11.8 Fixed cycle times for loaders.

Loader size, heaped bucket capacity (cy)	Wheel loader cycle time* (sec)	Track loader cycle time* (sec)
1.00–3.75	27–30	15–21
4.00–5.50	30–33	—
6.00–7.00	33–36	—
14.00–23.00	36–42	—

Source: Peurifoy, Schexnayder, and Shapira, 2006.
*Includes load, maneuver with four reversals of direction (minimum travel), and dump.

FIGURE 11.10 Loading travel cycle for a loader.

Consider a wheel loader with a 2½-cy heaped capacity bucket, handling well-blasted rock weighing 2,700 lb per lcy, for which the swell is 25%. This unit, equipped with a torque converter and a power-shift transmission, has these speed ranges, forward and reverse:

Low	0–3.9 mph
Intermediate	0–11.1 mph
High	0–29.5 mph

The average speeds [in feet per minute (fpm)] should therefore be about

Hauling, all distances	0.8×3.9 mph $\times$ 88 fpm per mph	= 274 fpm
Returning, 0–100 ft	0.6×11.1 mph $\times$ 88 fpm per mph	= 586 fpm
Returning, over 100 ft	0.8×11.1 mph $\times$ 88 fpm per mph	= 781 fpm

The effect of increased haul distance on production is shown by the calculations provided in Table 11.9.

Example 11.2 demonstrates the process for estimating loader production.

EXAMPLE 11.2

A 4-cy wheel loader will be used to load trucks from a quarry stockpile of processed aggregate having a maximum size of 1 in. The haul distance will be negligible. The aggregate has a loose unit weight of 3,100 lb per cy. Estimate

TABLE 11.9　Effect of haul distance on production.

Cycle Element	Haul distance (ft)				
	25	50	100	150	200
Fixed time	0.45	0.45	0.45	0.45	0.45
Haul time	0.09	0.18	0.36	0.55	0.73
Return time	0.04	0.09	0.13	0.19	0.26
Cycle time, (min)	0.58	0.72	0.94	1.19	1.44
Trips per 50-min-hour	86.2	69.4	53.2	42.0	34.7
Production (tons)*	262	210	161	127	105

*0.9 bucket fill factor.

the loader production in tons based on a 50-min-hour efficiency factor. Use a conservative fill factor.

Step 1.　Size of bucket, 4 cy

Step 2.　Bucket fill factor (Table 11.7) aggregate over 1 in., 85–90%; use 85%, conservative. Check tipping:

Load weight:

$$4 \text{ cy} \times 0.85 = 3.4 \text{ lcy}$$

$$3.4 \text{ lcy} \times 3100 \text{ lb/lcy (loose unit weight of material)} = 10{,}540 \text{ lb}$$

From Table 11.5: 4-cy machine static tipping load at full turn is 25,000 lb

Therefore, operating load (50% static tipping at full turn) is

$$0.5 \times 25{,}000 \text{ lb} = 12{,}500 \text{ lb}$$

10,540 lb actual load < 12,500 lb operating load, therefore *okay*

Step 3.　Typical fixed cycle time (Table 11.8) 4-cy wheel loader, 30 to 33 sec; use 30 sec.

Step 4.　Efficiency factor, 50-min-hour

Step 5.　Class of material, aggregate 3,100 lb per lcy

Step 6.　Probable production:

$$\frac{3600 \text{ sec/hr} \times 4 \text{ cy} \times 0.85}{30 \text{ sec/cycle}} \times \frac{50 \text{ min}}{60 \text{ min}} \times \frac{3{,}100 \text{ lb/lcy}}{2{,}000 \text{ lb/ton}} = 527 \text{ ton/hr}$$

EXAMPLE 11.3

The loader in Example 11.2 will also be used to charge the aggregate bins of an asphalt plant that is located at the quarry. The one-way haul distance from the 1¼ in. aggregate stockpile to the cold bins of the plant is 220 ft. The asphalt plant uses 105 tons per hour of 1¼-in. aggregate. Can the loader meet this requirement?

Step 3. Typical fixed cycle time (Table 11.8) 4-cy wheel loader, 30 to 33 sec; use 30 sec.

From Table 11.5: Travel speeds forward

First, 4.3 mph; second, 7.7 mph; third, 13.3 mph

Travel speeds reverse

First, 4.9 mph; second, 8.6 mph; third, 14.9 mph

Travel loaded: 220 ft; because of short distance and time required to accelerate and brake, use 80% of first gear maximum speed.

$$\frac{4.3 \text{ mph} \times 80\% \times 88 \text{ fpm/mph}}{60 \text{ sec/min}} = 5.04 \text{ ft per sec}$$

Return empty: 220 ft; because of short distance and time required to accelerate and brake, use 80% of second gear maximum speed.

$$\frac{7.7 \text{ mph} \times 80\% \times 88 \text{ fpm/mph}}{60 \text{ sec/min}} = 9.03 \text{ ft per sec}$$

1. Fixed time   30 sec   4-cy wheel loader
2. Travel with load   44 sec   220 ft, 80% first gear
3. Return travel   24 sec   220 ft, 80% second gear

     Cycle time   98 sec

Step 6. Probable production:

$$\frac{3600 \text{ sec/hr} \times 4 \text{ cy} \times 0.85}{98 \text{ sec/cycle}} \times \frac{50 \text{ min}}{60 \text{ min}} \times \frac{3{,}100 \text{ lb/lcy}}{2{,}000 \text{ lb/ton}} = 161 \text{ ton/hr}$$

161 tons per hour > 105 tons per hour required

The loader will meet the requirement.

TRENCH SAFETY

Time and again noncompliance with trench safety guidelines and a lack of common sense result in the loss of life from cave-ins and entrapments. The death rate for trench-related accidents is nearly double that for any other type of construction accident. The first line of defense against cave-ins is a basic knowledge of soil mechanics combined with knowledge as to the type of material that will be excavated.

A critical question, often neglected, is previous disturbance of the material being excavated. In this regard, OSHA standard 1926.651(b)(1) states: "The estimated location of utility installations, such as sewer, telephone, fuel, electric, water lines, or any other underground installations that reasonably may be expected to be encountered during excavation work, shall be determined prior to opening an excavation."

Any trench measuring 5 feet or more in depth must be sloped, shored, or shielded (see Fig. 11.11).

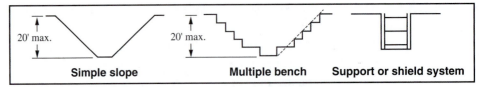

FIGURE 11.11 Methods for protecting workers in trenches.

Sloping is the most common method employed to protect workers in trenches. The trench walls are excavated in a V-shaped manner so that the angle of repose prevents the cave-in. Required slope angles vary depending on the specific soil type and soil moisture.

Benching is a subsidiary class of sloping that involves the formation of "steplike" horizontal levels. Both sloping and benching require ample right-of-way space. When the work area is restricted, shoring, sheathing, or shielding is necessary.

Shoring is a structural system that applies pressure against the walls of a trench to prevent collapse of the soil. *Sheathing* is a barrier driven into the ground to provide support to the vertical sides of an excavation. **Shields** or trench boxes (Fig. 11.12) are designed to protect the workers, not the excavation, from collapse.

shields

A structural system designed to protect the workers should an excavation in which they are working collapse.

Means of Egress

In trenches 4 feet or more deep, "A stairway, ladder, ramp or other safe means of egress shall be located . . . so as to require no more than 25 ft of lateral travel for

FIGURE 11.12 Positioning a trench box to protect workers in a trench.

employees" [OSHA standard 1926.651(c)(2)]. Spoil piles, tools, equipment, and materials must be kept at least 2 feet from the excavation's edge.

Other Requirements

OSHA is serious about trenching accident prevention and every construction manager should likewise be serious. In 2003, OSHA proposed $99,400 in penalties against a Louisiana contractor for failure to protect employees from potential trenching and excavation hazards. Three alleged repeat violations were issued for failing to provide a ladder for employees to get into and out of the trench, and placing soil from the trench within 2 feet from the trench's edge. Another regulation states that in the case of trenches 20 feet or more in depth, a registered professional engineer must design the excavation protection [1926.651(i)(2)(iii) and 1926.652(b)(4)].

Visit www.osha.gov/SLTC/trenchingexcavation/index.html for excellent tools that aid in identifying and controlling trenching hazards.

CRANES

MAJOR CRANE TYPES

Cranes are a broad class of construction equipment used to hoist and place material and machinery. Each type of crane is designed and manufactured to work economically in specific job conditions.

Construction cranes are generally classified into two major families: (1) mobile cranes and (2) tower cranes. The most common mobile crane types are

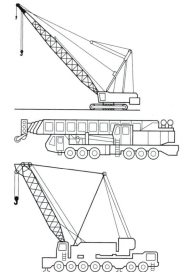

1. Crawler

2. Telescoping-boom truck mounted

3. Lattice-boom truck mounted

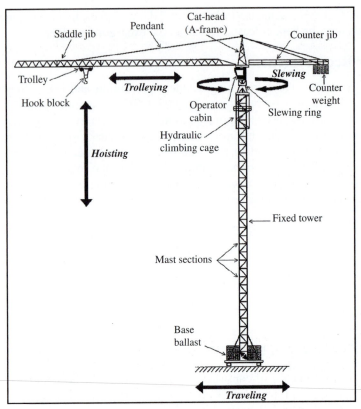

FIGURE 11.13   Nomenclature for a top-slewing tower crane.

The common tower crane types include

1. Top **slewing** *(fixed tower)* tower cranes (Fig. 11.13) have a fixed tower and a swing circle ("slewing ring" or "crown") mounted at the top, allowing only the jibs (main-jib and counter-jib), tower top, and operator cab to rotate. The tower is assembled from modular sections, and hence the term *sectional tower crane* is often used in reference to this type of crane. The crane is stabilized partly at its base (by ballasts or other means of ground anchoring) and partly by ballasts on the counter-jib.

2. Bottom slewing *(slewing tower)* tower cranes have the swing circle located under a slewing platform, and both the tower and jib assembly rotate relative to the base chassis. The tower is essentially a telescoping mast, and hence the term *telescopic tower crane* is often applied to this type of crane. The mast is usually a lattice-boom type (always for the midsize and larger cranes), but smaller cranes of this type may have a hollow-section mast similar to the telescopic boom of a mobile crane. The entire ballast is placed on the revolving base platform.

CRAWLER CRANES

The *full revolving superstructure* of this type of unit is mounted on a pair of continuous, parallel crawler tracks. Many manufacturers have different option packages available that enable configuration of the crane to a particular application, standard lift, tower unit, or duty cycle. Units in the low to middle range of lift capacity have good lifting characteristics and are capable of **duty-cycle work** such as handling a concrete bucket. Machines of 100-short-ton* capacity and above are built for lift capability and do not have the heavier components required for duty-cycle work. The **universal machines** incorporate heavier frames, have heavy-duty or multiple clutches and brakes, and have more powerful swing systems. These designs allow for quick changing of drum laggings that vary the torque/speed ratio of cables to the application. Figure 11.14 illustrates a large crawler crane on a building project.

The crawlers provide the crane with travel capability around the job site. The crawler tracks provide such a large ground contact area that soil failure under these machines is only a problem when operating on soils having a very

duty-cycle work
A repetitive lifting assignment of relatively short cycle time.

universal machines
The base machine can be used as a crane or dragline and for pile driving or other such application.

---

*A short ton (or U.S. ton) equals 2,000 lb as opposed to a metric or long ton (termed *ton* in the text), which equals 2,240 lb.

FIGURE 11.14 Crawler crane on a building project in Washington, DC.

low bearing capacity. Before hoisting a load the machine must be leveled and ground settlement considered. If soil failure or ground settlement is possible, the machine can be positioned and leveled on mats. The distance between crawler tracks affects stability and lift capacity.

To relocate a crawler crane between projects requires transport by truck, rail, or barge. As the size of the crane increases, the time and cost to dismantle, load, investigate haul routes, and reassemble the crane also increases. The durations and costs can become significant for large machines. Relocating the largest machines can require 15 or more truck trailer units. These machines usually have lower initial cost per rated lift capability, but movement between jobs is more expensive. Therefore, crawler-type machines should be considered for projects requiring long duration usage at a single site.

Many new models use modular components to make dismantling, transporting, and assembling easier. Quick-disconnect locking devices and pin connectors have replaced multiple-bolt connections.

Most crawler cranes have a fixed-length lattice boom (Fig. 11.14), which is also a crane type discussed in this section. A lattice-boom is cable-suspended, and therefore acts as a compression member, *not* a bending member like a telescoping hydraulic boom. There exist, however, new small-sized crawler models that are hydraulic boom equipped.

TELESCOPING-BOOM TRUCK-MOUNTED CRANES

There are truck cranes (see Fig. 11.15) that have a self-contained telescoping boom. The multisection-telescoping boom is a permanent part of the full revolving superstructure. In this case, the superstructure is mounted on a multiaxle truck/carrier. Most of these units can travel on the public highways between projects under their own power with a minimum of dismantling. Once the crane is leveled at the new work site, it is ready to work without setup delays. These machines, however, have higher initial cost per rated lift capability. If a job requires crane utilization for a few hours to a couple of days, a telescoping truck crane should be given first consideration because of its ease of movement and setup.

Telescoping-boom truck cranes have extendable **outriggers** for stability. In fact, many units cannot be operated safely with a full reach of boom unless the outriggers are fully extended and the machine raised so that the tires are clear of the ground. In the case of the larger machines, the width of the outriggered vehicle may reach 40 feet, which necessitates careful planning of the operation area. Additionally, these heavy machines transfer, through the outriggers, extremely high loads to the ground. This high-ground loading must be considered vis-à-vis the soil-bearing capacity. Large-sized timber or steel mats are used to spread the load over a larger ground area, further increasing the overall width requirement. These outrigger space considerations are also a concern when using large lattice-boom truck cranes.

outriggers
Movable beams that can be extended laterally from a mobile crane to stabilize and help support the unit.

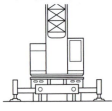

FIGURE 11.15 Telescoping truck crane on a construction site.

LATTICE-BOOM TRUCK-MOUNTED CRANES

As with the telescoping-boom truck crane, a full-revolving superstructure is mounted on a multiaxle truck/carrier. The advantage of this machine is the *lattice-boom.* The lattice-boom structure is lightweight. This reduction in boom weight means additional lift capacity as the machine predominately handles hoist load and less weight of boom. The lattice-boom does take longer to assemble than simply extending a telescoping boom. The lightweight boom will give a less expensive lattice-boom machine (Fig. 11.16) the same hoisting capacity as a larger telescoping-boom unit.

The disadvantage of these units is the time and effort required disassembling them for transport. In the case of the larger units, it may be necessary to remove the entire superstructure. Additionally, a second crane is often required for this task. Some newer models are designed so that the machine can separate itself without the aid of another crane.

FIGURE 11.16   Lattice-boom truck crane.

sheave
Grooved pulley-wheel for changing the direction of a wire rope's pull.

LIFTING CAPACITIES OF CRANES

Because cranes are used to hoist and move loads from one location to another, it is necessary to know the lifting capacity and working range of a crane selected to perform a given service. Individual manufacturers and suppliers will furnish machine-specific information in the literature describing their machines.

When a crane lifts a load attached to the hoist line that passes over a **sheave** located at the boom point of the machine, there is a tendency to tip the machine over. This introduces what is defined as the *tipping condition*. With the crane on a firm, level, supporting surface in calm air, it is considered to be at the point of tipping when a balance is reached between the overturning moment of the load and the stabilizing moment of the machine.

During tests to determine the tipping load for wheel-mounted cranes, the outriggers should be lowered to relieve the wheels of all weight on the supporting surface or ground. The radius of the load is the horizontal distance from the axis of rotation of the crane to the center of the vertical hoist line or tackle with the load applied. The *tipping load* is the load that produces a tipping condition at a specified radius. The load includes the weight of the item being lifted plus the weights of the hooks, hook blocks, slings, and any other items used in hoisting the load, including the weight of the hoist rope located between the boom-point sheave and the item being lifted.

RATED LOADS FOR LATTICE- AND TELESCOPIC-BOOM CRANES

The rated load for a crane as published by the manufacturer is based on ideal conditions. Load charts can be complex documents listing numerous booms, jibs, and other components that can be employed to configure the crane for various tasks. It is critical that the chart being consulted be for the actual crane configuration that will be used.

> Interpolation between the published values IS NOT permitted; use the next lower value. Rated loads are based on ideal conditions, a level machine, calm air (no wind), and no dynamic effects.

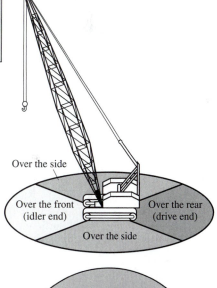

A partial safety factor in respect to tipping is introduced by the PCSA rating standards that state that the rated load of a lifting crane shall not exceed the following percentages of tipping loads at specified radii [11]:

1. Crawler-mounted machines, 75%
2. Rubber-tire-mounted machines, 85%
3. Machines on outriggers, 85%

It should be noted that there are other groups that recommend rating criteria. The Construction Safety Association of Ontario [7, 10] recommends that for rubber-tire-mounted machines a factor of 75% be used.

One manufacturer is producing rubber-tire-mounted cranes having intermediate outrigger positions. For intermediate positions greater than one-half the fully extended length, the manufacturer is using a rating based on 80% of the tipping load. For intermediate positions less than one-half the fully extended length, a rating based on 75% is used. At this time, there is no standard for this type machine.

Load capacity will vary depending on the quadrant position of the boom with respect to the machine's undercarriage. In the case of crawler cranes, the three quadrants that should be considered are

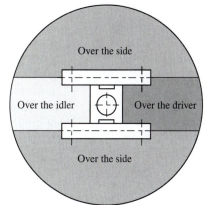

1. Over the side
2. Over the drive end of the tracks
3. Over the idler end of the tracks

Crawler crane quadrants are usually defined by diagonal lines through the crane's center of rotation and the driver or idler at the ends of the track. However, the quadrants are sometimes defined by the longitudinal centerline of the machine's crawlers. The area between the centerlines of the two crawlers is considered over the end and the area outside the crawler centerlines is considered over the side.

In the case of wheel-mounted cranes, the quadrants of consideration will vary with the configuration of the outrigger locations. If a machine has only four outriggers, two on each side, one set located forward and one set to the rear, the quadrants are usually defined by imaginary lines running from the superstructure center of rotation through the position of the outrigger support. In such a case, the three quadrants to consider are

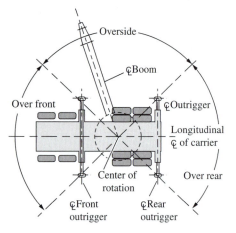

1. Over the side
2. Over the rear (of the carrier)
3. Over the front (of the carrier)

Some wheel-mounted cranes have an outrigger directly in the front, or there can be other machine-specific outrigger configurations. Therefore, the best practice is to carefully consult the manufacturer's specifications.

The important point is that the rated load should be based on the direction of minimum stability for the mounting, unless otherwise specified. The minimum stability condition restricts the rated load because the crane must both raise and swing loads. The swinging motion will cause the boom to move through various quadrants, changing the load's effect on the machine. Further, it should be remembered that the rating is based on the fact that the outriggers are fully extended.

Rated loads are based on the assumption that the crane is in a level position (for the full 360° of swing). When a crane is not level, even small variations significantly affect lifting capacity. In the case of a short-boom machine operating at minimum radius, 3° out of level can result in a 30% loss in capacity. For long-boom machines, the loss in capacity can be as great as 50% [10].

Another important consideration with modern cranes is that tipping is not always the critical capacity factor. At short radii, capacity may be dependent on boom or outrigger strength and structural capacity, and at long radii, pendant tension can be the controlling element. Manufacturers' load charts will limit the rated capacity to values below the minimum critical condition taking into account all possible factors.

Table 11.10 illustrates the kind of information issued by the manufacturers of cranes. The crane in this example is described as a 200-ton, nominal rating, crawler-mounted cable-controlled crane with 180 feet of boom. It is important to realize that the lifting capacity by which mobile cranes are identified (although not necessarily named) does not represent a standard classification method. That is why the word *nominal* is added to the description. In most cases, if a ton rating is used, it refers to lifting capacity with a basic boom and at minimum

TABLE 11.10 Lifting capacities in pounds for a 200-ton, nominal rating, crawler crane with 180 ft of boom.

Radius (ft)	Capacity (lb)	Radius (ft)	Capacity (lb)	Radius (ft)	Capacity (lb)
32	146,300	80	39,200	130	17,900
36	122,900	85	35,800	135	16,700
40	105,500	90	32,800	140	15,500
45	89,200	95	30,200	145	14,500
50	76,900	100	27,900	150	13,600
55	67,200	105	25,800	155	12,700
60	59,400	110	23,900	160	11,800
65	53,000	115	22,200	165	11,100
70	47,600	120	20,600	170	10,300
75	43,100	125	19,200	175	9,600

Source: Manitowoc Engineering Co.
Note: Specified capacities based on 75% of tipping loads.

radius. Some manufacturers use a load moment rating classification system. The designation would be "tm." As an example, a Demag AC 650 (reflecting 650-ton lifting capacity, in this case referring to a metric ton) is named by a big crane rental company AC 2000 (reflecting 2000 tm max. load moment).

The capacities located in the upper portion of a load chart (Table 11.11), usually defined by either a bold line or by shading, represent structural failure conditions. Operators can feel the loss of stability prior to a tipping condition.

TABLE 11.11 Lifting capacities in pounds for a 25-ton truck-mounted hydraulic crane.

Load radius (ft)	Lifting capacity (lb)[†], boom length (ft)						
	31.5	40	48	56	64	72	80
12	50,000	45,000	38,700				
15	41,500	39,000	34,400	30,000			
20	29,500	29,500	27,000	24,800	22,700	21,100	
25	19,600	19,900	20,100	20,100	19,100	17,700	17,100
30		14,500	14,700	14,700	14,800	14,800	14,200
35			11,200	11,300	11,400	11,400	11,400
40			8,800	8,900	9,000	9,000	9,000
45				7,200	7,300	7,300	7,300
50				5,800	5,900	6,000	6,000
55					4,800	4,900	4,900
60					4,000	4,000	4,000
65						3,100	3,300
70							2,700
75							2,200

Note: Specified crane capacities based on 85% of tipping loads.

[†]The loads appearing below the solid line are limited by the machine stability. The values appearing above the solid line are limited by factors other than machine stability.

But in the case of a structural failure, there is no sense of feeling to warn the operator; therefore, load charts must be understood and all lifts must be in strict conformance with the ratings.

While the manufacturer will consider crane structural factors when developing a capacity chart for a particular machine, operational factors affect absolute capacity in the field. The manufacturer's ratings can be thought of as valid for a *static* set of conditions. A crane on a project operates in a *dynamic* environment, lifting, swinging, and being subjected to air currents and temperature variations. The load chart provided by the manufacturer does not take into account these dynamic conditions. Factors that will greatly affect actual crane capacity on the job are

1. Wind forces on the boom or load
2. Swinging the load
3. Hoisting speed
4. Stopping the hoist

These dynamic factors should be carefully considered when planning a lift.

Rated Loads for Hydraulically Operated Cranes

The rated tipping loads for hydraulic cranes are determined and indicated as for cable-controlled cranes. However, in the case of hydraulic cranes, the critical load rating is sometimes dictated by hydraulic pressure limits instead of tipping. Therefore, load charts for hydraulic cranes represent the controlling-condition lifting capacity of the machine, and the governing factor may not necessarily be tipping. The importance of this is that the operator cannot use physical sense of balance (machine feel) as a gage for safe lifting capability.

TOWER CRANES

Tower cranes provide high lifting height and good working radius, while taking up a very limited area. These advantages are achieved at the expense of lifting capacity and mobility, as compared to mobile cranes. The three common tower crane configurations are (1) a special vertical boom arrangement on a mobile crane, (2) a mobile crane superstructure mounted atop a tower, or (3) a vertical tower with a jib (see Fig. 11.17). The latter description is often referred to in the United States as the European type, but it is the type perceived elsewhere as a "tower crane" when this term is used with no further details. Tower cranes of jib type usually fall within one of two categories:

1. *Top-slewing (fixed tower)* tower cranes (Fig. 11.17) have a fixed tower and a swing circle ("slewing ring" or "crown") mounted at the top, allowing only the jibs (main-jib and counter-jib), tower top, and operator cab to rotate. The tower is assembled from modular sections, and hence the term *sectional tower crane* is often used in reference to this type of crane. The crane is stabilized partly at its base (by ballasts or other means of ground anchoring) and partly by ballasts on the counter-jib.

FIGURE 11.17 Top slewing (fixed tower) tower crane.

2. *Bottom-slewing (slewing tower)* tower cranes (Fig. 11.18) have the swing circle located under a slewing platform, and both the tower and jib assembly rotate relative to the base chassis. The tower is essentially a telescoping mast, and hence the term *telescopic tower crane* is often applied to this type of crane. The mast is usually a lattice-boom type (always for the midsized and larger cranes), but smaller cranes of this type may have a hollow-section mast similar to the telescopic boom of a mobile crane. The entire ballast is placed on the revolving base platform.

The main differences between these two categories are reflected in setup and dismantling procedures and in lifting height. Bottom slewing tower cranes essentially erect themselves using their own motors, in a relatively short and simple procedure. They are often referred to as, "self-erecting" or "fast-erecting" cranes. This is achieved at the expense of service height, as dictated by the telescoping tower (mast) that, because of its revolving base, cannot be braced to a permanent structure. On the other hand, setting up and dismantling top-slewing tower cranes requires more time, is more complicated, and can be a

FIGURE 11.18 Bottom-slewing tower crane.

costlier procedure. Erection of a top-slewing tower crane requires the assistance of other equipment, but the crane can reach extreme heights. Consequently, the bottom-slewing models are suitable mainly for shorter-term service of low-rise buildings, while top-slewing cranes commonly serve high-rise buildings on jobs requiring a crane for a long duration.

TOWER CRANE SELECTION

The use of a tower crane requires considerable planning because the crane is a fixed installation on the site for the duration of the major construction activities. From its fixed position, it must be able to cover all points from which loads are to be lifted and to reach the locations where the loads must be placed. Therefore, when selecting a crane for a particular project, the engineer must ensure that the weight of the loads can be handled at their corresponding required radius. Individual tower cranes are selected for use based on

1. Weight, dimension, and lift radii of the heaviest loads
2. Maximum free-standing height of the machine
3. Maximum braced height of the machine
4. Machine-climbing arrangement
5. Weight of machine supported by the structure
6. Available head room that can be developed
7. Area that must be reached
8. Hoist speeds of the machine
9. Length of cable the hoist drum can carry

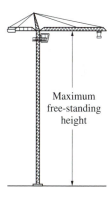

Maximum free-standing height

The vertical movement of material during the construction process creates an *available headroom* clear-distance requirement. This distance is defined as the vertical distance between the maximum achievable crane-hook position and the uppermost work area of the structure. The dimensions of those loads that must be raised over the upper most work area during the building process set the requirement. For practical purposes and safety, hook height above the serviced building top should never be less than 20 feet. When selecting a tower crane for a very tall structure, a climbing-type crane may be the only choice capable of meeting the available headroom height requirement.

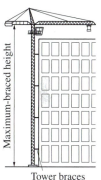

Maximum-braced height

Tower braces

RATED LOADS FOR TOWER CRANES

While hook or lift height does not affect capacities directly, there is a relation when hoist speed is considered. Tower cranes are usually powered by alternating current (ac) electric motors, producing only low-level noises for city-friendly operation. A crane having a higher motor horsepower can achieve higher operating speeds. When considering the production capability of a crane for duty-cycle work, hoist line speed and the effect motor size has on speed are very important. This is especially true for high-rise construction, where travel time of the hook between loading and unloading areas is the most significant part of the crane's cycle time, as opposed to low-rise construction, where travel time is insignificant compared to load rigging and unrigging times. If a project requires operating speeds that are higher than provided by an existing crane, replacing the crane's motors with more powerful ones is an alternative option to bringing in another crane.

Hoist-cable configuration is another factor affecting lifting speed. Tower cranes can usually be rigged with one of two hoist-line configurations, a two-part line or a four-part line. The four-part-line configuration provides a greater lifting capacity than a two-part line within the structural capacity constraints of the tower and jib configuration. The maximum lifting capacity of the crane will be increased by 100% with the four-part-line configuration. However, the increased lifting capacity is acquired with a resulting loss of 50% in vertical hoist speed. Table 11.12 is a capacity chart for a climbing tower crane rigged with a two-part line and having a maximum reach of 218 feet. This particular

TABLE 11.12 Lifting capacities in pounds for a tower crane.

Jib model Maximum hook reach	L1 104' 0"	L2 123' 0"	L3 142' 0"	L4 161' 0"	L5 180' 0"	L6 199' 0"	L7 218' 0"	Hook reach
	27,600	27,600	27,600	27,600	27,600	27,600	27,600	10' 3"
	27,600	27,600	27,600	27,600	27,600	27,600	27,600	88' 2"
	27,600	27,600	27,600	27,600	27,600	27,600	25,800	94' 6"
	27,600	27,600	27,600	27,600	27,600	25,800	24,200	101' 0"
	27,600	27,600	27,600	27,600	26,800	24,900	23,400	104' 0"
		27,600	27,600	27,600	25,200	23,600	22,200	109' 8"
		27,600	27,600	25,600	23,300	21,800	20,500	117' 8"
		27,000	27,000	25,100	22,800	21,300	20,100	120' 0"
Lifting capacities		26,300	26,300	24,300	22,200	20,700	19,500	123' 0"
in pounds,			24,800	22,800	20,800	19,300	18,300	130' 0"
two-part line			22,400	20,700	18,700	17,400	16,400	142' 0"
				19,500	17,600	16,300	15,400	150' 0"
				18,800	16,800	15,700	14,800	155' 0"
				17,900	16,200	15,100	14,200	161' 0"
					15,200	14,200	13,300	170' 0"
					14,200	13,200	12,400	180' 0"
						12,300	11,600	190' 0"
						11,700	10,800	199' 0"
							10,200	210' 0"
							9,700	218' 0"

Source: Morrow Equipment Company, L.L.C.

crane can have a stationary free-standing height such that there is 212 feet of clear under-hook height.

Tower-crane load charts are usually structured assuming that the weight of the hook-block is part of the crane's dead weight. But the rigging system is taken as part of the lifted load. When calculating loads, the Construction Safety Association of Ontario recommends that a 5% working margin be applied to the computed weight.

EXAMPLE 11.4

Can the tower crane, whose load chart is shown in Table 11.12, lift a 15,000-lb load at a radius of 142 ft? The crane has an L7 jib and a two-part hoist line. The slings that will be used for the pick weigh 400 lb.

Weight of load	15,000 lb
Weight of rigging	400 lb (slings)
	15,400 lb
	× 1.05 working margin
Required capacity	16,170 lb

From Table 11.12, the maximum lifting capacity at a 142-ft hook reach is 16,400 lb.

$$16,400 \text{ lb} > 16,170 \text{ lb}$$

Therefore, the crane can safely make the lift.

CRANE SAFETY

On-site crane safety requires, first and foremost, that all involved parties (project manager, general superintendent, crane operator, etc.) be well aware of safety hazards—factors that increase the chance of an accident—on the particular job. Potential safety hazards, by which the expected safety level at a given site may be evaluated, preferably before actual construction has started, fall under three categories (factors associated primarily with tower cranes are marked by an asterisk—*):

1. The "human factor" is reflected mainly in the experience and competence of the operator as well as the signalers, the mode of operator employment (i.e., whether you use your own company operator or hire one through a manpower company)*, and the attitude of all personnel involved with on-site crane work.

2. "Project factors" are the presence of power lines and the compactness of the site; overlapping of crane work envelopes and oversailing of the crane's jib*; the length of the workday and working night shifts; work conditions inside the operator cab and the use of optional, advanced operator aids; various visibility interruptions, particularly hidden work zones; and hazardous loads and lifting assignments.

3. Typical "environmental (i.e., non-project-specific) factors" are winds and severe weather, maintenance standards of the cranes and lifting accessories, and corporate policy toward safety management.

Safety should be a major concern not only when the crane is in operation, but also in other phases of its presence on the project site. This is particularly true for tower cranes during erection and dismantling, climbing, and after-duty hours. During all of these periods, the crane is not in its full or "natural" working state. "Natural" working state is when the crane is doing what it is designed and built to do, namely, lift loads. During after-duty hours, when no loads are lifted, the balance of forces is shifted while no operator is in the cab. A gust of wind, a local structural failure, or a disengagement of brakes that went unnoticed while the cab was unmanned may develop into an accident.

Even more hazardous are erection, climbing, and dismantling operations of tower cranes. A considerable number of tower-crane accidents that involved these operations have been reported. These operations differ from the routine state of employment first and foremost because the structure of the crane and its various operation and control systems are not fully configured; the crane is thus in a delicate and constantly changing balance of forces. Additionally, these operations are unique in that they involve other equipment in close proximity, as well as other personnel. Since these operations are commonly subcontracted (i.e., their executioners are out of the direct control of the construction company and not subject to its internal safety plans and programs), utmost caution should be exercised in the prequalification and selection process of these subcontractors (e.g., thorough examination of experience and accident records).

OTHER CONSTRUCTION EQUIPMENT

Most equipment is designed to efficiently perform specific tasks. The manager should visualize how best to build the project and select the appropriate machines based on their mechanical capabilities.

DOZERS

A dozer is a tractor-powered unit that has a blade attached to the machine's front (Fig. 11.19). It is designed to provide tractive power for drawbar work. A dozer has no set volumetric capacity. The amount of material the dozer moves is dependent on the quantity that will remain in front of the blade during the push. Crawler dozers equipped with special clearing blades are excellent machines for land clearing.

SCRAPERS

Tractor-pulled scrapers (Fig. 11.20) are designed to load, haul, and dump loose material. The greatest advantage of tractor-scraper combinations is their versatility. The key to a pusher-scraper spread's economy is that both the pusher and

FIGURE 11.19 Crawler dozer.

FIGURE 11.20 Tractor scraper.

the scraper share in the work of obtaining the load. Scrapers can be used in a wide range of material types (including shot rock) and are economical over a wide range of haul lengths and haul conditions. Scrapers are best suited for haul distances greater than 500 feet but less that 3,000 feet, although with larger units the maximum distance can approach a mile.

FINISHING EQUIPMENT

Finishing, finish grading, and fine grading are all terms used in reference to the process of shaping materials to the required line and grade. Graders are multipurpose machines used for finishing and shaping (Fig. 11.21). The components of the grader that actually do the work are the moldboard (blade) and the scarifier. Graders may also be equipped with lightweight rear-mounted rippers.

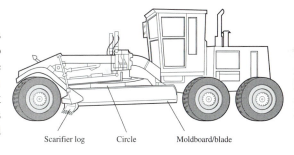

Scarifier log   Circle   Moldboard/blade

FIGURE 11.21 Grader working a fill.

TRUCKS AND HAULING EQUIPMENT

Trucks are hauling units that provide relatively low hauling costs because of their high travel speeds (Fig. 11.22). There are at least three methods of rating the capacities of trucks and wagons:

1. *Gravimetric*—the load it will carry, expressed as a weight

2. *Struck volume*—the volumetric amount it will carry, if the load is water level in the body (bowl or dump box)

3. *Heaped volume*—the volumetric amount it will carry, if the load is heaped on a 2:1 slope above the body (bowl or dump box)

The productive capacity of a truck depends on the size of its load and the number of trips it can make in an hour. The number of trips completed per hour is a function of cycle time. Truck cycle time has four components: (1) load time, (2) haul time, (3) dump time, and (4) return time.

COMPACTION EQUIPMENT

The effectiveness of different compaction methods is dependent on the individual soil type being manipulated. Appropriate compaction methods based on soil type are identified in Table 11.13.

Manufacturers have developed distinct compactors that incorporate at least one of the compaction methods and in some cases more than one into their performance capabilities. Many types of compacting equipment are available, including

FIGURE 11.22 Rigid-frame rear-dump truck.

TABLE 11.13 Soil types versus method of compaction.

Material	Impact	Pressure	Vibration	Kneading
Gravel	Poor	No	Good	Very good
Sand	Poor	No	Excellent	Good
Silt	Good	Good	Poor	Excellent
Clay	Excellent with confinement	Very good	No	Good

 1. Tamping rollers

 2. Smooth drum vibratory soil compactors

 3. Pad drum vibratory soil compactors

 4. Pneumatic-tired rollers

Tamping Rollers   Tamping foot compactors (Fig. 11.23) are high-speed, self-propelled, nonvibratory rollers. These rollers usually have four steel-padded wheels and can be equipped with a small blade to help level the lift. The pads are tapered with an oval or rectangular face. The pad face is smaller than the base of the pad at the drum.

FIGURE 11.23 Self-propelled tamping roller with blade.

FIGURE 11.24 Dual-drum vibratory compactor on asphalt.

Smooth Drum Vibratory Compactors The smooth drum compactors, whether single or dual drum (see Fig. 11.24), generate three compactive forces: (1) pressure, (2) impact, and (3) vibration. These rollers are most effective on granular materials with particle sizes ranging from large rocks to fine sand, and for compacting asphalt pavements.

Padded Drum Vibratory Soil Compactors These rollers are effective on soils with up to 50% of the material having a plasticity index (PI) of 5 or greater.

Pneumatic-Tired Rollers These are surface rollers that apply the principle of kneading action to effect compaction below the surface. Pneumatic-tired rollers are used in compacting asphalt, chip seals, recycled pavement, and base and sub-base materials.

ASPHALT PAVERS

An asphalt paver consists of a tractor, either track or rubber-tired, and a screed (see Fig. 11.25). The tractor-powered unit has a receiving hopper in the front and a system of slat conveyors to move the mix through a tunnel under the power plant to the rear of the tractor unit. At the rear of the tractor unit, the mix is deposited on the surface to be paved and augers are used to spread the asphalt evenly across the front of the trailing screed. Two tow arms, pin connected to the tractor unit, draw the screed behind the tractor. The screed controls the asphalt placement width and depth, and imparts the initial finish and compaction to the material.

CONCRETE EQUIPMENT

Increasingly, concrete is proportioned at a central location and transported to the purchaser in a fresh state, mixed en route. This type of concrete is termed *ready-mixed concrete* or *truck-mixed concrete.*

FIGURE 11.25 Wheel-mounted asphalt paver.

Concrete Transit-Mix Trucks   Transit mixers (Fig. 11.26) are available in several sizes up to about 14 cy, but the most popular size is 8 cy. They are capable of thoroughly mixing the concrete with about 100 revolutions of the mixing drum. Mixing speed is generally 8 to 12 rpm.

Concrete Pumps   By applying pressure to a column of fresh concrete in a pipe, the concrete can be moved through the pipe if a lubricating outer layer is

FIGURE 11.26 Transit-mix trucks delivering concrete to the job site.

provided and if the mixture is properly proportioned for pumping. To work properly, the pump must be fed concrete of uniform workability and consistency. In the United States, perhaps as much as one-fourth of all concrete is placed by pumping (a pump can be seen behind the transit-mix trucks in Fig. 11.26). Pumps are available in a variety of sizes, capable of delivering concrete at sustained rates of 10 to 150 cy per hour. Effective pumping distance varies from 300 to 1,000 feet horizontally, or 100 to 300 feet vertically, although occasionally pumps have moved concrete more than 5,000 feet horizontally and 1,000 feet vertically.

SUMMARY

The same elements govern the production of all types of excavators. The first issue is how much material is actually loaded into the bucket. This is a function of the bucket volume and the type of material being excavated; a bucket fill factor is applied to the rated heaped capacity of the bucket. The second issue is cycle time. In the case of hoes, cycle time is a function of depth of cut and swing angle. When travel distance is more than minimum for a loader, the production cycle time is influenced by travel speed.

The rated load for a crane, as published by the manufacturer, is based on ideal conditions. Load charts can be complex documents listing numerous booms, jibs, and other components that may be employed to configure the crane for various tasks. It is critical that the chart being consulted be for the actual crane configuration that will be used. Rated loads are based on ideal conditions, a level machine, calm air (no wind), and no dynamic effects.

Critical learning objectives include

- An ability to adjust bucket volume based on appropriate fill factors
- An ability to ascertain bucket cycle time as a function of machine type and job conditions to include angle of swing, depth or height of cut, and, in the case of loaders, travel distance
- An ability to select an efficiency factor to use in the production calculation by considering both project-specific physical conditions and contractor management ability
- An understanding of the basic mobile and tower-crane types
- An understanding of load charts and their limitations

These objectives are the basis for the questions in this chapter.

REVIEW QUESTIONS

11.1 A crawler hoe with a 2-cy bucket and a cost per hour, including the wages to an operator, of $67 will excavate and load haul units under each of the stated conditions. The maximum digging depth of the machine is 20 ft. Determine the cost per bank cubic yard for each condition (Condition 1, 300 bcy/hr, $0.223/bcy; condition 2, 319 bcy/hr, $0.210/bcy; condition 3, 128 bcy/hr, $0.523/bcy).

Condition	(1)	(2)	(3)
Material	Moist loam, earth	Sand and gravel	Rock, well blasted
Depth of excavation (ft)	12	16	14
Angle of swing (degrees)	60	80	120
Percent swell		14	
Efficiency factor (min-hr)	45	50	45

11.2 A crawler hoe with a 3-cy bucket and a cost per hour, including the wages to an operator, of $86 will excavate and load haul units under each of the stated conditions. The maximum digging depth of the machine is 25 ft. Determine the cost per bank cubic yard for each of the stated conditions.

Condition	(1)	(2)	(3)	(4)
Material	Sandy clay	Hard clay	Rock, well blasted	Sand and gravel
Depth of excavation (ft)	10	18	15	20
Angle of swing (degrees)	60	90	90	150
Percent swell	20			13
Efficiency factor (min-hr)	45	50	50	55

11.3 A 3-cy wheel loader will be used to load trucks from a quarry stockpile of processed aggregates having a maximum size of ¼ in. The haul distance will be negligible. The aggregate has a loose unit weight of 2,950 lb per cy. Estimate the loader production in tons based on a 50-min-hour efficiency factor. Use an aggressive cycle time and fill factor. (467 ton/hr)

11.4 A 7-cy wheel loader will be used to load a crusher from a quarry stockpile of blasted rock (average breakage) 180 ft away. The rock has a loose unit weight of 2,700 lb per cy. Estimate the loader production in tons based on a 50-min-hour efficiency factor.

11.5 High Lift Construction Co. has determined that the heaviest load to be lifted on a project they anticipate bidding weighs 14,000 lbs. From the proposed tower crane location on the building site, the required reach for this lift will be 150 ft. The crane will be equipped with an L5 jib and a two-part hoist line. From the load chart, the rated capacity is 17,600 lbs. This critical lift is of a piece of limestone facing and will require a 2,000-lb spreader bar attached to a 300-lb set of slings. If assembled in the proposed configuration, can the crane safely make the pick?

REFERENCES

1. Barnes, Jonathan (2005). "OSHA May Update Warning on Quick Excavator Attachments," *ENR,* McGraw-Hill Construction, New York, February 28, p.12.

2. *Articulating Boom Cranes* (1994). ASME B30.22-1993, an American National Standard, The American Society of Mechanical Engineers, New York.

3. Bates, Glen E., and Robert M. Hontz (1998). *Exxon Crane Guide, Lifting Safety Management System,* Specialized Carriers & Riggers Association, Fairfax, VA.

4. *Below-the-Hook Lifting Devices* (1994). ASME B30.20-1993, an American National Standard, The American Society of Mechanical Engineers, New York.

5. *Caterpillar Performance Handbook,* Caterpillar Inc., Peoria, IL. (published annually) www.cat.com.

6. *Construction Standards for Excavations* (1990). AGC publication No. 126, promulgated by the Occupational Safety and Health Administration, The Associated General Contractors of America, Washington, DC.

7. *Crane Handbook,* 10th printing (1999). Construction Safety Association of Ontario, Toronto, Ontario, Canada.

8. *Crane Safety on Construction Sites* (1998). ASCE Manuals and Reports on Engineering Practice No. 93, American Society of Civil Engineers, Reston, VA.

9. *Hammerhead Tower Cranes* (1992). ASME B30.3-1990, an American National Standard, The American Society of Mechanical Engineers, New York.

10. *Mobile Crane Manual* (1993). Construction Safety Association of Ontario, Toronto, Ontario, Canada.

11. *Mobile Power Crane and Excavator and Hydraulic Crane Standards, PCSA Standard 1* (1968). Power Crane and Shovel Association, a Bureau of Construction Industry Manufacturers Association, Milwaukee, WI.

12. Nichols, Herbert L., Jr., and David A. Day (1998). *Moving the Earth, the Workbook of Excavation,* 4th ed., McGraw-Hill, New York.

13. O'Brien, James J., John A. Havers, and Frank W. Stubbs, Jr. (1996). *Standard Handbook of Heavy Construction,* 3rd ed., McGraw-Hill, New York.

14. Peurifoy, Robert L., Clifford J. Schexnayder, and Aviad Shapira (2006). *Construction Planning, Equipment, and Methods,* McGraw-Hill Companies, New York.

15. *Rigging Manual* (1992). Construction Safety Association of Ontario, Toronto, Ontario, Canada.

16. Schexnayder, Cliff, Sandra L. Weber, and Brentwood T. Brooks (1999). "Effect of Truck Payload Weight on Production," *Journal of Construction Engineering and Management,* ASCE, 125(1), January–February: 1–7.

17. Shapira, Aviad, and Jay D. Glascock (1996). "Culture of Using Mobile Cranes for Building Construction," *Journal of Construction Engineering and Management,* ASCE, 122(4): 298–307.

18. Shapira, Aviad, and Clifford J. Schexnayder (1999). "Selection of Mobile Cranes for Building Construction Projects," *Construction Management & Economics* (UK), 17(4): 519–27.

19. Shapiro, Howard I., Jay P. Shapiro, and Lawrence K. Shapiro (1991). *Cranes and Derricks,* 2nd ed., McGraw-Hill Book Company, New York.

EXCAVATOR WEBSITE RESOURCES

1. www.aem.org   The Association of Equipment Manufacturers (AEM) is the international trade and business development resource for companies that manufacture equipment, products, and services used worldwide in the construction, agricultural, industrial, mining, forestry, materials-handling, and utility industries. AEM was formed on January 1, 2002, from the consolidation of the Construction Industry Manufacturers Association (CIMA) and the Equipment Manufacturers Institute (EMI).

2. www.caterpillar.com   Caterpillar Inc. is the world's largest manufacturer of construction and mining equipment.

3. www.coneq.com/index.asp   Construction Equipment magazine online.

4. www.deere.com/deerecom/Contractors/default.htm   Deere & Company, the John Deere construction equipment site provides specifications for many machines and has equipment videos.

5. www.equipmentworld.com/   Equipmentworld.com is an online magazine featuring construction industry and equipment news.

6. www.hitachi-c-m.com   Hitachi Construction Machinery Co., Ltd, manufacture and sell excavators, wheel loaders, off-road dump trucks, tunnel boring machines, and other products.

CRANE AND LIFTING WEBSITE RESOURCES

1. www.csao.org   Construction Safety Association of Ontario, 74 Victoria St., Toronto, Ontario, Canada M5C 2A5, provides health and safety education, consultation, and information to workers and management in the construction industry.

2. www.cranestodaymagazine.com   Cranes Today. Cranes Today on the Web is a gateway to the lifting industry online and bringing industry news—when it happens.

3. www.aem.org/CBC/ProdSpec/PCSA   The Power Crane & Shovel Association (PCSA) of the Association of Equipment Manufacturers explores business issues, technological questions and legislative and regulatory concerns (domestic and worldwide) that affect manufacturers of cranes. It also promotes the standardization and simplification of terminology and and classification of cranes for worldwide harmonization. They are located at 6737 W. Washing ton Street, Suite 2400, Milwaukee, WI 53214-5647.

4. www.sae.org/servlets/index   Society of Automotive Engineers (SAE). SAE World Headquarters, 400 Commonwealth Drive, Warrendale, PA 15096-0001 USA; phone 1-877-606-7323 (United States and Canada only) or 724-776-4970 (outside the United States and Canada); fax Customer Service: 724-776-0790; Headquarters: 724-776-5760.

5. www.scranet.org   Specialized Carriers and Rigging Association (SC&RA), 2750 Prosperity Avenue, Suite 620, Fairfax, Virginia 22031-4312; phone 703-698-0291, fax 703-698-0297, e-mail, info@scranet.org. Specialized Carriers and Rigging Association (SC&RA) provides information about safe transporting, lifting, and erecting oversized and overweight items.

12

Equipment Costs

A correct and complete understanding of the costs that result from equipment ownership and operation provide companies a market advantage that leads to greater profits. Ownership cost is the cumulative result of those cash flows an owner experiences whether or not the machine is productively employed on a project. Operating cost is the sum of those expenses an owner experiences by working a machine on a project. The process of selecting a particular type of machine for use in constructing a project requires knowledge of the cost associated with operating the machine in the field. There are three basic methods for securing a particular machine to use on a project: (1) buy, (2) rent, or (3) lease.

INTRODUCTION

Equipment cost is often one of a contractor's largest expense categories, and it is a cost fraught with variables and questions. To be successful, equipment owners must carefully analyze and answer two separate cost questions about their machines:

1. How much does it cost to operate the machine on a project?
2. What is the optimum economic life and optimum manner to secure a machine?

The first question is critical to bidding and operations planning. The only reason for purchasing equipment is to perform work that will generate a profit for the company. This first question seeks to identify the expense associated with productive machine work, and is commonly referred to as ownership and operating costs (**O&O**). It is usually expressed in dollars per equipment operating hour.

The second question seeks to identify the optimum point in time to replace a machine and the optimum way to secure a machine. This is important in that it will reduce O&O cost and thereby lower production expense, enabling a

O&O
The ownership and operating costs of a machine.

415

contractor to achieve a better pricing position. The process of answering this question is known as replacement analysis. A complete replacement analysis must also investigate the cost of renting or leasing a machine.

The economic analyses, which answer these cost questions, require the input of many expense and operational factors. These input costs will be discussed first and a development of the analysis procedures will follow.

EQUIPMENT DOCUMENTATION

Data on both machine utilization and costs are the keys to making rational equipment decisions, but collection of individual pieces of data is only the first step. The data must be assembled and presented in usable formats. Many contractors recognize this need and strive to collect and maintain accurate equipment records for evaluating machine performance, establishing operating cost, analyzing replacement questions, and managing projects.

Realizing the advantages to be gained therefrom, owners are directing more attention to accurate record keeping. Advances in computer technology have reduced the effort required to implement record systems. Several computer companies offer record-keeping packages specifically designed for contractors. In many cases, the task is simply the retrieval of equipment cost data from existing accounting files.

Automation introduces the ability to handle more data economically and in shorter time frames, but the basic information required to make rational decisions is still the critical item. A commonly used technique in equipment costing and record keeping is the standard rate approach. Under such a system, jobs are charged a standard machine utilization rate for every hour the equipment is employed. Machine expenses are charged either directly to the piece of equipment or to separate equipment cost accounts. This method is sometimes referred to as an internal or company rental system. Such a system usually presents a fairly accurate representation of *investment consumption* and it properly assigns machines expenses. In the case of a company replacing machines each year and continuing in the construction business, this system allows a check at the end of each year on estimate rental rates, as the internally generated rent should equal the expenses absorbed.

The first piece of information necessary for rational equipment analysis is not an expense but a record of the machine's use. One of the implicit assumptions of a replacement analysis is that there is a continuing need for a machine's production capability. Therefore, before beginning a replacement analysis, the disposal–replacement question must be resolved. Is this machine really necessary? A projection of the ratio between total equipment capacity and utilized capacity provides a quick guide for the dispose–replace question.

The level of detail for reporting equipment use varies. Both independent service vendors and equipment companies offer data collection devices that provide accurate real-time information about machine use. These devices are installed in the machine and transmit data via the most cost-effective wireless network (satellite or cellular networks). As a minimum, data should be collected on a daily basis to record whether a machine worked or was idle. A more sophisticated system will seek to identify use on an hourly basis, accounting for actual production time and categorizing idle time by classifications such as standby, down weather, and down repair. The input for either type of system is easily incorporated into regular personnel timekeeping reports, with machine time and operator time being reported together.

Most of the information required for ownership and operating or replacement analyses is available in the company's accounting records. All owners keep records on a machine's initial purchase expense and final realized salvage value as part of the accounting data required for tax filings. Maintenance expenses can be tracked from mechanics' time sheets, purchase orders for parts, or shop work orders. Service logs provide information concerning consumption of consumables. Fuel amounts can be recorded at fuel points or with automated systems. Fuel amounts should be cross-checked against the total amount purchased. When detailed and correct reporting procedures are maintained, the accuracy of equipment cost analyses is greatly enhanced.

Many discussions of equipment economics include *interest* as a cost of ownership. Sometimes the authors make comparisons with the interest rates that banks charge for borrowed funds or with the rate that could be earned if the funds were invested elsewhere. Such comparisons imply that these are appropriate rates to use in an equipment cost analysis. A few authors appear to have perceived the proper character of interest by realizing that a company requires capital funds for all of its operations. It is not logical to assign different interest costs to machines purchased wholly with retained earnings (cash) as opposed to those purchased with borrowed funds. A single corporate interest rate should be determined by examination of the combined costs associated with all sources of capital funds: debt, equity, and internal. For a complete treatment of cost of capital, see Modigliani and Miller's classic paper published in *The American Economic Review* [4].

OWNERSHIP COST

ELEMENTS OF OWNERSHIP COST

Ownership cost is the cumulative result of those cash flows an owner experiences whether or not the machine is productively employed on a job. It is a cost related to finance and accounting exclusively, and it does not include the wrenches, nuts and bolts, and consumables (fuel) necessary to keep the machine operating.

Most of these cash flows are expenses (outflows), but a few are cash inflows. The most significant cash flows affecting *ownership cost* are

1. Purchase expense
2. Salvage value
3. Major repairs and overhauls
4. Property taxes
5. Insurance
6. Storage and miscellaneous

Purchase Expense

The cash outflow the firm experiences in acquiring ownership of a machine is the purchase expense. It is the total delivered cost (drive-away cost), including amounts for all options, shipping, and taxes, less the cost of tires if the machine has rubber tires. The machine will show as an asset on the books of the firm. The firm has exchanged money (cash or borrowed funds), liquid assets, for a machine, a fixed asset with which the company hopes to generate profit. As the machine is used on projects, wear takes its toll and the machine can be thought of as being used up or consumed. This consumption reduces the machine's value because the revenue stream it can generate is likewise reduced. Normally, an owner tries to account for the decrease in value by prorating the consumption of the investment over the *service life* of the machine. This prorating is known as depreciation.

It can be argued that the amount that should be prorated is the difference between the initial acquisition expense and the expected future salvage value. Such a statement is correct to the extent of accounting for the amounts involved, but it neglects the timing of the cash flows. Therefore, it is recommended that each cash flow be treated separately to allow for a time value analysis and to allow for ease in changing assumptions during sensitivity analyses.

Salvage Value

Salvage value is the cash inflow a firm receives if a machine still has value at the time of its disposal. This revenue will occur at a future date.

Used equipment prices are difficult to predict. Machine condition (see Fig. 12.1), the movement of new machine prices (see Fig. 12.2), and the machine's possible secondary service applications affect the amount an owner can expect to receive. A machine having a diverse and layered service potential will command a higher resale value. Medium-sized dozers, which often exhibit rising salvage values in later years, can have as many as seven different levels of useful life. These may range from an initial assignment as a high-production machine on a dirt spread to an infrequent land-clearing assignment by a farmer.

Historical resale data can provide some guidance in making salvage value predictions and can be fairly easily accessed from auction price books. By studying such historical data and recognizing the effects of the economic environment, the magnitude of salvage value prediction errors can be minimized and the accuracy of an ownership cost analysis improved.

FIGURE 12.1 Salvage value is dependent on machine condition.

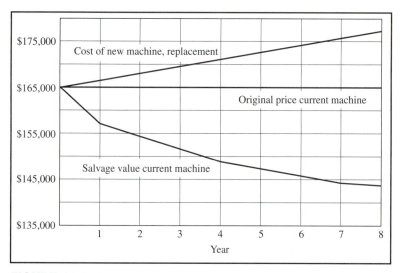

FIGURE 12.2 The movement of new machine prices; cost is one factor affecting salvage value.

Tax Saving from Depreciation

The tax saving from **depreciation** are a phenomenon of the tax system in the United States. (This may not be an ownership cost factor under the tax laws in other countries.) Under the tax laws of the United States, depreciating a machine's loss in value with age will lessen the net cost of machine ownership. The cost saving, the prevention of a cash outflow, afforded by tax depreciation is a result of shielding the company from taxes. This is an applicable cash flow factor only if a company is operating at a profit. There are carry-back features in

depreciation
An accounting method used to describe the loss in value of an asset.

the tax law so that the saving can be preserved even though there is a loss in any one particular year, but the long-term operating position of the company must be at a profit for tax saving from depreciation to come into effect.

The rates at which a company can depreciate a machine are set by the revenue code. These rates usually have no relation to actual consumption of the asset (machine). Therefore, many companies keep several sets of depreciation numbers, one for depreciation tax purposes, one for corporate earnings tax accounting purposes, and one for internal and/or financial statement purposes. The first two are required by the revenue code. The last tries to accurately match the consumption of the asset based on work application and company maintenance policies.

Under the current tax laws, tax depreciation accounting no longer requires the assumption of a machine's future salvage value and useful life. The only piece of information necessary is basis. Basis refers to the cost of the machine for purposes of computing gain or loss. Basis is essential. To compute tax depreciation amounts, fixed percentages are applied to unadjusted basis. The terminology adjusted/unadjusted refers to changing the book value of a machine by depreciation.

The tax law allows the postponement of taxation on gains derived from the exchange of like-kind depreciable property. If there is a gain realized from a like-kind exchange, the depreciation basis of the new machine is reduced by the amount of the gain. However, if the exchange involves a disposal sale to a third party and a separate acquisition of the replacement, the gain from the sale is taxed as ordinary income.

EXAMPLE 12.1

A tractor with an adjusted basis (from depreciation) of $25,000 is traded for a new tractor that has a fair market value of $400,000. A cash payment of $325,000 is made to complete the transaction. Such a transaction is a nontaxable exchange and no gain is recognized on the trade-in. The unadjusted basis of the new tractor is $350,000, even though the cash payment was $325,000 and the apparent gain in value for the traded machine was $50,000 [($400,000 − $325,000) − $25,000].

Cash payment	$325,000
Adjusted basis of the trade-in tractor	25,000
Basis of the new tractor	$350,000

If the owner had sold the old tractor to a third party for $75,000 and then purchased a new tractor for $400,000, the $50,000 profit on the third-party sale would have been taxed as ordinary income and the unadjusted basis of the new tractor would be $400,000.

The current tax depreciation law establishes depreciation percentages that can be used based on the specific year of machine life. These are usually the optimum depreciation rates in terms of tax advantages. However, an owner can still utilize the straight-line method of depreciation or methods that are not expressed in terms of time duration (years). Unit-of-production would be an example of a depreciation system that is not time based.

Straight-Line Depreciation   Straight-line depreciation is easy to calculate. The annual amount of depreciation D_n, for any year n, is a constant value, and thus the book value (BV_n) decreases at a uniform rate over the useful life of the machine. The equations are

$$\text{Depreciation rate, } R_n = \frac{1}{N} \qquad \text{[12.1]}$$

where N = number of years

$$\text{Annual depreciation amount, } D_n = \text{Unadjusted basis} \times R_n$$

Substituting Eq. [12.1] yields

$$D_n = \frac{\text{Unadjusted basis}}{N} \qquad \text{[12.2]}$$

$$\text{Book value year } m, \ BV_n = \text{Unadjusted basis} - (n \times D_n) \qquad \text{[12.3]}$$

EXAMPLE 12.2

Consider the new tractor in Example 12.1 and assume it has an estimated useful life of 5 years. Determine the depreciation and the book value for each of the 5 years using the straight-line method.

$$\text{Depreciation rate, } R_n = \frac{1}{5} = 0.2$$

$$\text{Annual depreciation amount, } D_n = \$350{,}000 \times 0.2 = \$70{,}000$$

Years m	BV_{n-1}	D_n	BV_n
0	$ 0	$ 0	$350,000
1	350,000	70,000	280,000
2	280,000	70,000	210,000
3	210,000	70,000	140,000
4	140,000	70,000	70,000
5	70,000	70,000	0

Declining-Balance Depreciation Methods   Declining-balance (DB) depreciation methods are accelerated depreciation methods that provide for larger portions of the cost of a piece of equipment to be written off in the early years. Often these methods more closely approximate the actual loss in market value

TABLE 12.1 Tax code specified depreciation rates.

Year of life	3-yr property	5-yr property
1	0.33	0.20
2	0.45	0.32
3	0.22	0.24
4	—	0.16
5	—	0.08

with time. Declining methods range from 1.25 times the current book value divided by the life to 2.00 times the current book value divided by the life (the latter is termed double-declining balance). Note that although the estimated salvage value (S) is not included in the calculation, the book value cannot go below the salvage value. The following equations are necessary for using the declining-balance methods.

The symbol R is used for the declining-balance depreciation rate:

1. For 1.25 declining-balance (1.25 DB) method, $R = 1.25/N$.
   For 1.50 declining-balance (1.5 DB) method, $R = 1.50/N$.
   For 1.75 declining-balance (1.75 DB) method, $R = 1.75/N$.
   For double-declining-balance (DDB) method, $R = 2.00/N$.

2. The allowable depreciation D_n, for any year n and any depreciation rate R is

$$D_n = (BV_{n-1}) \times R \qquad \text{[12.4]}$$

3. The book value for any year n is

$$BV_n = BV_{n-1} - D_n \text{ provided that } BV_n \geq S \qquad \text{[12.5]}$$

Since the book value can never go below the estimated salvage value, the DB method must be forced to intersect the value S at time N.

Tax Code Depreciation Schedules   Modified Accelerated Cost Recovery System (MACRS) is a property depreciation system adopted by the Internal Revenue Service (IRS). Under the tax code, machines are classified as 3-, 5-, 10-, or 15-yr real property. Cars and light-duty trucks (under 13,000 lb unloaded) are classified as 3-yr property. Most other pieces of construction equipment are 5-yr property. The appropriate depreciation rates are given in Table 12.1. MACRS provides for a slightly larger write-off in the earlier years of the cost recovery period. The full set of depreciation tables showing the MACRS percentages are available in the IRS's free publication 946, *How to Depreciate Property*, available on the Internet at http://www.irs.ustreas.gov.

If a machine is disposed of before the depreciation process is completed, no depreciation can be recovered in the year of disposal. Any gain, as measured against the depreciated value or adjusted basis, is treated as ordinary income.

EXAMPLE 12.3

A 5-yr life class machine is purchased for $125,000. It is sold in the third year after purchase for $91,000. Using the tax code specified depreciation rates, what are the depreciation amounts and what is the book value of the machine when it is sold? Will there be income tax and, if so, on what amount?

$125,000 × 0.20 = $25,000   depreciation at end of the first year
$125,000 × 0.32 = $40,000   depreciation at end of the second year
                 $65,000   total depreciation
Book value when sold = $125,000 − $65,000 = $60,000.

The amount of gain on which there will be a tax is
    $91,000 − $60,000 = $31,000.

The tax savings from depreciation are influenced by

1. Disposal method for the old machine
2. Value received for the old machine
3. Initial value of the replacement
4. Class life
5. Tax depreciation method

Based on the relationships between these elements, three distinct situations are possible:

1. No gain on the disposal—no income tax on zero gain.
2. A gain on the disposal:
   a. Like-kind exchange—no added income tax, but basis for the new machine is adjusted.
   b. Third-party sale—the gain is taxed as income; the basis of the new machine is fair market value paid.
3. A disposal in which a loss results—the basis of the new machine is the same as the basis of the old machine, decreased by any money received.

Assuming a corporate profit situation, the applicable tax depreciation shield formulas are

1. For a situation where there is no gain on the exchange:

$$\text{Total tax shield} = \sum_{n=1}^{N} t_c D_n \qquad \text{[12.6]}$$

where

n = individual yearly time periods within a life assumption of N years
t_c = corporate tax rate
D_n = annual depreciation amount in the nth time period

2. For a situation where a gain results from the exchange
    a. Like-kind exchange—Eq. [12.6] is applicable. It must be realized that the basis of the new machine will be affected.
    b. Third-party sale.

$$\text{Total tax shield} = \left(\sum_{n=1}^{N} t_c D_n \right) - \text{gain} \times t_c \qquad \text{[12.7]}$$

Gain is the actual salvage amount received at the time of disposal minus the book value.

The implication of the basis is that in making analysis calculations, the actual salvage derived from the machine directly affects the depreciation saving. To perform a valid analysis, the depreciation accounting practices for tax purposes and the methods of machine disposal and acquisition the company chooses to use must be carefully examined. These dictate the appropriate calculations for the tax effects of depreciation.

Major Repairs and Overhauls

Major repairs and overhauls are included under ownership cost because they result in an extension of a machine's service life. They can be considered as an investment in a new machine. Because a machine commonly works on many different projects, considering major repairs as an ownership cost prorates these expenses to all jobs. These costs should be added to the book value of the machine and depreciated.

Property Taxes

In this context, taxes refer to those equipment ownership taxes that are charged by any government subdivision. They are commonly assessed at a percentage rate applied against the book value of the machine. Depending on location, property taxes can range up to about 4.5% of book value. In many locations, there will be no property tax on equipment. Over the service life of the machine, they will decrease in magnitude as the book value decreases.

Insurance

Insurance, as considered here, includes the cost of policies to cover fire, theft, and damage to the equipment either by vandalism or construction accident. Annual rates can range from 1% to 3% of book value. This cost can be actual premium payments to insurance companies, or it can represent allocations to a self-insurance fund maintained by the equipment owner.

Storage and Miscellaneous

Between jobs or during bad weather, a company will require storage facilities for its equipment. The cost of maintaining storage yards and facilities should be prorated to those machines that require such harborage. Typical expenses include space rental, utilities, and the wages for laborers or watchmen. These expenses are all combined in an overhead account and then allocated on a proportional basis to the individual machines. The rate may range from nothing to perhaps 5% of book value.

OPERATING COST

ELEMENTS OF OPERATING COST

Operating cost is the sum of those expenses an owner experiences by working a machine on a project. Typical expenses include

1. Fuel
2. Lubricants, filters, and grease
3. Repairs
4. Tires
5. Replacement of high-wear items

Operator wages are rarely included under operating costs; this is because of wage variance between jobs, the general practice is to keep operator wages as a separate cost category (see Tables 6.4 and 6.9). Such a practice aids in estimation of machine cost for bidding purposes as the differing project wage rates can readily be added to the total machine O&O cost. In applying operator cost, all benefits paid by the company must be included—direct wages, fringe benefits, and insurance. This is another reason wages are separated. Some benefits are based on an hourly basis, some on a percentage of income, some on a percentage of income to a maximum amount, and some are paid as a fixed amount. The assumptions about project work schedule will therefore affect wage expense.

Fuel

Fuel expense is best determined by measurement on the job. Good service records tell the owner how many gallons of fuel a machine consumes over what period of time and under what job conditions. Hourly fuel consumption can then be calculated directly.

When company records are not available, manufacturer's consumption data can be used to construct fuel use estimates. The amount of fuel required to power a piece of equipment for a specific period of time depends on the brake horsepower of the machine and the work application. Therefore, most tables of hourly fuel consumption rates are divided according to the machine type and the working conditions. To calculate hourly fuel cost, a consumption rate is found in the tables (see Table 12.2) and then multiplied by the unit price of fuel. The cost of fuel for vehicles used on public highways will include applicable taxes.

TABLE 12.2 Average fuel consumption—wheel loaders.

Horsepower (fwhp)	Type of utilization		
	Low (gal/hr)	Medium (gal/hr)	High (gal/hr)
90	1.5	2.4	3.3
140	2.5	4.0	5.3
220	5.0	6.8	9.4
300	6.5	8.8	11.8

However, in the case of off-road machines used exclusively on project sites, there is usually no fuel tax. Therefore, because of the tax laws, the price of gas or diesel will vary with machine usage.

Fuel consumption formulas have been published for both gasoline and diesel engines. The resulting values from such formulas must be adjusted by *time and load factors* that account for working conditions. This is because the formulas are derived assuming that the engine is operating at maximum output. Working conditions that must be considered are the percentage of an hour that the machine is actually working (*time factor*) and the percentage of rated horsepower (*load factor*) at which it is working. When operating under standard conditions, a *gasoline engine* will consume approximately 0.06 gal of fuel per flywheel horsepower hour (fwhp-hr). A *diesel engine* will consume approximately 0.04 gal per fwhp-hr.

Lubricants, Filters, and Grease

The cost of lubricants, filters, and grease will depend on the maintenance practices of the company and the conditions of the work location. Some companies follow machine manufacturer's guidance concerning time periods between lubricant and filter changes. Other companies have established their own preventive maintenance change period guidelines. In either case, the hourly cost is arrived at by (1) considering the operating hour duration between changes and the quantity required for a complete change plus (2) a small consumption amount representing what is added between changes.

Many manufacturers provide quick cost-estimating tables (Table 12.3) for determining the cost of these items. Whether using manufacturer's data or past experience, notice should be taken about whether the data match expected field

TABLE 12.3 Lubricating oils, filters, and grease cost for crawler tractors.

Horsepower (fwhp)	Approximate cost per hour	
	Materials ($)	Labor ($)
< 100	0.22	0.15
100 to < 200	0.49	0.24
200 to < 300	0.65	0.24

conditions. If the machine is to be operated under adverse conditions, such as deep mud, water, severe dust, or extreme cold, the data values will have to be adjusted.

A formula that can be used to estimate the quantity of oil required is

Quantity consumed gph (gal per hour) =

$$\frac{\text{hp} \times f \times 0.006 \text{ lb/hp-hr}}{7.4 \text{ lb/gal}} + \frac{c}{t} \qquad \textbf{[12.8]}$$

where

$$\begin{aligned}
\text{hp} &= \text{rated horsepower of the engine} \\
c &= \text{capacity of the crankcase in gallons} \\
f &= \text{operating factor} \\
t &= \text{number of hours between oil changes}
\end{aligned}$$

This formula contains the assumption that the quantity of oil consumed per rated horsepower hour between changes will be 0.006 lb.

Repairs

Repairs, as referred to here, mean normal maintenance-type repairs (Fig. 12.3). These are the repair expenses incurred on the jobsite where the machine is

FIGURE 12.3 Normal repairs are included in operating cost.

operated and include the costs of parts and labor. Major repairs and overhauls are accounted for as ownership cost.

Repair expenses increase with machine age. The U.S. Army has found that 35% of its equipment maintenance cost is directly attributable to the oldest 10% of its equipment. Instead of applying a variable rate, an average is usually calculated by dividing the total expected repair cost, for the planned service life of the machine, by the planned operating hours. Such a policy builds up a repair reserve during a machine's early life. That reserve will then be used to cover the higher costs experienced later. As with all costs, company records are the best source of expense information. When such records are not available, manufacturers' published guidelines can be used.

Tires

Tires for wheel-type equipment (Fig. 12.4) are a major operating cost because they have a short life in relation to the "iron" of a machine. Tire cost includes repair and replacement charges. These costs are very difficult to estimate because of the variability in tire wear with project site conditions and operator skill. Tire and equipment manufacturers both publish tire life guidelines based on tire type and job application. Manufacturers' suggested life periods can be used with local tire prices to obtain an hourly tire cost. It must be remembered, however, that the guidelines are based on good operating practices and do not account for abuses such as overloading haul units.

Replacement of High-Wear Items

The cost of replacing those items that have very short service lives with respect to machine service life can be a critical operating cost. These items will differ depending on the type of machine, but typical items include cutting edges, rip-

FIGURE 12.4   Tires are a major operating cost.

FIGURE 12.5 Bucket teeth are a high-wear-item replacement cost.

per tips, bucket teeth (Fig. 12.5), body liners, and cables. By using either past experience or manufacturer life estimates, the cost can be calculated and converted to an hourly basis.

All machine-operating costs should be calculated per working hour. That way it is easy to sum the applicable costs for a particular class of machines and obtain a total operating hour cost.

COST FOR BIDDING

GENERAL INFORMATION

The process of selecting a particular type of machine for use on a construction project requires knowledge of the cost associated with operating the machine in the field. In selecting the proper machine, a contractor seeks to achieve unit production at the least cost. This cost for bidding is the sum of the O&O expenses. O&O costs are stated on an hourly basis (e.g., $90/hr for a dozer) because it is used in calculating the cost per unit of machine production. If a dozer can push 300 cy per hour and it has a $90/hr O&O cost, the cost for bidding is $0.30/cy ($90/hr ÷ 300 cy/hr). The estimator/planner can use the cost per cubic yard figure directly in unit-price work. On a lump sum job, it will be necessary to multiply the cost/unit price by the estimated quantity to obtain the total amount that should be charged.

Ownership Cost

The outflow of cash when a machine is purchased and the inflow of money in the future (when the machine is retired from service) are the two most significant components of ownership cost. The net result of these two cash flows, which defines the machine's decline in value across time, is termed *depreciation*. As used in this section, depreciation is the measuring system used to account for purchase expense at time zero and salvage value after a defined period of time. Depreciation is expressed on an hourly basis over the service life of a machine. Do not confuse the depreciation discussed here with tax depreciation. Tax depreciation has nothing to do with consumption of the asset; it is simply an artificial calculation for tax code purposes.

> Because tires are a high-wear item that will be replaced many times over a machine's service life, their cost will not be included in these calculations but will be addressed as a part of operating cost.

The depreciation portion of ownership cost can be calculated by either of two methods: (1) time value or (2) average annual investment.

Depreciation—Time Value Method   The time value method will recognize the timing of the cash flows (i.e., the purchase at time zero and the salvage at a future date). The cost of the tires is deducted from the total purchase price, which includes amounts for all options, shipping, and taxes (total cash outflow—cost of tires). A judgment about the expected service life and a corporate cost of capital rate are both necessary input parameters for the analysis. These input parameters are entered into the uniform series capital recovery factor formula to determine the machine's purchase price equivalent annual cost.

To account for the salvage cash inflow, the uniform series sinking fund factor formula is used. The input parameters are the estimated future salvage amount, the expected service life, and the corporate cost of capital rate.

EXAMPLE 12.4

A company having a cost of capital rate of 8% purchases a $300,000 loader. This machine has an expected service life of 4 years and will be utilized 2,500 hours per year. The tires on this machine cost $45,000. The estimated salvage value at the end of 4 years is $50,000. Calculate the depreciation portion of the ownership cost for this machine using the time value method.

Initial cost	$300,000
Cost of tires	− 45,000
Purchase price less tires	$255,000

Now it is necessary to calculate the uniform series required to replace a present value of $255,000. Using the uniform series capital recovery factor formula

$$A = \$225,000 \left[\frac{0.08(1 + 0.08)^4}{(1 + 0.08)^4 - 1} \right]$$

$$A = \$255,000 \times 0.3019208 = \$76,990 \text{ per year}$$

Next calculate the uniform series sinking fund factor to use with the salvage value.

$$A = \$50,000\left[\frac{0.08}{(1 + 0.08)^4 - 1}\right]$$

$$A = \$50,000 \times 0.02219208 = \$11,096 \text{ per year}$$

Therefore, using the time value method, the depreciation portion of the ownership cost is

$$\frac{\$76,990/\text{yr} - \$11,096/\text{yr}}{2,500 \text{ hr/yr}} = \$26.358/\text{hr}$$

Depreciation—Average Annual Investment Method A second approach to calculating the depreciation portion of ownership cost is the average annual investment (AAI) method.

$$\text{AAI} = \frac{P(n + 1) + S(n - 1)}{2n} \qquad \text{[12.9]}$$

where
$P =$ purchase price less the cost of the tires
$S =$ the estimated salvage value
$n =$ expected service life in years

The AAI is multiplied by the corporate cost of capital rate to determine the cost of money portion of the ownership cost. The straight-line depreciation of the cost of the machine less the salvage and less the cost of tires, if a pneumatic-tired machine, is then added to the cost of money portion (interest) to arrive at the total ownership depreciation.

EXAMPLE 12.5

Using the same machine and company information as in Example 12.4, calculate the ownership depreciation using the AAI method.

$$\text{AAI} = \frac{\$255,000(4 + 1) + \$50,000(4 - 1)}{2 \times 4}$$

$$= \$178,125/\text{yr}$$

$$\text{Cost of money portion} = \frac{\$178,125/\text{yr} \times 8\%}{2,500 \text{ hr/yr}} = \$5.700/\text{hr}$$

Straight-line depreciation

Initial cost	$300,000
Cost of tires	− 45,000
Salvage	− 50,000
	$205,000

$$\frac{\$205,000}{4 \text{ yr} \times 2,500 \text{ hr/yr}} = \$20.500/hr$$

Total depreciation portion of the ownership cost using the AAI method

$$\$5.700/hr + \$20.500 = \$26.200/hr$$

For Examples 12.4 and 12.5, the difference in the calculated total depreciation portion of the ownership cost is $0.158/hr ($26.358/hr − $26.200/hr). The choice of which method to use is strictly a company preference. Basically, either method is satisfactory, especially considering the impact of the unknowns concerning service life, operating hours per year, and expected future salvage. There is no single solution to calculating ownership cost. The best approach is to perform several analyses using different assumptions and to be guided by the range of solutions.

Tax Saving from Tax Code Depreciation   To calculate the tax saving from depreciation, the government tax code depreciation schedules (Table 12.1) must be used. The resulting depreciation amounts are then multiplied by the company's tax rate to calculate specific savings, using Eq. [12.6] or Eq. [12.7]. The sum of the yearly saving must be divided by the total anticipated operating hours to obtain an hourly cost saving.

EXAMPLE 12.6

Using the same machine and company information as in Example 12.4, calculate the hourly tax saving resulting from tax code depreciation. Assume that under the tax code the machine is a 5-year property and that there had been no gain on the exchange that procured the machine. The company's tax rate is 37%.

First, calculate the annual depreciation amounts for each of the years. In this case, the tax code depreciation rate must be used to calculate the depreciation.

Year	5-yr property rates	BV_{n-1}	D_n	BV_n
0	0	$ 0	$ 0	$300,000
1	0.20	300,000	60,000	240,000
2	0.32	240,000	96,000	144,000
3	0.24	144,000	72,000	72,000
4	0.16	72,000	48,000	24,000
5	0.08	24,000	24,000	0

Using Eq. [12.6], the tax shielding effect for the machines' service life would be

Year	D_n	Shielded amount*
1	$60,000	$22,200
2	96,000	35,520
3	72,000	26,640
4	48,000	17,760
	Total	$102,120

*$D_n \times 37\%$

$$\text{Tax saving from depreciation} = \frac{\$102,120}{4 \text{ yr} \times 2,500 \text{ hr/yr}} = \$10.21/\text{hr}$$

Major Repairs and Overhauls   When a major repair and overhaul takes place, the machine's ownership cost will have to be recalculated. This is done by adding the cost of the overhaul to the book value at that point in time. The resulting new adjusted book value is then used in the depreciation calculation, as already described. If there are separate calculations for true depreciation and for tax depreciation, both will have to be adjusted.

Taxes, Insurance, and Storage   To calculate the taxes, insurance, and storage costs, common practice is to simply apply a percentage value to either the machine's book value or its AAI amount. The expenses incurred for these items are usually accumulated in a corporate overhead account. That value divided by the value of the equipment fleet and multiplied by 100 will provide the percentage rate to be used.

$$\text{Taxes, insurance, and storage portion of ownership cost} =$$
$$\text{rate } (\%) \times BV_n \text{ (or AAI)} \qquad \textbf{[12.10]}$$

EXAMPLE 12.7

Using the same machine and company information as in Example 12.4 and 12.5, calculate the hourly owning expense associated with taxes, insurance, and storage. Annually, the company pays, as an average, 1% in property taxes on equipment and 2% for insurance, and allocates 0.75% for storage expenses.

Total percentage rate for taxes, insurance, and storage

1% + 2% + 0.75% = 3.75%

From Example 12.5, the average annual investment for the machine is $178,125/yr.

Taxes, insurance, and storage expense

$$\frac{\$178,125/\text{yr} \times 3.75\%}{2,500 \text{ hr/yr}} = \$2.672/\text{hr}$$

Operating Cost

Figures based on actual company experience should be used to develop operating expenses. Many companies, however, do not keep good equipment operating and maintenance records; therefore, many operating costs are estimated as a percentage of a machine's book value. Even companies that keep records often accumulate expenses in an overhead account and then prorate the total back to individual machines using book value.

Fuel   The amount expended on fuel is a product of how a machine is used in the field and the local cost of fuel. In past years, fuel could be purchased on long-term contracts at a fixed price. Today fuel is usually offered with a *time of delivery price.* A supplier will agree to supply the fuel needs of a project, but the price will not be guaranteed for the duration of the work. Therefore, when bidding a long-duration project, the contractor must make an assessment of future fuel prices.

To calculate hourly fuel expense, a consumption rate is multiplied by the unit price of fuel. Service records are important for estimating fuel consumption.

EXAMPLE 12.8

A 220-fwhp dozer will be used to push an aggregate stockpile. This dozer is diesel powered. It is estimated that the work will be steady at an efficiency equal to a 50-min-hour. The engine will work at full throttle while loading the bucket (30% of the time) and a three-quarter throttle to travel and position. Calculate the fuel consumption using the engine consumption averages. If diesel cost $2.07/gal, what is the expected fuel expense?

Fuel consumption diesel engine 0.04 gal per fwhp-hr

Throttle load factor (operating power):

Push load	1.00 (power) × 0.30 (% of the time) = 0.30
Travel and position	0.75 (power) × 0.70 (% of the time) = 0.53
	0.83

Time factor (operating efficiency): 50-min hour: 50/60 = 0.83

Combined factor: 0.83 × 0.83 = 0.69

Fuel consumption = 0.69 × 0.04 gal/fwhp-hr × 220 fwhp = 6.1 gal/hr

---

Lubricants   The quantity of lubricants used by an engine will vary with the size of the engine, the capacity of the crankcase, the condition of the piston rings, and the number of hours between oil changes. For extremely dusty conditions, it may be desirable to change oil every 50 hours, but this is an unusual condition. It is common practice to change oil every 100 to 200 hours. The quantity of the oil consumed by an engine per change will include the amount added during the change plus the makeup oil between changes.

EXAMPLE 12.9

Calculate the oil required, on a per hour basis, for the 220-fwhp dozer in Example 12.8. The operating factor will be 0.69, as calculated in that example. The crankcase capacity is 8 gal and the company has a policy to change oil every 150 hr.

$$\frac{220 \text{ fwhp} \times 0.69 \times 0.006 \text{ lb/hp-hr}}{7.4 \text{ lb/gal}} + \frac{8 \text{ gal}}{150 \text{ hr}} = 0.18 \text{ gal/hr}$$

The cost of hydraulic oil, filters, and grease will be added to the expense of engine oil. The hourly cost of filters is simply the actual expense to purchase the filters divided by the hours between changes. If a company does not keep detailed machine-servicing data, it is difficult to accurately estimate the cost of hydraulic oil and grease. The usual solution is to refer to manufacturers' published tables of average usage or expense.

Repairs   The cost of repairs is normally the largest single component of machine cost (see Fig. 12.6). Some general guidelines published in the past by the Power Crane and Shovel Association (PCSA) estimated repair and maintenance expenses at 80% to 95% of depreciation for crawler-mounted excavators, 80% to 85% for wheel-mounted excavators, 55% for crawler cranes, and 50% for wheel-mounted cranes. The lower figures for cranes reflect the work they perform and the intermittent nature of their use. In the case of mechanical machines, it was assumed that half of the cost was materials and parts, and half was labor. For hydraulic machines, two-thirds of the cost is for materials and parts, and one-third for labor.

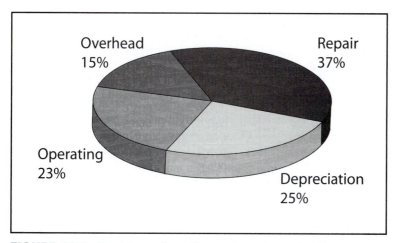

FIGURE 12.6   Breakdown of machine cost over its service life.

Equipment manufacturers supply tables of average repair costs based on machine type and work application. Repair expenses will increase with machine usage (age). The repair cost to establish a machine rate for bidding should be an average rate.

Tires   Tire expenses include both tire repair and tire replacement. Tire maintenance is commonly handled as a percentage of straight-line tire depreciation. Tire hourly cost can be derived simply by dividing the cost of a set of tires by their expected life, and this is how many companies prorate this expense. A more sophisticated approach is to use a time-value calculation, recognizing that tire replacement expenses are single-point-in-time outlays that take place over the life of a wheel-type machine.

EXAMPLE 12.10

Calculate the hourly tire cost that should be part of the machine operating cost if a set of tires can be expected to last 5,000 hours. Tires cost $38,580 per set of four. Tire repair cost is estimated to average 16% of the straight-line tire depreciation. The machine has a service life of 4 years and operates 2,500 hours/year. The company's cost of capital rate is 8%.

Not considering the time value of money:

$$\text{Tire repair cost} = \frac{\$38,580}{5,000 \text{ hr}} \times 16\% = \$1.235/\text{hr}$$

$$\text{Tire use cost} = \frac{\$38,580}{5,000 \text{ hr}} = \$7.716/\text{hr}$$

Therefore, tire operating cost is $8.951/hr ($1.235/hr + $7.716/hr).

Considering the time value of money:
Tire repair cost is the same $1.235/hr.
Calculate the number of times the tires will have to be replaced.

$$\left(\frac{4 \text{ yr} \times 2,500 \text{ hr/yr}}{5,000 \text{ hr per set of tires}} \right) = 2 \text{ sets}$$

Will have to purchase a second set at the end of the second year.

First set: Calculate the uniform series required to replace a present value of $38,580. Using the uniform series capital recovery factor formula:

$$A = \$38,580 \left[\frac{0.08(1 + 0.08)^4}{(1 + 0.08)^4 - 1} \right]$$

$$\frac{\$38,580 \times 0.301921}{2,500 \text{ hr/yr}} = \$4.659/\text{hr}$$

Second set: The second set will be purchased 2 yr in the future. Therefore what amount at time zero is equivalent to $38,580, 2 yr in the future? Using the present worth compound amount factor the equivalent time zero amount is calculated.

$$P = \frac{\$38,580}{(1 + 0.08)^2} = \$33,076$$

Calculate the uniform series required to replace a present value of $33,076.

$$A = \$33,760 \left[\frac{0.08(1 + 0.08)^4}{(10.08)^4 - 1} \right]$$

$$\frac{\$33,076 \times 0.301921}{2,500 \text{ hr/yr}} = \$3.995/\text{hr}$$

Therefore, considering the time value of money, tire operating cost is $9.889/hr ($1.235/hr + $4.659/hr + $3.995/hr).

High-Wear Items   Because the cost of high-wear items is dependent on job conditions and machine application, the cost of these items is usually accounted for separately from general repairs.

EXAMPLE 12.11

A dozer equipped with a three-shank ripper will be used in a loading and ripping application. Actual ripping will take place only about 20% of total dozer operating time. A ripper shank consists of the shank itself, a ripper tip, and a shank protector. The estimated life for the ripper tip is 30 hours. The estimated life of the ripper shank protector is three times tip life. The local price for a tip is $40 and $60 for shank protectors. What hourly high-wear item charge should be added to the operating cost of a dozer in this application?

Tips: $\frac{30 \text{ hr}}{0.2}$ = 150 hr of dozer operating time

$\frac{3 \times \$40}{150 \text{ hr}}$ = $0.800/hr for tips

Shank protectors: 3 times tip life $\times$ 150 hr = 450 hr of dozer operating time

$\frac{3 \times \$60}{450 \text{ hr}}$ = $0.400/hr for shank protectors

Therefore, the cost of high-wear items is $1.200/hr ($0.800/hr tips + $0.400/hr shank protectors)

BUY, RENT, OR LEASE

GENERAL INFORMATION

There are three basic methods for securing a particular machine to use on a project: (1) *buying* (direct ownership), (2) *renting*, or (3) *leasing*. Each method has inherent advantages and disadvantages. Ownership guarantees control of machine availability and mechanical condition, but it requires a continuing sequence of projects to pay for the machine. Ownership may at times force a company into using obsolete equipment. The calculations applicable for determining the cost of direct ownership have been developed.

RENTAL

The rental of a machine is a short-term alternative to direct equipment ownership. With a rental, a company can pick the machine that is exactly suited for the job at hand. This is particularly advantageous if the job is of short duration or if the company does not foresee a continuing need for the particular type of machine in question. Rentals are very beneficial to a company in such situations, even though the rental charges are higher than *normal* direct ownership expense. The advantage lies in the fact that direct ownership costing assumes a continuing need for and utilization of the machine. If that assumption is not valid, a rental should be considered. Another important point to consider is the fact that with a rental, the company loses the tax depreciation shield of machine ownership but gains a tax deduction because rental payments are treated as an expense.

It must be remembered that rental companies only have a limited number of machines and, during the peak work season, all types are not always available. Additionally, many specialized or custom machines cannot be rented.

Firms oftentimes use rentals as a way to test a machine prior to a purchase decision. A rental provides the opportunity for a company to operate a specific make or model machine under actual project conditions. Profitability of the machine, based on the company's normal operating procedures, can then be evaluated before a major capital expenditure is approved to purchase the machine.

The general practice of the industry is to price rental rates for equipment on either a daily (8 hour), weekly (40 hour), or monthly (176 hour) basis. In the case of larger pieces of equipment, rentals may be available only on a monthly basis. Cost per hour usually is less for a longer-term rental (i.e., the monthly rate figured on a per hour basis would be less than the daily rate on an hourly basis).

Responsibility for repair cost is stated in the rental contract. Normally, in the case of tractor-type equipment, the renter is responsible for all repairs. If it is a pneumatic-tired machine (on rubber), the renting company will measure tread wear and charge the renter for tire wear. In the case of cranes and shovels, the renting company usually bears the cost of normal wear and tear. The user must provide servicing of the machine while it is being used. The renter is almost always responsible for fuel and lubrication expenses. Industry practice is that rentals are payable in advance. The renting company will require that the user furnish certificates of insurance before the machine is shipped to the job site.

Equipment cost is very sensitive to changes in use hours. Fluctuations in maintenance expenses or purchase price barely affect cost per hour. But a decrease in use hours per year can make the difference between a cost-effective machine ownership versus renting.

The basic cost considerations that need to be examined when considering a possible rental can be illustrated by a simple set of circumstances. Consider a small wheel loader with an ownership cost of $10.96 per hour. Assume that the cost is based on the assumption that the machine will work 2,400 hours each year of its service life. If $10.96/hour is multiplied by 2,400 hours/year, the yearly ownership cost is found to be $26,304.

Checking with the local rental company, the construction firm receives rental quotes of $3,558 per month, $1,182 per week, and $369 per day for this size loader. By dividing by the appropriate number of hours, these rates can be expressed as hourly costs. Likewise, by dividing the calculated hourly rental rates into the construction firm's yearly ownership cost figure ($26,304), the operating hour break-even points can be determined (see Table 12.4).

If the loader will be used for less than 1,300 hours but more than 890 hours, the construction company should consider a monthly rental instead of ownership. When the projected usage is less than 1,300 hours but greater than 120 hours, a weekly rental would be appropriate. In the case of very limited usage, that is, less than 26 hours, the daily rate is optimal.

The point is that when a company rents, it pays for the equipment only when project requirements dictate a need. The company that owns a machine must continue to make the equipment payments even when the machine sits idle. When investigating a rental, the critical question is usually *expected hours of usage.*

LEASE

A **lease** is a long-term agreement for the use of an asset. It provides an alternative to direct ownership. During the lease term, the leasing company (lessor) always owns the equipment and the user (lessee) pays the owner to use the equipment. The lessor must retain ownership rights in order for the contract to be considered a true lease by the Internal Revenue Service. The lessor will receive lease payments in return for providing the machine. The lease payments do not have to be uniform across the lease period. The payments can be structured in the agreement to best fit the situation of the lessee or the lessor. In the lessee's case, cash flow at the

lease
A long-term agreement for the use of an asset.

TABLE 12.4 Rental versus ownership, operating hour break-even points.

Rental duration	Rate ($)	Hours	Rental rate ($/hr)	Operating hour break-even point [$26,304 (hr)]	Operating point break-even point [$3,558 (hr)]
Monthly	3,558	176	20.22	1,300	—
Weekly	1,182	40	29.55	890	120
Daily	369	8	46.13	570	77

beginning may be low, so the lessee wants payments that are initially less. Because of tax considerations, the lessor may agree to such a payment schedule. Lease contracts are binding legal documents, and most equipment leases are noncancelable by either party.

A lease pays for the use of a machine during the most reliable years of a machine's service life. Sometimes the advantage of a lease is that the lessor provides the maintenance and servicing. This frees the contractor from hiring mechanics and service personnel and allows the company to concentrate on the task of building.

Long-term, when used in reference to lease agreements, is a period of time that is long relative to the life of the machine in question. An agreement that is for a very short period of time, as measured against the expected machine life, is a rental. A conventional—true—lease will have one of three different end-of-lease options: (1) buy the machine at fair market value, (2) renew the lease, or (3) return the equipment to the leasing company.

As in the case of a rental, a lessee loses the tax depreciation shield of machine ownership but gains a tax deduction because lease payments are treated as an expense. The most important factor contributing to a decision to lease is reduced cost. Under specific conditions, the actual cost of a leased machine can be less than the ownership cost of a purchased machine. This is caused by the different tax treatments for owning and leasing an asset. An equipment user must make a careful examination of the cash flows associated with each option to determine which results in the lowest total cost.

Working capital is the cash that a firm has available to support its day-to-day operations. This *cash asset* is necessary to meet the payroll on Friday, to pay the electric bill, and to purchase fuel to keep the machines running. To be a viable business, working capital assets must be greater than the inflow of bills. A machine is an asset to the company, but it is not what the electric company will accept as payment for its bill.

A commonly cited advantage of leasing is that working capital is not tied up in equipment. This statement is only partly true. It is true that when a company borrows funds to purchase a machine, the lender normally requires that the company establish an equity position in the machine, a *down payment*. Additionally, the costs of delivery and initial servicing are not included in the loan and must be paid by the new owner. Corporate funds are therefore tied up in these up-front costs of a purchase. Leasing does not require these cash outflows and is often considered as 100% financing. However, most leases require an advance lease payment. Some even require a security deposit and charge other up-front costs.

Still another argument is that because borrowed funds are not used, credit capacity is preserved. Leasing is often referred to as off-balance-sheet financing. A lease is considered an operating expense, not a liability, as is the case with a bank loan. With an operating lease (used when the lessee does not ultimately want to purchase the equipment), leased assets are expensed. Therefore, such assets do not appear on the balance sheet. Standards of accounting, however, require disclosure of lease obligations. It is hard to believe that *lenders* would be so naive as to not consider all of a company's fixed obligations, including both loans and leases. But the off-balance-sheet lease typically will not hurt *bonding capacity,* which is important to a company's ability to bid work.

Before entering into a contract with a construction company, most owners require that the company post a bond guaranteeing that it will complete the project. A third-party surety company secures this bond. The surety closely examines the construction company's financial position before issuing the bond. Based on the financial strength of the construction company, the surety typically restricts the total volume of work that the construction company can have under contract at any one time. This restriction is known as bonding capacity. It is the total dollar value of work under contract that a surety company will guarantee for a construction company.

Owners should make a careful examination of the advantages of a lease situation. The cash flows that should be considered when evaluating the cost of a lease include

1. Inflow, initially, of the equivalent value of the machine
2. Outflow of the periodic lease payments
3. Tax shielding provided by the lease payments (This is allowed only if the agreement is a true lease. Some "lease" agreements are essentially installment sale arrangements.)
4. Loss of salvage value when the machine is returned to the lessor

These costs all occur at different points in time, so present-value computations must be made before the costs can be summed. The total present value of the lease option should be compared to the minimum ownership costs, as determined by a time-value replacement analysis. In most lease agreements, the lessee is responsible for maintenance. If, for the lease in question, maintenance expense is the same as for the case of direct ownership, then the maintenance expense factor can be dropped from the analysis. A leased machine would exhibit the same aging and resulting reduced availability as a purchased machine.

SUMMARY

Equipment owners must carefully calculate machine ownership and operating costs. These costs are combined and expressed in dollars per operating hour. The most significant cash flows affecting *ownership cost* are (1) purchase expense; (2) salvage value; (3) major repairs and overhauls; and (4) property taxes, insurance, storage, and miscellaneous expenses. *Operating cost* is the sum of those expenses an owner experiences by working a machine on a project: (1) fuel; (2) lubricants, filters, and grease; (3) repairs; (4) tires; and (5) replacement of high-wear items. *Operator wages* are sometimes included under operating costs, but because of wage variance between jobs, the general practice is to keep operator wages as a separate cost category. Critical learning objectives would include

■ An ability to calculate ownership cost
■ An ability to calculate operating cost
■ An understanding of the advantages and disadvantages associated with
  direct ownership, renting, and leasing machines

These objectives are the basis for the questions in this chapter.

REVIEW QUESTIONS

12.1 What is the largest single equipment cost?

12.2 Regardless of how much a machine is used, is it true or false that the owner must pay owning cost?

12.3 A machine's owning cost includes which of the following

   a. Tires

   b. Storage expenses

   c. Taxes

   d. General repair

12.4 Asphalt Pavers, Inc., purchases a loader to use at its asphalt plant. The purchase price delivered is $235,000. Tires for this machine cost $24,000. The company believes it can sell the loader after 7 years (3,000 hr/yr) of service for $79,000. There will be no major overhauls. The company's cost-of-capital is 6.3%. What is the depreciation part of this machine's ownership cost? Use the time value method to calculate depreciation. ($9.236/hr)

12.5 Using the AAI method to calculate depreciation and the Question 12.4 information, what is the depreciation part of the machine's ownership cost?

12.6 Pushem Down clearing contractors purchases a dozer with a delivered price of $275,000. The company believes it can sell the used dozer after 4 years (2,000 hours/yr) of service for $56,000. There will be no major overhauls. The company's cost-of-capital is 9.2% and its tax rate is 33%. Property taxes, insurance, and storage will run 4%. What is the owning cost for the dozer? Use the time value method to calculate the depreciation portion of the ownership cost. ($40.379/hr)

12.7 Earthmovers Inc. purchases a grader to maintain haul roads. The purchase price delivered is $165,000. Tires for this machine cost $24,000. The company believes it can sell the grader after 6 years (15,000 hr) of service for $26,000. There will be no major overhauls. The company's cost-of-capital is 7.3% and its tax rate is 35%. There are no property taxes, but insurance and storage will run 3%. What is the owning cost for the grader? Use the time value method to calculate the depreciation portion of the ownership cost.

12.8 A 140-fwhp diesel-powered wheel loader will be used at an asphalt plant to move aggregate from a stockpile to the feed hoppers. The work will be steady at an efficiency equal to a 55-min hour. The engine will work at full throttle while loading the bucket (32% of the time) and a three-quarter throttle to travel and dump. Calculate the fuel consumption using the engine consumption averages and compare the result to a medium rating in Table 12.2. (4.3 gal/hr, 4.0 gal/hr)

12.9 A 60-fwhp gasoline-powered pump will be used to dewater an excavation. The work will be steady at an efficiency equal to a 60-min hour. The engine will work at half throttle. Calculate the fuel consumption using the engine consumption.

12.10 A 260-fwhp diesel-powered wheel loader will be used to load shot rock. This loader was purchased for $330,000. The estimated salvage value at the end of 4 years is $85,000. The company's cost-of-capital is 8.7%. A set of tires costs $32,000. The work will be at an efficiency equal to a 45-min hour. The engine will work at full throttle while loading the bucket (33% of the time) and at three-quarter throttle to travel and dump. The crankcase capacity is 10 gallons and the company has a policy to change oil every 100 hours on this job. The annual cost of repairs equals 70% of the straight-line machine depreciation. Fuel cost is $1.07/gallon, and oil is $2.50/gallon. The cost of other lubricants and filters is $0.45/hour. Tire repair is 17% of tire depreciation. The tires should give 3,000 hours of service. The loader will work 1,500 hours/year. In this usage, the estimated life for bucket teeth is 120 hours. The local price for a set of teeth is $640. What is the operating cost for the loader in this application? ($48.483/hr)

12.11 A 400-fwhp diesel-powered dozer will be used to support a scraper fleet. This dozer was purchased for $395,000. The estimated salvage value at the end of 4 years is $105,000. The company's cost-of-capital is 7.6%. The work will be at an efficiency equal to a 50-min hour. The engine will work at full throttle while push loading the scrapers (59% of the time) and at three-quarter throttle to travel and position. The crankcase capacity is 16 gallons and the company has a policy to change oil every 150 hours. The annual cost of repairs equals 68% of the straight-line machine depreciation. Fuel costs $1.03/gallon and oil is $2.53/gallon. The cost of other lubricants and filters is $0.65/hour. The dozer will work 1,800 hours/year. In this usage, the estimated life for cutting edges is 410 hours. The local price for a set of cutting edges is $1,300. What is the operating cost for the dozer in this application?

12.12 Using data found on the Web, create a salvage value plot of Caterpillar D6H (crawler dozer) resale prices. From the plot, make an estimate of the loss in value that can be expected for a crawler tractor that is replaced after a 5-year life. (*Hint:* one website to use to obtain data is Web ref. 7.)

12.13 Use the Web to determine if the Goodyear RL-3J (E-3) scraper tire gives superior lateral traction.

12.14 Using Michelin's earthmover website, determine what tires could be used on a CAT 12G grader.

REFERENCES

1. *Caterpillar Performance Handbook,* Caterpillar Inc., Peoria, IL, issued annually.

2. Equipment Leasing Association of America, 4301 N. Fairfax Drive, Suite 550, Arlington, VA. www.elaonline.com

3. Lewellen, Wilbur G. (1976). *The Cost of Capital,* Kendall/Hunt Publishing Co., Dubuque, IA.

4. Modigliani, Franco, and Merton H. Miller (1958). "The Cost of Capital, Corporate Finance and the Theory of Investment," *The American Economic Review,* 48(3), June.

5. *Rental Contract Checklist,* Associated Equipment Distributors, Inc., 615 W. 22nd Street, Oak Brook, IL. www.aednet.org

6. Schexnayder, C. J., and Donn E. Hancher (1981). "Contractor Equipment Management Practices," *Journal of the Construction Division, Proceedings, ASCE,* 107(CO4), Dec.

7. Schexnayder, C. J., and Donn E. Hancher (1982). "Inflation and Equipment Replacement Economics," *Journal of the Construction Division, Proceedings, ASCE,* 108(CO2), June.

8. Schexnayder, C. J., and Donn E. Hancher (1981). "Interest Factor in Equipment Economics," *Journal of the Construction Division, Proceedings, ASCE,* 107(CO4), Dec.

9. *The Cost of Renting Construction Equipment,* Associated Equipment Distributors, Inc., 615 W. 22nd St., Oak Brook, IL. www.aednet.org

WEBSITE RESOURCES

1. www.aednet.org   The Associated Equipment Distributors, Inc. (AED) is an international trade association representing construction equipment distributors, manufacturers, and industry-service firms.

2. www.aem.org   The Association of Equipment Manufacturers (AEM) is a trade and business development resource for companies that manufacture equipment. AEM was formed on January 1, 2002, from the consolidation of the Construction Industry Manufacturers Association (CIMA) and the Equipment Manufacturers Institute (EMI).

3. www.caterpillar.com   Caterpillar Inc. is the world's largest manufacturer of construction and mining equipment.

4. www.constructionequipment.com *Construction Equipment* magazine online.

5. www.deere.com Deere and Company, John Deere's Construction and Forestry Equipment Division.

6. www.elaonline.org Equipment Leasing Association (ELA) is a national organization made up of member companies within the equipment leasing and finance industry.

7. www.equipmentworld.com Equipmentworld.com is an online magazine featuring construction industry and equipment news.

8. www.goodyearotr.com The Goodyear Tire & Rubber Company off-the-road tire information.

9. www.hcmac.com Hitachi Construction Machinery Co., Ltd., is an earthmoving construction machinery manufacturer.

10. www.machinerytrader.com Machinery Trader is a marketplace for buying and selling heavy construction equipment.

11. earthmover.webmichelin.com/na_eng Michelin tire company's construction equipment tire information site.

Building Materials

There are several kinds of buildings, with the most obvious distinction being commercial and residential. Commercial buildings vary greatly in size, type of construction, number of stories, and intended use. One-story buildings often look like large boxes on concrete slabs, as this provides maximum floor space and minimum structural frame. There are many kinds of residential buildings, varying greatly in style, grandeur, and number of individual residences in each building. Engineers should appreciate the impact of selecting different materials and different construction methods. They must also understand the impact materials and methods have on the overall cost of a structure, and on its appearance and life expectancy. The four major building materials are concrete, structural steel, wood, and masonry.

CONCRETE

Let us begin our discussion of the world's oldest human-made mass-produced construction material—concrete. Picture for a moment the many kinds of structures you have seen in your life built with concrete. Any such list must include highway pavements (Fig. 13.1), sidewalks, bridge decks, traffic barriers, foundations, building frames (Fig. 13.2), building curtain walls, dams (Fig 13.3), and retaining walls. Most people could take a few minutes and double the list. The point is that concrete is a versatile, workable material with many construction applications. It is not proper to refer to a highway as a "cement pavement" or to a sidewalk as a "cement sidewalk." They are both made of *concrete*. Cement is one of the basic ingredients in concrete—aggregate, sand, and water are the other ingredients.

The earliest known use of cement to make concrete dates back to the ancient Romans, about 2,000 years ago. They discovered a sand or mineral on Mount Vesuvius that was high in silica and alumina, which they used to make cement.

FIGURE 13.1 Concrete highway pavement under construction.

FIGURE 13.2 Concrete-framed building.

FIGURE 13.3   Concrete dam under construction.

There are still concrete structures today that were constructed by the Romans. Modern cement was developed in the 18th century. Combining the proportioned amounts of lime, iron, silica, and alumina, and burning them in a rotating oven or kiln (Fig. 13.4) at about 3,000°F creates today's Portland cement. The product from the kiln is known as clinker. To achieve the final usable product cement, the clinker is ground to a very fine consistency. Cement is bagged in 94-lb bags or sold by the ton in bulk. The types of cement are shown in Table 13.1.

Concrete is a mixture that contains cement, sand, gravel, water, and usually additives. The cement is the material that causes the other ingredients to adhere to one another. The water is needed to cause the cement to hydrate. Concrete does not harden by drying, but by hydration. Hydration is the chemical reaction that causes the setting up or hardening of concrete. The aggregate used to make concrete should be uniformly graded to provide a good fill of all the spaces between the individual pieces of aggregate. The sand will fill the small spaces in the mix. Gap-graded aggregates have certain particle sizes entirely or mostly absent. The lack of two or more successive sizes can create segregation problems with the concrete mix. The amount of water used in the mix should not be excessive, because excess water eventually evaporates, leaving small voids in the spaces it once occupied. The amount of water is controlled by specifying the water/cement ratio, the amount of water in relation to the amount of cement. The water/cement ratio is defined as the ratio by weight of the free water content

FIGURE 13.4 Rotary kiln for making cement clinker.

to the cement content. It long has been accepted that a low free water/cement ratio in a concrete mix is essential for the concrete's subsequent strength and durability. There are many different additives and they are used for a variety of purposes from changing the color of the concrete to shortening or lengthening the time for the concrete to harden or set up.

Concrete is characterized by strong compressive strength but is relatively weak in tension and shear. It can support large axial loads and is used extensively in the foundations of large structures such as bridges and tall buildings, but it breaks easily under bending without reinforcing steel. Concrete with compressive strength of 3,000–6,000 pounds per square inch (psi) is common. The

TABLE 13.1 Cement types listed by the American Concrete Institute.

Cement type	Characteristics
Type I	Normal
Type IA	Normal with air-entraining
Type II	Moderate resistance to sulfate
Type IIA	Moderate resistance with air-entraining
Type III	High early strength
Type IIIA	High early strength with air-entraining
Type IV	Low heat of hydration
Type V	High resistance to sulfate

American Concrete Institute defines any concrete over 6,000 psi as high-strength concrete.

For structures that require flexural strength for resistance to bending moments, reinforcing steel—rebar—is embedded in the concrete. In some cases, steel cables are placed in the concrete to make prestressed or posttensioned concrete. The tension to resist the bending moment comes from the reinforcing steel or the cables.

ultimate strength of concrete

The compressive strength of concrete continues to increase slowly for many months. The ultimate strength may be higher than the 28-day strength, which is used for design calculations.

For a concrete mix to reach its ultimate strength, a long period of time is needed. The **ultimate strength of concrete** and its 28-day strength are not the same. For design purposes, engineers use the 28-day strength, and the concrete should be subjected to only minimum loads until the 28 days have passed. That rule is very rigidly enforced in heavy construction. It is not at all uncommon for work in residential and medium-rise commercial building construction to be processed on a 3-day floor cycle. Place concrete one day, form and shore the next floor over the next two days (this puts minimum load on the new slab) and then place concrete on the third day. This placement increases the load on the previously placed floor, but the load is carried primarily by the shoring, which passes continually down for several floors, thereby shifting the load to concrete that has had time to reach its ultimate strength. Note the shoring and reshoring in Fig. 13.5 is in place to a depth of four floors below the slab that is currently being placed.

FIGURE 13.5   Concrete-frame building under construction.

The constructor's needs and considerations relative to concrete are very different from the designer's. The designer is concerned with strength of the material. The size of the concrete member and the configuration and size of its internal reinforcing steel are determined by the designer, based on the projected live loads and dead loads on the structure. The constructor, on the other hand, is concerned with building the forms (Figs. 13.6, 13.7, and 13.8), placing and tying the reinforcing steel (Figs. 13.9 and 13.10), placing the plastic concrete in the forms, stripping the forms, and curing the concrete. The designer is not concerned about the distance between the jobsite and the concrete plant. The constructor is very much concerned about this issue. The designer is not concerned with the time and cost impacts of curing concrete for 28 days. The constructor must figure out how to do this economically. The designer is not concerned about whether the concrete can be delivered by truck directly to the forms, or needs to be pumped. Again, these are the concerns that determine the job methods that must be used, what the job costs are, and what the impacts are on the placement schedule—all concerns of the constructor.

It is important to understand when different concrete placement methods would be used. Concrete trucks are designed to receive concrete directly from the discharge chute of the batch plant, continue mixing while enroute to the job site, and discharge concrete directly into the forms. As it is placed in the forms, the concrete must be vibrated sufficiently to ensure there are no voids, but not so much that the large aggregates all migrate to the bottom of the forms. Above the first floor of buildings, the wet or plastic concrete must be lifted or pumped up

FIGURE 13.6 Concrete formwork for a reservoir water intake tower.

FIGURE 13.7 Concrete formwork for a highway bridge.

FIGURE 13.8 Concrete formwork for a multistory building.

FIGURE 13.9 Placing and tying reinforcing steel.

FIGURE 13.10 Bridge deck reinforcing steel ready for concrete placement.

to the forms. When a crane is used, the concrete is dumped from the truck into a bucket, which is lifted by the crane to the forms on the upper floors (Fig. 13.11). The cost of placing concrete on the upper floors of a building is obviously much greater than the cost of placing concrete at ground level.

The concrete components of buildings include foundations, structural frames, and walls. Foundations may be drilled piers, grade beams, and slabs or mats. The building frame, if it is concrete, may be reinforced concrete or pre-stressed concrete, such as would be designed for a parking garage. In some areas of the country, it is common for large, one-story buildings for stores and other commercial buildings to have concrete tilt-up walls, which are cast in place on the floor of the building then lifted into place and set on a concrete **grade beam**. This method of construction is economical and rapid, and offers the opportunity to build very attractive buildings by using different exposed aggregates on the outside of the concrete panels.

grade beam

Portion of reinforced concrete foundation that supports walls or the exterior edges of a slab. Often constructed without forms by excavating the earth to the required width and depth. Also referred to as a footing, which transfers the weight from walls and columns to the soil or bedrock.

FIGURE 13.11 Placement of concrete into the formwork by means of a bucket and crane.

REINFORCED CONCRETE

The purpose of reinforcing steel, or rebar, is to give the concrete flexural strength. Without reinforcement, concrete cracks very easily when placed under tension or moment. Reinforcing bars are placed in the areas of greatest tension. In columns, reinforcing bars also help carry the vertical, compressive load, and make it possible to reduce the diameter of the column. In that case, the bars must be tied with stirrups or spiral ties to prevent buckling. Reinforcing bars must be covered by enough concrete to prevent exposing them to the elements. Exposure would cause the steel to corrode. As steel corrodes, it increases in volume, forcing cracks in the concrete and allowing additional water and other corrosives to reach the steel, thereby accelerating the oxidizing process.

The amount of reinforcing steel required in any concrete member, such as a floor slab or a grade beam, is calculated by determining the cross-sectional area of the steel bars needed to resist the tension. A small number of large bars or a large number of small bars, properly placed, has essentially the same effect.

Reinforcing bars are available as billet-steel, rail-steel, axle-steel, and low-alloy steel. The available grades include Grade 40, which has a yield strength of 40,000 psi, Grades 50, 60, and 75, each with its corresponding yield strength. Some bars have deformations rolled into their surface to improve the bond between the concrete and steel. The **American Society of Testing Materials (ASTM)** specifications require that identification marks be rolled into the bars, which identify the producing mill, the bar size, the type of steel, and the grade (see Fig. 13.12). Bars are sized according to their diameter in multiples of ⅛ in. For instance, a #3 bar has a nominal diameter of ⅜ in. and a #8 bar has a nominal diameter of ⅝ or 1 in. Coated bars are used to reduce the effects of corrosion. The most common coated bars are epoxy coated (see Fig. 13.10) or

American Society of Testing Materials (ASTM)
Tests engineering materials and writes standard test methods and specifications.

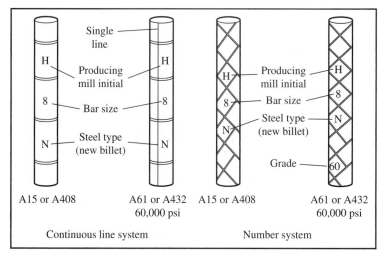

FIGURE 13.12 Reinforcing steel mill marks.

galvanized. They are more expensive than normal reinforcing bars, and therefore are used only in structures that are exposed to highly corrosive environments, such as roads in northern climates where salt is used on the roads in the winter to melt snow and ice.

The location of the reinforcing steel in the concrete is important, and is controlled carefully during the process of placing and tying the bars (see Fig. 13.13). The spacing between the bars is not as critical as the spacing between the forms and bars. For example, a mat of steel in a 6-in. floor slab may require that the bars be spaced 6 in. on-center (OC), and at least 1 in. from the concrete surface. The 6-in. spacing of the bars may vary, but the total number of bars cannot. The general rule is that the spacing of the bars cannot exceed three times the slab thickness at any point. On the other hand, the space of the bars from the

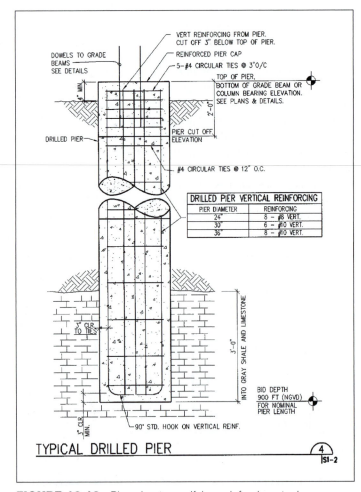

FIGURE 13.13 Plan sheet specifying reinforcing steel placement.

concrete formwork must be maintained within a $\pm\frac{3}{8}$ in. tolerance ($\pm\frac{1}{2}$ in. for slabs thicker than 8 in.). The space between the steel and the forms must be great enough to allow the aggregate to pass between the bars and the forms and not hang on the bars, causing segregation, and the steel must be tied firmly in place to prevent movement during concrete placement.

In horizontal structures, such as floor slabs and grade beams, supports such as chairs, saddles, and dowel blocks are used to support the bars and to maintain the required spacing above the bottom of the form. It is not good practice to walk on the reinforcing bars before the concrete is placed, but the reality is that workers walk on them all the time. Therefore, high-quality supports for the bars are important because the bars must be placed with accuracy. They cannot be allowed to bend or otherwise be moved out of their correct alignment. The supports may be covered with plastic or epoxy, or galvanized to prevent rusting. There are many kinds of supports, some designed to support only one bar, some designed to support several bars (see Fig. 13.14).

In vertical building elements, such as walls and columns, the reinforcing steel must also be held rigidly in place until the concrete is placed. The bars are normally tied to the forms using tie wire to maintain the correct spacing.

Reinforcing steel may be prebent at an off-site fabricating facility. The steel is then shipped, usually by truck, to the site. Each load of reinforcement steel is accompanied by a "bar list," which lists all bars in the shipment, with their quantity, grade, and length. Bars that have been prebent are divided into "heavy bending" for bars size #4 and larger, and "light bending" for #3 bars and smaller. Light bending includes items such as stirrups and spirals.

At the site, the bars should be delivered directly to the point where they are needed, or stored in a clean, dry storage area. If the material can be scheduled

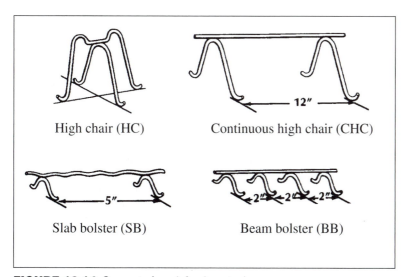

High chair (HC)   Continuous high chair (CHC)

Slab bolster (SB)   Beam bolster (BB)

FIGURE 13.14 Supports for reinforcing steel.

to arrive at the site as it is needed, and lifted directly to the location where it is needed, considerable costs are saved by eliminating double handling. Some contracts are governed by labor agreements that require that all reinforcing bars must be bent to shape on the site, and no prefabrication shop may be used. This agreement represents an added expense that must be accounted for in the bid price.

STRUCTURAL STEEL

Iron and steel production is a very old craft. Scientists have discovered evidence that ancient African tribes have known how to make iron in small quantities for several thousand years. During the 1800s, most iron products were either wrought iron or cast iron. In 1856 Sir Henry Bessemer invented the converter that forces air into the molten iron ore mix. Before the Bessemer converter, the production of steel was expensive and inefficient and could not have supported the extensive uses of steel we see today.

Steel is iron with a small amount of carbon dissolved in its chemical structure. Modern steel contains small amounts of manganese or silicon to promote hardness and uniformity of internal grain structure, sulfur to retain malleability, and phosphorus to increase strength and corrosion resistance. Other elements, such as aluminum, chromium, copper, lead, nickel, tungsten, tellurium, and cobalt, are considered alloying elements. The oxygen burns most of the carbon from the iron and produces steel. Bessemer's process greatly increased production of steel by cutting the time required to make steel, and thus cut the price of steel in half. In 1850, annual steel production in England amounted to about 50 thousand tons. By 1880 the production was up to 130 thousand tons. Today, the United States produces about 800 million tons per year, and about 40% of that is from recycled steel. The rate of growth and the ready availability of steel products are attributed to the development of the Bessemer converter. Structural steel is one of the most useful construction materials in the world (see Fig. 13.15).

The efficient production of structural steel is one of the developments that led to the construction of tall buildings. Another important development making tall buildings practical was the elevator, needed for vertical transportation. Steel provides the strength and rigidity to build tall buildings. The elevator makes use of the upper floors acceptable and even desirable. There are many other technological advances that are needed to make tall buildings comfortable and convenient, such as plumbing and water supply on the upper floors, fire protection on floors above the fire department's capability, and vibration damping; but none of these would have even been needed without steel and the elevator.

There are four basic steps involved in the production of steel:

1. The raw materials—iron ore, limestone, and coal—are loaded continuously into the top of a blast furnace to make pig iron. The raw materials must be heated to a temperature of about 1,700°F, which separates the molten pig iron from the slag. The molten pig iron is removed by drawing

FIGURE 13.15 The Eiffel Tower in Paris, France, is one of the most famous steel structures.

it off through a tap at the bottom of the blast furnace. The slag consists of impurities such as melted coke, rock, and soil, and tends to float on the molten pig iron. A tap slightly higher up in the blast furnace is used to remove the slag. Burning coal with the mix produces carbon monoxide and carbon dioxide, which increases the carbon content of the steel. Blasting air or oxygen into the furnace increases the temperature and reduces the time the materials must spend in the blast furnace.

2. The pig iron is remelted in small furnaces to produce cast metals and ingots. The common cast metals used in construction are gray and white cast iron, malleable cast iron, and wrought iron.

3. Steel is produced by oxidizing pig iron in another furnace at about 3,000°F. The composition of the steel is controlled by the addition of recycled steel and other chemical additives such as silicon or manganese.

Steel is defined as the chemical union of iron and carbon. The carbon accounts for less than 2% of the weight of the steel.

4. Steel is moved from the furnace to a soaking pit, where it is lowered to a uniform temperature of about 2,300°F before it is taken to a rolling mill and rolled into its final shape: long products such as wire, bars, H sections, angles; or flat products such as plates and hot rolled coils.

Hot working steel increases the strength of the steel by breaking up grains and forcing tiny air pockets out of the steel. Cold rolling increases strength and hardness by elongating the grains. It also produces smoother surfaces because there is no cooling process.

The Society of Automotive Engineers (SAE) has developed a numbering system that describes the content and type of steel, using four or more digits. The American Iron and Steel Institute (AISI) adopted the system but modified it by adding a letter prefix to designate the type of furnace that produced the steel. The American Institute of Steel Construction, Inc. (AISC) has its own system in which each type of steel is known by the number of the ASTM standard that describes it. For example, A36 is the all-purpose carbon grade steel widely used in building and bridge construction.

Rusting is caused by the oxygen in the air reacting with the iron in the steel to form iron oxide. Rusting is a slow process, but over time it reduces the cross section of the steel member and therefore reduces its strength. Exposure to more caustic atmospheres, such as could occur in a manufacturing plant, can increase the rate of rusting. The formation of rust can be prevented or slowed by covering the steel with a coating, such as paint or galvanizing (zinc). Paint has the disadvantage that the steel must be repainted periodically, but galvanizing is not acceptable for all applications because of its color. Stainless steel is formed by the addition of chromium or nickel, but it is too expensive and too difficult to work with for general usage.

piling
May be steel H-beams, wood poles, or concrete caissons, driven or drilled into the ground, often to bedrock, to support a foundation of a structure.

There are many common structural uses for rolled shapes (see Fig. 13.16) and plates on a construction project: **piling** for foundations, sheathing for excavation and trenches, I-beams for the building frame, floor and roof joists, to name a few. In the cold-rolled steel production process, a typical I-beam makes 40 passes through the rollers to be formed into its final shape. The accepted designation for wide flange shapes consists of the letter W (for wide flange), a number representing the nominal depth in inches, an x, and a number denoting pounds per linear foot. A W 12 × 50 would be a wide flange beam, 12 in. deep, weighing 50 lb per ft. Structural angles have a different, but similar, designation. An L 4 × 3 × ½ is an angle, 4 in. by 3 in. by ½ in. thick.

Yield stress is the basis for design of all steel structures. It is the point that produces a yield on the stress-strain curve, the point at which the steel frame of the building would begin to change its shape through elongation or distortion of its steel members.

Steel is shipped from the steel mill to a fabrication shop (Fig. 13.17), where the steel is cut to the correct shape, connector plates are shaped, and the bolt

FIGURE 13.16 Rolled steel in the fabrication shop.

FIGURE 13.17 Steel fabrication shop.

holes are punched. The steel fabricator will do as much as possible in the shop to reduce the work required to erect the steel in the field. The fabrication shop is where the shop drawings are produced (and it is the reason they are called "shop drawings"). The shop drawings detail every piece of steel, every bolt hole, and every weld in the structure.

Shapes of steel items in buildings (Fig. 13.18) include wide flange beams, C channels, and H piles for the foundation pilings; reinforcing bar, discussed in the previous section, to reinforce the concrete foundation; wide flange beams for the columns and major floor beams; and open floor and roof joists, to name just a few. On structures such as bridges, steel girders may support the deck, or the deck may be supported on concrete prestressed or posttensioned beams. Even

FIGURE 13.18 Steel-frame building under construction.

individual residential homes are sometimes constructed with steel beams supporting the floors and steel studs replacing the usual lumber 2 × 4s. Termites do not eat steel but metal studs do not perform as well in a fire.

When exposed to intense heat, steel loses its strength. It becomes like a noodle after it's cooked. Therefore, steel members in buildings are coated with fireproofing (Fig. 13.19). The fireproofing is not designed to completely prevent collapse of the steel in a major fire but to give people time to get out of buildings. Columns (with proper fireproofing) should withstand 1,600°F heat for 1 to 2 hours.

Steel connections (Fig. 13.20) are made by rivets, welds, or bolts. Rivets are not as common in construction as they once were. To make a riveted connection, a hole is drilled through the two pieces of steel, a rivet that is slightly larger than the drilled hole is heated to 1,000°F, pounded into the hole, and braced while the end is hammered to a flattened head. As the rivet cools, it contracts to tighten the connection. Welding has an aesthetically pleasing appearance, and can produce a connection that is stronger than the surrounding steel. The welding rod

FIGURE 13.19 Applying fireproofing material to building steel.

FIGURE 13.20 The intricacies of steel member connections.

used must be matched to the type of steel. Heat is produced by passing either alternating current or direct current electricity through the welding rod into the steel surrounding the joint. Fillet welds connect steel plates oriented at an angle to one another. Butt welds connect plates in line with one another. Welds must be protected from corrosion because they absorb air and therefore corrode more easily than the surrounding steel.

Bolts are manufactured with a head on one end and threads for the nut on the other. If washers are used, they fit loosely on the bolt. Common bolts are generally not strong enough and high-strength bolts must be used to withstand both the tensions and shear stresses found in the connection. Bolts are tightened by electric or compressed air impact wrenches and then tightened with a calibrated torque wrench until bolt tension reaches that specified by the American Institute of Steel Constructors.

WOOD AND WOOD PRODUCTS

Wood is the very common building material. People enjoy the appearance of high-quality wood. It has a warm pleasant appearance and it is used for framing, floors, walls, paneling, and even ceilings. As wood ages, it often becomes even more attractive. Finally, when we are finished with it, wood will decay easily and become part of the soil again. Wood is abundant, and will continue to be abundant as long as we manage the forests properly.

The wood used for framing (see Fig.13.21) is classified as soft wood because it comes from evergreen trees. In fact, some softwood is harder than some of the hardwoods, which come from deciduous trees (those that lose their leaves in the winter). Evergreen trees grow rapidly and produce great quantities of wood used in the construction industry.

After wood is harvested, it must be sawn and seasoned (dried). After a log is cut into boards, the mill dries the boards either in a kiln or in the open air. Natural wood contains water that makes up between 30% and 300% of the dry weight of the wood. Structural wood is usually dried to about 19% water content. Below about 15%, there is no advantage to drying wood any further because the wood will simply absorb water to match the surrounding atmosphere. Some wood will warp because of the high humidity of its surroundings.

When logs are brought to the mill, the mill master must determine what sawing pattern to use to saw the log. Plain sawing produces the greatest quantity of salable lumber and is the way most wood for framing is sawed (see Fig. 13.22). Plain sawing produces boards with grain variation from nearly parallel to the width of the board to nearly perpendicular to the width of the board. This variation causes the boards to vary in strength and to shrink differently as they dry.

Quarter sawing (see Fig. 13.23) does not produce as much salable wood, but it produces more uniform boards. The log is first cut into quarters, and then the boards are all cut across the grain. Quarter sawing is used for wood to be used in furniture and cabinetwork. The wood is more expensive.

FIGURE 13.21 Wood-frame building under construction.

Large logs are sawed into patterns that most economically use the log. Large sizes can be cut from the center portion of the log, and smaller sizes from the outer layers.

Wood is not straight, it is not perfectly uniform, and it has random defects. For all of these reasons, wood must be graded. It is graded for structural strength and for appearance. Framing lumber is graded for structural strength, with most sizes graded as either Select Structural No. 1, No. 2, or No. 3, with No. 1 being the best and No. 3 being the worst. Lumber is also graded according to appearance. Appearance grades are selects, finish, paneling, and bevel or bungalow siding. These grading systems allow engineers and architects to select the quality of wood best suited to the building requirements.

In recent years, the wood products industry has developed laminated wood products, collectively referred to as "**Glulam**," such as **oriented strand board (OSB)**, laminated veneer lumber (LVL), and I-joists, to name a few. LVL is an engineered wood product created by layering dried and graded wood veneers with waterproof adhesive into blocks of material called billets. In LVL billets, the grain of each layer runs in the same direction producing a lumber that outperforms conventional lumber. OSB is manufactured from waterproof,

Glulam
Trade name for laminated wood products.

oriented strand board (OSB)
Glued wood product (Glulam), manufactured by gluing together wood chips that are all oriented in the same plane. Typical dimensions are 4' × 8'. Used in place of more expensive plywood for shear walls and roofs on homes and small commercial buildings.

FIGURE 13.22 A plain cut log.

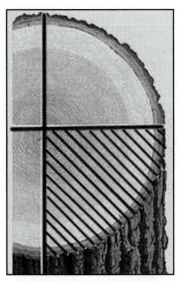

FIGURE 13.23 Quarter sawn log.

heat-cured adhesives and rectangularly shaped wood strands arranged in cross-oriented layers, similar to plywood. OSB shares many of the characteristics of plywood. These products are more uniform than natural lumber, and they offer exceptional strength and reliability. I-joists are I-shaped engineered beams that offer strength and versatility for use in residential construction and other similar-sized buildings. I-joists can be manufactured using either solid sawn lumber or structural composite lumber for the flange of the beam and plywood or OSB for the web.

The variety of engineered wood products is increasing every year and is driven by efforts to make more efficient use of new-growth trees and protect the old-growth trees. Engineered wood products make efficient use of smaller trees.

Plywood is a **laminated wood product.** It is made up of three or more thin veneers of wood. There is always an odd number of veneers to balance the effects of atmospheric moisture. The outer veneer may be a high-quality wood selected for its appearance, and the inner layers are selected for their strength. The American Plywood Association veneer grades for plywood are given in Table 13.2.

The two major weaknesses of wood are its combustibility and susceptibility to decay and insects. Several treatments have been developed in an effort to reduce the effects of these two weaknesses. The cost of fire-retardant treated wood is too high to justify its use in residential construction. Wood that is used in or near the ground must be treated to prevent decay and insect damage. Creosote, which may be environmentally harmful, is sometimes used to treat lumber used in marine applications, but it is not acceptable for use in residential construction or office buildings because of its smell and inability to be painted.

laminated wood product

Includes products such as plywood, oriented strand board, and glued beams.

TABLE 13.2 Plywood veneer grades.

Grade	Characteristics
N	Smooth surface "natural finish" veneer. Select, all heartwood or all sapwood. Free of open defects. Allows not more than six repairs, wood only, per 4 × 8 panel, made parallel to grain and well matched for grain and color.
A	Smooth, paintable. Not more than 18 neatly made repairs, boat, sled, or router type, and parallel to grain, permitted. May be used for natural finish in less-demanding applications.
B	Solid surface. Shims, circular repair plugs and tight knots 1" across grain permitted. Some minor splits permitted.
C Plugged	Improved C veneer with splits limited to ⅛-in. width and knotholes and borer holes limited to ¼ in. and ½ in. Admits some broken grain. Synthetic repairs permitted.
C	Tight knots to 1 ½ in. Knotholes to 1 in. across grain and some to 1 ½ in. if total width of knots and knotholes is within specified limits. Synthetic or wood repairs. Discoloration and sanding defects that do not impair strength permitted. Limited splits allowed. Stitching permitted.
D	Knots and knotholes to 2-in. width across grain and ½ in. larger within specified limits. Limited splits are permitted. Stitching permitted. Limited to interior (Exposure 1 or 2) panels.

The United States began mandating the use of nonarsenic containing wood pre-servatives for virtually all residential use timber in 2004. The most widely used treatment for wood is alkaline copper quaternary (ACQ). ACQ is a preservative made up of copper, a fungicide, and quaternary ammonium compound (quat), an insecticide that also augments the fungicidal treatment. Since it contains high levels of copper, ACQ-treated timber is five times more corrosive to common steel, according to American Wood Preservers Association (AWPA) test results. Therefore, it is necessary to use double-galvanized or stainless steel fasteners in ACQ timber. This treatment allows the wood to be left exposed or painted or stained. It is widely used for outdoor structures such as decks, porches, and sheds.

MASONRY

Masonry is found everywhere in the world in some form. Archeologists have found evidence dating back 8,000 years of masonry construction, using sun-baked clay bricks. The process of kiln or oven drying bricks was developed around 2,500 years ago. The art of masonry construction therefore is not a new method (see Fig. 13.24).

Consider some examples. The majestic cathedrals in Europe were built from stone (Fig. 13.25), some requiring construction periods of up to 300 years. Adobe brick construction in Mexico and the Southwest, which features

FIGURE 13.24 Roman masonry aqueduct.

FIGURE 13.25 Five-hundred-year-old cathedral in Portugal.

sun-dried bricks, has been practiced for generations. There is probably no way to know when the Eskimo developed the igloo, which is ice masonry, but it must be counted among the world's earliest forms of masonry construction. Some of the world's greatest construction projects are masonry projects—the ancient pyramids of Egypt, the Roman aqueducts, the Parthenon, the Great Wall of China, and the Inca cities of South America.

Masonry building blocks are quarried stone (Fig. 13.26), brick (Fig. 13.27), tile, or **concrete masonry units** (CMUs). Mortar is used to provide the bond between the blocks and to ensure even contact in spite of irregularities. The requirements of any masonry structure are that it be pleasing in appearance, strong and durable enough for its design purpose, and watertight.

Mortar is a mixture of cement, lime, sand, and water. Most mortar is made using Type I Portland cement, but Types II or III also may be used in special circumstances such as exposure to more highly caustic atmospheres. Lime, which is made by heating seashells or from other sources, is needed because mortar without lime is too harsh, which means that it does not flow or adhere to the surfaces of the bricks properly. There are no tests for determining when the mortar is too harsh but the brick mason will know. Sand must be sound, clean, and well graded. Either natural or manufactured sand is acceptable. The water used

concrete masonry units

Concrete units used in masonry for outdoor walls, external building walls, and other such purposes.

FIGURE 13.26 Stone building under construction.

FIGURE 13.27 Brick veneer wall under construction.

should be clean and free of organic impurities. Mortar is usually mixed on the construction site.

Brick and *tile* are both made of clay. Tile is essentially the same product as brick, but tiles have larger internal openings, such as would be needed for the inside of chimneys. *Surface clay* can be excavated on or near the surface of the earth, and it is used for most ordinary brick. Brick for furnaces is formed using *fire clay,* which is mined from deep within the earth and has fewer impurities. Excavated natural clays are blended for color and uniformity. They are crushed to break up any large chunks, and they are screened to remove any stones. The clay is then mixed with water in a pugmill in a process known as tempering. The product of the pugmill is a plastic mix of clay and water that is ready for forming into bricks. The three methods of forming bricks are known as stiff-mud, soft-mud, and dry-press.

The most common method, which is used for most ordinary brick, is the stiff-mud method. The clay is mixed with about 15% water, by weight, and dewatered by a vacuum process. The mixture is forced through a die to form a ribbon, which is then cut to the desired size by wire.

Soft-mud bricks are made from clay that has a high natural water content. The clay is forced into forms, and water or sand is used to break the bond with the form.

The dry-press process is used with a mix of clay and 10% water content by weight. The mix is forced into forms under pressure to form the bricks.

Bricks or tiles are fired for a period of 2 to 5 days depending on the type of kiln used. The firing of brick controls its color and strength (bricks can have ultimate compressive strengths as low as 1,600 psi, whereas well-burned bricks have compressive strengths exceeding 20,000 psi), and therefore the temperature of the kiln as the bricks move through must be very carefully controlled. At some point during the firing process, the brick may reach temperatures as high as 2,400°F. The cooling down process is also carefully controlled to prevent cracking, and can require 2 or 3 days.

The most important engineering property of brick is its compressive strength, which can vary from 1,500 to 20,000 psi.

There are no standard brick sizes, although most bricks are about 4 in. wide by 8 in. long, but may be as much as 12 in. long. Masonry units have nominal dimensions based on the 4-in. module. The height varies from about 2 to 4 in. Bricks are sized to fit comfortably into the mason's hand. Larger bricks not only are less comfortable in the hand, but also are heavier and therefore more difficult to handle. Most bricks have a hollowed out interior, with either two or three holes formed into them. The purpose of the holes is to make the brick lighter. They do not significantly reduce the compressive strength of the brick. The bricks commonly used in construction are delineated in Table 13.3.

Concrete masonry units are concrete blocks made from a concrete mix rather than from clay. CMUs are used for the interior wythe on buildings, or

TABLE 13.3 Classes of bricks commonly used in construction.

Brick classification	Description
Building bricks (common, hard, or kiln-run)	These are made from ordinary clays or shales and fired in kilns. They have no special scorings, markings, surface texture, or color.
Face bricks	These are better quality and have better durability and appearance than building bricks because they are used in exposed wall faces.
Clinker bricks	These bricks are overburned in the kiln. They are usually rough, hard, durable, and sometimes irregular in shape.
Pressed bricks	These are made by the dry-press process rather than by kiln-firing. They have regular smooth faces, sharp edges, and perfectly square corners.
Glazed bricks	These have one surface coated with ceramic glazing. This surface has a glasslike texture.
Fire bricks	These are made from a special type of fire clay to withstand the high temperatures of fireplaces, boilers, and similar conditions without cracking or decomposing.
Cored bricks	These have 10 holes—two rows of five holes each—extending through them to reduce weight.
Sand-lime bricks	These are made from a lean mixture of slaked lime and fine sand containing lots of silica. They are molded under steam pressure and hardened under steam pressure.

for exterior walls that do not require the same high standard of appearance as a masonry wall. Originally they were made as solid blocks. They then progressed to hollow core blocks that were hand packed in forms. Today, a modern CMU-forming machine can turn out 20,000 units per day. The raw materials are Portland cement, sand, aggregate, and water. Admixtures may be used, essentially as they would be used in concrete, to provide color or high early strength. As in concrete, the aggregate makes up a high percentage of the weight of the block. The quality of the aggregate is very important. The aggregate should be sound and tough, with sufficient strength to provide the required compressive strength.

Concrete masonry units are manufactured by weighing the materials from a weigh bin into the mix, forcing the mix into the forms, and finally steam curing to prevent breakage during the curing process. Finished blocks may be stored outside without adverse impact.

The standard size of concrete blocks is 8 in. × 8 in. × 16 in. The thickness may vary to 3 or 8 in. Most blocks are hollow to reduce the weight and provide improved curing with less cracking. The normal blocks are called stretcher blocks. There are several other types of blocks, such as corner blocks and bull-nose blocks. Additionally, special-shaped blocks are used for lintels, door jambs, and other applications. Strength and absorption requirements are given in ASTM C90.

SUMMARY

Concrete is a mixture that contains cement, sand, gravel, water, and usually additives. The cement is the material that causes the other ingredients to adhere to one another. Concrete does not harden by drying but by a chemical reaction called hydration. The aggregate should be uniformly graded to provide a good fill of all the spaces between the individual pieces of aggregate. The sand will fill the small spaces in the mix. The amount of water used in the mix should not be excessive because excess water eventually evaporates, leaving small voids in the spaces it once occupied.

The constructor is concerned with the construction of the forms, tying the reinforcing steel, transporting the mix to the site, placing the concrete in the forms, and the curing process.

The purpose of reinforcing steel is to give the concrete flexural strength. In columns, reinforcing steel also helps carry the vertical compressive load, and makes it possible to reduce the diameter of the column. The ASTM specifications require that identification marks be rolled into the bars, which identify the producing mill, the bar size, the type of steel, and the grade. Reinforcing steel may be prebent at an off-site fabricating facility.

In recent years, the wood products industry has developed laminated wood products, collectively referred to as "Glulam," such as oriented strand board (OSB), laminated veneer lumber (LVL), and I-joists.

Masonry building blocks are normally quarried stone, brick, tile, or concrete masonry units. Tile and brick are clay products fired in a furnace. Concrete masonry units are a concrete product. Mortar is used to provide the bond and to ensure even contact between the masonry blocks.

REVIEW QUESTIONS

13.1 Describe the process of manufacturing wide flange beams.

13.2 What is the meaning of the designation W 24 × 204?

13.3 What are the advantages of using structural steel in building construction? What are the disadvantages?

13.4 What is the meaning of modular-sized masonry units?

13.5 What is the water/cement ratio? What is its significance?

13.6 Why is gap-graded aggregate not acceptable?

13.7 What is the basic purpose of reinforcing steel in concrete slabs and grade beams?

13.8 Why are bricks fired in a kiln? What are the characteristics of bricks that make them such a universally used construction material?

13.9 What is the difference between brick and concrete masonry units?

13.10 What are the basic uses of CMUs?

REFERENCES

1. Allen, Edward, and Joseph Iano (1998). *Fundamentals of Building Construction Materials and Methods,* 3rd ed., John Wiley and Sons, New York.

2. Concrete Reinforcing Steel Institute (1990). *Manual of Standard Practice,* 25th ed. Schaumburg, IL.

3. Concrete Reinforcing Steel Institute (1992). *Placing Reinforcing Bars,* 6th ed. Schaumburg, IL.

4. Marotta, Theodore M., and Charles A. Herubin (2001). *Basic Construction Materials,* 6th ed., Prentice Hall, Upper Saddle River, NJ.

5. Simmons, H. Leslie (2001). *Construction Principles, Materials, and Methods,* 7th ed. John Wiley & Sons, New York.

WEBSITE RESOURCES

1. www.apawood.org   The Engineered Wood Association web page is one of the most informative and complete web pages available for information regarding wood and wood products. It offers general information and design specifications for a complete line of engineered wood products including Glulam, I-joists, laminated veneer lumber, oriented strand board, plywood, and other specialty wood products. APA offers a complete line of low cost reference material and links to other websites.

2. www.aitc-glulam.org   The American Institute of Timber Construction.

3. www.awc.org   American Wood Council.

4. www.pbmdf.com   National Particleboard Association.

5. www.imiweb.org   This International Masonry Institute site contains technical notes relating to masonry materials. Online versions are free.

6. www.bia.org   The Brick Institute of America site. This site features an idea gallery, technical notes, and the brick university courses.

7. www.steel.org   The American Iron and Steel Institute presents industry data, consumer guides, public policy issues papers, and information about steel applications.

8. www.aisc.org   American Institute of Steel Constructors. Contains technical data relating to structural steel.

9. www.sweets.com   Sweets online catalog offers essentially all materials for buildings.

10. www.portcement.org   Website of the Portland Cement Association offers instructional materials, catalogs, member services, and information about concrete.

14

Building Construction Methods

This chapter contains a description of the construction of three different kinds of buildings—residential, commercial, and multistory (up to about 20 stories). Most people are familiar with houses and may even have an understanding of how they are built. The activities required to build a larger commercial building are similar to the activities required to build a house, but more complicated—in some cases more repetitious. Most commercial buildings have essentially the same design—a one-story open shell with a concrete slab floor. Multistory buildings are a completely separate class of buildings and are addressed separately.

MOBILIZATION

The work of constructing a building begins long before a constructor arrives on the site. Some of the work begins before the constructor is even sure of being selected to do the work. Specifically, bidding or negotiating and taking bids from subcontractors takes place before the contract is signed. This is the "get work" portion of the constructor's model.

Once selected to build a project, the constructor has several tasks:

■ Arrange financing for accomplishing the work. This involves an estimate of expenses and a cash flow analysis so the constructor will know before the work starts what the monthly cash requirements will be. (See Chapters 3 and 6.)

■ Select subcontractors. Award subcontracts, or at least develop a plan of when the subcontracts will be awarded.

■ Prepare a detailed project schedule showing the work of each subcontractor so the project manager can ensure there is no conflict over work space and tasks.

■ Select material suppliers and finalize as many purchase agreements as possible.

■ Set up the project office. The size of the office generally reflects the size of the project. Construction of a residence would probably be managed out of the superintendent's pickup truck. A multimillion-dollar shopping mall would require an on-site field office.

■ Provide secure storage for tools, equipment, and materials.

■ Attend a preconstruction meeting with the owner and interested public agencies. The preconstruction meeting has several purposes, but the most important is to open the lines of communication between the constructor and owner. The topics discussed would normally include safety requirements, permits, maintenance of proactive relationships with the neighbors, payment schedule, and actions that will be required by the owner during construction. Each project is different, and each preconstruction conference covers different topics.

■ Begin the preparation of submittals and shop drawings, particularly shop drawings for the long-lead-time materials. Often the construction of a house does not require the submission of shop drawings because the builder has a standard list of materials and supplies used in the house. For construction of other kinds of facilities, shop drawings are required. The first shop drawings submitted are for items that would be installed first and the items that have the longest lead time, such as structural steel and nonstandard mechanical equipment.

BUILDING SYSTEMS

A building structure is composed of numerous subsystems. The key subsystems are:

■ *Foundation.* The foundation supports the building and carries the live loads and the dead loads. Live loads consist of the people and movable equipment that will occupy the building as well as environmental loads such as wind, snow, etc. Dead loads are the structure itself and the nonmoving equipment in the building.

■ *Framing.* In the case of most residential construction, the framing material is 2 × 4 or 2 × 6 lumber. In a commercial building, the frame could be steel or concrete.

■ *Building envelope.* The exterior walls and roof make up the envelope that encloses the building. The building envelope is basically everything outside the frame.

■ *Interior walls.* The interior walls may be part of the structural frame of the building, as they normally are in a house, or they might be completely separate from the building frame as they are in a high-rise commercial building.

- *Utilities.* These include electricity, water, wastewater disposal, telephone, television cable, and data and communication systems. Sophisticated buildings, such as hospitals or computer chip manufacturing facilities, have many more utilities serving the building. An example would be an oxygen system in a hospital or an argon gas system in a computer chip manufacturing facility.

- *Environmental control systems.* These include heating, lighting, air-conditioning and ventilation, and acoustics, along with the control systems that operate each.

- *Transportation systems.* Most building transportation is vertical—elevators, escalators, and stairs. Only buildings that are large in horizontal dimensions such as airport terminals have horizontal transportation systems (people movers).

- *Fire suppression.* Fire sprinklers are normally required in most public buildings and some cities now require fire suppression systems in new homes.

SITE PREPARATION

Consider the construction of a residence, a one-story, single-family house with a deck and an attached two-car garage. The house has three bedrooms, a great room with a fireplace, a dining room, a kitchen, and a poured concrete basement. The walls have 2×4 or 2×6 studs. The exterior is wood shingles and the roof is asphalt shingles. Water is provided by a well and the house has a septic system. A gas furnace provides heat and there is a central air-conditioning system.

The first on-site activity is usually site preparation or site clearing. This involves clearing only those trees that are in the footprint of the structure (while saving as many as possible) and removing unwanted debris. Surveying to establish the exact location of the building will follow the clearing work. The well can be drilled at this time; however, the contractor will delay construction of the septic system until later in the project to ensure that the system is not damaged by heavy equipment coming onto the site. The concrete trucks will need to have access to the foundation, and the septic system would create an unnecessary obstacle to their operations. Other service entrances such as electric power and telephone will also be constructed later for the same reason. The electric meter and interior wiring can be installed without having power brought to the building. The same is true for television cable, telephone, and any other similar services.

FOUNDATION

Homes typically have shallow foundations. The most common types of residential foundations are concrete slab-on-grade and grade beams. A basement is simply a special case of a grade beam. The space between the grade beams

FIGURE 14.1 Basement grade beam forms being constructed.

is excavated to the level of the basement floor and a concrete floor is provided. The home in this example has a basement, so the contractor must excavate to the correct elevation for the foundation footings, and place the drainage pipes in the ground immediately inside the footings. Such drainage systems normally drain into a sump. Concrete footings are required to support the basement walls.

The concrete basement walls are constructed by placing the rebar (probably #4, 9 in. on-center), setting the forms (reusable), and then placing the concrete in the forms. Tie-down bolts will be placed in the concrete to hold the frame of the structure in place. For most homes, these operations can be completed in 1 to 2 days. Figure 14.1 shows the house foundation being constructed.

Following the grade beam construction, the basement walls are constructed. The forms and reinforcing steel are placed, and the concrete is placed into the forms.

FRAMING

Today, much of the framing for residential construction is prefabricated off site and trucked to the site (see Fig. 14.2). The scrap lumber then remains at the prefabrication site and there is minimal scrap generated at the construction site. Framing is becoming more and more a job of erecting wall sections, rather than field measuring, cutting, and fabricating.

The first step is bolting the sill to the top of the foundation concrete. This is followed by the installation of the floor joists for the first floor and placing the subfloor. The size and type of the floor joists and the type of subfloor are

FIGURE 14.2 Prefabricated framing for residential construction.

determined based on the free span between the basement walls. Joists are either dimension lumber such as 2 × 10s or manufactured joists made from plywood or oriented strand board (OSB). The subfloor is typically either plywood or OSB. The exterior and interior walls are raised on top of the subfloor. Homes without basements have frames bolted directly to the concrete floor slab (Fig. 14.3). Homes with basements have the floor joists and subfloor on the sill.

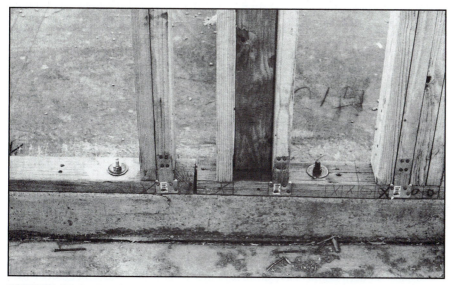

FIGURE 14.3 Framing for a residence bolted directly to the concrete floor slab.

FIGURE 14.4   The beginning of the framing for the residence.

The first-floor frame is built on the subflooring. The floor for the second story is built onto the top of the first-floor frame. Constructing the frame for the second story is the same process as constructing the frame of the first floor. The roof framing system, consisting of rafters and joists, is erected on top of the second-story frame. Figures 14.4 through 14.6 depict the house as framing progresses.

BUILDING ENVELOPE

The building envelope consists of the insulation, sheathing, moisture barrier, exterior siding and roofing, windows, and doors. Sheathing, which is the plywood or OSB nailed directly onto the frame to create a shear wall, is attached to both the walls and roof.

Most roofs have a layer of roofing paper nailed to the roof sheathing. The roof is shingled immediately following the placing of the roofing paper. The builder must know the required weight and required overlap of the roofing paper, in accordance with the local building code.

The sheathing is placed onto the frame as soon as the work progresses enough to allow its installation. Nailing the wall sheathing in place helps increase the stability of the building. Following the wall sheathing, the windows and doors can be installed and the exterior of the building applied.

Many modern homes have chimneys that consist of a sheet metal duct with a wood frame built on the roof to look like a chimney. If there is a real masonry

FIGURE 14.5 Most of the exterior framing, sheathing, and vapor barrier are complete. Interior framing can be seen through the patio door.

FIGURE 14.6 Roof framing and sheathing partially completed.

fireplace with a real brick chimney, it must be constructed prior to roofing the building because the flashing must be placed properly around the chimney at the roof level to prevent leakage.

Insulation is placed under the first floor, in the exterior walls, and above the ceiling of the top floor, below the attic space. This completes the insulation envelope around all six sides of the climate-controlled spaces in the residence. The basement, attic, and garage will be outside the insulation envelope. There is no reason to insulate between floors. There is also no reason to insulate the exterior walls of the garage unless the garage itself is heated. Insulation is rated in terms of thermal resistance, or **R value**, which indicates the insulation's resistance to heat flow. Higher R values indicate greater resistance to heat flow.

There are several types of insulation, not all of which are used in residential construction. The wall insulation (Fig. 14.7) would normally be blanket insulation in the form of batts or rolls, which are made from mineral fibers. They are available in widths suited to standard spacing of wall studs and attic or floor joists, with or without **vapor barrier** facing. Styrofoam sheets or panels may be used on the inside of the concrete basement walls to reduce heat loss into the concrete. Styrofoam sheets may also be installed outside the sheathing on homes that are to receive a stucco finish. The insulation below the first floor is placed between the floor joists. Normally, the joist braces are adequate to hold the insulation, but it may be necessary to add holders.

Vapor barriers are used in most parts of the country to keep the insulation dry because it does not insulate as effectively if it becomes wet, and the moisture can cause damage to the structure if it condenses inside the walls. The warm air inside the home contains moisture, so vapor barriers are placed inside the insulation in cold climates. The vapor barrier would be placed outside the insulation

R value

Numerical measure of the insulating characteristics of materials.

vapor barrier

Plastic sheets placed either inside or outside the insulation, depending on the climate, to protect the insulation and building from excessive moisture.

FIGURE 14.7   Blanket insulation used in a framed wall.

in a crawl space because the moist air would be in the space below the structure. In hot, moist climates, the vapor barrier may be placed outside the insulation. Builders must determine what practice works best in their local area.

The ceiling insulation is another matter. Here, **blown-in insulation** is usually used. Blown-in loose fill insulation is placed above the ceiling of the top story using special pneumatic equipment. It obviously cannot be blown in until the sheetrock is completed. The insulation will be placed in the attic on top of the ceiling of the upper story. The blown-in attic insulation is normally thicker and has a higher R value than the insulation in the walls because heat rises.

Heat naturally flows toward an area of lower temperature. The purpose of the insulation is to retard that flow of heat. Homes in cold climates need insulation to keep the warmth inside the structure during the cold winters. Homes in warmer climates need insulation to keep the outdoor heat from flowing into the cool air-conditioned air during the heat of the summer. The amount of insulation needed for a house depends on the climate, building materials, heating system, size and shape of the house, and habits of the occupants. Typical R values are R-18 for walls and R-30 for ceilings.

The electric and plumbing rough-in should be done before the insulation is installed to prevent damage to the insulation during wiring, and to allow the insulation installer to place the insulation around and behind the wiring and electric boxes. The electric boxes and wiring must be placed accurately to meet code requirements. Insulation has more flexibility and can be made to fit around obstructions such as electric wiring and plumbing.

Windows can be installed any time after the framing is in place. Window units are designed to be installed by attaching directly onto the framing of the building prior to the application of the building exterior material such as shingles, siding, or masonry.

Most contractors do not want to install the exterior doors until they are ready to secure the building, but the doorframes must be installed immediately after framing. When the exterior doors are in place, entry becomes more of a nuisance, but there is less likelihood of pilferage.

blown-in insulation
Type of insulation placed over ceilings. Placed by blowing in through hoses with compressed air.

INTERIOR WALLS

In a house, framing the interior walls is an integral part of the framing operation. In large commercial buildings, interior walls may not be part of the original construction process. Each building tenant has interior partitions constructed to suit their own needs. Interior framing is usually covered with sheetrock.

Sheetrock consists of a plaster core with a paper covering on each side. The edges are slightly thinner than the rest of the board, to accommodate the tape and joint compound, which is used to hide the joints. Placing sheetrock on the ceiling and walls is a matter of either nailing or screwing the sheets to the studs. As sheetrock is installed, cutouts are made for electric switches and wall outlets, light fixtures, windows, and any other items that protrude through the sheetrock (Fig. 14.8).

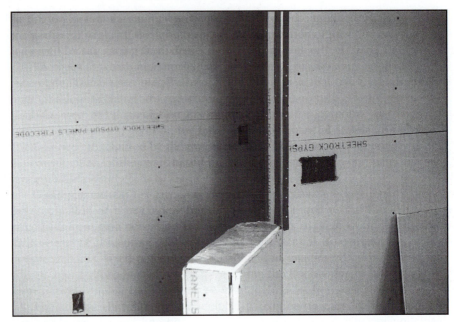

FIGURE 14.8 Sheetrock installed and cutouts made for electric switches and plugs.

The nail or screw heads are covered by a layer of joint compound to create a surface that is smooth and completely free of dents and blemishes. Covering the nails, screws, and joints is an operation called tape and paste, and must be done at least twice, preferably three times, to produce a quality surface. The final product must dry for several days to produce a hard surface capable of accepting paint or other wall coverings.

UTILITIES

utility entrances

The points where utilities such as electricity, water, and sewer penetrate the foundation or outer wall of the building.

The utilities in a building are all constructed following the same general principles. The **utility entrances** to the building do not need to be completed early in the construction process. It is much better to place the entrances for water lines, telephone lines, and electrical cable to the meter late in the construction process to avoid damage during construction. Inside the building, utilities are generally built in two phases—rough-in and finish. Utility rough-in can begin when the framing is nearly complete.

Electric rough-in is a matter of identifying the locations for all items such as the main electrical panel, light fixtures, switches, wall receptacles, and appliance outlets. There may be other systems such as central sound systems, a security system, or computers that need to be wired. All of these systems are wired at this stage of construction. Everything that resides inside the building walls, such as wiring, wall receptacles, and switches, is installed. The electricians will return to the project after the sheetrock is in place and painted for the finish electric work.

FIGURE 14.9 Part of the electric rough-in for the residence.

They will install the lights, speakers, and building control systems. The electric entrance cable is installed and connected at this time. Part of the electric rough-in is shown in Fig. 14.9.

Plumbing rough-in is like electrical rough-in, except that slab-on-grade construction requires that the water supply and wastewater drain lines be installed prior to the construction of the slab. All above-slab pipes for the sinks and showers, washer, and outdoor faucets are installed during the framing process or as soon as the framing is complete (Fig. 14.10). Some items such as bathtubs must be installed during framing. Pipes are not connected to the water supply at this point to prevent any accidental discharge of water into the building. The water service line, meter, and backflow prevention valve will be installed in the final stages of the project. The drain lines can be connected to the sewer system at any time. Finish plumbing work will be completed very late in the construction process as the building is nearing completion.

Mechanical rough-in consists of installation of the ductwork (Fig. 14.11) for the forced air heating and air-conditioning, and wiring for the mechanical system controls. At the same time, other minor items such as the exhaust ducts for the dryer are installed. Finish mechanical work will be accomplished at the same time as finish plumbing and finish electric.

Typically, electric, plumbing, mechanical, and telephone rough-ins can be accomplished in 1 or 2 days each, with crews of two to four workers, depending on the expertise of the crews and the size of the residence.

Water supply for a home will be from either a well or the municipal water system. A well can be drilled at any convenient time during construction, since

FIGURE 14.10   Plumbing rough-in, copper water pipes, and ABS drain pipes.

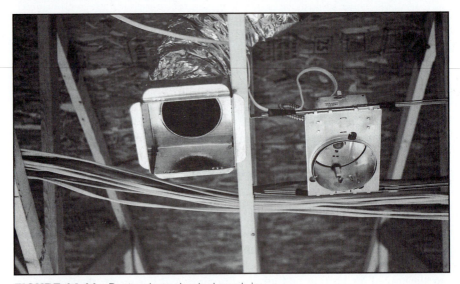

FIGURE 14.11   Ductwork mechanical rough-in.

most wells are not located close to the building. A municipal water line would be built as the building is nearing completion, again to avoid damage.

Waste disposal can also be accomplished by either a septic system or the municipal waste system. This home has a septic system. It will not be built until there is no longer any need for heavy equipment operations in the area.

FIGURE 14.12 View of the completed residence.

ENVIRONMENTAL CONTROL SYSTEMS

Examples of environmental control systems are lighting, heating and cooling, ventilation, and the systems that control them. The wiring and ductwork for these systems are installed during the utilities rough-in stage. As the building nears completion, the final work on these systems is completed. For instance, the furnace and air-conditioning equipment do not need to be installed until the project is nearly complete unless there is an access problem that will be created by later stages of construction. Most heating systems for homes are physically quite small and do not require early installation. Air-conditioning compressor systems are installed outside the building. Lights and appliances are among the last items installed in the house.

INTERIOR FINISH

The interior needs to be painted before the finish electric and finish plumbing. The very last item will be the floor covering. Figure 14.12 shows the completed residence.

COMMON COMMERCIAL BUILDINGS

There are many chain restaurants, department stores, grocery stores, office buildings, and warehouses that are essentially the same standardized one-story building design. The floor is a concrete slab-on-grade. The exterior walls are made of either **tilt-up concrete panel construction** or concrete blocks. The front of the building may be partially glass if the building is occupied by a retail outlet. The

tilt-up concrete panel construction
Walls, usually for commercial buildings, formed in place on the building's concrete floor, and raised by crane into place.

roof is approximately 30 feet above the floor and only heavy enough to support the local snow loads or wind forces in areas having high winds or hurricanes. Typically, the roof is flat and is supported by rows of steel columns in a grid approximately 30-feet square. These buildings are economical and they can be erected quickly. They are built as an empty shell, and then the interior work is added according to the use of the building. These are not low-quality buildings, but they are quite simple and easily built.

Figures 14.13 through 14.15 show a selection of these buildings. Note that they are all different, yet they are all constructed in the same basic way. Each is a big box.

The high cost of formwork is the primary driver of the tilt-up concrete construction technique. The process is used to reduce the cost of materials and labor for wall construction. The tilt-up wall panels can be designed to be load bearing, which is usually the case for the exterior walls, or non–load bearing. This building process works very well for shell-type buildings. The panels are formed, constructed, erected, and supported very quickly, as compared to other building methods.

The slab-on-grade building floor serves as the casting yard for the tilt-up wall panels (Fig. 14.16). A commercial concrete curing bond-breaking compound is used as the curing compound on the floor slab. A second coat would be applied later to serve as the bond breaker between the floor slab and the tilt-up panels.

The floor is constructed so there is a 2- to 3-foot gap between its outside dimension and the final interior face of the exterior tilt-up panels (Fig. 14.17).

FIGURE 14.13   Fuel and food quick stop.

FIGURE 14.14 Single-story strip mall buildings.

FIGURE 14.15 Offices with warehouse space in the rear.

This gap provides space for construction of a strip foundation around the perimeter of the building. After the wall panels are erected, the gap between the outside dimensions of the floor slab and the final interior face of the exterior tilt-up panels is filled with concrete to complete the slab out to the wall. Reinforcing bars extend outward from the previously placed casting slab and the tilt-up panels so

FIGURE 14.16 Slab-on-grade building floor serves at the casting yard for the tilt-up wall panels.

FIGURE 14.17 A 2- to 3-foot gap between the outside dimension of the building's floor slab and the final interior face of the exterior tilt-up panels.

that when the concrete is placed in the space between the slab and the interior of the panel wall, the structural systems are tied together (see Fig. 14.17).

Blockouts for specified doors and windows are created in the panels. In these areas, there is no reinforcing steel or electrical conduit. Knockouts are also created. The outline of knockout areas are grooved about 1 in. into the panel sections. The knockouts are created so that in the future if a tenant should chose to place a door in that panel, it is possible to easily remove the concrete section by saw cutting without striking any reinforcing steel or electrical conduit. The concrete panels are designed to be structurally sound with the knockouts removed.

With tilt-up panel construction, the layout of the panels can be accomplished in many ways. All panels can be cast on the slab-on-grade if there is sufficient room. At that stage, the slab-on-grade has no joints in it so the wall panels will be smooth. The slab joints will be saw cut after the panels are lifted into place. If site conditions greatly restrict the available slab area, the panels can be constructed on top of one another. Panels can be successfully stacked up to eight panels on top of each other.

Each panel has a particular function and position in the overall structure. Contractors try to cast the tilt-up wall panels as close to their final position as possible. They do not have to be directly adjacent to their erected location, but they are normally laid out on the slab in a pattern that minimizes crane time during the erection. Panels are usually erected in a clockwise rotation, starting from a chosen building corner. To ensure a proper placement sequence, the erection of certain panels at the correct place and time may require some panels to be rigged and hoisted over or around others that remained in their horizontal cast position on the slab.

The wall panel forms are installed on the building's slab-on-grade as soon as that concrete work is completed. This operation includes the panel forming and the placement of reinforcing steel, blockouts, lift and brace points, ledger plates, and conduit for the electric utilities (see Fig. 14.16).

A panel placement map is used to lay out the panels on the building slab. A spacing of 2 feet is maintained between the edge forms. The wall panel edge forms are positioned accordingly. Steel concrete edge forms having an angle shape are used, and the forms are nailed to the floor slab. The nails are hammered into predrilled holes in the concrete floor. Chamfer strips are positioned along each panel edge so that when the panels are lifted and placed, concrete edges are not damaged. Once the forms are in place, the floor is cleaned and the bond breaker is applied. At that point, the crews must be careful not to contaminate the clean surface. The electric conduit, reinforcing steel, lifting and anchoring inserts, and ledgers are added. The inserts are needed to provide a means to lift the panels and to support them temporarily during construction. The number of inserts depends on the size of the panel. The placement of the concrete must be accomplished as quickly as possible.

The final phase is the erection of the concrete tilt-up panels. Hoisting the panels requires the use of a crane (Fig. 14.18). It is imperative to follow the proper lifting sequence so the crane can erect as many panels as possible to

FIGURE 14.18   Hoisting a tilt-up wall panel.

minimize the total cost of the crane. Each panel must be leveled and plumbed before the temporary braces are attached to maintain the panel in an upright position. The braces should be designed for a 60-mph wind; note the braces in Fig. 14.18.

After the panels are erected, the roof system can be placed on the structure (Fig. 14.19). At the same time, the space between the floor slab and the wall panels can be filled with concrete, securing the wall panels to the foundation and the floor. Once these operations have been completed, the temporary braces can be removed and the joints can be saw cut into the floor slab to prevent random cracking.

The horizontal dimensions of the buildings may change depending on the space requirements for the owner. If two stories are required in part of the building, the second story is inserted into the existing building shape. The roof profile does not necessarily change. A fast-food restaurant may be only 5,000 to 10,000 square feet, while a department store may be 40,000 square feet or more. The exterior finish may be anything from painted concrete masonry units to exposed aggregate concrete panels. When these buildings are used as office space or restaurants, a drop ceiling may be installed to create a finished look in the commercial space. When these buildings are used as department stores or warehouses, there is typically no ceiling, and the ducts for the air-conditioning are exposed, as is the underside of the roof. The elements of these buildings that

FIGURE 14.19 Erection of the interior structural steel and roof system for the tilt-up building.

are constant for the general structure are the utility systems, the roof system, and the economical nature of the construction process.

These buildings contain all the systems discussed in residential construction.

- *Foundation.* Grade beam to support the walls and roof. Floor is slab-on-grade.
- *Framing.* The exterior walls are the framing, typically concrete masonry units or tilt-up concrete panels. The walls and the interior columns support the roof joists (see Fig. 14.19).
- *Building envelope.* The insulation, exterior walls, and the roof make up the building envelope. Insulation is applied directly to the inside of the walls.
- *Interior partitions.* The interior of the building is built to suit the purpose of the building. For instance, a store would have a display area and a storage area, with smaller rooms for the administrative staff. A restaurant would have a kitchen, possibly a bar, and seating areas for patrons.
- *Building utilities.* Electricity, water, wastewater disposal, telephone, television cable, and communication systems such as sound systems or closed-circuit television. Utilities are installed in two stages, rough-in and finish, just as they are done in a home.
- *Environmental control systems.* These include heating, lighting, air-conditioning and ventilation, and acoustics, along with the control systems that operate them.
- *Transportation systems.* These buildings are one story and do not require transportation systems.
- *Fire suppression.* Fire sprinklers are required in commercial buildings.

MULTISTORY BUILDINGS

The nature of multistory building construction is very different from homes and one- or two-story commercial building construction. Multistory buildings contain all of the same systems and the same general sequence of construction operations, but the accomplishment of those operations is technically very different from the techniques and methods used in residential and small commercial buildings. The mechanical, electrical, transportation systems, and life safety systems are much more complex in multistory buildings.

Consider the construction of the building foundation. Small buildings such as houses have shallow foundations. Multistory buildings have deep foundations. The design of the foundation varies according to the kind of soils and bedrock at the location, but usually involves driving piles (Fig. 14.20) or drilling deep foundations. A new project engineer, one week into his very first construction project, was recently asked how large the building would be when finished. The reply was, "I don't know, but there are 1,180 steel H-piles in the foundation." This is a large building, indeed.

FIGURE 14.20 Driving concrete piling for the foundation of a multistory building.

Let us take an abbreviated look at the construction of multistory buildings. Construction of multistory buildings began in the United States around the end of the 19th century. The Monadnock Building, which was completed in Chicago in 1891, would not be an acceptable building by today's standards because too much of the lower floor space is taken up by heavy masonry walls that are more than 2 feet thick. Today, multistory buildings are designed with the same amount of usable floor space on each floor.

The 60-story Woolworth Building in New York was built in 1913, and the most famous of the early skyscrapers, the Empire State Building, was completed in 1931. The twin towers of the unfortunate World Trade Center, now familiar to everyone in the world because of the September 11, 2001, attack, used a framing design called the framed-tube design, meaning that the exterior walls formed a tube made of vertical I-beams. The 1,450-ft-tall Sears Tower in Chicago, which was completed in 1974, is a variation of the framed-tube design called the bundled-tube. It is an attempt to make the framed-tube even more stable.

During the first half of the 20th century, there was basically one way to design frames for multistory buildings—a very heavy steel skeleton. During the second half of the century, structural steel and reinforced concrete design changed. Now multistory buildings are constructed using different designs such as rigid frame, **shear wall**, wall-frame, framed-tube, and bundled-tube. Very tall buildings may use a system of suspending each story or floor from the rigid core of the building.

The design of buildings has changed to meet the basic requirements of the building occupants. Today, multistory buildings must have large areas of open space on each floor that can be divided easily into individual work spaces. A building must provide an environment that is both structurally sound and architecturally pleasing. The architect and the structural engineer must work together to develop the form they want for the building. At its best level, structural engineering is as much an art form as is architecture, making the engineer and the architect a team in the development of the building.

It may be easier to consider the impact of the building's requirements on the structural engineer after a short discussion of the kinds of building structures in use today. In each case, the purpose of the structure is the same, to economically support the dead and live loads and to support the external loads caused by wind, weather, temperature change, and earthquakes or hurricanes. The building must be structurally strong enough to meet all these support requirements and also keep vibration to an acceptable level, keep a high percentage of floor space available, provide convenient access to each floor, and possess a pleasing exterior appearance. Figures 13.18 and 14.21 show the erection of the steel skeleton for a new building. In Fig. 14.22, the contractor is placing the concrete floor decks on the steel frame.

The buildings in Figs. 13.5 and 14.23 are concrete frame. In these buildings, there are typically no diagonal braces. The building rigidity comes from the strength of the connections between the frame members.

shear wall
Walls designed to resist horizontal shear or distortion from horizontal loads, such as wind loads.

FIGURE 14.21 Erecting the steel skeleton for a multistory building.

FIGURE 14.22 Concrete floor decks on the steel frame of a multistory building.

Braced-Frame Buildings

The braced-frame structure is one of the more economical methods of designing the steel structure of a building. The vertical members are the columns, the horizontal members are the girders, and the diagonals are the braces. The diagonal braces work in both tension and compression. Any force that tends to distort the square shape of the frame, pulls one of the diagonals and pushes on the other in each square opening of the frame. Without the diagonal braces, the structure could distort very easily. The wind, for instance, would push the building out of its square shape.

A variation of the braced frame is achieved by changing the length and placement of the diagonal braces so they span more than one floor. This design has a disadvantage of increasing the difficulty of hiding the diagonal braces. They interfere with window placement, and they change the appearance of the building if they are external. A braced-frame building is suitable for buildings up to about 50 stories.

Rigid-Frame Structures

Rigid-frame structures have no diagonal braces (Fig. 14.23). There are only vertical columns and horizontal girders. Figure 14.24 is a parking garage, which

FIGURE 14.23 Concrete frame building.

FIGURE 14.24   Rigid-frame parking garage.

is a common use of a rigid-frame building. The columns and girders are large, heavy members that support a concrete floor slab. This kind of building is best suited for reinforced concrete columns. The connections of the columns and girders must be so strong and massive that they do not distort and do not require any diagonal braces to keep the shape of the building square.

This frame system is not suitable for buildings with more than around 10 stories. The structural members are too large and too heavy. The girders take up too much of the ceiling space and the columns take up too much of the floor space. This kind of design is commonly used for buildings that require a great deal of strength and rigidity, such as parking garages that are only a few stories tall. For buildings over 10 stories, a more efficient, lighter design is needed.

Shear Wall Buildings

Shear wall buildings work by replacing the diagonal bracing with walls that are rigid enough to resist distortion. Masonry walls do not work for this purpose because the mortar lacks the required tensile strength. Most shear walls are made of concrete panels. Many smaller buildings have shear walls made of plywood or OSB.

Framed-Tube Structures

A framed-tube structure (Fig. 14.25) works like the center tube in a roll of paper towels, except that there is also a tube on the outside, the walls are more massive

FIGURE 14.25 Example of a framed-tube structure. Most of the exterior of the building is glass, a common practice.

than the paper towel roll, and the frames are either square or rectangular. The external walls of the building make up the tube, which consists of large columns and large horizontal girders placed close together. The columns and girders are rigid and no diagonal bracing is needed.

The interior floor girders do not need to be as massive as the large steel girders in the exterior walls, so they take up less of the ceiling space and allow the floors to be spaced closer together vertically. Interior columns may be required to carry some of the building's weight if the building is large. There is always a structurally heavy central core in framed-tube buildings to accommodate the elevators and utilities.

There are several variations of the framed-tube structure, such as tubes inside tubes, several tubes bundled together, and tubes with diagonal braces. They all begin with the same principle, the large supporting steel frame is located around the perimeter of the building where it will not occupy too much of the interior floor space.

Suspended Structures

The final type of building frame system is the suspended frame. In such a building, a central structural steel core is built, and the floors are suspended from frames that are cantilevered out from the central core. These buildings are sometimes referred to as cantilever structures. This type of building is too complicated for use in the kind of 10- to 20-story buildings common in most cities.

Building Frame Selection

For a building of 20 stories, it makes sense to select a shear wall design. The building will have a structural steel skeleton built around the central core with steel columns and steel girders for each floor. Instead of steel diagonal bracing that is difficult to hide, the building will use concrete shear walls.

At the time the upper floors are being framed, exterior shell and utilities rough-in of the lower floors can proceed. The core of the building carries the elevators and the utilities. Most multistory commercial buildings are constructed without any interior finish work because the tenants will build their own interior to suit their own requirements.

Multistory buildings present many engineering challenges. The challenge of providing water, fire suppression, and wastewater disposal on the upper floors is a major challenge. Electrical and communication systems are also major installations. Vertical transportation for both people and materials is a greater challenge for a 20-story than for a 2-story building, and often requires banks of elevators.

The remaining construction challenge of multistory buildings is their inherent lack of work space. Although there are some exceptions, multistory buildings are nearly always built in the downtown area of large cities (Fig. 14.26). The movement of large equipment such as cranes and trucks carrying structural steel becomes a major coordination issue. Movement of such equipment must be done during times of minimum traffic. There is little or no storage space at the site, so all materials are delivered to the site when needed. The contractor's coordination problem is to ensure the steady arrival of materials to keep the work progressing on schedule.

As with all buildings, construction begins with the foundation, probably a deep foundation. The building will be supported on piles or caissons. As soon as the foundation is complete, the erection of the steel frame of the building and the placing of the concrete floor slabs begins, and continues in a repetitious manner until all floors are completed. Since all floors are designed with the same shape, the erection of a 20-story building is a matter of repeating the same process 20 times. The placement of the exterior cladding follows the construction of the

FIGURE 14.26 Crowded multistory building project in a metropolitan center.

floor, and progresses upward from the ground level. When the roof and exterior cladding of the building are in place, the interior finish work can begin. It is not uncommon for companies to be moved into a building and to be open for business before all of the available office space has been completed.

SUMMARY

Most framing for residential construction is prefabricated off site and trucked to the site. The building envelope consists of the insulation, sheathing, moisture barrier, exterior siding and roofing, windows, and doors. The envelope is basically everything outside the frame.

In a house, framing the interior walls is an integral part of the framing operation. In large commercial buildings, interior walls may not be part of the original construction process. Each building tenant would construct interior partitions to suit their own needs.

FIGURE 14.27 Two-story masonry building under construction.

The utilities in a building are all constructed following the same general principles.

■ The utility entrances to the building do not need to be completed early in the construction process. It is much better to place entrances such as water lines from the well or municipal water line, telephone lines, and electrical entrance cable to the meter late in the construction process to avoid damage during construction.

■ Inside the building, utilities are generally built in two phases—rough-in and finish. Utility rough-in can begin when the framing is nearly complete.

There are many commercial buildings such as chain restaurants, department stores, grocery stores, office buildings, and warehouses that are essentially the same one- or two-story standardized building design. The floor is a concrete slab-on-grade. The exterior walls are usually concrete tilt-up panels, but may be masonry (Fig. 14.27). Typically the roof is flat and is supported by steel columns in a grid approximately 30-feet square. These buildings are economical and they can be erected quickly.

The nature of multistory buildings is so different from homes and one-story commercial buildings as to be another whole branch of the construction industry. The buildings contain all of the same systems, and the same general sequence of construction operations must be followed, but the accomplishment of those operations is a great departure from the techniques and methods used in residential and small commercial buildings.

REVIEW QUESTIONS

14.1 Describe the sequence for the construction of a simple one-story commercial building. How does this differ from the sequence of construction in a one-story residential building?

14.2 What are the common materials used for roofing and exterior siding on commercial buildings where you live? Why are those materials used while different materials are used for the same purposes in other areas?

14.3 Describe the rough-in activities of the plumber and electrician. How are these activities different in a residence as compared to a commercial building?

14.4 Make a tour of your local area. Note the number of big box buildings within a few miles of your home or school.

14.5 Why are shear wall type multistory buildings a preferred form of construction for buildings up to about 20 stories? What is a shear wall?

14.6 Make a list of the specific building codes that govern construction in your area.

14.7 Research several different kinds of roof insulation methods used in commercial buildings. How do these compare to roof insulation used in residential construction?

14.8 Does the insulation in residential construction have a vapor barrier in your area? If so, is the barrier inside or outside the insulation? Why?

14.9 Explain why electrical, mechanical, and plumbing rough-in usually begins before framing is complete.

14.10 What role do shop drawings play in the steel framing of a small commercial building?

REFERENCES

1. Allen, Edward (1998). *Fundamentals of Building Construction Materials and Methods,* 3rd ed., John Wiley & Sons, New York.

2. Billington, David P. (1983). *The Tower and the Bridge,* Basic Books, Inc., New York.

3. Dobrowolski, Joseph A. (1998). *Concrete Construction Handbook,* 4th ed., McGraw-Hill, New York.

4. Feldman, Patti, and William Feldman (1996). *Construction and Computers,* McGraw-Hill, New York.

5. Hornbostel, Caleb, and William J. Hornung (1982). *Materials and Methods for Contemporary Construction,* 2nd ed., Prentice-Hall, Englewood Cliffs, NJ.

6. Space, William R. (2000). *Encyclopedia of Construction Methods and Materials,* Sterling Publishing Co., New York.

WEBSITE RESOURCES

1. www.sweets.com/sdff05.htm   This resource for construction professionals offers information on metal roofing, framing, molding, and building materials. Provided by a division of McGraw-Hill Companies.

2. www.timberframe.org/intro.html   A trade organization of timber frame, log home, and post and beam builders, home designers, architects, tool suppliers, and manufacturers of structural insulated panels.

3. www.ornl.gov/roofs+walls/insulation   Department of Energy fact sheet details how insulation works, what to know when buying a house, and how to select insulation.

4. www.insulation.org   National Insulation Association. Offers insulation industry news, a resource library and company guide, and a directory of local and regional associations.

5. www.pbs.org   Contains videos on projects such as tall buildings. Nova produced an excellent video "Skyscraper" but it is no longer listed as available on the website. However the book on which it was based (*Skyscraper: The Making of a Building,* by Karl Sabbagh) can still be purchased in bookstores. It is about the construction of the Worldwide Plaza building in midtown Manhattan.

15

Quality and Productivity

Quality and productivity are closely related. Quality is defined as meeting or exceeding the needs of the customer. Quality in construction is achieved by meeting the customer's requirements in the best possible way. Productivity is expressed as a measure of output per level of effort, or dollar value of output per level of effort. Contractors often define productivity in terms of hourly output, and then complain about how human nature causes people to be less productive late in the day when they are tired. Recent studies in productivity indicate that productivity should be measured against weekly hours worked, and not individual hours or events. The more important issue is how productive crews are that work, say, 45 hours per week, and not how productive crews are in their ninth hour of work. In some cases, construction worker productivity is improved by working more than 8 hours per day or more than 40 hours per week, a notion that may be contrary to commonly accepted ideas about productivity.

QUALITY

Quality in construction is defined as "meeting or exceeding the needs of the customer." There may be individuals or organizations that feel they were the first to use that definition, and that they should be given credit for being its author, but the definition has become so common that now everyone uses it in common conversation about construction quality. It has become the definition. Everyone should recognize that intrinsic quality does still exist in construction, in the sense that some products and workmanship are simply better than others. Some grades of lumber are straighter than other grades, for instance. Concrete can be made with different compressive strengths. It might seem that straighter lumber or higher compressive strength concrete would be judged to be higher quality; however, the contract normally specifies the grade of lumber, the compressive strength of the concrete, and the "quality" of all workmanship and materials to be used in the facility. These matters are seldom left to the contractor's choice. The shop drawing process is used to control the "quality" of the materials used by ensuring each installed product meets the requirements of the specifications. In spite of that, some contractors have a reputation for being high-quality contractors, and others are still working on their quality reputation. One must conclude that quality in construction is more than supplying the right materials. Quality is also about a high level of workmanship, finishing on time, safely, within budget, and without claims and litigation. Quality is about meeting or exceeding the needs of the customer. Quality in construction includes ensuring the facility will perform its intended purpose.

There has been a growing worldwide emphasis on quality in all industries and in construction in particular. International quality standards are increasingly defined by the ISO 9000 standards. ISO is the *International Organization for Standardization,* founded in Switzerland in 1947. ISO 9000 consists of three quality standards that apply to all types of organizations. ISO 9001: 2000 contains requirements for companies that aspire to be ISO 9000 certified. **ISO 9000: 2000** and ISO 9004: 2000 contain guidelines. ISO is important because it offers an internationally recognized systemic approach, coupled with institutionalization of the attitudes, policies, procedures, record keeping, technologies, and resources for managing quality work. The principles advocated by ISO 9000 provide the framework for managing a quality system. Those principles are [Web 5].

ISO 9000: 2000
Quality standards published by the International Organization for Standardization.

1. *Focus on your customers.* Organizations rely on customers. Therefore,
   - Organizations must understand customer needs.
   - Organizations must meet customer requirements.
   - Organizations must exceed customer expectations.

2. *Provide leadership.* Organizations rely on leaders. Therefore,
   - Leaders must establish a unity of purpose and set the direction the organization should take.
   - Leaders must create an environment that encourages people to achieve the organization's objectives.

3. *Involve your people.* Organizations rely on people. Therefore,
   - Organizations must encourage the involvement of people at all levels.
   - Organizations must help people to develop and use their abilities.

4. *Use a process approach.* Organizations are more efficient and effective when they use a process approach. Therefore,
   - Organizations must use a process approach to manage activities and related resources.

5. *Take a systems approach.* Organizations are more efficient and effective when they use a systems approach. Therefore,

- Organizations must identify interrelated processes and treat them as a system.
- Organizations must use a systems approach to manage their interrelated processes.

6. *Encourage continual improvement.* Organizations are more efficient and effective when they continually try to improve. Therefore,
   - Organizations must make a permanent commitment to continually improve their overall performance.

7. *Get the facts before you decide.* Organizations perform better when their decisions are based on facts. Therefore,
   - Organizations must base decisions on the analysis of factual information and data.

8. *Work with your suppliers.* Organizations depend on their suppliers to help them create value. Therefore,
   - Organizations must maintain a mutually beneficial relationship with their suppliers.

Quality Assurance and Quality Control

Quality assurance and quality control are two different concepts that students should understand. **Quality assurance** (QA) refers to the management systems employed by construction companies to produce high-quality work consistently. Those systems consist of many programs such as hiring highly qualified employees, safety programs (safety is an integral part of any quality assurance program), training programs, incentives and recognition programs that reward high-quality performance, procurement systems designed to identify the best-quality suppliers, and personnel policies designed to reduce turnover and promote retention. All of those procedures should be institutionalized in a company QA manual. In manufacturing, QA also includes statistical control of processes, but in construction there are so few repetitive processes that we do not normally include statistical control in the QA procedures of a construction company. One exception is concrete and asphalt paving. Statistical analysis of these two construction activities is sometimes undertaken.

quality assurance
Management programs designed to produce high-quality work.

Quality control (QC) is another matter, although many people use the terms *QA* and *QC* interchangeably or together. Quality control is about the inspection of work to ensure it meets the quality standards specified in the contract. All government highway construction projects have "inspectors." Their job is quality control. They perform that function by accepting work that meets the specifications and rejecting work that does not. They perform such tests as the density of soils, the slump and compressive strength of concrete, and the temperature of asphalt delivered to the site. These are all quality control functions.

quality control
System of inspection to monitor the quality of completed work.

QA/QC is a common abbreviation used by engineers and contractors. It ought not to be used because QA and QC are different. QA is made up of good management practices. QC is an inspection or sampling process.

Consider the implications of a contractor with a reputation for producing low-quality construction, which would be the obvious result of ignoring QA, or cutting too many corners, or using only untrained low-paid workers. Not all contractors intend to be the highest quality contractor in town. Some believe they can make a better profit by working fast with untrained crews, and moving on to the next job as soon as possible. Such contractors ignore the complaints of their customers because they intend to get the next job by being the lowest bidder. The low bid system does not usually contain incentives for contractors to produce high-quality work.

commodity
Product differentiated by price only.

Too many owners consider construction to be a **commodity** distinguished only by price. Gasoline is an example of a commodity. Drivers look for the lowest price gas in their area because they believe price is the only difference. There is essentially no brand loyalty. Contractors work very hard to establish a reputation for quality—and to avoid being tagged with the commodity label. Contractors simply do not agree that they produce a commodity, but many owners believe that any contractor will produce the same facility, given the same set of plans and specifications. It is an attitude that is fed by the low bid system. It is in the best interests of all good contractors to produce high-quality work, and to make it known that their work is in fact of higher quality than the competition. This is not easy. Good contractors prefer to negotiate for work because it is difficult to distinguish themselves on low bid projects.

Total Quality Management

total quality management (TQM)
Quality system based on continuous improvement originated by W. Edwards Deming.

Total quality management (TQM) is a system of constant improvement first promoted by W. Edwards Deming. According to Deming, American companies require a transformation of management style and of government relations with industry. In his book, *Out of the Crisis,* Deming offers a theory of management based on his 14 points for management. Deming claims that American companies lose market share, and therefore jobs, because they fail to plan for the future. He maintains that companies should be judged by their planning for future dividends and their plans to ensure they stay in business, not by their quarterly reports to the stockholders. Deming was instrumental in helping the Japanese automobile manufacturers establish their quality control programs.

Deming's 14 points are [6]

1. Create and publish to all employees a statement of the aims and purposes of the company or other organization. The management must demonstrate constantly their commitment to this statement.
2. Learn the new philosophy, top management and everybody.
3. Understand the purpose of inspection, for improvement of processes and reduction of cost.

4. End the practice of awarding business on the basis of price tag alone.

5. Improve constantly and forever the system of production and service.

6. Institute training.

7. Teach and institute leadership.

8. Drive out fear. Create trust. Create a climate for innovation.

9. Optimize toward the aims and purposes of the company, the efforts of teams, groups, and staff areas.

10. Eliminate exhortations for the workforce.

11. Eliminate numerical quotas for production. Instead, learn and institute methods for improvement. Learn the capabilities of processes, and how to improve them.

12. Remove barriers that rob people of pride of workmanship.

13. Encourage education and self-improvement for everyone.

14. Take action to accomplish the transformation.

Many of Deming's ideas are very difficult for industry leaders to accept.

The key to understanding TQM is to first understand that the objective is continuous improvement. TQM is built on a foundation of

1. A well-defined vision and mission that everyone understands

2. Upper management commitment and leadership in quality matters

3. Training

From those three components, companies build better communications and teamwork, which leads to four important factors:

1. Customer satisfaction

2. Improved supplier management

3. Process improvement

4. Focus on employees

The umbrella over these factors is *continuous improvement.*

Quality and Contractor Selection

The owner's greatest impact on quality is through contractor selection. Private owners do not resort to using the low bid system because they know that the system can produce the poorest construction quality, with the greatest number of change orders, claims, and litigation. Most private owners negotiate from a short list of preferred contractors. Government agencies use the low bid system because it provides a convenient avenue to demonstrate that the contract was awarded fairly, and all contractors had a fair opportunity to compete. Those factors alone do not make low bid a good system.

Government agencies are searching for ways to improve on the low bid system so they can begin to reap some of the same benefits now reserved for private owners who select their contractors based on qualifications. Government contracting officers are not allowed to select bidders, except on highly technical projects that have a prequalification process. Projects that require prequalification are still open to all contractors who can prequalify.

To satisfy the government requirement that all qualified bidders must be allowed to submit bids, and yet to add some ability to select high-quality contractors, government agencies are beginning to try alternative selection systems such as construction management at risk and competitive negotiation. Some agencies are also trying different contracting methods such as A + B contracts that reward early completion. As with any change, there are proponents and opponents, but something has to replace the low bid system. As comfortable as we are with the current system, most observers agree that quality of construction cannot improve without eliminating the low bid system. Whether that change will be to competitive negotiation, construction manager at risk, or some other concept, remains to be sorted out.

"Do it right the first time" is a quality motto that is used in most good construction companies. Companies lose money when they start having to tear out work because they failed to do it right the first time. Replacing work changes the schedule, increases the cost of the work, and destroys pride and morale. Doing it right the first time is the connection between quality and productivity. Highly productive crews know this and they are productive because they "do it right the first time."

PRODUCTIVITY

There are two definitions of productivity that relate to construction, one that defines productivity in terms of the amount of work produced and one that defines productivity in terms of the dollar value of the work produced. Productivity is output per worker-hour or dollar value per worker-hour. Contractors usually prefer the work-output-related definition because they can make changes to affect the worker-hours of effort. They can change crew size or change the mix of equipment, so they prefer the work-output-related definition. At the same time, they recognize that in reality they are dealing with dollar output because all output relates to the amount the owner will be billed. Schedules are developed in crew days or worker-hours, so it is natural to define productivity as units of output in relation to effort. The government measures the country's productivity in dollar output per worker-hour or total cost per unit of output. Either definition is acceptable.

$$\text{Productivity} = \frac{\text{Units of output (or output dollar value)}}{\text{worker-hour}} \qquad \textbf{[15.1]}$$

Owner Effect on Productivity

What are the factors that affect productivity? The first and one of the most important is the project owner. The owner selects the designer and sets the standards for the design. The owner can impose a facility design that is difficult to construct or the owner can work with the designer to design a high-quality facility that can be constructed efficiently. The owner can select the contractor or at least the method that will be used to select the contractor. The owner can select the type of construction contract. For instance, the contract could be a low bid lump sum contract, or it could be a construction manager at risk contract with incentives. The impact on productivity will be significantly different.

Jobsite Management's Effect on Productivity

The jobsite management organization also has great impact on productivity. Is the construction management organization on the site set up in an efficient manner? Decisions about how work will be accomplished need to be made quickly. Does management have adequate information? Does the PM have the staff to adequately manage the subcontracts? The project manager, superintendent, foremen, and project engineers must understand their roles and be prepared to work together efficiently and harmoniously. The presence or absence of a project labor agreement can impact productivity because such agreements contain clauses, not only about hiring but also about work rules. Subcontracts need to be signed early. Subcontractors may need assistance with administrative matters such as shop drawings. Is the field organization ready for such challenges? Does the company use a system that integrates scheduling, accounting, and cost control? Are there agreements in place with the designer that ensure RFIs will receive immediate attention? Organization of a field staff may be a matter of understanding the "people problems," but it is also a matter of understanding the problems that are going to be faced and preparing for them before they occur. That requires experience and knowledge.

The layout of the site itself is one of the most obvious of the impacts on productivity. Materials placed too far from the point where they are needed cause wasted time. Should reinforcing steel be cut and bent on site or should it be precut and bent at an offsite fabrication plant (Fig. 15.1)? How is access to the

FIGURE 15.1 Reinforcing steel delivered precut and bent to the construction site.

FIGURE 15.2   Congested urban street construction site.

site controlled? Are deliveries made efficiently without interfering with ongoing work? The locations of the temporary storage facilities, the project manager trailers, the subcontractor trailers, and the site utilities are all important.

The location of the site is important (Fig 15.2). The suppliers need to be able to make deliveries efficiently. Large vehicles can be delayed by a need to travel through heavily populated residential or commercial areas. The distance from concrete or asphalt plants is important. In Washington, DC, for the construction of the Ronald Reagan Building, which is located at 14th Street and Pennsylvania Avenue, the contractor actually placed a concrete plant in the basement area of the building during construction. This was done because of the traffic problems in the area and how traffic would have affected delivery of concrete from local plants. Aggregates and cement were delivered to the on-site plant at night when traffic was minimal.

The quality of the local roads may impact the ability of suppliers to reach the site easily. In some areas, contractors are restricted from working during certain hours of the day because of the noise or dust pollution problems. Construction noise is a major community nuisance problem. The critical element in an urban environment is the close proximity of adjacent property owners. The identification of methods and techniques for mitigating construction noise nuisances is a crucial planning requirement for both project owners and contractors. Contractors may sometimes be restricted from working on highway projects during high traffic hours. All of these factors have impacts on productivity that are greater than the simple ability of human beings to work hard and steadily for a full day.

Productivity Studies

For many years, contractors have sponsored studies of the impact of overtime on productivity. Most contractors believe that after 40 hours per week productivity declines. This belief (of lower productivity at the end of the day) is often stated as the reason for an increase in the cost of change orders that require work that must be done on an overtime basis. It also causes contractors to avoid using overtime whenever possible. Every project manager, superintendent, foreman, and worker knows that crews do not produce the same amount of work every day. Each day is different. What makes the difference? Consider the following list:

- Combinations of labor and equipment
- Weather
- Constructability of the design
- Length of the workday
- Day shift versus night shift
- Efficiency of the tools and equipment
- Material availability
- Effort expended by the labor force
- Level of training of crews
- Number of crews working in the same space
- Government regulations

Consider the possible impact of requiring crews to work on an extended overtime schedule but keeping all support activities such as machine maintenance or engineer support on a normal schedule. If a machine should break down, there will be no maintenance support to perform the repair so the crew's productivity will be severely affected. Consider the possible impact of assigning crews to sites away from home for an extended period. Consider the comparison of union and nonunion crews. There are many factors that affect productivity; and, many times, to understand the effect of a factor, the entire construction process must be studied. The importance of the ISO 9000: 2000 principles 4 and 5 can be clearly understood when you consider some of the questions posed here.

Productivity studies are accomplished in several different ways. Work sampling is a technique of making random observations of crew members and noting their activity. Activities are classified as productive direct work, productive support work, or nonproductive work. Work sampling is used to determine if the size and composition of crews should be altered. Motion studies, which originated in the manufacturing industry, are designed to determine the best way to accomplish a task. Time studies are used to determine the amount of time devoted to individual activities in the completion of a task. The purpose is to figure out which activities might be shortened or eliminated. Individual activities required to place plywood sheets on a floor might be

- Walk to the plywood storage location
- Pick up a sheet of plywood
- Carry the plywood to its position on the floor
- Nail the plywood in place

A time study would determine the durations of each of these activities and provide the project manager with information that could be used to make crew composition decisions.

One of the issues with productivity studies is how to eliminate the effect of being watched, known as the **Hawthorne effect**. Crews have a natural tendency to be more productive simply because of the attention they are receiving (see *The Human Problems of an Industrial Civilization* by Elton Mayo [8]).

To study the impacts of overtime work, researchers have recently developed methods to evaluate crews with no on-site presence. Crew production is determined from field records, and the actual hours worked are determined from payroll records. The purpose of such a study would be to determine the most efficient number of hours for crews to work. Several studies done by organizations such as the Mechanical Contractors Association, the National Electrical Contractors Association, the Business Roundtable, and the Construction Industry Institute have established that crews become less productive after 40 hours per week. It still may be more efficient for crews to work more than 40 hours per week because there are many factors that contribute to productivity. It has been shown that it is sometimes more economically efficient to require crews to work overtime even given that their hourly productivity for those hours after the initial 8 hours may be somewhat reduced. The question that must be examined is—how much is produced in the total work duration? Some early work hours may be devoted to mobilization or setup, moving materials, training, or taking a break. These activities make it more productive overall to work the overtime hours, even if the later hours may be less productive than the hours earlier in the day. This is because mobilization, setup time, and other nonproductive time is spread over a greater number of hours.

The process of conducting such a study consists of collecting data to analyze crew productivity, and plotting the data against the percentage of overtime worked. This approach enables the researcher to use regression to calculate a theoretical productivity trend line for each type crew (see Fig. 15.3). The equation of the **productivity trend line** is used to determine theoretical optimum overtime and theoretical optimum productivity expressed as a multiple of the crew's productivity at 40 hours. From the productivity trend line, the most cost efficient level of overtime is calculated by taking the first derivative of the productivity trend line equation to determine the point of maximum productivity (i.e., the point of zero slope of the line).

The general shape of the productivity trend lines is the same for each type of crew. This methodology is applicable to many kinds of crews. Figure 15.3 shows a crew whose maximum productivity is attained at approximately 22% overtime. This particular type of crew is most productive working a 49-hour workweek. This amount of overtime may or may not produce the lowest labor costs (because of overtime premium pay), but it does produce the maximum productivity.

As a project manager, the more important data would be the number of hours in a week that would be most productive and would result in the lowest

Hawthorne effect

Increase in productivity caused by the presence of researchers.

productivity trend line

Mathematically calculated trend line depicting the productivity of crews working overtime, as a percentage of their productivity at 40 hours per week.

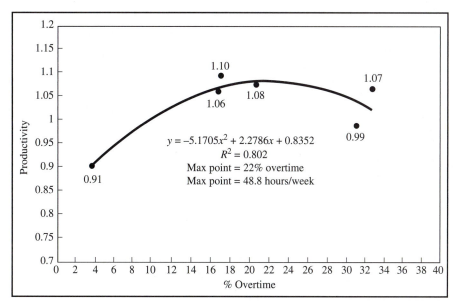

FIGURE 15.3   Effect of overtime on productivity.

overall job labor costs. This can be determined by developing a **productivity efficiency factor** using the following formula:

$$\text{Efficiency factor} = \frac{\text{Productivity at } x\% \text{ overtime}}{\text{Productivity at 40 hr per week}} \qquad \textbf{[15.2]}$$

productivity efficiency factor
Productivity at x% overtime divided by productivity at 40 hours per week.

The efficiency factor is then multiplied by the average hourly cost at $x\%$ overtime, providing a relative cost factor. Average hourly cost is computed by the formula:

$$\begin{array}{c}\text{Average}\\\text{hourly}\\\text{Cost}\end{array} = \frac{\left(40 \text{ hr} \times \dfrac{\text{Base}}{\text{rate}}\right) + \left[\left(\dfrac{\text{Hours}}{\text{per week}} - 40 \text{ hr}\right) \times \dfrac{\text{Overtime}}{\text{rate}}\right]}{\text{Hours per week}} \qquad \textbf{[15.3]}$$

A plot of **relative cost factors** produces a chart as shown in Fig. 15.4. Inspection of Fig. 15.4 shows that this crew is most cost effective at 19% (47.5 hr) overtime, if the cost differential (not the wage differential) between straight time and overtime is 50%. For purposes of illustration, Fig. 15.4 also contains cost plots for 0% and 25% cost differential to illustrate that it can be used for companies with different cost differentials for overtime work. The actual overtime-cost differential is determined differently for each company and depends primarily on how the company accounting system loads overhead costs onto labor costs. The minimum relative cost will determine the point of maximum cost efficiency. Operating at this number of overtime hours will result in the lowest overall job labor costs.

relative cost factors
Relate the cost of output to the average hourly cost.

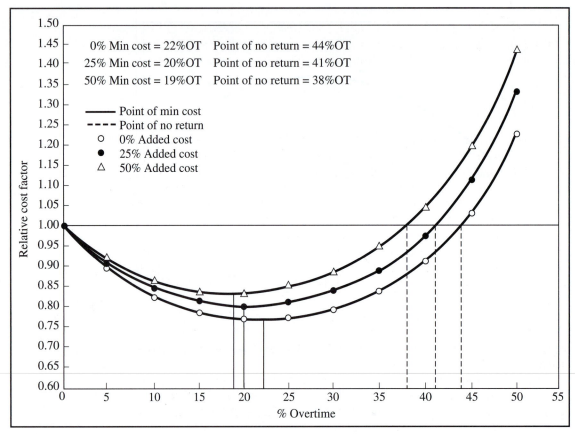

FIGURE 15.4 Effect of overtime on relative cost factor.

The information in Fig. 15.4 can be used to determine lowest overall labor costs, lowest total man-hours for a project, or most efficient use of crews based on projected workload. The figure can also be used to show the highest acceptable overtime amount and predict the costs associated with additional overtime. It must be noted that the data in Fig. 15.4 are not applicable to other crews. Each crew must be studied independently.

Consider a project estimated to require 1,000 hours of labor working 40 hours per week. If the cost of labor for this company is $50.00 per hour, the job total labor cost will be $50,000. If the cost differential for overtime work is 50%, and if the crews work 19% overtime (say 47.5 hours per week) at an average hourly cost computed as follows:

$$\text{Average hourly cost} = \frac{(40 \times \$50) + (7.5 \times \$75)}{47.5} = \$53.95$$

The number of hours worked will be 1,000/1.08 = 926 (from Fig. 15.3).

The job labor cost is projected to be 926 × \$53.95 = \$49,957.70.

The same crew at straight time would work 1,000 hours. The savings from reducing the crew's time on the project is not significant, but the crew's time on the job was reduced 74 crew-hours. Most contractors would consider that a worthy savings.

SUMMARY

Quality in construction is defined as "meeting or exceeding the needs of the customer." One must conclude that quality in construction is more than supplying the right materials. Quality is also about finishing on time, without shortcuts, safely, within budget, and without claims and litigation.

ISO is important because it offers an internationally recognized systemic approach, coupled with institutionalization of the attitudes, policies, procedures, record keeping, technologies, and resources for managing quality work.

Quality assurance refers to the management systems employed by construction companies to produce high-quality work consistently. Quality control is about the inspection of work to ensure it meets the quality standards specified in the contract.

Total quality management is a system of constant improvement, first promoted by W. Edwards Deming. According to Deming, American companies require a transformation of management style and of government relations with industry.

There are two definitions of productivity that relate to construction, one that defines productivity in terms of the amount of work produced and one that defines productivity in terms of the dollar value of the work produced. Contractors usually prefer the work-output-related definition because they can make changes to affect the worker-hours of effort.

Work sampling is used to determine if the size and composition of crews should be altered. Motion studies, which originated in the manufacturing industry, are designed to determine the best way to accomplish a task.

REVIEW QUESTIONS

15.1 As a project engineer, what actions will you take to help your construction company maintain its reputation as a "high-quality contractor"?

15.2 How do the teachings of W. Edwards Deming apply to construction quality?

15.3 What impact will ISO 9000 have on your management of construction projects?

15.4 Prepare a short class presentation from your own experience to illustrate the relationship between quality and productivity.

15.5 Define productivity for a concrete crew.

15.6 Your PM has asked you to conduct a time study of the concrete crews on your commercial project. Prepare a sequential list of activities for placing the reinforced concrete slab on grade. Start with setting the formwork and end with stripping the formwork.

15.7 Explain why the traditional tools of productivity such as Pareto diagrams and histograms have limited applicability to construction projects. What are the other traditional productivity tools?

15.8 Give an example of a project where work sampling would be an appropriate productivity measuring technique.

15.9 Elton Mayo (listed in the references here) described the "Hawthorne effect" as one of the problems of productivity studies. What is the Hawthorne effect and what is its impact? What measures are taken to avoid its effect?

15.10 Explain how a crew might actually be more productive by working overtime, when evaluating the week as a whole.

REFERENCES

1. Adrian, James J. (1987). *Construction Productivity Improvement,* Elsevier Science Publishing Co., New York.

2. "An Analysis of the Methods for Measuring Construction Productivity" (1986). Source Document 13, Construction Industry Institute, Austin, TX.

3. *Change Orders, Overtime, Productivity* (1994). Mechanical Contractors Association of America, Rockville, MD.

4. Chase, G. W. (1993). Implementing TQM in a Construction Company, The Associated General Contractors of America, 333 John Carlyle Street, Suite 200, Alexandria, VA.

5. Coburn, Ed (1999). "Fighting the Fatigue Factor," *Industrial Safety & Hygiene News.*

6. Deming, W. Edwards (1986). *Out of the Crisis,* MIT Press, Cambridge, MA.

7. "Effects of Scheduled Overtime on Labor Productivity: A Literature Review and Analysis" (1990). Source Document 60, Construction Industry Institute, Austin, TX.

8. Mayo, Elton (1946). *The Human Problems of an Industrial Civilization,* Harvard University, Cambridge, MA.

9. Oglesby, Clarkson, Henry Parker, and Gregory Howell (1989). *Productivity Improvement in Construction,* McGraw-Hill, New York.

10. *Overtime and Productivity in Electrical Construction,* 2nd ed. (1989). National Electrical Contractors Association, Bethesda, MD.

11. *Scheduled Overtime Effect on Construction Projects,* Report C-2, (1990). Business Round Table, New York.

12. Thomas, H. R., and D. F. Kramer (1987). *The Manual of Construction Productivity Measurement and Performance Evaluation,* Construction Industry Institute, Austin, TX.

13. Thomas, H. R., W. F. Maloney, G. R. Smith, S. R. Sanders, R. M. W. Horner, and V. K. Handa (1990). "Productivity Models for Construction," *Journal of Construction Engineering and Management,* American Society of Civil Engineers, 116(4): 705–726.

14. Thomas, H. R., and Jerald Rounds (1991). *Procedures Manual for Collecting Productivity and Related Data on Overtime Activities on Industrial Construction Projects: Electrical and Piping,* Construction Industry Institute, Austin, TX.

WEBSITE RESOURCES

1. www.asq.org   American Society for Quality. Resources for professionals include publications, education, conferences, standards, and certification, plus a virtual quality network to access.

2. www.psb.gov.sg   Productivity & Standards Board, Singapore. Features the board's services for small and medium enterprises such as manpower development and incentives. Includes labor statistics.

3. www.ppress.com   Productivity Press. Publishes books on quality management, corporate leadership, human resources development, and other business topics.

4. www.gpsqtc.com   Global Productivity Solutions Management Consultants specialize in the development and enhancement of human capital.

5. www.iso.ch/iso/en/iso9000-14000/iso9000/qmp.html   The international Organization for Standardization (ISO) is a worldwide federation of national standards bodies from more than 140 countries. The site contains information about ISO standards, in particular ISO 9000: 2000 Principles.

Safety

INTRODUCTION

The safety of employees is a paramount concern for all employers. This is especially true in the construction industry that continuously boasts one of the highest job injury rates of all professions. Construction employs 7.9 million people, 5.4% of the workforce, yet accounts for over 20% of fatalities and over 12% of the injuries in the workforce according to the 2005 Bureau of Labor and Statistics reports.

Safety on the jobsite is the responsibility of everyone on the site, but as the construction manager, your examples set the tone for the jobsite. If your attitude toward safety is carefree, your employees will exemplify that attitude. On the other hand, if safety is a priority to you and the company and clearly displayed, employees will be inclined to follow that example. To a construction manager, safety is a component of any project that is as important as estimating, scheduling, and equipment maintenance. It should be treated with the same concern and attention to detail that weekly job cost reports receive. It is not something that should be treated as an added responsibility to an employee's job. This means that everyone is responsible for the safety and well-being of all employees on the jobsite. It is up to the manager to plan this into the project and adapt it as part of the day-to-day culture on the jobsite.

Safety is the part of a project that has to take the front stage of the planning process and continue through all the way to the daily project operations. Safety is an investment, and a good safety program is part of a successful portfolio. The financial well-being of a company can be tied to the safety performance displayed by the company. While rates of return of $4 for every $1 spent on safety are quoted, the safety manager knows that this is not hard cash that is added into the bottom line of the company. A strong safety record benefits companies in several intangible and tangible ways. The most easily quantified is the experience modification rate (EMR).

The EMR is an excellent measure of how your loss prevention and control practices compare to others in the construction industry. The EMR compares your worker's compensation claims experience to other employers of similar size operating in the same type of construction business. Your EMR is calculated by the National Council on Compensation Insurance or in some states an independent agency. To calculate the EMR, each year your insurance company will report your payrolls and losses for the last 5 years. To complete the calculations, 3 complete years of data ending 1 year prior to the effective date of the rating period are used. For example, a rating in 2009 would not normally use 2008 data, but would include 2007, 2006, and 2005. The formula compares your specific payrolls and losses to industry averages. So, if your EMR is 1.0, you are at the industry average. If you are 25% better than the industry average, your EMR is 0.75, or if you are 25% worse, your EMR would be 1.25. Table 16.1 highlights how the EMR affects insurance premiums.

The backbone of a safety program is the written plan. Throughout this chapter we will highlight essentials of an effective safety program, providing examples that will help make your life easier as a project manager. The chapter will also highlight the regulatory component of safety on the job and the costs of accidents, and provide tips for you to practice on a daily basis. Safety is more than a slogan and it must be treated that way for it to be successful.

TECHNICAL INFORMATION

How much are accidents costing the construction industry on average each year? The construction industry boasts the highest spending on worker's compensation—$5.17 billion. It is a high risk industry so the premiums are higher. In addition to higher insurance expenditures, accident costs include lost wages, medical expenses, insurance administration, property damage, and indirect costs. The greatest impacts associated with accidents are the effect upon worksite morale, loss of productivity, and lawsuits.

A construction manager needs to be knowledgeable about programs and policies relating to daily jobsite operations that aim to protect employers and

TABLE 16.1 EMR example.

	ABC contractors	XYZ contractors
Payroll	$1,200,000	$1,200,000
Rate per $100	$10.25	$10.25
Premium	$123,000	$123,000
EMR	0.75	1.25
EMR dollars	($30,750)	$30,750
Modified Premium	**$92,250**	**$153,750**

their employees. Worker's compensation is an insurance program whose objective is to provide employers with immunity from lawsuits in exchange for payment to the affected worker, the replacement of income for injured employees, and rehabilitation of the injured employee. The program varies from state to state. Accident prevention plan strategies have been employed in the construction industry to reduce the high cost of worker's compensation rates. These include cultivating job satisfaction among employees, making safety a part of the company culture, and involving employees at all levels in the safety program and recognizing them for their efforts. While these efforts may ultimately lower the cost of worker's compensation, an effective safety program must be developed at the highest level within the company.

Williams-Steiger Act
Known as the Occupational Safety and Health Act. Passed in 1970.

While providing a safe working environment is the right thing to do, it is also the law. To assure safe and healthful working conditions for employees, the **Williams-Steiger Occupational Safety and Health Act (OSH Act)** was signed in 1970. The act authorized enforcement of the currently developed health and safety standards while assisting and encouraging States in their efforts to assure safe and healthful working conditions. The act also provides for research, information, education, and training in the field of occupational safety and health. Of the numerous standards promulgated by OSHA, the regulations that pertain specifically to the construction industry are contained within 29 CFR 1926.

The General Duty Clause of the OSHA regulations establishes the basic fundamental responsibilities of employers. The clause states,

> Each employer shall furnish to each of their employees employment and a place of employment which are free from recognized hazards that are causing or are likely to cause death or serious physical harm to their employees and shall comply with occupational safety and health standards promulgated under this Act.

Individual states are encouraged to develop and operate their own job safety and health programs. States must develop a plan that is as least as stringent as the federal standards, but they have the option to develop and promulgate standards that address hazards not covered within the federal standards. For example, California requires more stringent ergonomic standards, Nevada has different regulations regarding fatalities, Kentucky has a different way to report accidents, South Carolina revised the definition of a competent person, and Oregon has a revised method for hazard determination. Be certain to investigate the regulations that are pertinent to the state within which you are completing work.

FEDERAL OSHA REPORTING REQUIREMENTS

There are OSHA reporting requirements with which the construction manager must comply. You are required to keep records and reports of OSHA-recordable injuries and illnesses and submit this information to OSHA every year. These reports include a log of work-related injuries and illnesses (OSHA Form 300), an injury and illness incident report that includes data about how an injury or illness occurred (OSHA Form 301), and a summary of work-related injuries and illnesses (OSHA Form 300A). These forms and instructions for their completion can be downloaded from OSHA at the following link: www.osha.gov/recordkeeping/RKforms.html. At the end of each year, the OSHA 300 form must be reviewed to verify that the entries are complete and accurate, and a summary of the injuries and illnesses recorded on the OSHA 300 must be created and certified. This summary must be posted in a conspicuous place or placed where notices to employees are customarily posted. The summary must be posted no later than February 1 of the year following the year covered by the records and be kept in place until April 30. Additionally, you must save the OSHA 300 Log, the annual summary, and the OSHA 301 Incident Report forms for 5 years following the end of the calendar year that these records cover.

The OSHA 300 form requires each employer to keep records of fatalities, injuries and illnesses that are work related, or result in death, days away from work, restricted work or transfer to another job, medical treatment beyond first aid, or loss of consciousness. A work-related injury is an event or exposure in the work environment that caused or contributed to the resulting condition. All work-related incidents resulting in a fatality or the in-patient hospitalization of three or more employees must be reported to OSHA within 8 hours.

A question frequently asked of safety personnel is, what is normal first aid? For the purposes of OSHA, "first aid" means the following:

- Using a nonprescription medication
- Administering a tetanus shot
- Cleaning, flushing, or soaking wounds on the surface of the skin
- Using wound coverings such as bandages or Band-Aids
- Drilling of a fingernail or toenail to relieve pressure, or draining fluid from a blister
- Removing foreign bodies from the eye using only irrigation or a cotton swab
- Removing splinters or foreign material from areas other than the eye by irrigation, tweezers, cotton swabs, or other simple means
- Drinking fluids for relief of heat stress

If you are still having trouble deciding if an injury or illness is reportable, the decision tree illustrated in Fig. 16.1, adapted from OSHA for recording work-related injuries and illnesses, shows the steps involved in making this determination.

According to OSHA, record keeping is a critical part of an employer's safety and health efforts for several reasons:

- Keeping track of work-related injuries and illnesses can help prevent them in the future.
- Using injury and illness data helps identify problem areas. The more you know, the better you can identify and correct hazardous workplace conditions.

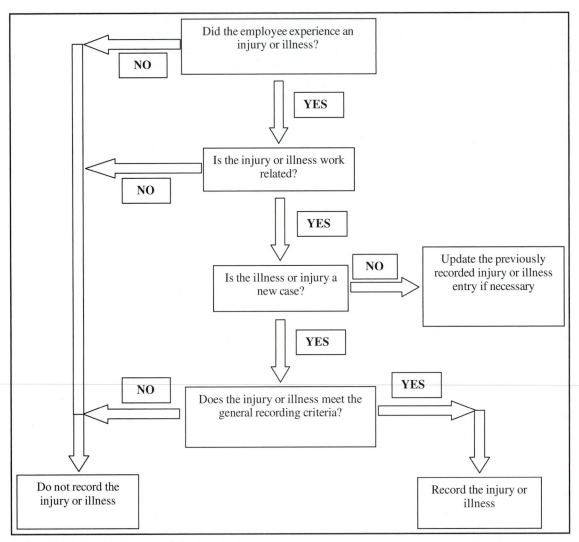

FIGURE 16.1 OSHA recordable injury decision tree.

- The recording process leads to improved administration of company safety and health programs with accurate records.
- As employee awareness about injuries, illnesses, and hazards in the workplace improves, workers are more likely to follow safe work practices and report workplace hazards.

OSHA JOBSITE VISITS

OSHA has the right of entry to any site of work performance to investigate or inspect the level of compliance with the safety and health standards. Knowing what to expect when an OSHA official visits your jobsite will help you better respond to the visit and ensure that your company is prepared in the event of a visit.

OSHA will inspect a jobsite for several reasons, but the order of priority for inspections is as follows: reports of imminent danger to employees; a fatality or catastrophe investigation; response to a complaint or referral; a planned inspection. A planned inspection may include a follow-up or monitoring visit due to a previous citation that required remediation. Inspections are unannounced because many investigation conditions can be easily and quickly altered if an employer knows investigators are coming. In fact, in Section 8(a) of the OSH Act, it prohibits unauthorized advance notice, except in special circumstances.

The Inspection

Inspections are made during regular working hours so working conditions may be observed. The first phase of the inspection is the opening conference. Upon arrival, the OSHA inspector will locate the general contractor and present his or her credentials. The inspector will note the conditions of the worksite upon arrival as well as any changes that may occur during the opening conference. So, fixing things during the opening conference is not a strategy to employ. Be certain that you have the necessary people present at the opening conference. If you would like an owner's representative present or another official from your company, for example the company safety officer, you may delay the inspection for no longer than 1 hour to await the arrival of the necessary party.

The inspector will alert you to the reason for inspecting the jobsite, to include the scope of the inspection and the rationale. If the inspection is due to a report of imminent danger, the inspector must provide a description of the alleged imminent danger situation, the date received, and the source of the information. If the inspection is being conducted as a result of a complaint, a copy of the complaint is to be furnished to the general contractor and any affected subcontractors. If the visit is a follow-up or monitoring visit, the inspector must provide the date of the initial inspection and detail the reasons for the follow-up or monitoring based upon prior employer agreement.

As part of the opening conference, the inspector will review the injury and illness records to the extent necessary to determine compliance and identify trends. Other OSHA programs, such as the Hazardous Communication Plan and the Fall Protection Plan, and records will be reviewed at the inspector's professional discretion. The inspector may also determine whether the employees of any other employers are working on the jobsite. If these employers may be affected by the inspection, the scope may be expanded to include others or a referral made at the discretion of the inspector. At multiemployer sites, copies of complaint(s), if applicable, shall be provided to all employers affected by the alleged hazard(s), and to the general contractor.

Walk-around Inspection

The main purpose of the walk-around inspection is to identify potential safety and health hazards in the workplace and to determine employer compliance.

Inspectors assess the employee and employer knowledge of any hazards on the jobsite through interviews and evaluate the good-faith efforts of the employer's safety and health program. The inspector will bring to your attention any apparent violations at the time they are documented. They will take photographs and measurements and may collect samples, such as air samples or surface samples. Be certain to take the same photographs as the inspector for your own records. A good rule of thumb is, if the inspector is taking photos, so should you.

Closing Conference

Upon completion of the inspection process, the inspector will meet with the general contractors and all appropriate subcontractors or their representatives and advise each of the apparent violations disclosed by the inspection. Employee representatives participating in the inspection shall also be invited to participate in the closing conference(s). The inspector will describe the apparent violations found during the inspection and other pertinent issues. The inspector will inform you of your rights regarding the citations, contesting procedures, abatement times, and a possible schedule for follow-up visits. Awareness of the OSHA site visit process, procedures, and policies is important knowledge for any construction manager.

Citation Types

OSHA inspectors can issue citations of the following types: serious, willful, repeated, and nonserious. A serious violation is a condition that has substantial probability to cause death or serious physical harm. The maximum fine for a serious violation is $7,000. A willful violation is issued when an employer commits an intentional and knowing violation of the OSH Act. The maximum fine for such a violation is $70,000. Repeated violations are issued for a hazardous condition that is substantially similar to a previously cited condition and must be issued within 3 years of the previous citation. Other than serious violations or nonserious violations are also possible, which net a maximum fine of $7,000. A failure to abate violation is issued if a noted violation is not corrected within the agreed upon period of time. This violation incurs a maximum fine of $7,000 per day for every day that the violation is not corrected. It is important to correct violations quickly and disseminate the information broadly to avoid this type of problem in the future on the site or on other projects.

Most-Violated Construction Standards

According to the 2006 data, the most violated OSHA standards in the construction industry center around a few main topics and include scaffolding, fall protection, ladders, excavations, hazard communication, personal protection, and general training requirements. Table 16.2 highlights the top 10 most cited standards for the reporting period from October 2005 through September 2006, based upon OSHA reporting information. For that period, there were a total of

TABLE 16.2 Top 10 most violated OSHA standards in construction.

Standard	Subsection	Number of citations	Dollars fined	Description
1926.451	L – Scaffolding	9,706	10,306,196	General requirements
1926.501	M – Fall Protection	6,820	8,242,454	Duty to have fall protection
1926.1053	X – Ladders	2,514	1,298,126	Ladders
1926.651	P – Excavations	2,078	2,500,316	Specific excavation requirements
1926.020	A – General	2,008	1,073,362	General safety and health provisions
1926.503	M – Fall Protection	2,002	966,063	Training requirements
1926.100	E – Personal Protection Equipment	1,709	867,654	Head protection
1910.1200 (1926.59)	HAZCOM	1,652	219,434	Hazard communication
1926.453	L – Scaffolding	1,489	1,325,076	Aerial lifts
1926.652	P – Excavations	1,466	4,167,996	Requirements for protective systems

FIGURE 16.2 Questionable scaffolding—no toeboards.

51,749 citations for a total of $46,526,993 in penalties in the construction industry. Figures 16.2 and 16.3 illustrate scaffolding and fall protection violations respectively.

FIGURE 16.3 Hole in safety fencing.

It is suggested that during the planning stages of any project the construction manager assess the project with these frequently cited standards in mind. This is a good starting point for jobsite safety planning.

WORKING WITH OSHA

As a construction manager, you may wish to investigate the opportunities that are available to assist you in the development of a site safety program or help in the identification and abatement of hazards. OSHA supports several programs that offer help and collaboration to employers to improve the health and safety of their jobsites. Examples include the Voluntary Protection Programs, alliances, and participation in a consultation program. The free health and safety consultation program is targeted to improve safety on the site level, and for a new construction manager, it is a recommended program to investigate.

CONSULTATION PROGRAM

This confidential program is completely separate from the OSHA compliance section and no citations are issued as a result of a consultation visit. The only obligation is committing to correcting serious job safety and health hazards in a timely manner.

The program provides a consultant to work with you and your employees to examine the conditions of your jobsite. The key piece to the consultation is the jobsite walkthrough. During the walkthrough, the consultant will work with you and your employees to identify and judge the nature and extent of specific hazards. This is another meaningful way that employees can be involved in the safety process.

A comprehensive consultation includes:

■ An appraisal of all mechanical and environmental hazards and physical work practices

■ An appraisal of the present job safety and health program or the establishment of one

■ A conference with management on findings from the consultation

■ A written report of recommendations and agreements

■ Training and assistance with implementing recommendations

Overall, the program will improve your knowledge of workplace hazards and ways to eliminate them. This is a proactive step toward establishing a vital health and safety program. Additional information about the consultation program and other OSHA sponsored safety programs can be found on the OSHA website. The site also contains several success stories. Reading and researching these programs are excellent ways to gain information about different approaches to safety and health that may be applicable to your jobsite. Figs. 16.4 and 16.5 are examples of worksite housekeeping issues that may be identified during a consultation visit. These sites can be cleaned up and easily organized to meet OSHA standards.

FIGURE 16.4 Poor housekeeping.

FIGURE 16.5 Improper material storage.

SAFETY PROGRAM DEVELOPMENT

For a safety program to be successful, it must be a systematic, orderly, all-inclusive management process that reaches beyond the minimum regulatory requirements providing the utmost safety and health protection for employees. In construction, safety programs must include industry-specific hazards. Construction work has the potential to be hazardous and occurs in several phases to include preplanning, planning, design, construction, and operation and maintenance. Safety needs to be considered at all stages of work on a project to ensure seamless functioning throughout a company and project.

Development of a safety program requires the input from all involved. Active involvement by both management and employees in implementation is critical to the success and support of a safety program. Creating a companywide atmosphere that safety is everyone's job ensures active involvement and ownership at every level and phase.

As a construction manager, your leadership provides the motivating force and the resources for organizing and controlling safety within the organization. Safety must be regarded as a fundamental and valuable activity within the organization. The emphasis placed upon safety should parallel the resources dedicated to scheduling, estimating, and cost control. Safety must be included as a fundamental operating principle of the company. This emphasis will send a strong message to all employees about the importance placed upon worker's health and well-being.

A commitment must be made to developing a company safety policy with goals and objectives that will be maintained, evaluated, and assessed on a recur-

FIGURE 16.6 Appropriate use of scaffolding signage.

ring basis. Included in this policy shall be a requirement for site-specific health and safety plans. While many owners require a project-specific safety plan as part of the contract, it should be an established company practice, as the hazards, demands, environment, and employees vary among all jobs. Unique plans that are founded upon the fundamental company policies are critical to sustaining a vibrant safety culture. The written plans should include all necessary aspects appropriate for the project undertaken. For example, personal protection equipment requirements may vary from project to project. Certain jobsites may be using building materials that can easily fray, fragment, or splinter. On these sites, safety gloves may be mandatory at all times. Figs. 16.6 and 16.7 illustrate simple, yet effective signage that emphasizes appropriate safety precautions that must be taken in specific areas.

As a manager, you must create an environment of trust and establish a clear line of communication with your employees. They should feel comfortable raising safety issues to any level within the company. In this role, you must actively demonstrate your commitment to their safety and health. This includes ensuring that all employees have access to the required safety equipment and that they know how to properly utilize that equipment. They also must have adequate training and knowledge of the dangers that are present on the jobsite.

While safety is everyone's responsibility, written clear definition of the responsibilities of all individuals on the site is necessary. Safety responsibilities should be clearly assigned. At all levels, each employee has his or her

FIGURE 16.7 Effective safety signage.

individual and assigned functional safety and health related activities. Employees are required to know their specific duties, be trained in how to perform them, and have the authority to act upon their relevant requirements. The ability to act when a situation is deemed unsafe is critical. For example, if the scaffolding superintendent notices that the crew trained in scaffolding erection is not wearing the appropriate personal protection equipment, he or she can stop the work and prevent the crew from continuing their work until the appropriate equipment is in place. The situation pictured in Fig. 16.8 is an excellent example of when an employee should be wearing fall protection equipment and is not. Anyone on the jobsite should have the authority to stop him and require him to wear the appropriate equipment.

Enforcement of safety policy is key to success. Employees must be held accountable for their actions and that accountability should be from top management, to field supervisors, to site employees. Employee performance should be tracked. Reward or recognize it when it is done well and correct it when it is not. Everyone is accountable for meeting their responsibilities for completing a job safely.

Company safety and health policy must be communicated to and understood by your employees. Safety should be integrated into the day-to-day management system within the company and on the jobsite. Employees understand the priority placed upon safety—they live it day to day and nothing interferes with it as their top priority. If this is the case, then enforcement of the policy will become easier over time.

Involving employees in the safety planning at all phases of a construction project not only is a smart decision, but also will help to increase the success of your program. Employees develop and express a commitment to safety and health protection for themselves and their fellow workers. They develop a sense

FIGURE 16.8 Unsafe worker—no fall protection.

of ownership for the success of the program because they were involved in the planning and in decisions that affect their safety and health. Their participation in the program must be meaningful and their input valued. Some examples of where employee participation should be sought include analyzing job or process hazards, participating in accident or incident investigations, and the documentation of near misses. Consideration should also be given to the development of an employee-suggestion program for safety improvements or initiatives. One company initiated a contest soliciting safety slogans from employees. The winning slogans were printed on banners and posted on the jobsite (Fig. 16.9).

FIGURE 16.9 Employee safety slogan sign.

Last, consider engaging employees in training other employees. They possess intimate knowledge of the jobs they perform and may be better equipped to educate others who will be completing the same work.

Another aspect that is important for construction managers is the presence of subcontractors on the site. Be certain to have a system in place for subcontractors when they come onto the jobsite. Prequalify your subcontractors and make sure they have a safety plan or better yet, require a written safety plan prior to award of the contract. It should be a requirement that they develop and operate an effective safety program. Be certain that the plan is job specific and not a generic plan. Their plan must be equal to or better than the plan in place by the general contractor.

To quantify the success of a safety program requires the implementation of a self-evaluation protocol. The protocol should examine the overall program and evaluate strengths, weaknesses, areas in need of improvements, and achievements. Determining where you are meeting your goals will help you to refocus your efforts and energies for the next year or project. The evaluation process must include interviewing employees at all levels for knowledge, awareness, and perceptions of the program. These people can help provide workable solutions to improving areas of weaknesses. The assessment should also review site conditions and, where hazards are found, determine the weaknesses that allowed the hazards to occur. Be certain to conduct evaluations immediately upon completing a project. This will provide valuable information that can be used for improvement in the future and be more competitive in the next bid.

JOB HAZARD ANALYSIS

A major component of a project-specific safety plan is the analysis of the work site for hazards that may impact safety and health. A job-hazard analysis should be conducted prior to even bidding a project or at the latest before completion of a task. Comprehending the hazards ahead of time may lead to prevention or elimination of any accidents or injuries. It should be the first step in completing a site-specific safety program.

To begin, identify and document common hazards associated with the project and the site. This should be accomplished for all phases of the construction project and the processes associated with each of the phases. Use OSHA regulations, building codes, industry standards, and equipment manufacturer recommendations as guidelines. Establish procedures to identify hazards and their prevention and control. Be certain to analyze routine jobs, tasks, and processes and identify uncontrolled hazards. Document the process by which the analysis takes place and how these will lead to hazard elimination or control.

In addition to completing a preliminary job hazard analysis for the entire project, individual procedures should be developed for specific task hazard analysis. The procedure should describe the process to identify uncontrolled hazards prior to an activity and detail the precautionary measures for the day-to-day operations. These should be developed for all areas of the project, but should focus upon areas that are considered high risk for the project.

In construction operations, employees are exposed to potential hazards as part of their daily work. Operations that may require additional attention for identifying hazards include scaffolding erection, work that could involve a fall from height, trench excavation, electrical work, or other areas that have a history of high accidents or injuries. In 2006, OSHA recorded 13,267 citations for violations of scaffolding regulations. These citations resulted in fines totaling over $13 million. Establishing a method for identifying and analyzing safety hazards for routine and nonroutine processes will help you to avoid being one of these statistics. Fig. 16.10 illustrates a simple but highly effective way to document and communicate daily hazards to employees. This simple message board conveys the dangers and increases employee awareness.

SELF-INSPECTIONS

As part of a vigilant safety program, jobsite inspections should be conducted on a regular basis, at least weekly during the project. Development of a systematic approach with checklists will help streamline the inspections and ensure a

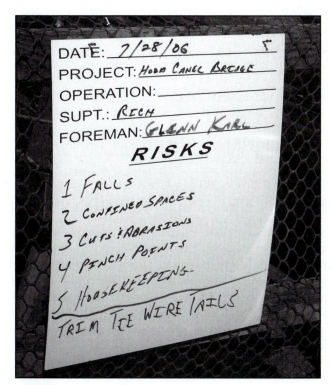

FIGURE 16.10 Daily risk signage.

consistent process. The inspection guidelines should include what will be inspected, who is responsible for conducting the inspection, what the schedule is for inspections, and also include a system that tracks any required actions to completion. A verification system to ensure that the inspection took place should also be implemented. This documentation will help to determine trends and possibly aid in correction of such issues.

ACCIDENT/INCIDENT INVESTIGATIONS

Investigating the root cause of an accident is critical to the success of a health and safety program. As part of the written program, the development of procedures that detail the process by which an investigation will take place is required. This procedure should outline the type of incident that is investigated. Is it for reportable incidences and fatalities only? Or will there be an investigation for near-misses, first-aid cases, and other incidents? Who is authorized to conduct the investigation? What are the criteria for an investigation and what is the appropriate level of investigation? How will the findings be represented? It is ideal to have written findings that aim to identify all contributing factors and corrective actions required, tracking actions to completion with a summation of what action was taken to prevent similar events in the future.

Why are these investigations so important to a safety and health program? It is all in how the data collected is utilized. A system should be in place to analyze injury, illness, and related data that include investigation results. The goal of the investigation is to identify common causes and accident trends over a year or for the duration of a project and enact corrections in the appropriate systems, equipment, or programs. For example, if there is a high rate of nail gun injuries to employees during the first 3 weeks of employment, it may indicate the need for more training on the equipment or possibly investigation of the use of another manufacturer's nail gun. Awareness of this issue can save a project time and money and improve the safety of the employee.

WRITTEN EMERGENCY RESPONSE PLAN

When it comes to safety in construction, most energy is spent on prevention. Safety programs focus upon the elimination of hazards and reducing injuries and illnesses. However, do the employees on your jobsite know what to do if an accident occurs? A written emergency response plan is another vital aspect of a safety program, but many jobsites do not have a detailed plan. Why? A possible reason may be due to the lack of federal standards. The standards merely mention a rescue component and do not define what is meant by a "prompt rescue." In the ANSI standards, Section A.10, which details the safety requirements for Construction and Demolition Operations, section A.10.26 Emergency Procedures for Construction Sites, address the NEED for emergency procedures on construction sites, but go no further. Without regulations requiring these types of plans, not much effort is placed upon preparedness.

When developing a plan, begin with the end in mind, and that is decreasing the number of fatalities, injuries, and illnesses that may result if an accident occurs. The plan must include training for employees in rescue procedures and be site specific. Key elements of the plan include knowing when and how to rescue workers, making sure that the necessary life-saving tools are available, understanding the roles of various emergency responders, and effective coordination with local authorities.

The plan should detail the local resources available to the jobsite, to include emergency response teams, HAZMAT teams, and disaster relief specialty teams. It should set procedures with preplanned objectives to include defining the "golden hour." This is the longest amount of time that should pass between trauma and treatment. In addition to having the right tools and equipment available, employees must be trained in how to use them. One of the hardest things to train and plan for is the human emotional responses to the incident. Brief your employees on what they may expect. Advanced planning with local officials can greatly improve the effectiveness of the plan. At a minimum, the local officials should be familiar with the jobsite location, know who is part of the rescue team, and have detailed contact information for all company representatives and related personnel.

Rescue planning and training is critical. As mentioned earlier, falls account for a large number of fatalities and injuries each year as do trenching accidents. In 2005, 394 construction personnel lost their lives as a result of falls and 39 fatalities were recorded due to trench and excavation cave-ins. When analyzing falls, if an employee is wearing a safety harness, it will do its job. The rescue is the vital component. To ensure a successful rescue, training and timing are critical. The training program must include a briefing that covers hazards daily and daily inspection of related equipment. Time is of the essence in a rescue, because the harness restricts blood flow and brain damage can occur in as little as 3–4 minutes after loss of consciousness and stoppage of breathing.

Also, coworkers should remain calm, so as not to cause any further trauma to the injured worker. Rescues from trenches are no less difficult. Most deaths in trenches are caused by medical asphyxia, due to the immense weight of the soil crushing the ribs, or compartment syndrome, where blood is cut off from the muscles and tissue necrosis can occur in as little as 10 minutes. Once a trench has collapsed, certain precautions must be taken to prevent another collapse. These include turning off equipment, halting traffic near the trench, and keeping people back at least 50 feet. The last precaution may be the most difficult to enforce, as most coworkers want to help. However, coworkers can do very little and statistics show that 75% of would-be rescuers become victims themselves. The trench shown in Fig. 16.11 is correctly shored and has a readily accessible method of egress—the ladder.

The success of an emergency response plan involves the amount of preparation and how effective communication is during a crisis. A highly coordinated

FIGURE 16.11   Example trench bracing.

management and communications strategy is the foundation for an emergency response plan, and training is the cornerstone.

SAFETY AND HEALTH TRAINING

While a majority of the OSHA regulations focus on the creation of safe work environments, some regulations concentrate on the responsibility of management to maintain safe worksites by ensuring that appropriate safety training is received by employees. Subsection 1926.21, entitled "Safety Training and Education," outlines the requirements of employers to provide employee training, but does not describe the details of how to meet these requirements. At a minimum, the training programs must meet the guidelines and regulations promulgated by OSHA. These requirements are that all employees in the construction industry are to be trained in the recognition, avoidance, and prevention of unsafe conditions.

Despite the fact that training is a requirement, in the period from October 2005 to September 2006, there were 1,105 citations relating to the education and training standard that resulted in 1,066 investigations by OSHA officials. Of these incidences, the fines assessed for violation of §1926.21 were over $950,000.

QUANTIFY TRAINING NEEDS

On each jobsite, employees will be conducting different types of work requiring various levels of safety training to meet regulatory guidelines. The first step in development of an effective company training program is to determine the regulatory requirements. §1926.21 outlines these training and education requirements in general terms. Other subsections address specific training requirements for the safe operation of various equipment and tools.

The most extensive section regarding training and education of employees in the OSHA standards can be found in §1926.59, Hazard communication. There was a time when employees had no legal means to force their employers to provide them with information about the hazardous substances they used at work. They demanded a "Right-to-Know" or the right to have access to such information. The federal OSHA standards detailed within 29 CFR 1910.12 and 1926.59, better known as the "hazard communications" or "right-to-know" standards, require employers to do just that. This section of the standards requires employers to develop and implement employee training programs that inform them of the hazards of chemicals used in the workplace and the appropriate protective measures available for hazard avoidance.

When identifying the needs of a program, determine the type of work to be performed and then review the applicable OSHA standards. Different types of work and the use of equipment associated with that work have specific training requirements according to the OSHA regulations. The education of your employees regarding these standards is required by law, so be sure these topics are included within the training program.

In addition to the regulatory and specific jobsite requirements, a determination must be made regarding the solution of a safety problem via training. Do safety problems currently exist? Will training solve the unsafe practices of employees or are there other actions that must be taken? Not every problem may be solved through training, such as shoddy equipment or unavailability of materials. Problems that can be addressed effectively by training include:

- lack of knowledge of a work process
- unfamiliarity with equipment
- incorrect execution of a task

In determining what training is needed, you must identify the work the employee is expected to complete and if the employee's performance is deficient. To obtain this information, an analysis of what an employee needs to know in order to perform a job in a safe manner must be conducted. With this information, specific training requirements will be identified.

When defining the general goals of your program, consider the following areas where adequate training and education can really affect the bottom line.

- Regulatory or performance based goals (i.e., meet OSHA standards)
- Reduction in accidents, injuries, and illnesses

- Reduction in lost workdays, hours, etc.
- Improved EMR rating
- Strong safety records

In addition to overreaching company goals and improvement of the bottom line, goals should be set that involve the level of training, the frequency of training, and the quality of training. Consider goals such as:

- Corporate managers and top executives will receive annual safety training.
- Supervisors, project managers, and superintendents will receive project-specific training.
- Every employee receives initial company safety orientation within 1–3 days of starting work.
- Every employee receives refresher training at regular intervals.
- Every employee is instructed in safe practices and the hazards associated with the activity prior to the commencement of that activity.

In order to effectively administer a safety training program, guidelines and objectives must be determined for the training. The goal of instructional objectives is to clearly state what you want your employees to do, to do better, or to stop doing as a result of the training. These objectives must be definitive and measurable, precisely indicating the skills or knowledge that must be demonstrated by an employee upon completion of the training. The learning objectives should also state the conditions for acceptable employee performance and the level of competency required to satisfactorily complete training.

The critical elements of a training program are the content of the program, the people delivering the training, and the employees receiving the training. After development of safety goals and the objectives of the training program, employers can begin to define the elements of the program. The content of the program should be outlined with precise activities that the employee will take part in during training. These activities should simulate the actual job environment as closely as possible to best ensure that the employee will transfer the skills learned in training to the job. By arranging the course material to mimic the sequence of events that would be used on-site, a specific safe practice can be developed.

For example, if an employee is learning the process of safe operation of a power tool, the sequence might be:

1. Ensure that the power source cord is free of defects.
2. Check that the power source is connected.
3. Identify and acknowledge that safety devices are in place and operating.
4. Understand the manufacturer's recommendations for safe operation.
5. Identify the power switch and its operation.
6. Determine and use the appropriate personal protective equipment, etc.

When determining the actual content of the training program, examine the needs that were previously identified.

TABLE 16.3 Top five training standard violations.

Standard	Subsection	Number of citations	Dollars fined	Description
1926.503	M – Fall Protection	2,002	966,063	Training requirements
1926.454	L – Scaffolding	1,281	1,073,088	Training requirements
1926.21	C – General Requirements	1,105	954,189	Safety training and education
1926.1060	X – Ladders	455	113,738	Training requirements
1926.761	R – Steel Erection	122	101,916	Training

Each of these areas has specific requirements with respect to training that can be cited. In the period from October 2005 to September 2006, there were 1,281 citations and 1,140 investigations leading to over $1,073,000 in fines for violation of the scaffolding training requirements outlined in §1926.454. Falls are one of the leading causes of death on construction sites. During the same time period, there were 2,002 citations and 1,881 investigations for failure to comply with §1926.503, fall protection training. These citations yielded over $966,000 in fines. Table 16.3 highlights the top five most violated OSHA training standards. Is it just a coincidence that falls from heights are the leading cause of death in the construction industry and that the most violated OSHA standard is §1926.451—General Requirements for All Types of Scaffolding? Based upon the above facts, every safety training program should begin the development of its content by focusing on those standards.

KEY COMPONENTS TO A SUCCESSFUL PROGRAM—BEYOND REGULATORY REQUIREMENTS

In addition to the regulatory requirements that you must meet in relation to training programs, other important components contribute to the success of a training program. Research completed on safe workplaces indicates that certain actions can influence the overall success or failure of a safety program. These components are easy to implement, most only take time and planning prior to completion of work. However, what makes implementation difficult is that these steps are attempting to change individual attitudes and habits. Although the steps are tangible methods that can be outlined on paper, the real effect and delivery is somewhat intangible. The focus of these components is making the shift from written policy to action at all levels.

ESTABLISHING THE SAFETY FOUNDATION

Safety demands a constant effort from all parties involved. The degree of success of any training program is contingent upon field implementation of the knowledge acquired during the instruction. Practicing the guidelines detailed within the training on a daily basis must be expected from all employees. When establishing a strong safety foundation, a message must be sent to employees indicating not only that they are expected to follow the safety regulations, but also that they should immediately take care of any safety problems that they encounter. If they are unable to correct the problem by themselves, they should contact their immediate supervisor to get the appropriate help.

All employees need to know that they are empowered to correct any unsafe conditions on a site and that they can stop working if they feel they are in an unsafe environment. With this concept in mind and knowing that it is supported by the employer, employees will begin to act safely. Employees must see that compliance with safe work practices is for their benefit and that violation of these practices will not be tolerated. To enforce that belief, it must be clear that noncompliance with the safety regulations will result in immediate action. For example, some employers have adopted a three-strikes system that deals with safety violations in stages. The employee is issued a warning regarding his or her unsafe behavior for the first infraction. Next, the employee is given time off without pay. Finally, if behavior fails to improve, the employee is terminated. This system, if followed consistently, will send a clear message regarding the company's policy toward safety compliance.

IMPROVING THE OVERALL ATTITUDE

In order to increase the success of a training program, the attitude of the worker must be changed and apathetic behavior must be modified. People can be trained so that acting safely and performing work in a safe manner is second nature. The atmosphere that needs to be created is one where unsafe acts are not tolerated, and that atmosphere needs to permeate every level of employment. Regardless of the level of employment, whether top management, project managers, field personnel, or office administrative staff, the environment created must be one where unsafe practices do not exist. If the environment exists where safety is the first consideration given to every task by every employee, then the "safety first" sign at the project entrance is more than just a sign—it has real significance.

The attitude toward safety is even more important during the actual training sessions. The focus of the training should be positive. Safety should not be viewed as something that has to be done to avoid a punishment. It should be addressed in a favorable light, with enthusiasm and a positive attitude. Training must also possess a level of seriousness since the goal of the program is to ultimately save lives and reduce injuries. Make sure employees understand what could happen if the practices are not followed. Part of a safety training program should include images of what has happened as a result of unsafe practices. Although sometimes graphic, these images will leave a lasting impression. Additionally, OSHA publishes documents easily accessible on its website that highlight fatal accidents. Employees can learn from others' mistakes. Employees need to know that the company cares about their safety and well-being. Safety training should take center stage regularly throughout the month. It should be viewed as an invest-ment and not as an expenditure. As the old saying goes, "People don't care how much you know, until they know how much you care." Keep that phrase in mind as training is conducted. Focus on issues that are important to workers and listen to their feedback. Try something novel with a safety twist, such as a safety essay or poster contest for employee's children. This is an easy and fun way to involve employees. Last, remember during training that the attitude of trainees is often a reflection of the trainer's attitude.

PROMOTION OF THE SAFETY CULTURE

Safe people develop a mindset and when they work with other people of the same mind set, a new safety culture is achieved. Employees need to be an active and integral part of the safety team. To help promote this sense of inclusion and responsibility, have your employees sign a personal commitment to safety and positive interaction. This document is a powerful tool that noted safety profes-sional Art Fettig developed several years ago. People need to learn to interact with each other in positive ways to promote safe work environments and prac-tices. Part of the training program must address this issue. On many jobsites where accidents occur, you hear coworkers say, "I knew they were going to hurt themselves or someone, they were always taking chances." Why don't cowork-ers get involved? A positive interaction program can work only if you first have the participant's signed permission. It should state that they give their permis-sion for other employees to point out to them when they are acting unsafely. Workers need to admit that they are human and that they do make mistakes. Injuries and accidents most often occur when individuals become complacent with their work and let their guards down. If part of the training program empha-sizes that everyone is an equal player in the game of workplace safety, everyone will benefit.

OTHER SAFETY RESOURCES

There are several resources available to the construction manager to help in developing a safety program and navigating the complex issues surrounding safety. The Construction Industry Institute (CII) as well as the National Safety Council and other local organizations provide safety training programs. Addi-tionally, trade and professional organizations have actively participated in improving safety education and practice. Consider involvement or contacting these agencies before you begin developing a program.

SUMMARY

Comprehending the impact of safety on your bottom line, what is required by law, what to expect during an OSHA jobsite visit, and the development of a safety program are either directly or indirectly part of your responsibilities as a construction manager. Effective planning for safety at all phases of a project will help you to provide a safe and healthy jobsite for all of your employees. Safety is a combination of education, training, and enforcement, but do not forget the human element. Leading by example and providing a positive attitude toward safety will help establish you as a safety leader.

REVIEW QUESTIONS

16.1 What is the purpose of an Experience Modification Rate?

16.2 When calculating the Experience Modification Rate for a company in 2009, what years of data would be included?

16.3 What act created federal OSHA?

16.4 What are the record keeping requirements for OSHA?

16.5 How long do you have to keep OSHA logs of injuries and illnesses?

16.6 What is the leading cause of fatalities in the construction industry?

16.7 Why do you need to have worker's compensation insurance?

16.8 Summarize the OSHA site inspection process.

16.9 What makes an injury reportable according to OSHA standards?

16.10 Describe the consultancy program supported by OSHA.

16.11 Identify three tasks that are commonly performed on a jobsite and perform a job hazard analysis on each.

16.12 Develop a mitigation plan for the tasks identified in problem 16.11.

16.13 Write an emergency response plan for a fall from heights.

16.14 Outline the key components of a safety plan for a residential and highway construction project. How do they differ?

16.15 Develop a list of emergency contacts for jobsite rescue and emergency response teams in your local area.

REFERENCES

1. Diether, J., and G. Loos. (2002). "Advancing Safety and Health Training," *Occupational Health and Safety,* 69: 28–34.

2. Fettig, A. (1998). *Winning the Safety Commitment.* Growth Unlimited Inc., Battle Creek, MI.

3. Hinze, J. W. (2003). "Safety Plus: Making Zero Accidents a Reality," Construction Industry Institute Research Report 160-11.

4. Occupational Safety and Health Administration (OSHA). (2007). *OSHA Standards for the Construction Industry,* 29 CFR 1926. U.S. Department of Labor, Washington, DC.

WEBSITE RESOURCES

1. www.osha.gov/recordkeeping/RKforms.html
2. www.osha.gov

17

Trends

In the face of growing populations and expanding economies, the need for more facilities, and therefore for more design and construction services, is inevitable. In response to these demands, the purchasers of construction services— owners—are investigating and experimenting with nontraditional contracting and innovative financing techniques, and contractors are devoting greater attention to new technologies that can improve their competitive positions. The business of construction will be changed in the near term by innovative contracting methods and technologically advanced tools. Construction will also be impacted further in the future by emerging technologies being developed in other disciplines.

COST-EFFECTIVE AND TIMELY CONSTRUCTION

Population growth and expanding economies are driving the need for more facilities and therefore for more design and construction services. Traffic congestion, especially in metropolitan areas, places a significant and serious burden on the public and threatens the economic vitality of the nation.

Transportation systems must be greatly expanded and improved; the Panama Canal Authority is pushing ahead with its third lane locks project in order to meet the needs of global shipping (Fig. 17.1). To link their two countries, Brazil and Peru are building a highway that traverses the jungles of the Amazon, the heights of the Andean Mountains, and the sands of the Peruvian desert. In the United States, the growth in roadway capacity has not kept pace with the increased demand for travel and there has been a lack of maintenance for existing structures.

Manufacturing and commercial facilities must be constructed: building owners are demanding sustainable designs and "green buildings." The use of the

FIGURE 17.1 Dredging work to expand the Panama Canal.

Leadership in Energy and Environmental Design (LEED) Rating System is providing the industry with standards for what constitutes a green building.

Housing inventory must be increased; the affordable housing deficit is a major problem in all countries, and in developing countries the inability to provide adequate housing to the population is a critical problem. There is a quantitative deficit and a qualitative deficit, meaning that a large number of existing residences do not provide an adequate quality of life for the inhabitants.

The question is, how will the construction industry deliver the needed facilities and maintain existing facilities? The key to addressing intensified demand lies not only in accomplishing the work but also in accomplishing the work while at the same time enhancing the general population's quality of life. Designers and constructors must perform the work in a more efficient manner while minimizing construction impacts such as noise, odor, and vibrations.

Building information modeling (BIM), a digital representation of physical and functional characteristics of a facility, is poised to fundamentally change the way projects are built. Such model-based technology linked with a database of project information will change the way project stakeholders communicate with each other. There is a shift to offsite prefabrication of structural and finish elements that are then installed or assembled rather than produced on site. Consequently, production equipment is being replaced on the construction site by *transportation* equipment. Material handling and lifting equipment now dominates building construction sites more than ever before and constitutes the critical element in achieving productivity.

Communities are demanding that construction projects and operations not degrade environmental quality. There are many research efforts directed at what is commonly referred to as sustainable development. Sustainable development can be broadly defined as meeting the needs of the present without compromising the ability of future generations to meet their own needs. It seeks to accomplish required construction while recognizing the interdependence of environmental, social, and economic systems.

The purchasers of construction services—owners—are investigating and experimenting with nontraditional contracting and innovative financing techniques, and contractors are devoting greater attention to new technologies that can improve their competitive positions. In the near term, innovative contracting methods and technologically advanced tools will change the business of construction. Further in the future, construction will be impacted by emerging technologies being developed in other disciplines.

Construction automation, better equipment, new processes, and new techniques will be employed to perform quality work with fewer workers. Tower cranes, common in Europe for decades, are globally gaining in popularity with surging real estate developments. Ideal for dense urban environments and coming with a small footprint, they are available in a growing diversity of sizes and configurations. Sophisticated electronic controls and operator assistance devices are enhancing their safe and productive operation.

The U.S. Department of Labor reported that an additional one million workers would be needed in the construction industry before 2012. The lack of skilled workers is a global problem impacting the industry. Additionally, the workforce of the future will need to have a very different set of skills as more automation is introduced into building processes. To address the issue, the industry is developing programs to attract new workers for high-skill positions in construction. These workers will include more women and minorities.

Those construction companies that embrace the improved techniques will prosper and the others that cling to old systems will become uncompetitive and cease to exist. Both engineering and construction firms are facing global competition and that competition will intensify in the future. Already the more sophisticated owners and contractors are experimenting with (1) innovative contracting methods, (2) technologies for guiding and monitoring, (3) smart structures, (4) simulation technologies, and (5) total electronic integration.

Innovative Contracting Methods

The facility owners have been experimenting with innovative contracting techniques such as qualifications-based procurement, design-build, and best-value procurement in the delivery of their projects. In Europe, there is greater use of nontraditional contracting and innovative financing techniques. Many owners are now beginning to use quality of prior work performed as the major qualification criteria in the selection of contractors. These methods of evaluation recognized the importance of the organizational structure that supports project work. One very successful approach is to colocate project-level owner and contractor personnel, and to empower those managers with the authority to solve problems on the jobsite.

Technologies for Guiding and Monitoring

Construction equipment or automated/preprogrammed machinery (robots) will incorporate advanced guiding, monitoring, and coordinating technologies. Many of these technologies were developed for the purposes of aeronautic navigation. They can be classified as inertial navigation systems, active beacon navigation systems, and **radio frequency identification** systems. After the machine position is known, a computer would control the mechanical action of the machine based on a laser scan of an existing surface (natural ground surface or wall surface in a building) compared against a calculated virtual surface. The electronic plan files would provide the virtual surface information.

Real-time monitoring of equipment can provide reliable data about performance of the work and alert management situations where the performance fails to meet plan expectations. The goal is to fully automate the measuring of various indirect parameters with the subsequent conversion of that information into project performance indicators. Work has been done on automatically monitoring construction lifting equipment as a means of gathering input information about progress [2]. Almost all major components and material of a vertical construction project are delivered by lifting equipment. Tracking daily lifting equipment activities can be described by two data flows: (1) gross weight on the hook and (2) hook location (Fig. 17.2). This information together with the planned

Radio frequency identification
A wireless system that permits noncontact reading of information.

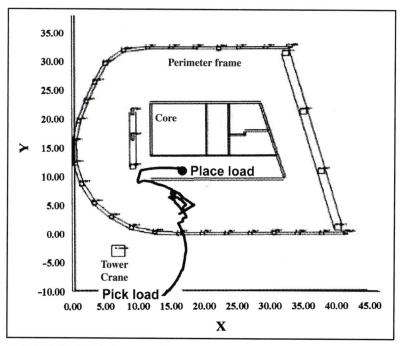

FIGURE 17.2 Motion trace of the tower crane's hook imposed on a building floor plan view.

construction activities permits identification of performed activities through crane monitoring.

Smart Structures

Advances in smart materials such as piezoelectric ceramics and shape memory alloys together with developments in fiber optics detector technologies will make low-cost, long-term, distance monitoring of strains, temperatures, and chemistry possible. This technology provides the ability to build smart structures that facilitate construction quality. Further enhancement of these technologies could lead to the embedment of materials that make judgments—information processing and intelligence—within the structure itself.

Simulation Technologies

Construction simulation tools can capture three-dimensional (3-D) data sets (space) and add a fourth dimension—time (3-D plus time). Visualization allows for better planning of construction sequencing and can improve coordination between owners, contractors, subcontractors, and suppliers. The technologies provide better analysis of many aspects of the construction process and support computer-based analysis of project schedules. The modeling functionality of these tools is constantly improving and being accepted by the industry.

Total Electronic Integration

There will be total electronic integration of the entire process of planning, design, and construction, linking together each step of the process in universally accessible electronic files. The intent of BIM is to create a repository of facility information that can be used throughout the life of a facility. The National Institute for Building Sciences has organized a committee to create an open national standard to enhance the efficiency of BIM. Site surveying, design, equipment/robot instructions, quality assurance/quality control, schedule reporting, payments, and as-built drawings will all be integrated in the future. This will be an integrated environment where information is passed electronically from step to step with minimal human handling of paper. Direct linking of electronic-design plans to construction equipment and robot computers will minimize errors, improve quality, and speed construction operations.

INNOVATIVE CONTRACTING METHODS

Innovative contracting practices present challenges to both owners and industry members. But it is evident that nontraditional contracting methods are being used more frequently. Some of these contracting methods significantly increase contractor responsibilities and contractors will also be assuming more liability—risk. In some cases, the use of these methods will require revisions to state statutes governing construction contracts, but in some states this has already happened.

Design-Build Contracting

Under the design-build approach (Fig. 17.3), a single company or a joint venture performs design and construction of a project. The contract can be awarded on either a low bid or best-value basis. This approach provides the owner with one source of responsibility for the project. Design-build has been used for large projects, the $1.2 billion I-15 project in Utah is an example, but is it an appropriate alternative contracting method for smaller projects. Many owners are seriously considering the use of design-build contracting for smaller projects.

Typically, design-build projects show a significant advantage in lowering the duration of the project, with a broad range of 4% to 60% reduction relative to design-bid-build [Web 2]. Research on design-build has shown that while each project should be considered on an individual basis, no more than 30% of preliminary design should be completed before design-build contract award, with lower percentages as the contracting agency gains more experience with design-build contracting. Additionally, the method relies more on performance-based specifications.

Public/Private Partnerships

As public owners look for innovative ways to finance and deliver their programs, a few have enacted public-private partnership statutes. The state of Virginia enacted the Public Private Partnership Act in 1995. The Texas Turnpike Commission has used exclusive development agreements to finance, design, construct, operate, and maintain approximately 100 miles of toll roads on new alignment.

There has been a backlash, however, in several states including Texas. In April 2007, both houses of the Texas legislature passed a two-year moratorium measure on the concept of private toll roads in the state. The legislature was bowing to a grassroots antiprivation movement. The eagerness to utilize private investment to create public works stems from financial constraints of governmental budgets.

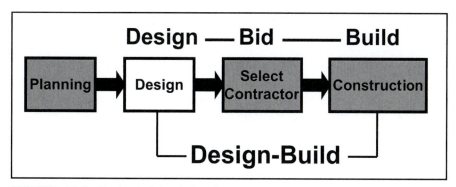

FIGURE 17.3 Design-build task time line.

Use of Electronic Contracting

Many contractors prepare bids using computerized spreadsheet and databases, but a paper copy still must be submitted to the owner at a set time. This requirement can cause a hardship for contractors at a distance from where bids are received and for those that receive subcontractor prices at the last minute. The use of electronic bidding should benefit both contractors and owners. The major concern with electronic bidding, however, is security. How can the owner and contractor be certain that the files sent or received have not been tampered with and have been sent by the correct person? Electronic signature standards are continuing to mature, but the U.S. Congress enacted the Electronic Signatures in Global and National Commerce Act ("ESIGN") on June 30, 2000, to facilitate the use of electronic records and signatures in interstate and foreign commerce by ensuring the validity and legal effect of contracts entered into electronically.

Performance-Related Specifications

Performance-related specification language has been developing for nearly 30 years but still needs work to be accepted by the construction industry and owners. Performance language must describe how a product should perform in service. It should reduce the prescriptive requirements found in many specifications and concentrate on measurement of factors critical to the performance of the final product.

The Strategic Highway Research Program II (SHRP 2) is making a major push to advance performance specifications for highway renewal projects. The desired *specification* language must work effectively and properly in all types of contracts, from traditional design-bid-build to design-build, and with other innovative contract types. The objectives are to (1) reduce the completion time of projects while maintaining or improving quality and (2) encourage further innovation by reducing mandatory method specification requirements.

GUIDING AND MONITORING MACHINES

Navigation of equipment and machines is a broad topic, covering a large spectrum of different technologies and applications. It draws on some very ancient techniques, as well as some of the most advanced in space science and engineering. The new field of geospatial engineering is rapidly expanding and a spectrum of technologies is being developed for the purposes of aeronautic navigation, mobile robot navigation, and geodesy. Equipment manufacturers are developing dynamic geofencing of machines using these technologies. With such fencing, an owner can create sight or area boundaries for individual machines. Electronically, a crane could be restricted from power transmission lines or an excavator from a gas main.

Physical Scales

The physical scale of navigation requirements can be measured by the accuracy to which the mobile unit needs to navigate—this is the resolution of navigation.

These requirements vary greatly with application; however, a first-order approximation of the accuracy required can be taken from the dimensions of the unit itself. Any autonomous device must be able to determine its position to a resolution within at least its own dimensions to be able to navigate and interact with its environment correctly. To help in categorizing this scale of requirements, three terms are used:

■ *Global navigation,* which is the ability to determine one's position in absolute or map-referenced terms and to move to a desired destination point

■ *Local navigation,* which is the ability to determine one's position relative to objects (stationary or moving) in the environment and to interact with them correctly

■ *Personal navigation,* which involves being aware of the positioning of the various parts that make up the unit, in relation to its other parts and in handling objects

The full range of scale is depicted in Fig. 17.4.

A few of the more important navigation technologies with respect to construction applications are (1) inertial navigation systems, (2) active beacon navigation systems, (3) global positioning systems, (4) ground-based frequency systems, and (5) radio frequency identification systems.

Global Positioning System

In 1973, the American Defense Navigation Satellite System was formed as a joint service between the U.S. Navy and Air Force, along with other departments including the Department of Transportation, with the aim of developing a highly precise satellite-based navigation system—the **global positioning system.** Since its conception, GPS has firmly established itself into many military and civilian uses across the world.

global positioning system
A highly precise satellite-based navigation system.

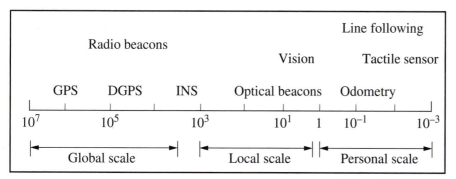

FIGURE 17.4 Summary of the scales of navigation (scale is in meters).

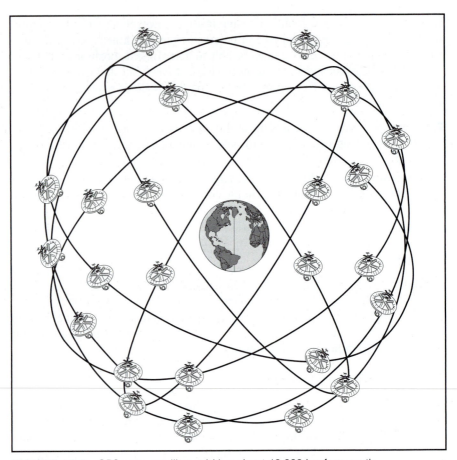

FIGURE 17.5 GPS uses satellites orbiting about 19,000 km from earth.

A typical GPS computes ground positions using a network of 24 satellites orbiting about 19,000 km from earth (Fig. 17.5). The system uses two pseudo-microwave signals transmitted from the satellites. The position of an object is calculated based on the trilateration—a measurement of the distance between a number of beacons and the user. The main advantages of GPS are declining cost, increasing accuracy, and the fact that it can be accessed by multiple users—an important point for the purpose of monitoring construction equipment and tools. The main disadvantage is that a GPS cannot be used inside a building or under roof structures.

There are two levels of GPS precision when using an individual receiver and only getting the signals from the satellites and nothing else. For such receivers, the only modus operandi is referred to as standalone operation. There are two standalone levels: one available to the civilian world, called the standard positioning service (SPS), and the military's precise positioning service (PPS). The

(a) Reference station

(b) System mounted on a CAT 613 scraper

(c) CAES equipped scraper working

(d) CAES equipped scraper working at night

FIGURE 17.6 Corps of Engineers field test of a computer-aided earthmoving system.

civilian (SPS) level was set by Department of Defense (DOD) policy. When the intentional degradation of accuracy was in place (called selective availability, or SA), SPS accuracy was 100 m horizontal 5% of the time. On May 1, 2000, SA was set to a level of 0—effectively being turned off. There are now several manufacturers achieving 10-cm accuracy levels with devices for civilian use.

The U.S. Army Corps of Engineers conducted a field test of a computer-aided earthmoving system (CAES) developed by Caterpillar Inc. in 2001 (see Fig. 17.6). From that limited test, the corps reported CAES-equipped scrapers:

- Moved 5.4% more earth in the 20-hr test period
- Reduced preconstruction and restaking time by 28 hours
- Reduced manpower requirements by 54%
- Achieved an accuracy of 2.3 inches vertical, 9.6 inches horizontal

Active Beacon Systems

Navigation using **active beacon systems** (ABSs) has been with us for many centuries. Using the stars for navigation is one of the oldest examples of global referenced navigation. Technology has produced many other systems, such as lighthouses and more recently radio navigation.

Active beacon systems can be used for surveying both before design and during construction; these will speed project layout and reduce errors. Using an ABS to guide and control construction equipment directly will reduce construction surveying requirements. Active beacon systems are based on three or more transmitters at known locations, and one receiver on the mobile unit whose location is to be determined. There could also be only one transmitter on the mobile unit and three or more receivers at the known locations. An ABS can operate on two principles: trilateration or triangulation.

- *Trilateration* uses a measurement of distance between a number of beacons and the user.
- *Triangulation* measures and uses the bearing between the user's heading and a number of beacons.

One significant challenge to deploying ABS for construction automation is horizontal, vertical, and longitudinal datums. The issue here is that there are very few (if any) as-built surveys of existing infrastructure that have the relative accuracy necessary for implementing autonomous construction equipment in a systematic manner. However, once there is a commitment to obtaining (and maintaining) these surveys, it is easy to imagine how plans can be downloaded to autonomous or semiautonomous equipment for execution.

Ground-Based Radio Frequency Systems

Ground-based radio frequency (RF) systems use the same principles as a GPS, with two differences. The first difference is that instead of satellites as a reference, they use ground stations. The second difference is the transmitting frequency. Here, too, the position of an object is calculated based on the trilateration principles. These systems are typically of two types: (1) active radar-like systems that have several fixed stations and (2) passive systems in which all the fixed stations are connected to a control center.

There are global scale systems, such as Omega and Loran, that allow for an unlimited number of users from a finite number of beacons. Commercial localized beacon systems are available in some locations. These allow many users in one area, but some are more suitable for autonomous mobile control because the position information is calculated at the mobile end. Others are more suited to tracking applications.

The main advantages of RF systems are (1) they can serve a large number of mobile units, (2) a desirable accuracy is achievable, (3) they can be relatively small, and (4) they are affordable. The major shortcomings of these systems are that the commercially available systems still have electromagnetic shielding and reflection problems.

Radio Frequency Identification

Radio frequency identification (RFID) first appeared in tracking and access applications during the 1980s. These wireless systems allow for noncontact reading of information. RFID has already established itself in a wide range of markets, including automated vehicle identification systems because of its ability to track moving objects. The technology will become an important construction industry management tool by allowing automated data collection, identification, and analysis of operations worldwide.

RFID technology provides the means to wirelessly transfer information via a tag or transponder and a reader (Fig. 17.7). RFID tags can be attached to almost any object including construction materials or equipment. In an RFID system, an identification transponder sends and receives bidirectional radio signals to a reader. RFID has advantages over other ID technologies by being able to perform in locations where vision is blocked or where surfaces become dirty. Current applications involve

- Reading tags as an object passes a fixed scanner, to record the movement of the object past the scanner
- Writing information on the tag that can be retrieved later
- Retrieving information from tags (at a distance of up to 73 m) using a mobile scanner. This expedites information gathering, or finding misplaced objects, because the operator does not have to go to the exact location of the object.

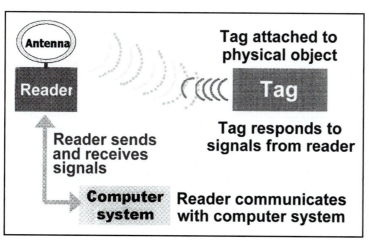

FIGURE 17.7 Model of a RFID system.

Construction materials supplier Granite Rock Co. has installed RFID tags on its trucks so that it can record and track material orders and speedily send trucks to the proper loading location in its quarries. RFID technology has been tested for material control on large process plant projects when there is a multitude of piping parts.

SMART STRUCTURES AND INTELLIGENT BUILDINGS

The use of sensors has been common to researchers building prototypes and models of structural systems. But now they are making a transition to mainstream engineering applications. Sensors are being used to create smart structures. A smart structure is a nonbiological physical structure having a definite purpose, a means to achieve that purpose, and a functioning pattern that mimics that of biological materials. A smart structure has three tasks: (1) receiving information, (2) processing the information, and (3) acting on the decision. There is not really a standard definition for an intelligent building.

Intelligent building systems use data from design, together with sensed data, to automatically configure controls and commission (i.e., start-up and check out) and operate buildings. Control systems will use robust techniques and will be based on smaller, cheaper, and more abundant sensors. Intelligent devices will use this wealth of data to ensure optimal building performance by continuously controlling building systems through the use of automated tools that detect and diagnose performance anomalies. The detector/controls will optimize operation across building systems, inform and implement energy purchasing, guide maintenance activities, and report building performance at the lowest possible cost.

The ultimate dream in the design of an intelligent building is to integrate four operating areas into one single computerized system:

- Energy efficiency
- Life safety systems
- Telecommunications systems
- Workplace automation

The constructor will build these systems or things into the structure for the use of the owner and occupants. There are also "smart structure" systems that support the construction effort. During the construction of structural members, fiber-optic strain gauge systems can be used to monitor prestressing during and after casting of the members as well as concrete temperature history and shrinkage in both precast and cast-in-place elements. At the completion of construction, the system can be used to measure deformations under load to verify design calculations. In addition to the benefits during construction, these fiber-optic systems remain in place for operational needs such as weighing in motion, incident detection, and overall structural health monitoring.

The Federal Highway Administration has developed a system of fiber-optic sensors but expects significant improvements in sensor technologies that will advance the ability of engineers to monitor the performance and "health" of bridges and structures. This technology is especially attractive because a large number of sensors can be placed on a single fiber, reducing the physical space required for cabling and installation labor costs. Furthermore, the sensors are very durable and are not subject to drift or electromagnetic interference. Prototype fiber-optic sensor systems that have been embedded in concrete bridges in the United States and in other countries have performed well. The sensors can be embedded directly in the concrete or can be attached to reinforcements before the concrete is placed.

Self-Healing Structures

Researchers are working to develop smart structural composite materials that self-heal when damaged. As an early example, researchers at the University of Illinois, who were inspired by biological systems in which damage triggers an autonomic healing response, have developed a synthetic material that can heal itself when cracked or broken. Their self-healing polymers hold the promise to mitigate the costly and inconvenient weakening effects of material "fatigue."

SIMULATION TECHNOLOGIES

By using computer graphics, engineers can form three-dimensional and four-dimensional visual simulations of projects, creating a virtual building process on their computer screens. Users get the impression they are moving through a particular design before the design is even implemented. Engineers can use what they learn from this virtual construction process to improve the actual process. Four-D visualization enables:

- Linking model objects with scheduled activities
- Simulating construction
- Communicating in a visual environment

The 3-D computer images are generated from two-dimensional (flat) computer-aided design files. The third dimension adds depth and texture to an image. The fourth dimension is the element of the simulation that allows the user to "walk through" the design.

Walt Disney Imagineering is the creative development, master planning, design, engineering, production, project management, and research and development area of the Walt Disney Company. Imagineering has been working with university researchers to develop a 4-D visualization system. The driver behind this effort is the cost of errors and omissions in project plans.

The process of moving from a blue-sky idea to a constructed project can be described as a series of *over-the-wall deliverables* (Fig. 17.8).

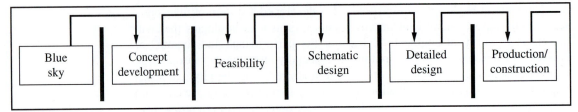

FIGURE 17.8 Over-the-wall deliverables.

The thinking that drove the Disney project was a dedication to improving project communication. Today, construction project communication can be described as

- Mostly by 2-D computer-aided design (CAD) files
- Many constituencies with different nomenclatures and standards
- Thousands of sheets and pages
- Thousands of paper-based transactions

The construction process is highly dependent on people. Because of this and the fact that there is a multitude of parties involved in the design and construction process, it is difficult if not impossible to share data and the reasoning.

> Our current communication procedures and systems do not capture the why and how behind construction project decisions.

The vision of construction planning and construction in the future is a process where there is *concurrent design and engineering:*

- Virtual and rapid prototyping
- Model-based processes in all phases of the life cycle
- Broadly shared data
- Upstream migration of constructor and operator input

Research work in this area is being pursued at Stanford University. Some of the work is demonstrated on the website www.stanford.edu/group/4D/index.shtml. The researchers there candidly state, "The difficulty and cost of creating and using such models is currently blocking their widespread adoption." But their work holds much promise to improve communication of design and schedule information.

TOTAL ELECTRONIC INTEGRATION

Computers are playing an increasing role in the construction process, especially in the design phase. Designers still use computer-aided design systems as drafting tools, mimicking manual drafting to a large extent. Automatic integration of design and construction is not possible without a radical change in this procedure. There are now integrated models that allow automatic extraction of construction-related data from the design database. This is a very different approach to design. These models describe a CIC (computer-integrated construction) CAD model integrated with systems that implement the model. The object-oriented model and the CAD/CIC interface eliminate the need for manual interfacing between the two computer-aided phases—design and computer-aided construction planning as well as computer-driven construction (e.g., robots and computer-controlled construction equipment). The model and the implementation systems improve the construction process by

- Reducing the number of steps involving manual data processing, thereby reducing many error sources and their consequent economic implications
- Automating and improving the design process, consequently lowering design costs and improving the design product
- Increasing compatibility of design
- Improving communication

Someday the design process will include the electronic integration of automatically collected physical information. For the constructor, the systems will generate construction quantities and bills of material, will simulate the actual assembly processes, and will even provide guidance control for construction equipment.

Wireless Networks on Construction Sites

Wireless local area networks (WLANs) provide the ability to interface various Web-based project systems from designer CAD systems to project planning and control tools. Most important, they allow speedy transfer of information to and from the field. Data are transmitted from devices by radio frequency communications. A basic system has a (1) mobile radio frequency terminal, (2) a base station, and (3) a network controller. WLAN systems will give construction managers and staff real-time access to data on Internet sites and the ability to directly enter new data.

Technical issues still to be resolved include:

- Interoperability amount systems and components
- Suitability of the equipment for use under construction site conditions
- Frequency interference and conflict

EMERGING TECHNOLOGIES

Not next year or even the year after, but in the future, the emerging technologies discussed here, and some yet to be considered, could greatly improve construction quality and productivity. These technologies, which are mainly being investigated in other disciplines, could significantly impact how the construction industry conducts business:

- Proactive computing
- Nonmonolith documents
- Artificial life schedule optimization
- Audit agents
- Holographic storage
- Nanotechnology
- Tacit knowledge extraction

Proactive Computing

All computing to date has been reactive; it responds to human commands explicitly or as a consequence of a programmed set of actions. The next generation of computing will be proactive; it will be given our goals and concerns and will be capable of reciprocating with advice. In so doing, it will learn the paths or processes by which it can assist its human user. A situation can easily be conceptualized where a project manager receives information about events that affect the schedule: examples would be delayed materials, equipment out of service, or an approaching storm system. Actions or responsive notifications would be generated automatically with a proactive system that understands the effects of these events. The computer would search for previous historical situations or for similar situations or events, assess the situation, and communicate to affected parties or programs such as the project schedule.

Intel Corporation believes there are seven key challenges that must be addressed to make proactive computing a reality.

1. Making it personal—empowering individuals and addressing their concerns over security and privacy
2. Closing the loop—bridging the gap between anticipating and acting on needs—predictably, and under human supervision
3. Anticipation—creating proactive software that anticipates our needs and produces answers before they are required
4. Dealing with uncertainty—using statistical modeling to deal with uncertainty inherent in the physical world
5. Planetary scale systems—developing software that works across a wide range of diverse platforms and networks
6. Deep networking—locally networking billions of embedded nodes; driving computing deeper into the infrastructure that surrounds us
7. Getting physical—connecting computers directly to the physical world around them

Nonmonolith Documents

Vannevar Bush's notion of hypertext, in the 1930s, quickly led to the idea of automatically linking one piece of information to another. Bush is the pivotal figure in hypertext research. His conception of the Memex introduced, for the first time, the idea of an easily accessible, individually configurable storehouse of knowledge. This idea is slowly being exploited. Bush saw the ability to navigate the enormous data store as a more important development than futuristic computer hardware. He described building a path to connect information of interest:

> When the user is building a trail, he names it, inserts the name in his codebook, and taps it out on his keyboard. Before him are the two items to be joined, projected onto adjacent viewing positions. At the bottom of each there are a number of blank code spaces, and a pointer is set to indicate one of these on each item. The user taps a single key, and the items are permanently joined. . . .
>
> Thereafter, at any time, when one of these items is in view, the other can be instantly recalled merely by tapping a button below the corresponding code space. Moreover, when numerous items have been thus joined together to form a trail, they can be reviewed in turn, rapidly or slowly, by deflecting a lever like that used for turning the pages of a book. It is exactly as though the physical items had been gathered together from widely separated sources and bound together to form a new book.

This passage is an apt description of the process of forming a link between nodes in today's hypertext packages.

The next step will be the situational assembly of information. The computer (our computer) would investigate workflow or a process, or would be given a goal and in response would automatically provide relevant data into spontaneous documents that would not otherwise exist.

Artificial Life Schedule Optimization

Traditional scheduling has an inescapable element that limits its effectiveness— the schedule is confined by the imagination and prior experience of its human creators. The addition of artificial life—systems that are modeled, copied, or adapted from biological life—to the computing process will allow mutations that could produce solutions not confined by intuition. Living systems have evolved small self-organizing construction systems to make hardware via self-referential information processing. If this artificial development of solutions technique were applied to construction, new approaches would surface that might yield operational improvements.

Holographic Storage

Researchers are working to develop systems of data storage and retrieval based not on two-dimensional digital methods, but using holographic techniques. Theoretical projections suggest that **holographic storage** will eventually provide the ability to store trillions of bytes in a piece of crystalline material the size of a sugar cube.

Instead of seeking information orthogonally (organized in row and column formats), it would be possible to search and retrieve from any angle and use multiple combinations of constituents. Unlike other technologies that record one

holographic storage
Systems of data storage and retrieval based on holographic techniques.

data bit at a time, holography records and reads over a million bits of data with a single flash of light. Light from a single laser beam is split into two beams, the signal beam (which carries the data) and the reference beam. The hologram is formed where these two beams intersect in the recording medium.

Holography has long held promise as a data storage technology. However, the major challenge has been the development of a suitable storage medium. In the summer of 2007, InPhase Technologies, Inc., of Colorado, which is a spin off of Bell Labs, began offering tapestry in the data storage marketplace. This new product could be the first step toward new initiatives in data mining.

There has been a growing interest in *data mining,* which requires sifting through extremely large warehouses of data to find relationships or patterns that might guide decision making and business process refinements. Mining techniques seek unusual patterns and relationships that are either counterintuitive or simply hidden in complex associations. Holographic memory's associative retrieval capabilities are increasingly attractive in support of management techniques such as data mining.

Nanotechnology

Nobel laureate and physicist Richard Feynman presented a vision of extreme miniaturization in his December 29, 1959, lecture "There's Plenty of Room at the Bottom." The lecture was given at the California Institute of Technology (Caltech) during the annual meeting of the American Physical Society and was first published in the February 1960 issue of Caltech's *Engineering and Science,* which owns the copyright. The full text of the lecture can be viewed at www .zyvex.com/nanotech/feynman.html.

nanotechnology
Precision construction having characteristic dimensions less than about 1,000 nanometers.

The word **nanotechnology** has become very popular since Feynman presented that idea and it is used to describe many types of research. One nanometer is about one ten-thousandth the width of a human hair or about the length of three atoms set side by side. This technology provides the opportunity to build within the structure itself, or within the machines we use, sensory tags for tracking quality or environment changes. Atomic precision construction could produce metal structures devoid of microimperfections, dramatically increasing strength. Bearings made to atomic precision would last far longer, run cooler, and bear greater loads. Crane components could have safety limit monitors incorporated into the boom structure. Similarly, performance-monitoring tags could be placed within the structural components of a building.

A National Nanotechnology Initiative (NNI) federal R&D program has now been established to coordinate the multiagency efforts in nanoscale science, engineering, and technology. NNI is managed within the framework of the National Science and Technology Council (NSTC), the cabinet-level council by which the president coordinates science, space, and technology policies across the federal government.

Researchers at the Idaho National Energy and Environment Laboratories are working with a new material they call nanosteel. Nanosteel is the result of applying a coating powder at high temperature to standard steels. The powder of particles approximately 50 nm in diameter is made from steel alloys. The result is a material about 30% stronger than standard steels.

Tacit Knowledge Extraction

Valuable information is commonly compartmentalized and buried in a very unorganized manner making it difficult if not impossible to retrieve. The explicit expression, tacit knowledge, was introduced by Polányi Mihály (1891–1976), in the 1930s when he wrote about the knowledge scientists carry in their brains but which is not easily articulated. Tacit knowledge is often described as hunches, intuition, and know-how. It is knowledge people have as a result of personal experience and culture.

Today e-mail is one of our largest repositories of information and it is our least organized data depository. There is some skepticism toward the extent to which this often highly subjective knowledge and learning can actually be made explicit. But maybe the issue is that the data source that caused the hunch is not recognized.

Techniques are being developed that allow the extraction of tacit information. Once that is accomplished, it is possible to organize the information, restructure it, and even mine it.

DISTANCE LEARNING

Today the World Wide Web is the fastest growing information environment. As computer technology becomes a common appliance and Internet connectivity becomes a ubiquitous utility, the ability to deliver learning programs to learners in their workplace and in their homes is increasing dramatically. Distance learning will expand and use new multimedia and simulation capabilities. These capabilities will complement and enhance the effectiveness of traditional forms of teaching without replacing them. There will be an increase in visual and inductive learning. There will be significant changes in engineering and construction education because of these technological changes, but it must be remembered that education is not merely making information available.

There will also be a requirement that constructors gain a better understanding of sociopolitical and economic issues affecting the building of projects. Communication capability was always important and will continue to be a critical skill that must be developed during the education process.

SUMMARY

These advanced technologies, innovative methods, and other concepts yet to be conceived must be developed and employed in construction to meet the public's demand for quality infrastructure without the inconvenience of the construction process. Innovative and more flexible contracting practices such as design-build,

build-operate-transfer, or bonuses for early completion are already being used around the world. These are improved business methods. It is also necessary to develop improved construction methods for actually putting the bricks and mortar in place. Many of the future technologies that address better building practices will be derivatives of research conducted to support other fields of endeavor, as is the case with many of the navigational control ideas coming from the aerospace industry. Designers and constructors must make full use of these advanced technologies to competitively complete projects quickly with minimum disruption to the public. The future is speed, quality, and lifetime economics.

REVIEW QUESTIONS

17.1 List the four basic components of internal navigation systems.

17.2 Ground-based radio frequency navigation systems use the same principles as what other positioning system?

17.3 What are the complete names for the technologies to which we apply the abbreviations ABS, INS, and RFID?

17.4 A lighthouse would be classified as what type of navigation system?

17.5 Name the main advantages of GPS.

17.6 Ground-based radio frequency systems use the same principles as GPS. Name two global-scale RF systems.

17.7 What are two major shortcomings of commercial ground-based radio frequency systems?

17.8 In reference to construction management, name three important capabilities of RFID technology.

17.9 The ultimate dream in the design of an intelligent building is to integrate four operating areas into a single computerized system. Name the four operating areas.

17.10 Four-D visualization enables engineers to improve the construction process in three specific ways. Name these three process improvements.

17.11 Name four emerging technologies that could significantly impact how the construction industry conducts business.

17.12 Vannevar Bush is associated with what idea that could support information recovery?

17.13 Is it true or false that holographic data storage techniques arrange information orthogonally?

17.14 Here are 25 technologies that John Voeller has identified as likely to impact construction in the next decade. Investigate one of these technologies in detail and prepare a two-page report.

■ XML variants
■ Nonmonolith documents

- Decision indexing
- Genetic schedulers
- Audit agents
- Virtual proximity
- Wearable computers
- Multiphysics analysis
- Optical cognition
- Exoskeleton robot
- Quantum encryption
- Observation-based training
- Digital signature/seal
- Proactive computing
- Site-hardened voice recognition
- Artificial life schedule optimization
- Goal-based search
- Situation simulation
- Holographic storage
- Molecular mechanicals/electricals
- Nanotechnology
- Human/robot collaboration
- Tacit knowledge extraction
- Virtual hall
- Communications convergence at broadband wireless

REFERENCES

1. Borg, R. F., J. Gambatese, K. Haines, Jr., C. T. Hendrickson, J. Hinze, A. Horvath, E. Koehn, and S. L. Moritz (2002). *Rebuilding the World Trade Center,* Technical Report. Construction Institute, American Society of Civil Engineers.

2. Brodetskaia, Irina (2004). "An Examination of the Potential for Monitoring Construction Project Performance by Monitoring Lifting Equipment," Research Thesis, The Technion—Israel Institute of Technology, Haifa, July.

3. Burnell, Scott R. (2002). "Nanocoating Could Strengthen Buildings," *Small Times,* April.

4. Dry, C. (1996). "Procedures Developed for Self-Repair of Polymeric Matrix Composite Materials," *Composite Structures,* 35.

5. Fang, H., S. Azarm, and L. E. Bernold (1994). "Multilevel Multiobjective Optimization in Precast Concrete Wall Panel Design," *J. Engineering Optimization,* Gordon and Breach Science Publishers, S.A., Vol. 22.

6. Finch, E., R. Flanagan, and L. Marsh (1996). "Auto-ID Application in Construction," *Construction Management and Economics,* 14(2):121–129.

7. Goodrum, P. M., and Haas, C. T. (2004). "Long-term Impact of Equipment Technology on Labor Productivity in the U.S. Construction Industry at the Activity Level." *Journal of Construction Engineering and Management,* ASCE, 130(1), 124–133.

8. Hendrickson, Chris T., Arpad Horvath, Satish Joshi, and Lester B. Lave (1998). "Use of Economic Input-Output Models for Environmental Life Cycle Assessment," *Environmental Science & Technology,* April.

9. Horvath, Arpad, and Chris T. Hendrickson (1998). "A Comparison of the Environmental Implications of Asphalt and Steel-Reinforced Concrete Pavements," *Transportation Research Record.*

10. Horvath, Arpad, and Chris T. Hendrickson (1998). "Steel vs. Steel-Reinforced Concrete Bridges: An Environmental Assessment," *Journal of Infrastructure Systems,* ASCE, 4(3):111–117.

11. Jaselskis, E. J., M. R. Anderson, C. T. Jahren, Y. Rodrigues, and S. Njos (1995). "Radio-Frequency Identification Applications in Construction Industry," *Journal of Construction Engineering and Management,* ASCE, 121(2):189–196.

12. "New Mexico Airport Project Utilizes Stringless Paving" (2001). *ENR,* 23 April, 14.

13. Parish, D., and R. Grabbe (1993). "Robust Exterior Autonomous Navigation," *Proceedings of the 1993 SPIE Conference on Mobile Robots,* Boston: 280–291.

14. Schexnayder, Cliff. J., and James Ernzen (1999). *Mitigation of Nighttime Construction Noise, Vibrations, and Other Nuisances,* Synthesis of Highway Practice 218, Transportation Research Board, National Research Council, Washington, DC.

15. Schexnayder, Cliff. J., and Scott A. David (2002). "Past and Future of Construction Equipment—Part IV," *Journal of Construction Engineering and Management,* ASCE, 128(4):279–286.

16. Spillman, W. B. Jr., J. S. Sirkis, and P. T. Gardiner (1995). "The Field of Smart Structures as Seen by Those Working in It: Survey Results," *Smart Structures and Materials 1995: Smart Sensing, Processing, and Instrumentation,* Proceedings of SPIE, Vol. 2444, The International Society for Optical Engineering, Bellingham, WA.

17. Tadashi, Kanzki, Nishizawa Shuiti, and Toida Hiroshi (1993). "Survey Robot Using Satellite GPS," *Proceedings of the 10th International*

Symposium on Automation and Robotics in Construction, Houston: 479–486.

18. Tingle, Jeb S., and Travis A. Mann (2001). *Expedient Airfield Construction Using the Computer-Aided Earthmoving System,* ERDC/GSL TR-01-20, U.S. Army Corps of Engineers, Washington, DC.

19. Voeller, John (2001). "Technologies to Watch," *Civil Engineering,* ASCE, Aug.

20. White, S. R., et al. (2001). "Autonomic Healing of Polymer Composites," *Nature* 409.

21. Wiezel, Avi, and Hugo Arellano (2001). "Contractors Are Using Data Mining to Increase Profits," *Proceedings 2001 Conference,* Canadian Society for Civil Engineering, CD paper C54, June.

WEBSITE RESOURCES

1. www.ibuilding.gr/index.html European Intelligent Building Group communicates information concerning intelligent buildings to owners, designers, and constructors.

2. www.fhwa.dot.gov/reports/designbuild/designbuild.htm *Design-Build Effectiveness Study* (2006). Final Report, Prepared for: USDOT—Federal Highway Administration, January.

3. www.post-gazette.com/pg/06169/698927-84.stm Amy Goldstein (2006). "Privatization Backlash in Indiana," *The Washington Post,* June 18.

4. www.leica-geosystems.com/us Leica Geosystems is a global surveying technology group. It provides systems for highly accurate 3-D data capturing, visualization, and modeling of space-related data (GIS).

5. cic.nist.gov Computer-Integrated Construction Group (CICG) at the National Institute of Standards and Technology (NIST). NIST is an agency of the U.S. Commerce Department's Technology Administration and the CICG is part of the Building Environment Division.

6. www.fiatech.org/about.htm Fully Integrated and Automatic Project Processes (FIATECH) is an industry-led, collaborative, not-for-profit research consortium serving the construction industry from asset creation through operation and maintenance. Its mission is to achieve significant cycle time and life-cycle cost reductions and efficiencies in capital projects from concept to design, construction, operation, and even to decommissioning and dismantling of capital facilities.

7. www.smartcommunities.ncat.org Center for Excellence in Sustainable Development, a project of the U.S. Department of Energy offers strategies for sustainability.

8. nano.gov   National Nanotechnology Initiative (NNI). It emphasizes long-term, fundamental research aimed at discovering novel phenomena, processes, and tools; addressing NNI Grand Challenges; supporting new interdisciplinary centers and networks of excellence including shared user facilities; supporting research infrastructure; and addressing research and educational activities on the societal implications of advances in nanoscience and nanotechnology.

Glossary

A

accounting   The practice of classifying, reporting, and interpreting the financial data of an organization.

active beacon systems   Determination of a mobile unit's position using three or more transmitters at known locations, and one receiver on the unit.

actual costs of work performed   A graph showing the real costs of work accomplished plotted against project time.

additive or deductive   Amounts that are added to or subtracted from the base bid on a project to ensure the bid is within the capital budget.

affirmative action   Actions required of employers to maintain a workforce that accurately represents the racial and ethnic makeup of the available hiring population.

AFL-CIO   American Federation of Labor and Congress of Industrial Organizations. Federation of America's 13 million union members.

AIA Document A201   The most common general conditions. May be purchased from the American Institute of Architects for use in contracts.

American Society of Testing Materials (ASTM)   Tests engineering materials and writes standard test methods and specifications.

Americans with Disabilities Act   Requires reasonable accommodations for persons with disabilities. Requirements extend to employees and the general public.

aqueduct   A bridge or channel for conveying water, usually over long distances.

architect   A person who designs structures; must have the ability to conceptualize and communicate ideas effectively to clients, engineers, government officials, and construction crews.

as-built drawings   Completed by the contractor as the work progresses and submitted to the owner at the completion of the work as a permanent record of the as-built conditions and dimensions.

asphalt macadam   Compressed layers of broken rock held together by spraying the layers with liquid asphalt.

asphalt shingles   Composition roof shingles made from asphalt-impregnated felt covered with mineral granules.

asset   A property or economic resource owned by a company.

Associated General Contractors of America, The   National association of general contractors devoted to the improvement of construction quality in America.

average end area   A calculation method for determining the volume of material bounded by two cross sections or end areas.

B

backward pass   A schedule calculation that determines the late start and late finish times of the precedence diagram activities under the condition that the project's minimum duration be maintained.

balance line   A horizontal line of specific length that intersects the mass diagram in two places.

balance sheet   A financial report showing the assets, liabilities, and owner's equity of a company on a specific date.

balloon framing   The building-frame construction in which each of the studs are one piece from the foundation to the roof of a two-story house.

bar chart   A graphical representation of planned construction activities, the estimated activity durations, and the planned sequence of activity performance.

base bid   Used for projects that may be in danger of exceeding the capital budget. Provides a mechanism for the owner to accept bids on a portion of the facility and not construct some of the features that would cause the budget to be exceeded.

baseboard   Finish board covering the interior wall where the wall and floor meet.

beam   A horizontal structural member that carries a load.

bearing wall or partition   A wall supporting any vertical load other than its own weight.

benchmark   Mark on some permanent object fixed to the ground from which land measurements and elevations are taken.

best value procurement   Procurement system based on criteria such as performance, quality of work, timeliness, and customer satisfaction, in addition to price.

bid   The offer of a bidder, submitted in the prescribed manner, to furnish all labor, equipment, and materials to complete the specified work.

bid bond   A bond that assures that the contractor will accept the contract when awarded and will supply the required performance and payment bonds.

bid depository   An organization used by owners, general contractors, subcontractors, and materials suppliers to help ensure fair bidding.

bid opening   The opening and tabulation of bills at the prescribed time and place.

bid shopping   Unethical procedure of requesting preferred subcontractors to lower bids to meet or beat bids submitted to the GC by another firm.

black powder   An explosive mixture composed of potassium nitrate, sulfur, and charcoal.

blown-in insulation   Type of insulation placed over ceilings. Placed by blowing in through hoses with compressed air.

blueprints   Architectural drawings used to graphically depict construction work. The original drawing is transferred to a sensitized paper that turns blue with white lines when printed.

board measure   System of lumber measurement. The unit is 1 BF, which is 1 ft square by approximately 1 in. thick.

bonding company   Provides bid, performance, and payment bonds to contractors. *See also* Surety.

budgeted cost of work performed   A graph showing the planned (estimated) cost of the work allocated to completed activities.

budgeted cost of work schedule   A graph showing how the project is budgeted to be accomplished according to the original master schedule.

building codes   Community or regional codes to establish minimum standards of design and construction.

built-up bituminous roof   Roofing for low-slope roofs composed of several layers of felt and hot asphalt or coal tar, usually covered with small aggregate.

Bureau of Labor Statistics   Part of the Department of Labor. Publishes statistics relating to labor in all industries, including construction.

C

caisson   A watertight, dry chamber in which people can work below the level of the water table.

cash flow analysis   Process for analyzing a project to determine the amount of working capital that will be required each month.

cast iron   A hard, brittle, nomalleable iron-carbon alloy. It is cast into shapes.

change orders   Directives issued by the owner to change the contract by adding or subtracting features within the scope of work.

coefficient of traction   The factor that determines the maximum possible tractive force between the wheels or tracks of a machine and the surface on which it is traveling.

commodity   Product differentiated by price only.

comparative statement   A financial statement with data for two or more successive accounting periods placed in columns side by side to illustrate changes.

conceptual estimate   A cost estimate prepared from only a conceptual description of a project. This is done before plans, specifications, and other project details have been fully developed.

concrete masonry units   Concrete units used in masonry for outdoor walls, external building walls, and other such purposes.

construction documents   The drawings and specifications setting forth the requirements for constructing the project.

construction management at risk Construction manager who also acts as the general contractor, with a guaranteed maximum price.

construction manager A person who coordinates the construction process on behalf of the owner.

construction mortgage Bank loan to finance construction, secured by the facility.

contractor An individual or firm undertaking the execution of construction work under the terms of a contract.

Copeland Act The Copeland Act makes it illegal for an employer to require a "kickback."

cost accounting The phase of accounting that deals with collecting and controlling the cost of producing a product or service.

course A continuous row of stone or brick of uniform height.

critical-cost activity Any single activity that has the potential to impact the estimated cost of the project by more than ½ of 1%.

critical path Path of activities on a critical path diagram that defines the duration of the project. Critical path activities have no float.

critical path method A project scheduling method where activities are arranged based on interrelationships and the longest time path through the network called the critical path is determined.

cross sections Earthwork drawings created by combining the project design with field measurements of existing conditions. These are typically views at right angles to the centerline of the project or a major project feature.

cross section view A construction drawing depicting a vertical section of earthwork at right angles to the centerline of the work. Used together with centerline distances to calculate earthwork quantities.

CSI The Construction Specifications Institute provides technical information to enhance communication among all disciplines of nonresidential building design and construction.

current ratio The relation of a company's current assets to its current liabilities.

D

Davis Bacon Act Requires workers on federally financed projects to be paid the "prevailing wage."

delayed payment Sometimes because of credit or cash flow problems, owners fail to pay contractors for services according to the schedule set by the contract. To justify it, they may claim the contractor's work is defective or substandard.

depreciation An accounting method used to describe the loss in value of an asset.

design-bid-build Conventional procurement system in which the owner contracts separately with a designer and then the contractor.

design-build Procurement system in which the owner contracts with a contractor/designer as a single unit, under one contract.

design-build contracting *See* Design-build.

dimensioned lumber Framing lumber that is of nominal thickness.

direct costs All cost elements that can be associated with a specific item of project work.

drawbar pull The available pull that a crawler tractor can exert on towed load.

dressed lumber Lumber machined and smoothed at the mill, to usually ½ in. less than nominal (rough) size.

drywall construction Interior wall covering other than plaster, usually referred to as "gypsum board" or "wallboard."

dummy activities A pseudoactivity with duration of zero. A dummy is a dotted line arrow and used solely to indicate sequence.

E

earned value approach Analysis of the contract-completed activities to determine the cost to date and compare with the revenue to date. The difference is the earned value.

elevation A drawing showing either the front, sides, or rear face of a building.

engineer's estimate A forecast of construction costs prepared on the basis of a detailed analysis of materials, equipment, and labor for all items of work.

equity A right, claim, or interest in property.

estimator Construction company employee whose primary assignment is to estimate the cost of projects and change orders for bids and negotiations.

ethics All construction and design organizations publish a code of ethics to guide the behavior of their members. Ethics involve doing the right thing and protecting the public.

expense Goods or services consumed in operating a business.

experience modifier rating (EMR) Factor used to calculate a contractor's cost of worker's compensation insurance, based on the accident experience of the company.

F

Fair Labor Standards Act (Wage and Hour Law), 1935   This legislation contains minimum wage, maximum hours, overtime pay, equal pay, and child-labor standards.

feasibility estimate   An estimate of project cost that is prepared before complete construction documents are available.

fill factor   A numerical value used to adjust rated heaped excavator bucket capacity based on the type of material being handled and the type of excavator.

float   The time flexibility of activity performance that states the maximum allowable for not delaying a following activity or the project.

foreman   Supervisor of a specific trade group of workers.

forward pass   A schedule calculation that determines the earliest start and finish time of the precedence diagram activities and the minimum project duration.

four-dimensional (4-D) visualization   A process for capturing both 3-D data sets (space) and the fourth dimension of time.

free float   The time duration that an activity can be allayed without delaying the project's completion and without delaying the start of any succeeding activity.

fringe benefits   Employee benefits that an employer must pay in addition to an employee's base pay. Typically these may be paid directly to the employee or they may be paid to various agencies on behalf of the employee. They would include health insurance, pension plans, and certain taxes.

front loading   Practice of assigning high values to activities that are completed early and compensated for by assigning low values to activities that are completed late in the project. The result is higher monthly pay estimates.

G

general and administrative overhead   The contractor's general operating expenses that are not related to a specific project but are incurred in operating the business.

general conditions   That part of the contract documents that defines the rights, responsibilities, and relationships of all parties to the construction contract.

general contractor   Contractor who accepts cost estimates from the subcontractors, and then signs the general contract with the owner. By doing so, the GC assumes responsibility for the entire project.

global positioning system (GPS)   A highly precise satellite-based navigation system.

Glulam   Trade name for laminated wood products.

grade assistance   The effect of gravitational force in aiding movement of a vehicle down a slope.

grade beam   Portion of reinforced concrete foundation that supports walls or the exterior edges of a slab. Often constructed without forms by excavating the earth to the required width and depth.

grade resistance   The force-opposing movement of a machine up a frictionless slope.

guaranteed maximum price (GMP)   Type of contract generally used by construction managers, who are selected during or before the design process. The GMP is used because the incomplete design does not permit detailed estimating.

H

Hawthorne effect   Increase in productivity caused by the presence of researchers.

holographic storage   Systems of data storage and retrieval based on holographic techniques.

hydraulic fill   Constructed by transporting in a water suspension the soil material for an embankment or dam. Placement of the material is accomplished by allowing it to settle out of suspension in a confined area.

I

incentive contracts   Contracts with a bonus for early completion, A + B contracts that reward contractors who know to accomplish the work most efficiently, or contracts that offer a higher pay based on quality.

income statement   A financial statement showing revenues earned by a business, the expenses incurred in earning the revenues, and the resulting net income or net loss.

indirect costs   Overhead or auxiliary costs incurred in completing a project but not attributable to any specific tasks.

inertial navigation system   Using a system of gyroscopes, accelerometers, a navigation computer, and a clock, the autonomous navigation of the mobile unit is accomplished from a known initial position.

inspector   Employee of the owner whose function is to inspect the work as it progresses, calculate the quantity of materials used, and assess the quality of the work.

interfering float The time available to delay an activity without delaying the projects completing but delaying the start of following noncritical activities.

International Code Council The ICC was established in 1994 as a nonprofit organization dedicated to developing a single set of comprehensive and coordinated national model construction codes.

iron A cheap, abundant, useful, and important metal but pure iron is only moderately hard. Iron is used to produce other alloys, including steel.

ISO 9000: 2000 Quality standards published by the International Organization for Standardization.

J

jackhammer An air- (originally) or electric- (today) driven reciprocating percussion tool used for drilling, or breaking rock, concrete, brick, or asphalt.

job order contracts Contracts used mainly for maintenance work. Contractors submit a bid in the form of a decimal multiplier for the owner-published book of standard rates for work. The contractor then performs work against a series of work orders. This process is designed to avoid a design contract for each work order.

just-in-time procurement Procurement process for common items to guarantee delivery within 24 hours and thereby reduce the need for storage at the construction site.

L

labor cost Wages paid directly to on-site personnel, including fringes.

laminated wood products Includes products such as plywood, oriented strand board, and glued beams.

lease A long-term agreement for the use of an asset.

liability A debt owed.

linear schedule A simple diagram that shows the relationships between activities, their duration, and the space where they take place at a given time.

liquidated damages Daily amount, specified by the contract, paid by the contractor to the owner to compensate the owner for late completion. Liquidated damages are not penalties. They represent compensation for expenses and lost income.

load-bearing wall Any wall that carries some of the weight of the structure above.

lock At a location where the water elevations up- and downstream in a waterway are different (at a dam), a structure closed off by gates in which, by controlling its inner water level, it is possible to raise and lower vessels.

lump sum contract Type of contract commonly used for buildings where the contractor can accurately estimate the costs.

M

mass diagram In earthwork calculations, a graphical representation of the algebraic cumulative quantities of cut and fill along the centerline, where cut is positive and fill is negative. Used to calculate haul in distance.

mass excavation The requirement to excavate substantial volumes of material, usually at considerable depth or over a large area.

millwright A carpenter having knowledge of gear ratios so that he could construct the machinery of a mill.

multiple prime contractors Contracting system that eliminates the general or prime contractor. The owner contracts individually with the specialty contractors.

N

nanotechnology Precision construction having characteristic dimensions less than about 1,000 nanometers.

National Labor Relations Act Established the right of workers to join a union without fear of management reprisal.

National Labor Relations Board The NLRB was established by the National Labor Relations Act. It has two major functions: (1) it oversees representation elections to determine whether groups of employees wish to be represented by a union and (2) it prevents and remedies unfair labor practices, as defined by the Labor Management Relations Act (Taft-Hartley Act).

negotiated contracts Contracts that are awarded through a process of negotiation rather than through a bid process. Commonly used in the private sector.

node On an activity-on-arrow schedule, the nodes mark the start and end points of an activity. With precedence diagramming, the nodes represent the activities of the schedule.

nominal size Size of lumber before dressing, rather than its actual or finished size.

Norris-LaGuardia Act The Norris-LaGuardia Act was the first legislation to limit the power of the federal courts to issue injunctions against union activities in labor disputes.

Notice to Proceed Document issued to the contractor after award of the contract. Formally authorizes access to the site and establishes the project start date.

O

O&O The ownership and operating costs of a machine.

on center Method of indicating spacing of framing members by stating the distance from the center of one to the center of the next.

open shop Contracting business that does not require union membership.

oriented strand board (OSB) Glued wood product (Glulam), manufactured by gluing together wood chips that are all oriented in the same plane. Typical dimensions are 4'x 8'. Used in place of more expensive plywood for shear walls and roofs on homes and small commercial buildings.

OSHA Safety and Health Standards Code of Federal Regulations, Title 29, Part 1910. Contains the safety and health regulations that pertain to construction.

outriggers Movable beams that can be extended laterally from a mobile crane to stabilize and help support the unit.

outriggers Movable beams that can be extended laterally from a mobile crane to stabilize and help support the unit.

overbilling Placing a higher value on work performed early in the project and lower values on work to be completed near the end of the project so that the contractor does not experience out of pocket on cash flow—have to finance the project.

owner's equity The equity of the owner or owners of a business in the assets of that business.

P

partnering Management process based on trust.

payment bonds Guarantee that the contractor will pay the subcontractors and material suppliers.

performance bonds Guarantee that the contractor will perform the work in accordance with the plans and specifications.

performance charts Graphical representations of the power and corresponding speed the engine and transmission of a machine can deliver.

periodic pay estimates Payment to the contractor during the work, normally monthly, and based on the progress to date.

permanent materials Materials that are installed permanently in the completed project.

pilaster Rectangular pier attached to a wall for the purpose of strengthening the wall; also a decorative column attached to a wall.

piling May be steel H-beams, wood poles, or concrete caissons, driven or drilled into the ground, often to bedrock, to support a foundation of a structure.

pitch Slope of a roof usually expressed as a ratio.

plan view A construction drawing representing the horizontal alignment of the work.

plans and specifications Contract drawings and written materials descriptions, which convey the detailed requirements to the contractor.

platform framing Framing in which each story is built upon the other.

prebid conference A meeting held prior to bid opening for the purpose of explaining the project and answering questions that bidders have with respect to the contract documents and the work.

predecessor activity An activity that must be completed before a given activity can be started.

preliminary schedule A schedule prepared to support development of the project cost estimate.

prevailing wage Determined by the Department of Labor and published in periodic wage decisions. Apply to federally funded contracts.

productivity A measure of the progress of a crew or contractor, measured against a standard such as the project schedule.

productivity efficiency factor Productivity at $x\%$ overtime divided by productivity at 40 hr per week.

productivity trend line Mathematically calculated trend line depicting the productivity of crews working overtime, as a percentage of their productivity at 40 hr per week.

profile view A construction drawing depicting a vertical plane cut through the centerline of the work. It shows the vertical relationship of the ground surface and the finished work.

profit The amount of money, if any, that a contractor retains after completing a project and paying all costs for materials, equipment, labor, and overhead.

profit (loss) to date Calculation of the profit to date based on the progress and expenditures to date, compared to the expected progress and expenditures to date.

project engineer Entry-level contractor management position at the site. Many times the first assignment for a new graduate is as project engineer.

project labor agreement   Agreement between the contractor and owner on a large project to hire workers through the union hiring hall in exchange for a no-strike agreement.

project manager   Senior construction company manager at the site, in overall charge of the project.

public works   Publicly funded construction.

Q

quality assurance   Management programs designed to produce high-quality work.

quality control   System of inspection to determine the quality of completed work.

quantity survey   The process of calculating the quantity of materials required to build a project.

quote   To make an offer at a guaranteed price.

R

R value   Numerical measure of the insulating characteristics of materials.

radio frequency identification   A wireless system that permits noncontact reading of information.

relative cost factors   Relate the cost of output to the average hourly cost.

request for information (RFI)   Formal document used by a contractor to clarify contract requirements.

resident engineer   Senior representative of the owner, usually on public works projects.

retainage   A project owner, in order to have protection that the contractor will complete the work, often holds back (retains) a portion of what a contractor earns until the end of the project. Ten percent is a common retainage percentage value.

retarder chart   A graph that identifies the speed that can be maintained when a machine is descending a grade having a slope such that the magnitude of the grade resistance is greater than the rolling resistance.

revenue   An inflow of assets, not necessarily cash, in exchange of goods and services sold.

rimpull   The tractive force between the rubber tires of a machine's driving wheels and the surface on which they travel.

risk management   Process of sharing risks, usually based on insurance for high-risk areas such as equipment loss or liability matters.

rolling resistance   The resistance of a level surface to constant-velocity motion across it.

S

scaffolding   Temporary structure, designed and built by the contractor, to provide elevated working space.

scope definition   Owner responsibility to develop and convey to the architect in sufficient detail to allow the design to progress.

shear wall   Walls designed to resist horizontal shear, or distortion from horizontal loads, such as wind loads.

sheathing   Barrier driven into the ground to provide support to the vertical sides of an excavation.

sheave   Grooved pulley-wheel for changing the direction o a wire rope's pull.

shields   A structural system designed to protect the workers should an excavation in which they are working collapse.

shop drawings   Drawings that supplement the contract drawings to identify fabrication details, makes and models, colors, and other details as required by the contract.

shoring   A structural system that applies pressure against the walls of an excavation to prevent collapse of the soil.

slewing   To turn or rotate on an axis.

specialty contractors   *See* Subcontractor.

spread   A group of construction earthwork machines that work together to accomplish a specific construction task such as excavating, hauling, and compacting material.

square   In roofing, 100 sf of roofing.

station   A horizontal distance of 100 ft.

steel   Carbon steel is an alloy of iron with small amounts of Mn, S, P, and Si. Carbon is the major variable that distinguishes between the properties of iron and steel, a very strong metal.

strike   A work stoppage by a body of workers to enforce compliance with demands made on an employer.

stripping   The upper layer of organic material that must be removed before beginning an excavation or embankment.

structural excavation   Excavation undertaken in support of structural element construction, usually involves removing materials from a limited area.

studs   Vertical framing members in a wall spaced at 16" or 24" on center.

subcontractor   Specialty contractor performing under contract to the GC.

successor activity   An activity that cannot start until a given activity is competed.

superintendent Person in overall charge of the day-to-day detailed management of the construction process. Supervises the foremen.

supplies Materials used during the construction process that are not permanently incorporated into the completed facility.

surety A firm executing a bond, payable to the project owner, securing performance of the contract or securing payments for labor and materials.

T

tilt-up concrete panel construction Walls, usually for commercial buildings, formed in place on the building concrete floor, and raised by crane into place.

total float The amount of time that an activity can be delayed without delaying project completion.

total quality management (TQM) Quality system based on continuous improvement originated by W. Edwards Deming.

tri-nitroglycerin An extremely sensitive-to-shock explosive compound produced by mixing glycerine with sulphuric and nitric acid.

two-step procurement Procurement process based on submission of a design, selection of the preferred design by the owner, and then submission of a price.

U

ultimate strength of concrete The compressive strength of concrete continues to increase slowly for many months. The ultimate strength may be higher than the 28-day strength, which is used for design calculations.

unfair labor practices Under the National Labor Relations Act, both management and labor can be prosecuted for unfair labor practices, such as for restraint of labor's right to organize.

Uniform Building Code (UBC) First enacted by the International Conference of Building Officials in 1927. The code is dedicated to the development of better building construction and greater safety to the public by uniformity of building laws.

unit cost The estimated or actual cost of a single unit of work.

unit-price contract Contract used for heavy/highway projects that do not allow the contractor or the owner to know the exact quantities of materials before the start of work. Bidding is based on the engineer's estimated quantities.

universal machines The base machine can be used as a crane or dragline and for pile driving or other such applications.

utility entrances The points where utilities such as electricity, water, and sewer penetrate the foundation or outer wall of the building.

V

value analysis Used during construction to estimate the contractor earnings to date. Construction activities are assigned a value at the beginning of the project. Completed activities are added up.

value engineering Evaluating a project for the purpose of bettering the objective in terms of cost and functional parameters.

vapor barrier Plastic sheets placed either inside or outside the insulation, depending on the climate, to protect the insulation and building from excessive moisture.

velocity diagram A graphic for monitoring the relationship between time and the accomplishment of an activity.

W

Williams-Steiger Act Known as the Occupational Safety and Health Act. Passed in 1970.

work breakdown structure A hierarchical breakdown of the scope of work for a project.

worker's compensation insurance Insurance that pays employee medical bills and a portion of hourly wage in case of job-related injury.

wrought iron A tough and malleable metal. Wrought iron contains only a few tenths of a percent of carbon compared to steel, which contains about 1% carbon.

Selected Unit Equivalents

Unit	Equivalent
1 acre	43,560 square feet
1 atmosphere	14.7 lb per square inch
1 Btu	788 foot-pounds
1 Btu	0.000393 horsepower-hour
1 foot	12 inches
1 cubic foot	7.48 gallons liquid
1 square foot	144 square inches
1 gallon	231 cubic inches
1 gallon	4 quarters liquid
1 horsepower	550 foot-pounds per second
1 metric ton	2,204 pounds
1 mile	5,280 feet
1 mile	1,760 yards
1 square mile	640 acres
1 pound	16 ounces avoirdupois
1 quart	32 fluid ounces
1 long ton	2,240 pounds
1 short ton	2,000 pounds

APPENDIX C

AIA Document
A101-2007

Document A101™ – 2007

Standard Form of Agreement Between Owner and Contractor *where the basis of payment is a Stipulated Sum*

AGREEMENT made as of the   day of
in the year
(In words, indicate day, month and year)

BETWEEN the Owner:
(Name, address and other information)

and the Contractor:
(Name, address and other information)

for the following Project:
(Name, location, and detailed description)

The Architect:
(Name, address and other information)

The Owner and Contractor agree as follows.

1

TABLE OF ARTICLES

ARTICLE 1   THE CONTRACT DOCUMENTS

The Contract Documents consist of this Agreement, Conditions of the Contract (General, Supplementary and other Conditions), Drawings, Specifications, Addenda issued prior to execution of this Agreement, other documents listed in this Agreement and Modifications issued after execution of this Agreement, all of which form the Contract, and are as fully a part of the Contract as if attached to this Agreement or repeated herein. The Contract represents the entire and integrated agreement between the parties hereto and supersedes prior negotiations, representations or agreements, either written or oral. An enumeration of the Contract Documents, other than a Modification, appears in Article 9.

ARTICLE 2   THE WORK OF THIS CONTRACT

The Contractor shall fully execute the Work described in the Contract Documents, except as specifically indicated in the Contract Documents to be the responsibility of others.

ARTICLE 3   DATE OF COMMENCEMENT AND SUBSTANTIAL COMPLETION

§ 3.1 The date of commencement of the Work shall be the date of this Agreement unless a different date is stated below or provision is made for the date to be fixed in a notice to proceed issued by the Owner.
(Insert the date of commencement if it differs from the date of this Agreement or, if applicable, state that the date will be fixed in a notice to proceed.)

If, prior to the commencement of the Work, the Owner requires time to file mortgages and other security interests, the Owner's time requirement shall be as follows:

§ 3.2 The Contract Time shall be measured from the date of commencement.

Init.

/

2

§ 3.3 The Contractor shall achieve Substantial Completion of the entire Work not later than
(   ) days from the date of commencement, or as follows:
(Insert number of calendar days. Alternatively, a calendar date may be used when coordinated with the date of commencement. If appropriate, insert requirements for earlier Substantial Completion of certain portions of the Work.)

, subject to adjustments of this Contract Time as provided in the Contract Documents.
(Insert provisions, if any, for liquidated damages relating to failure to achieve Substantial Completion on time or for bonus payments for early completion of the Work.)

ARTICLE 4 CONTRACT SUM
§ 4.1 The Owner shall pay the Contractor the Contract Sum in current funds for the Contractor's performance of the Contract. The Contract Sum shall be
Dollars ($), subject to additions and deductions as provided in the Contract Documents.

§ 4.2 The Contract Sum is based upon the following alternates, if any, which are described in the Contract Documents and are hereby accepted by the Owner:
(State the numbers or other identification of accepted alternates. If the bidding or proposal documents permit the Owner to accept other alternates subsequent to the execution of this Agreement, attach a schedule of such other alternates showing the amount for each and the date when that amount expires.)

§ 4.3 Unit prices, if any:
(Identify and state the unit price; state quantity limitations, if any, to which the unit price will be applicable.)

Item	Units and Limitations	Price Per Unit

§ 4.4 Allowances included in the Contract Sum, if any:
(Identify allowance and state exclusions, if any, from the allowance price.)

Item	Price

ARTICLE 5 PAYMENTS
§ 5.1 PROGRESS PAYMENTS
§ 5.1.1 Based upon Applications for Payment submitted to the Architect by the Contractor and Certificates for Payment issued by the Architect, the Owner shall make progress payments on account of the Contract Sum to the Contractor as provided below and elsewhere in the Contract Documents.

§ 5.1.2 The period covered by each Application for Payment shall be one calendar month ending on the last day of the month, or as follows:

§ 5.1.3 Provided that an Application for Payment is received by the Architect not later than the ( ) day of a month, the Owner shall make payment of the certified amount to the Contractor not later than the ( ) day of the ( ) month. If an Application for Payment is received by the Architect after the application date fixed above, payment shall be made by the Owner not later than ( ) days after the Architect receives the Application for Payment.
(Federal, state or local laws may require payment within a certain period of time.)

§ 5.1.4 Each Application for Payment shall be based on the most recent schedule of values submitted by the Contractor in accordance with the Contract Documents. The schedule of values shall allocate the entire Contract Sum among the various portions of the Work. The schedule of values shall be prepared in such form and supported by such data to substantiate its accuracy as the Architect may require. This schedule, unless objected to by the Architect, shall be used as a basis for reviewing the Contractor's Applications for Payment.

§ 5.1.5 Applications for Payment shall show the percentage of completion of each portion of the Work as of the end of the period covered by the Application for Payment.

§ 5.1.6 Subject to other provisions of the Contract Documents, the amount of each progress payment shall be computed as follows:
.1   Take that portion of the Contract Sum properly allocable to completed Work as determined by multiplying the percentage completion of each portion of the Work by the share of the Contract Sum allocated to that portion of the Work in the schedule of values, less retainage of percent (%). Pending final determination of cost to the Owner of changes in the Work, amounts not in dispute shall be included as provided in Section 7.3.9 of AIA Document A201™–2007, General Conditions of the Contract for Construction;
.2   Add that portion of the Contract Sum properly allocable to materials and equipment delivered and suitably stored at the site for subsequent incorporation in the completed construction (or, if approved in advance by the Owner, suitably stored off the site at a location agreed upon in writing), less retainage of percent (%);
.3   Subtract the aggregate of previous payments made by the Owner; and
.4   Subtract amounts, if any, for which the Architect has withheld or nullified a Certificate for Payment as provided in Section 9.5 of AIA Document A201–2007.

§ 5.1.7 The progress payment amount determined in accordance with Section 5.1.6 shall be further modified under the following circumstances:
.1   Add, upon Substantial Completion of the Work, a sum sufficient to increase the total payments to the full amount of the Contract Sum, less such amounts as the Architect shall determine for incomplete Work, retainage applicable to such work and unsettled claims; and
(Section 9.8.5 of AIA Document A201–2007 requires release of applicable retainage upon Substantial Completion of Work with consent of surety, if any.)
.2   Add, if final completion of the Work is thereafter materially delayed through no fault of the Contractor, any additional amounts payable in accordance with Section 9.10.3 of AIA Document A201–2007.

§ 5.1.8 Reduction or limitation of retainage, if any, shall be as follows:
(If it is intended, prior to Substantial Completion of the entire Work, to reduce or limit the retainage resulting from the percentages inserted in Sections 5.1.6.1 and 5.1.6.2 above, and this is not explained elsewhere in the Contract Documents, insert here provisions for such reduction or limitation.)

§ 5.1.9 Except with the Owner's prior approval, the Contractor shall not make advance payments to suppliers for materials or equipment which have not been delivered and stored at the site.

§ 5.2 FINAL PAYMENT

§ 5.2.1 Final payment, constituting the entire unpaid balance of the Contract Sum, shall be made by the Owner to the Contractor when

    .1   the Contractor has fully performed the Contract except for the Contractor's responsibility to correct Work as provided in Section 12.2.2 of AIA Document A201–2007, and to satisfy other requirements, if any, which extend beyond final payment; and

    .2   a final Certificate for Payment has been issued by the Architect.

§ 5.2.2 The Owner's final payment to the Contractor shall be made no later than 30 days after the issuance of the Architect's final Certificate for Payment, or as follows:

ARTICLE 6   DISPUTE RESOLUTION
§ 6.1 INITIAL DECISION MAKER

The Architect will serve as Initial Decision Maker pursuant to Section 15.2 of AIA Document A201–2007, unless the parties appoint below another individual, not a party to this Agreement, to serve as Initial Decision Maker.
(If the parties mutually agree, insert the name, address and other contact information of the Initial Decision Maker, if other than the Architect.)

§ 6.2 BINDING DISPUTE RESOLUTION

For any Claim subject to, but not resolved by, mediation pursuant to Section 15.3 of AIA Document A201–2007, the method of binding dispute resolution shall be as follows:
(Check the appropriate box. If the Owner and Contractor do not select a method of binding dispute resolution below, or do not subsequently agree in writing to a binding dispute resolution method other than litigation, Claims will be resolved by litigation in a court of competent jurisdiction.)

☐   Arbitration pursuant to Section 15.4 of AIA Document A201–2007

☐   Litigation in a court of competent jurisdiction

☐   Other *(Specify)*

ARTICLE 7   TERMINATION OR SUSPENSION

§ 7.1 The Contract may be terminated by the Owner or the Contractor as provided in Article 14 of AIA Document A201–2007.

§ 7.2 The Work may be suspended by the Owner as provided in Article 14 of AIA Document A201–2007.

ARTICLE 8 MISCELLANEOUS PROVISIONS

§ 8.1 Where reference is made in this Agreement to a provision of AIA Document A201–2007 or another Contract Document, the reference refers to that provision as amended or supplemented by other provisions of the Contract Documents.

§ 8.2 Payments due and unpaid under the Contract shall bear interest from the date payment is due at the rate stated below, or in the absence thereof, at the legal rate prevailing from time to time at the place where the Project is located.
(Insert rate of interest agreed upon, if any.)

§ 8.3 The Owner's representative:
(Name, address and other information)

§ 8.4 The Contractor's representative:
(Name, address and other information)

§ 8.5 Neither the Owner's nor the Contractor's representative shall be changed without ten days written notice to the other party.

§ 8.6 Other provisions:

ARTICLE 9 ENUMERATION OF CONTRACT DOCUMENTS

§ 9.1 The Contract Documents, except for Modifications issued after execution of this Agreement, are enumerated in the sections below.

§ 9.1.1 The Agreement is this executed AIA Document A101–2007, Standard Form of Agreement Between Owner and Contractor.

§ 9.1.2 The General Conditions are AIA Document A201–2007, General Conditions of the Contract for Construction.

§ 9.1.3 The Supplementary and other Conditions of the Contract:

Document	Title	Date	Pages

§ 9.1.4 The Specifications:
(Either list the Specifications here or refer to an exhibit attached to this Agreement.)

Section	Title	Date	Pages

§ 9.1.5 The Drawings:
(Either list the Drawings here or refer to an exhibit attached to this Agreement.)

Number	Title	Date

§ 9.1.6 The Addenda, if any:

Number	Date	Pages

Portions of Addenda relating to bidding requirements are not part of the Contract Documents unless the bidding requirements are also enumerated in this Article 9.

§ 9.1.7 Additional documents, if any, forming part of the Contract Documents:
.1   AIA Document E201™–2007, Digital Data Protocol Exhibit, if completed by the parties, or the following:

.2   Other documents, if any, listed below:
(List here any additional documents that are intended to form part of the Contract Documents. AIA Document A201–2007 provides that bidding requirements such as advertisement or invitation to bid, Instructions to Bidders, sample forms and the Contractor's bid are not part of the Contract Documents unless enumerated in this Agreement. They should be listed here only if intended to be part of the Contract Documents.)

ARTICLE 10 INSURANCE AND BONDS
The Contractor shall purchase and maintain insurance and provide bonds as set forth in Article 11 of AIA Document A201–2007.
(State bonding requirements, if any, and limits of liability for insurance required in Article 11 of AIA Document A201–2007.)

This Agreement entered into as of the day and year first written above.

_____   _____
OWNER *(Signature)*   **CONTRACTOR** *(Signature)*

_____   _____
(Printed name and title)   *(Printed name and title)*

Init.

/

7

AIA Documents
B102-2007
B201-2007
B201-2007 Exhibit A

AIA® Document B102™ – 2007

Standard Form of Agreement Between Owner and Architect without a Predefined *Scope of Architect's Services*

AGREEMENT made as of the   day of
in the year
(In words, indicate day, month and year)

BETWEEN the Owner:
(Name, address and other information)

and the Architect:
(Name, address and other information)

for the following Project:
(Name, location and detailed description)

The Owner and Architect agree as follows.

1

TABLE OF ARTICLES

ARTICLE 1 ARCHITECT'S RESPONSIBILITIES

§ 1.1 The Architect shall provide the following professional services:
(Describe the scope of the Architect's services or identify an exhibit or scope of services document setting forth the Architect's services and incorporated into this document in Section 9.2)

§ 1.2 The Architect shall perform its services consistent with the professional skill and care ordinarily provided by architects practicing in the same or similar locality under the same or similar circumstances. The Architect shall perform its services as expeditiously as is consistent with such professional skill and care and the orderly progress of the Project.

§ 1.3 The Architect shall identify a representative authorized to act on behalf of the Architect with respect to the Project.

§ 1.4 Except with the Owner's knowledge and consent, the Architect shall not engage in any activity, or accept any employment, interest or contribution that would reasonably appear to compromise the Architect's professional judgment with respect to this Project.

§ 1.5 The Architect shall maintain the following insurance for the duration of this Agreement. If any of the requirements set forth below exceed the types and limits the Architect normally maintains, the Owner shall reimburse the Architect for any additional cost:
(Identify types and limits of insurance coverage, and other insurance requirements applicable to the Agreement, if any.)

    .1   General Liability

    .2   Automobile Liability

    .3   Workers' Compensation

    .4   Professional Liability

ARTICLE 2 OWNER'S RESPONSIBILITIES

§ 2.1 Unless otherwise provided for under this Agreement, the Owner shall provide information in a timely manner regarding requirements for and limitations on the Project, including a written program which shall set forth the Owner's objectives, schedule, constraints and criteria, including space requirements and relationships, flexibility, expandability, special equipment, systems and site requirements. Within 15 days after receipt of a written request from the Architect, the Owner shall furnish the requested information as necessary and relevant for the Architect to evaluate, give notice of or enforce lien rights.

§ 2.2 The Owner shall identify a representative authorized to act on the Owner's behalf with respect to the Project. The Owner shall render decisions and approve the Architect's submittals in a timely manner in order to avoid unreasonable delay in the orderly and sequential progress of the Architect's services.

§ 2.3 The Owner shall coordinate the services of its own consultants with those services provided by the Architect. Upon the Architect's request, the Owner shall furnish copies of the scope of consulting services in the contracts between the Owner and the Owner's consultants. The Owner shall furnish the services of consultants other than those designated in this Agreement, or authorize the Architect to furnish them as an Additional Service, when the Architect requests such services and demonstrates that they are reasonably required by the scope of the Project. The Owner shall require that its consultants maintain professional liability insurance as appropriate to the services provided.

§ 2.4 The Owner shall furnish all legal, insurance and accounting services, including auditing services, that may be reasonably necessary at any time for the Project to meet the Owner's needs and interests.

§ 2.5 The Owner shall provide prompt written notice to the Architect if the Owner becomes aware of any fault or defect in the Project, including errors, omissions or inconsistencies in the Architect's Instruments of Service.

ARTICLE 3 COPYRIGHTS AND LICENSES

§ 3.1 The Architect and the Owner warrant that in transmitting Instruments of Service, or any other information, the transmitting party is the copyright owner of such information or has permission from the copyright owner to transmit such information for its use on the Project. If the Owner and Architect intend to transmit Instruments of Service or any other information or documentation in digital form, they shall endeavor to establish necessary protocols governing such transmissions.

§ 3.2 The Architect and the Architect's consultants shall be deemed the authors and owners of their respective Instruments of Service, including the Drawings and Specifications, and shall retain all common law, statutory and other reserved rights, including copyrights. Submission or distribution of Instruments of Service to meet official regulatory requirements or for similar purposes in connection with the Project is not to be construed as publication in derogation of the reserved rights of the Architect and the Architect's consultants.

§ 3.3 Upon execution of this Agreement, the Architect grants to the Owner a nonexclusive license to use the Architect's Instruments of Service solely and exclusively for the Project, provided that the Owner substantially performs its obligations, including prompt payment of all sums when due, under this Agreement. The Architect shall obtain similar nonexclusive licenses from the Architect's consultants consistent with this Agreement. The license granted under this section permits the Owner to authorize the Contractor, Subcontractors, Sub-subcontractors, and material or equipment suppliers, as well as the Owner's consultants and separate contractors, to reproduce applicable portions of the Instruments of Service solely and exclusively for use in performing services for the Project. If the Architect rightfully terminates this Agreement for cause as provided in Sections 5.3 and 5.4, the license granted in this Section 3.3 shall terminate.

§ 3.3.1 In the event the Owner uses the Instruments of Service without retaining the author of the Instruments of Service, the Owner releases the Architect and Architect's consultant(s) from all claims and causes of action arising from such uses. The Owner, to the extent permitted by law, further agrees to indemnify and hold harmless the Architect and its consultants from all costs and expenses, including the cost of defense, related to claims and causes of action asserted by any third person or entity to the extent such costs and expenses arise from the Owner's use of the Instruments of Service under this Section 3.3.1.

§ 3.4 Except for the licenses granted in this Article 3, no other license or right shall be deemed granted or implied under this Agreement. The Owner shall not assign, delegate, sublicense, pledge or otherwise transfer any license granted herein to

3

another party without the prior written agreement of the Architect. Any unauthorized use of the Instruments of Service shall be at the Owner's sole risk and without liability to the Architect and the Architect's consultants.

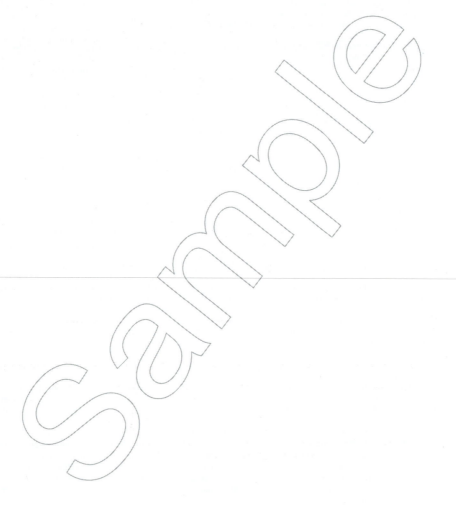

ARTICLE 4 CLAIMS AND DISPUTES

§ 4.1 GENERAL

§ 4.1.1 The Owner and Architect shall commence all claims and causes of action, whether in contract, tort, or otherwise, against the other arising out of or related to this Agreement in accordance with the requirements of the method of binding dispute resolution selected in this Agreement within the period specified by applicable law, but in any case not more than 10 years after the date of Substantial Completion of the Work. The Owner and Architect waive all claims and causes of action not commenced in accordance with this Section 4.1.1.

§ 4.1.2 To the extent damages are covered by property insurance, the Owner and Architect waive all rights against each other and against the contractors, consultants, agents and employees of the other for damages, except such rights as they may have to the proceeds of such insurance as set forth in AIA Document A201–2007, General Conditions of the Contract for Construction, if applicable. The Owner or the Architect, as appropriate, shall require of the contractors, consultants, agents and employees of any of them similar waivers in favor of the other parties enumerated herein.

§ 4.1.3 The Architect and Owner waive consequential damages for claims, disputes or other matters in question arising out of or relating to this Agreement. This mutual waiver is applicable, without limitation, to all consequential damages due to either party's termination of this Agreement, except as specifically provided in Section 5.7.

§ 4.2 MEDIATION

§ 4.2.1 Any claim, dispute or other matter in question arising out of or related to this Agreement shall be subject to mediation as a condition precedent to binding dispute resolution. If such matter relates to or is the subject of a lien arising out of the Architect's services, the Architect may proceed in accordance with applicable law to comply with the lien notice or filing deadlines prior to resolution of the matter by mediation or by binding dispute resolution.

§ 4.2.2 The Owner and Architect shall endeavor to resolve claims, disputes and other matters in question between them by mediation which, unless the parties mutually agree otherwise, shall be administered by the American Arbitration Association in accordance with its Construction Industry Mediation Procedures in effect on the date of the Agreement. A request for mediation shall be made in writing, delivered to the other party to the Agreement, and filed with the person or entity administering the mediation. The request may be made concurrently with the filing of a complaint or other appropriate demand for binding dispute resolution but, in such event, mediation shall proceed in advance of binding dispute resolution proceedings, which shall be stayed pending mediation for a period of 60 days from the date of filing, unless stayed for a longer period by agreement of the parties or court order. If an arbitration proceeding is stayed pursuant to this Section, the parties may nonetheless proceed to the selection of the arbitrator(s) and agree upon a schedule for later proceedings.

§ 4.2.3 The parties shall share the mediator's fee and any filing fees equally. The mediation shall be held in the place where the Project is located, unless another location is mutually agreed upon. Agreements reached in mediation shall be enforceable as settlement agreements in any court having jurisdiction thereof.

§ 4.2.4 If the parties do not resolve a dispute through mediation pursuant to this Section 4.2, the method of binding dispute resolution shall be the following:
(Check the appropriate box. If the Owner and Architect do not select a method of binding dispute resolution below, or do not subsequently agree in writing to a binding dispute resolution method other than litigation, the dispute will be resolved in a court of competent jurisdiction.)

☐ Arbitration pursuant to Section 4.3 of this Agreement

☐ Litigation in a court of competent jurisdiction

☐ Other *(Specify)*

§ 4.3 ARBITRATION

5

§ 4.3.1 If the parties have selected arbitration as the method for binding dispute resolution in this Agreement, any claim, dispute or other matter in question arising out of or related to this Agreement subject to, but not resolved by, mediation shall be subject to arbitration, which unless the parties mutually agree otherwise, shall be administered by the American Arbitration Association in accordance with its Construction Industry Arbitration Rules in effect on the date of this Agreement. A demand for arbitration shall be made in writing, delivered to the other party to this Agreement, and filed with the person or entity administering the arbitration.

§ 4.3.1.1 A demand for arbitration shall be made no earlier than concurrently with the filing of a request for mediation, but in no event shall it be made after the date when the institution of legal or equitable proceedings based on the claim, dispute or other matter in question would be barred by the applicable statute of limitations. For statute of limitations purposes, receipt of a written demand for arbitration by the person or entity administering the arbitration shall constitute the institution of legal or equitable proceedings based on the claim, dispute or other matter in question.

§ 4.3.2 The foregoing agreement to arbitrate and other agreements to arbitrate with an additional person or entity duly consented to by parties to this Agreement shall be specifically enforceable in accordance with applicable law in any court having jurisdiction thereof.

§ 4.3.3 The award rendered by the arbitrator(s) shall be final, and judgment may be entered upon it in accordance with applicable law in any court having jurisdiction thereof.

§ 4.3.4 CONSOLIDATION OR JOINDER

§ 4.3.4.1 Either party, at its sole discretion, may consolidate an arbitration conducted under this Agreement with any other arbitration to which it is a party provided that (1) the arbitration agreement governing the other arbitration permits consolidation; (2) the arbitrations to be consolidated substantially involve common questions of law or fact; and (3) the arbitrations employ materially similar procedural rules and methods for selecting arbitrator(s).

§ 4.3.4.2 Either party, at its sole discretion, may include by joinder persons or entities substantially involved in a common question of law or fact whose presence is required if complete relief is to be accorded in arbitration, provided that the party sought to be joined consents in writing to such joinder. Consent to arbitration involving an additional person or entity shall not constitute consent to arbitration of any claim, dispute or other matter in question not described in the written consent.

§ 4.3.4.3 The Owner and Architect grant to any person or entity made a party to an arbitration conducted under this Section 4.3, whether by joinder or consolidation, the same rights of joinder and consolidation as the Owner and Architect under this Agreement.

ARTICLE 5 TERMINATION OR SUSPENSION

§ 5.1 If the Owner fails to make payments to the Architect in accordance with this Agreement, such failure shall be considered substantial nonperformance and cause for termination or, at the Architect's option, cause for suspension of performance of services under this Agreement. If the Architect elects to suspend services, the Architect shall give seven days' written notice to the Owner before suspending services. In the event of a suspension of services, the Architect shall have no liability to the Owner for delay or damage caused the Owner because of such suspension of services. Before resuming services, the Architect shall be paid all sums due prior to suspension and any expenses incurred in the interruption and resumption of the Architect's services. The Architect's fees for the remaining services and the time schedules shall be equitably adjusted.

§ 5.2 If the Owner suspends the Project, the Architect shall be compensated for services performed prior to notice of such suspension. When the Project is resumed, the Architect shall be compensated for expenses incurred in the interruption and resumption of the Architect's services. The Architect's fees for the remaining services and the time schedules shall be equitably adjusted.

§ 5.3 If the Owner suspends the Project for more than 90 cumulative days for reasons other than the fault of the Architect, the Architect may terminate this Agreement by giving not less than seven days' written notice.

§ 5.4 Either party may terminate this Agreement upon not less than seven days' written notice should the other party fail substantially to perform in accordance with the terms of this Agreement through no fault of the party initiating the termination.

6

§ **5.5** The Owner may terminate this Agreement upon not less than seven days' written notice to the Architect for the Owner's convenience and without cause.

7

§ 5.6 In the event of termination not the fault of the Architect, the Architect shall be compensated for services performed prior to termination, together with Reimbursable Expenses then due and all Termination Expenses as defined in Section 5.7.

§ 5.7 Termination Expenses are in addition to compensation for the Architect's services and include expenses directly attributable to termination for which the Architect is not otherwise compensated, plus an amount for the Architect's anticipated profit on the value of the services not performed by the Architect.

§ 5.8 The Owner's rights to use the Architect's Instruments of Service in the event of a termination of this Agreement are set forth in Article 3 and Section 6.3.

ARTICLE 6 COMPENSATION

§ 6.1 The Owner shall compensate the Architect for services described in Section 1.1 as set forth below, or in the attached exhibit or scope document incorporated into this Agreement in Section 9.2.
(Insert amount of, or basis for, compensation or indicate the exhibit or scope document in which compensation is provided for.)

§ 6.2 Reimbursable Expenses are in addition to compensation for the Architect's professional services and include expenses incurred by the Architect and the Architect's consultants directly related to the Project, as follows:

   .1   Transportation and authorized out-of-town travel and subsistence;
   .2   Long distance services, dedicated data and communication services, teleconferences, Project Web sites, and extranets;
   .3   Fees paid for securing approval of authorities having jurisdiction over the Project;
   .4   Printing, reproductions, plots, standard form documents;
   .5   Postage, handling and delivery;
   .6   Expense of overtime work requiring higher than regular rates, if authorized in advance by the Owner;
   .7   Renderings, models, mock-ups, professional photography, and presentation materials requested by the Owner;
   .8   Architect's Consultant's expense of professional liability insurance dedicated exclusively to this Project, or the expense of additional insurance coverage or limits if the Owner requests such insurance in excess of that normally carried by the Architect's consultants;
   .9   All taxes levied on professional services and on reimbursable expenses;
   .10   Site office expenses; and
   .11   Other similar Project-related expenditures.

§ 6.2.1 For Reimbursable Expenses the compensation shall be the expenses incurred by the Architect and the Architect's consultants plus an administrative fee of   percent (%) of the expenses incurred.

§ 6.3 COMPENSATION FOR USE OF ARCHITECT'S INSTRUMENTS OF SERVICE

If the Owner terminates the Architect for its convenience under Section 5.5, or the Architect terminates this Agreement under Section 5.3, the Owner shall pay a licensing fee as compensation for the Owner's continued use of the Architect's Instruments of Service solely for purposes of the Project as follows:

§ 6.4 PAYMENTS TO THE ARCHITECT

§ 6.4.1 An initial payment of   Dollars
($) shall be made upon execution of this Agreement and is the minimum payment under this Agreement. It shall be credited to the Owner's account in the final invoice.

8

§ 6.4.2 Unless otherwise agreed, payments for services shall be made monthly in proportion to services performed. Payments are due and payable upon presentation of the Architect's invoice. Amounts unpaid
(   ) days after the invoice date shall bear interest at the rate entered below, or in the absence thereof at the legal rate prevailing from time to time at the principal place of business of the Architect.
(Insert rate of monthly or annual interest agreed upon.)

§ 6.4.3 The Owner shall not withhold amounts from the Architect's compensation to impose a penalty or liquidated damages on the Architect, or to offset sums requested by or paid to contractors for the cost of changes in the Work unless the Architect agrees or has been found liable for the amounts in a binding dispute resolution proceeding.

§ 6.4.4 Records of Reimbursable Expenses and services performed on the basis of hourly rates shall be available to the Owner at mutually convenient times.

ARTICLE 7 MISCELLANEOUS PROVISIONS

§ 7.1 This Agreement shall be governed by the law of the place where the Project is located, except that if the parties have selected arbitration as the method of binding dispute resolution, the Federal Arbitration Act shall govern Section 4.3.

§ 7.2 Terms in this Agreement shall have the same meaning as those in AIA Document A201–2007, General Conditions of the Contract for Construction.

§ 7.3 The Owner and Architect, respectively, bind themselves, their agents, successors, assigns and legal representatives to this Agreement. Neither the Owner nor the Architect shall assign this Agreement without the written consent of the other, except that the Owner may assign this Agreement to a lender providing financing for the Project if the lender agrees to assume the Owner's rights and obligations under this Agreement.

§ 7.4 If the Owner requests the Architect to execute certificates, the proposed language of such certificates shall be submitted to the Architect for review at least 14 days prior to the requested dates of execution. If the Owner requests the Architect to execute consents reasonably required to facilitate assignment to a lender, the Architect shall execute all such consents that are consistent with this Agreement, provided the proposed consent is submitted to the Architect for review at least 14 days prior to execution. The Architect shall not be required to execute certificates or consents that would require knowledge, services or responsibilities beyond the scope of this Agreement.

§ 7.5 Nothing contained in this Agreement shall create a contractual relationship with or a cause of action in favor of a third party against either the Owner or Architect.

§ 7.6 Unless otherwise required in this Agreement, the Architect shall have no responsibility for the discovery, presence, handling, removal or disposal of, or exposure of persons to, hazardous materials or toxic substances in any form at the Project site.

§ 7.7 The Architect shall have the right to include photographic or artistic representations of the design of the Project among the Architect's promotional and professional materials. The Architect shall be given reasonable access to the completed Project to make such representations. However, the Architect's materials shall not include the Owner's confidential or proprietary information if the Owner has previously advised the Architect in writing of the specific information considered by the Owner to be confidential or proprietary. The Owner shall provide professional credit for the Architect in the Owner's promotional materials for the Project.

§ 7.8 If the Architect or Owner receives information specifically designated by the other party as "confidential" or "business proprietary," the receiving party shall keep such information strictly confidential and shall not disclose it to any other person except to (1) its employees, (2) those who need to know the content of such information in order to perform services or construction solely and exclusively for the Project, or (3) its consultants and contractors whose contracts include similar restrictions on the use of confidential information.

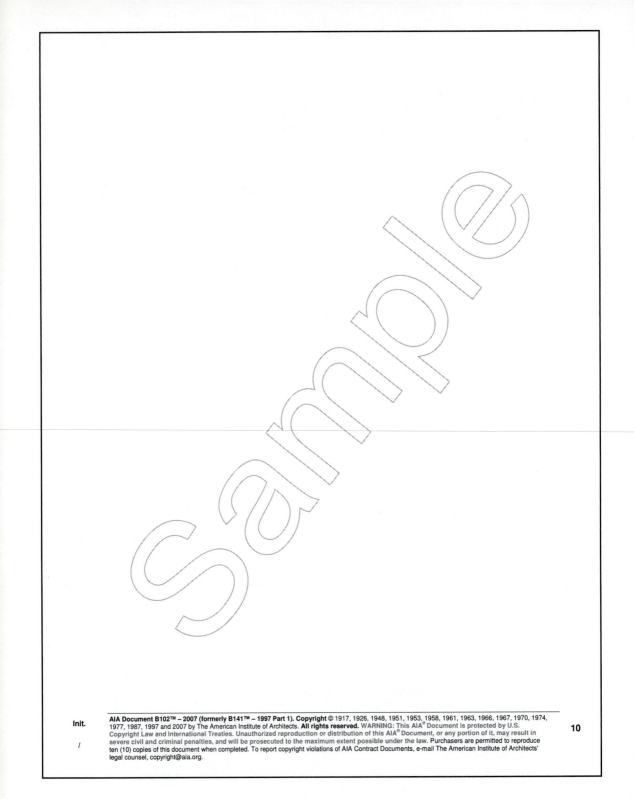

ARTICLE 8 SPECIAL TERMS AND CONDITIONS
Special terms and conditions that modify this Agreement are as follows:

ARTICLE 9 SCOPE OF THE AGREEMENT
§ 9.1 This Agreement represents the entire and integrated agreement between the Owner and the Architect and supersedes all prior negotiations, representations or agreements, either written or oral. This Agreement may be amended only by written instrument signed by both Owner and Architect.

§ 9.2 This Agreement is comprised of the following documents listed below:

   .1   AIA Document B102–2007, Standard Form Agreement Between Owner and Architect

   .2   AIA Document E201–2007, Digital Data Protocol Exhibit, if completed, or the following:

   .3   Other documents:
              (List other documents, including the Architect's scope of services document, hereby incorporated into the Agreement.)

This Agreement entered into as of the day and year first written above.

_____   _____
OWNER *(Signature)*   **ARCHITECT** *(Signature)*

_____   _____
(Printed name and title)   *(Printed name and title)*

AIA® Document B201™ – 2007

Standard Form of Architect's Services: *Design and Construction Contract Administration*

for the following **PROJECT**:
(Name and location or address)

THE OWNER:
(Name and address)

THE ARCHITECT:
(Name and address)

THE AGREEMENT
This Standard Form of Architect's Services is part of or modifies the accompanying Owner-Architect Agreement (hereinafter, the Agreement) dated the   day of   in the year
(In words, indicate day, month, and year.)

TABLE OF ARTICLES

1

ARTICLE 1 INITIAL INFORMATION

§ 1.1 This Agreement is based on the Initial Information set forth in Article 1 and in optional Exhibit A, Initial Information:

(*Complete Exhibit A, Initial Information and incorporate it into this services document at Section 7.1, or state below Initial Information such as details of the Project's site and program, Owner's contractors and consultants, Architect's consultants, Owner's budget for the Cost of the Work, authorized representatives, anticipated procurement method, and other information relevant to the Project.*)

§ 1.2 The Owner's anticipated dates for commencement of construction and Substantial Completion of the Work are set forth below:

   .1   Commencement of construction date:

   .2   Substantial Completion date:

§ 1.3 The Owner and Architect may rely on the Initial Information. Both parties, however, recognize that such information may materially change and, in that event, the Owner and the Architect shall appropriately adjust the schedule, the Architect's services and the Architect's compensation.

ARTICLE 2 SCOPE OF ARCHITECT'S BASIC SERVICES

§ 2.1 The Architect's Basic Services consist of those described in Article 2 and include usual and customary structural, mechanical, and electrical engineering services. Services not set forth in Article 2 are Additional Services.

§ 2.1.1 The Architect shall manage the Architect's services, consult with the Owner, research applicable design criteria, attend Project meetings, communicate with members of the Project team and report progress to the Owner.

§ 2.1.2 The Architect shall coordinate its services with those services provided by the Owner and the Owner's consultants. The Architect shall be entitled to rely on the accuracy and completeness of services and information furnished by the Owner and the Owner's consultants. The Architect shall provide prompt written notice to the Owner if the Architect becomes aware of any error, omission or inconsistency in such services or information.

§ 2.1.3 As soon as practicable after the date of this Agreement, the Architect shall submit for the Owner's approval a schedule for the performance of the Architect's services. The schedule initially shall include anticipated dates for the commencement of construction and for Substantial Completion of the Work as set forth in the Initial Information. The schedule shall include allowances for periods of time required for the Owner's review, for the performance of the Owner's consultants, and for approval of submissions by authorities having jurisdiction over the Project. Once approved by the Owner, time limits established by the schedule shall not, except for reasonable cause, be exceeded by the Architect or Owner. With the Owner's approval, the Architect shall adjust the schedule, if necessary, as the Project proceeds until the commencement of construction.

§ 2.1.4 The Architect shall not be responsible for an Owner's directive or substitution made without the Architect's approval.

§ 2.1.5 The Architect shall, at appropriate times, contact the governmental authorities required to approve the Construction Documents and the entities providing utility services to the Project. In designing the Project, the Architect shall respond to applicable design requirements imposed by such governmental authorities and by such entities providing utility services.

§ 2.1.6 The Architect shall assist the Owner in connection with the Owner's responsibility for filing documents required for the approval of governmental authorities having jurisdiction over the Project.

§ 2.2 SCHEMATIC DESIGN PHASE SERVICES

§ 2.2.1 The Architect shall review the program and other information furnished by the Owner, and shall review laws, codes, and regulations applicable to the Architect's services.

§ 2.2.2 The Architect shall prepare a preliminary evaluation of the Owner's program, schedule, budget for the Cost of the Work, Project site, and the proposed procurement or delivery method and other Initial Information, each in terms of the other, to ascertain the requirements of the Project. The Architect shall notify the Owner of (1) any inconsistencies discovered in the information, and (2) other information or consulting services that may be reasonably needed for the Project.

§ 2.2.3 The Architect shall present its preliminary evaluation to the Owner and shall discuss with the Owner alternative approaches to design and construction of the Project, including the feasibility of incorporating environmentally responsible design approaches. The Architect shall reach an understanding with the Owner regarding the requirements of the Project.

§ 2.2.4 Based on the Project's requirements agreed upon with the Owner, the Architect shall prepare and present for the Owner's approval a preliminary design illustrating the scale and relationship of the Project components.

§ 2.2.5 Based on the Owner's approval of the preliminary design, the Architect shall prepare Schematic Design Documents for the Owner's approval. The Schematic Design Documents shall consist of drawings and other documents including a site plan, if appropriate, and preliminary building plans, sections and elevations; and may include some combination of study models, perspective sketches, or digital modeling. Preliminary selections of major building systems and construction materials shall be noted on the drawings or described in writing.

§ 2.2.5.1 The Architect shall consider environmentally responsible design alternatives, such as material choices and building orientation, together with other considerations based on program and aesthetics, in developing a design that is consistent with the Owner's program, schedule and budget for the Cost of the Work. The Owner may obtain other environmentally responsible design services under Article 3.

§ 2.2.5.2 The Architect shall consider the value of alternative materials, building systems and equipment, together with other considerations based on program and aesthetics in developing a design for the Project that is consistent with the Owner's program, schedule and budget for the Cost of the Work.

§ 2.2.6 The Architect shall submit to the Owner an estimate of the Cost of the Work prepared in accordance with Section 5.3.

§ 2.2.7 The Architect shall submit the Schematic Design Documents to the Owner, and request the Owner's approval.

§ 2.3 DESIGN DEVELOPMENT PHASE SERVICES
§ 2.3.1 Based on the Owner's approval of the Schematic Design Documents, and on the Owner's authorization of any adjustments in the Project requirements and the budget for the Cost of the Work, the Architect shall prepare Design Development Documents for the Owner's approval. The Design Development Documents shall illustrate and describe the development of the approved Schematic Design Documents and shall consist of drawings and other documents including plans, sections, elevations, typical construction details, and diagrammatic layouts of building systems to fix and describe the size and character of the Project as to architectural, structural, mechanical and electrical systems, and such other elements as may be appropriate. The Design Development Documents shall also include outline specifications that identify major materials and systems and establish in general their quality levels.

§ 2.3.2 The Architect shall update the estimate of the Cost of the Work.

§ 2.3.3 The Architect shall submit the Design Development documents to the Owner, advise the Owner of any adjustments to the estimate of the Cost of the Work, and request the Owner's approval.

§ 2.4 CONSTRUCTION DOCUMENTS PHASE SERVICES
§ 2.4.1 Based on the Owner's approval of the Design Development Documents, and on the Owner's authorization of any adjustments in the Project requirements and the budget for the Cost of the Work, the Architect shall prepare Construction Documents for the Owner's approval. The Construction Documents shall illustrate and describe the further development of the approved Design Development Documents and shall consist of Drawings and Specifications setting forth in detail the quality levels of materials and systems and other requirements for the construction of the Work. The Owner and Architect acknowledge that in order to construct the Work the Contractor

will provide additional information, including Shop Drawings, Product Data, Samples and other similar submittals, which the Architect shall review in accordance with Section 2.6.4.

§ 2.4.2 The Architect shall incorporate into the Construction Documents the design requirements of governmental authorities having jurisdiction over the Project.

§ 2.4.3 During the development of the Construction Documents, the Architect shall assist the Owner in the development and preparation of (1) bidding and procurement information that describes the time, place and conditions of bidding, including bidding or proposal forms; (2) the form of agreement between the Owner and Contractor; and (3) the Conditions of the Contract for Construction (General, Supplementary and other Conditions). The Architect shall also compile a project manual that includes the Conditions of the Contract for Construction and Specifications and may include bidding requirements and sample forms.

§ 2.4.4 The Architect shall update the estimate for the Cost of the Work.

§ 2.4.5 The Architect shall submit the Construction Documents to the Owner, advise the Owner of any adjustments to the estimate of the Cost of the Work, take any action required under Section 5.5, and request the Owner's approval.

§ 2.5 BIDDING OR NEGOTIATION PHASE SERVICES
The Architect shall assist the Owner in establishing a list of prospective contractors. Following the Owner's approval of the Construction Documents, the Architect shall assist the Owner in (1) obtaining either competitive bids or negotiated proposals; (2) confirming responsiveness of bids or proposals; (3) determining the successful bid or proposal, if any; and, (4) awarding and preparing contracts for construction.

§ 2.5.2 COMPETITIVE BIDDING
§ 2.5.2.1 Bidding Documents shall consist of bidding requirements and proposed Contract Documents.

§ 2.5.2.2 The Architect shall assist the Owner in bidding the Project by
.1 procuring the reproduction of Bidding Documents for distribution to prospective bidders;
.2 distributing the Bidding Documents to prospective bidders requesting their return upon completion of the bidding process, and maintaining a log of distribution and retrieval and of the amounts of deposits, if any, received from and returned to prospective bidders;
.3 organizing and conducting a pre-bid conference for prospective bidders;
.4 preparing responses to questions from prospective bidders and providing clarifications and interpretations of the Bidding Documents to all prospective bidders in the form of addenda; and
.5 organizing and conducting the opening of the bids, and subsequently documenting and distributing the bidding results, as directed by the Owner.

§ 2.5.2.3 The Architect shall consider requests for substitutions, if the Bidding Documents permit substitutions, and shall prepare and distribute addenda identifying approved substitutions to all prospective bidders.

§ 2.5.3 NEGOTIATED PROPOSALS
§ 2.5.3.1 Proposal Documents shall consist of proposal requirements and proposed Contract Documents.

§ 2.5.3.2 The Architect shall assist the Owner in obtaining proposals by
.1 procuring the reproduction of Proposal Documents for distribution to prospective contractors, and requesting their return upon completion of the negotiation process;
.2 organizing and participating in selection interviews with prospective contractors; and
.3 participating in negotiations with prospective contractors, and subsequently preparing a summary report of the negotiation results, as directed by the Owner.

§ 2.5.3.3 The Architect shall consider requests for substitutions, if the Proposal Documents permit substitutions, and shall prepare and distribute addenda identifying approved substitutions to all prospective contractors.

§ 2.6 CONSTRUCTION PHASE SERVICES
§ 2.6.1 GENERAL
§ 2.6.1.1 The Architect shall provide administration of the Contract between the Owner and the Contractor as set forth below and in AIA Document A201–2007, General Conditions of the Contract for Construction. If the Owner and

Contractor modify AIA Document A201–2007, those modifications shall not affect the Architect's services under this Agreement unless the Owner and the Architect amend this Agreement.

§ 2.6.1.2 The Architect shall advise and consult with the Owner during the Construction Phase Services. The Architect shall have authority to act on behalf of the Owner only to the extent provided in this Agreement. The Architect shall not have control over, charge of, or responsibility for the construction means, methods, techniques, sequences or procedures, or for safety precautions and programs in connection with the Work, nor shall the Architect be responsible for the Contractor's failure to perform the Work in accordance with the requirements of the Contract Documents. The Architect shall be responsible for the Architect's negligent acts or omissions, but shall not have control over or charge of, and shall not be responsible for, acts or omissions of the Contractor or of any other persons or entities performing portions of the Work.

§ 2.6.1.3 Subject to Section 3.3, the Architect's responsibility to provide Construction Phase services commences with the award of the Contract for Construction and terminates on the date the Architect issues the final Certificate for Payment.

§ 2.6.2 EVALUATIONS OF THE WORK

§ 2.6.2.1 The Architect shall visit the site at intervals appropriate to the stage of construction, or as otherwise required in Section 3.3.3, to become generally familiar with the progress and quality of the portion of the Work completed, and to determine, in general, if the Work observed is being performed in a manner indicating that the Work, when fully completed, will be in accordance with the Contract Documents. However, the Architect shall not be required to make exhaustive or continuous on-site inspections to check the quality or quantity of the Work. On the basis of the site visits the Architect shall keep the Owner reasonably informed about the progress and quality of the portion of the Work completed, and report to the Owner (1) known deviations from the Contract Documents and from the most recent construction schedule submitted by the Contractor, and (2) defects and deficiencies observed in the Work.

§ 2.6.2.2 The Architect has the authority to reject Work that does not conform to the Contract Documents. Whenever the Architect considers it necessary or advisable, the Architect shall have the authority to require inspection or testing of the Work in accordance with the provisions of the Contract Documents, whether or not such Work is fabricated, installed or completed. However, neither this authority of the Architect nor a decision made in good faith either to exercise or not to exercise such authority shall give rise to a duty or responsibility of the Architect to the Contractor, Subcontractors, material and equipment suppliers, their agents or employees or other persons or entities performing portions of the Work.

§ 2.6.2.3 The Architect shall interpret and decide matters concerning performance under, and requirements of, the Contract Documents on written request of either the Owner or Contractor. The Architect's response to such requests shall be made in writing within any time limits agreed upon or otherwise with reasonable promptness.

§ 2.6.2.4 Interpretations and decisions of the Architect shall be consistent with the intent of and reasonably inferable from the Contract Documents and shall be in writing or in the form of drawings. When making such interpretations and decisions, the Architect shall endeavor to secure faithful performance by both Owner and Contractor, shall not show partiality to either, and shall not be liable for results of interpretations or decisions rendered in good faith. The Architect's decisions on matters relating to aesthetic effect shall be final if consistent with the intent expressed in the Contract Documents.

§ 2.6.2.5 Unless the Owner and Contractor designate another person to serve as an Initial Decision Maker, as that term is defined in AIA Document A201–2007, the Architect shall render initial decisions on Claims between the Owner and Contractor as provided in the Contract Documents.

§ 2.6.3 CERTIFICATES FOR PAYMENT TO CONTRACTOR

§ 2.6.3.1 The Architect shall review and certify the amounts due the Contractor and shall issue certificates in such amounts. The Architect's certification for payment shall constitute a representation to the Owner, based on the Architect's evaluation of the Work as provided in Section 2.6.2 and on the data comprising the Contractor's Application for Payment, that, to the best of the Architect's knowledge, information and belief, the Work has progressed to the point indicated and that the quality of the Work is in accordance with the Contract Documents. The foregoing representations are subject (1) to an evaluation of the Work for conformance with the Contract Documents upon Substantial Completion, (2) to results of subsequent tests and inspections, (3) to correction of minor deviations from the Contract Documents prior to completion, and (4) to specific qualifications expressed by the Architect.

§ 2.6.3.2 The issuance of a Certificate for Payment shall not be a representation that the Architect has (1) made exhaustive or continuous on-site inspections to check the quality or quantity of the Work, (2) reviewed construction means, methods, techniques, sequences or procedures, (3) reviewed copies of requisitions received from Subcontractors and material suppliers and other data requested by the Owner to substantiate the Contractor's right to payment, or (4) ascertained how or for what purpose the Contractor has used money previously paid on account of the Contract Sum.

§ 2.6.3.3 The Architect shall maintain a record of the Applications and Certificates for Payment.

§ 2.6.4 SUBMITTALS

§ 2.6.4.1 The Architect shall review the Contractor's submittal schedule and shall not unreasonably delay or withhold approval. The Architect's action in reviewing submittals shall be taken in accordance with the approved submittal schedule or, in the absence of an approved submittal schedule, with reasonable promptness while allowing sufficient time in the Architect's professional judgment to permit adequate review.

§ 2.6.4.2 In accordance with the Architect-approved submittal schedule, the Architect shall review and approve or take other appropriate action upon the Contractor's submittals such as Shop Drawings, Product Data and Samples, but only for the limited purpose of checking for conformance with information given and the design concept expressed in the Contract Documents. Review of such submittals is not for the purpose of determining the accuracy and completeness of other information such as dimensions, quantities, and installation or performance of equipment or systems, which are the Contractor's responsibility. The Architect's review shall not constitute approval of safety precautions or, unless otherwise specifically stated by the Architect, of any construction means, methods, techniques, sequences or procedures. The Architect's approval of a specific item shall not indicate approval of an assembly of which the item is a component.

§ 2.6.4.3 If the Contract Documents specifically require the Contractor to provide professional design services or certifications by a design professional related to systems, materials or equipment, the Architect shall specify the appropriate performance and design criteria that such services must satisfy. The Architect shall review shop Drawings and other submittals related to the Work designed or certified by the design professional retained by the Contractor that bear such professional's seal and signature when submitted to the Architect. The Architect shall be entitled to rely upon the adequacy, accuracy and completeness of the services, certifications and approvals performed or provided by such design professionals.

§ 2.6.4.4 Subject to the provisions of Section 3.3, the Architect shall review and respond to requests for information about the Contract Documents. The Architect shall set forth in the Contract Documents the requirements for requests for information. Requests for information shall include, at a minimum, a detailed written statement that indicates the specific Drawings or Specifications in need of clarification and the nature of the clarification requested. The Architect's response to such requests shall be made in writing within any time limits agreed upon, or otherwise with reasonable promptness. If appropriate, the Architect shall prepare and issue supplemental Drawings and Specifications in response to requests for information.

§ 2.6.4.5 The Architect shall maintain a record of submittals and copies of submittals supplied by the Contractor in accordance with the requirements of the Contract Documents.

§ 2.6.5 CHANGES IN THE WORK

§ 2.6.5.1 The Architect may authorize minor changes in the Work that are consistent with the intent of the Contract Documents and do not involve an adjustment in the Contract Sum or an extension of the Contract Time. Subject to the provisions of Section 3.3, the Architect shall prepare Change Orders and Construction Change Directives for the Owner's approval and execution in accordance with the Contract Documents.

§ 2.6.5.2 The Architect shall maintain records relative to changes in the Work.

§ 2.6.6 PROJECT COMPLETION

§ 2.6.6.1 The Architect shall conduct inspections to determine the date or dates of Substantial Completion and the date of final completion; issue Certificates of Substantial Completion; receive from the Contractor and forward to the Owner, for the Owner's review and records, written warranties and related documents required by the Contract

6

Documents and assembled by the Contractor; and issue a final Certificate for Payment based upon a final inspection indicating the Work complies with the requirements of the Contract Documents.

§ 2.6.6.2 The Architect's inspections shall be conducted with the Owner to check conformance of the Work with the requirements of the Contract Documents and to verify the accuracy and completeness of the list submitted by the Contractor of Work to be completed or corrected.

§ 2.6.6.3 When the Work is found to be substantially complete, the Architect shall inform the Owner about the balance of the Contract Sum remaining to be paid the Contractor, including the amount to be retained from the Contract Sum, if any, for final completion or correction of the Work.

§ 2.6.6.4 The Architect shall forward to the Owner the following information received from the Contractor: (1) consent of surety or sureties, if any, to reduction in or partial release of retainage or the making of final payment; (2) affidavits, receipts, releases and waivers of liens or bonds indemnifying the Owner against liens; and (3) any other documentation required of the Contractor under the Contract Documents.

§ 2.6.6.5 Upon request of the Owner, and prior to the expiration of one year from the date of Substantial Completion, the Architect shall, without additional compensation, conduct a meeting with the Owner to review the facility operations and performance.

ARTICLE 3   ADDITIONAL SERVICES

§ 3.1 Additional Services listed below are not included in Basic Services but may be required for the Project. The Architect shall provide the listed Additional Services only if specifically designated in the table below as the Architect's responsibility, and the Owner shall compensate the Architect as provided in Section 6.2.
(Designate the Additional Services the Architect shall provide in the second column of the table below. In the third column indicate whether the service description is located in Section 3.2 or in an attached exhibit. If in an exhibit, identify the exhibit.)

Additional Services		Responsibility *(Architect, Owner or Not Provided)*	Location of Service Description *(Section 3.2 below or in an exhibit attached to this document and identified below)*
§ 3.1.1	Programming		
§ 3.1.2	Multiple preliminary designs		
§ 3.1.3	Measured drawings		
§ 3.1.4	Existing facilities surveys		
§ 3.1.5	Site Evaluation and Planning (B203™–2007)		
§ 3.1.6	Building information modeling		
§ 3.1.7	Civil engineering		
§ 3.1.8	Landscape design		
§ 3.1.9	Architectural Interior Design (B252™–2007)		
§ 3.1.10	Value Analysis (B204™–2007)		
§ 3.1.11	Detailed cost estimating		
§ 3.1.12	On-site project representation		
§ 3.1.13	Conformed construction documents		
§ 3.1.14	As-designed record drawings		
§ 3.1.15	As-constructed record drawings		
§ 3.1.16	Post occupancy evaluation		
§ 3.1.17	Facility Support Services (B210™–2007)		
§ 3.1.18	Tenant-related services		
§ 3.1.19	Coordination of Owner's consultants		
§ 3.1.20	Telecommunications/data design		
§ 3.1.21	Security Evaluation and Planning (B206™–2007)		
§ 3.1.22	Commissioning (B211™–2007)		
§ 3.1.23	Extensive environmentally responsible design		
§ 3.1.24	LEED® Certification (B214™–2007)		

7

Additional Services		Responsibility *(Architect, Owner or Not Provided)*	Location of Service Description *(Section 3.2 below or in an exhibit attached to this document and identified below)*
§ 3.1.25	Fast-track design services		
§ 3.1.26	Historic Preservation (B205™–2007)		
§ 3.1.27	Furniture, Finishings, and Equipment Design (B253™–2007)		
§ 3.1.28	Other		

§ 3.2 Insert a description of each Additional Service designated above as the Architect's responsibility, if not further described in an exhibit attached to this document.

§ 3.3 Additional Services may be provided after execution of this Agreement, without invalidating the Agreement. Except for services required due to the fault of the Architect, any Additional Services provided in accordance with this Section 3.3 shall entitle the Architect to compensation pursuant to Section 6.3.

§ 3.3.1 Upon recognizing the need to perform the following Additional Services, the Architect shall notify the Owner with reasonable promptness and explain the facts and circumstances giving rise to the need. The Architect shall not proceed to provide the following services until the Architect receives the Owner's written authorization:

.1 Services necessitated by a change in the Initial Information, previous instructions or approvals given by the Owner, or a material change in the Project including, but not limited to, size, quality, complexity, the Owner's schedule or budget for Cost of the Work, or procurement or delivery method;

.2 Services necessitated by the Owner's request for extensive environmentally responsible design alternatives, such as unique system designs, in-depth material research, energy modeling, or LEED® certification;

.3 Changing or editing previously prepared Instruments of Service necessitated by the enactment or revision of codes, laws or regulations or official interpretations;

.4 Services necessitated by decisions of the Owner not rendered in a timely manner or any other failure of performance on the part of the Owner or the Owner's consultants or contractors;

.5 Preparing digital data for transmission to the Owner's consultants and contractors, or to other Owner authorized recipients;

.6 Preparation of design and documentation for alternate bid or proposal requests proposed by the Owner;

.7 Preparation for, and attendance at, a public presentation, meeting or hearing;

.8 Preparation for, and attendance at a dispute resolution proceeding or legal proceeding, except where the Architect is party thereto;

.9 Evaluation of the qualifications of bidders or persons providing proposals;

.10 Consultation concerning replacement of Work resulting from fire or other cause during construction; or

.11 Assistance to the Initial Decision Maker, if other than the Architect.

§ 3.3.2 To avoid delay in the Construction Phase, the Architect shall provide the following Additional Services, notify the Owner with reasonable promptness, and explain the facts and circumstances giving rise to the need. If the Owner subsequently determines that all or parts of those services are not required, the Owner shall give prompt written notice to the Architect, and the Owner shall have no further obligation to compensate the Architect for those services:

.1 Reviewing a Contractor's submittal out of sequence from the submittal schedule agreed to by the Architect;

.2 Responding to the Contractor's requests for information that are not prepared in accordance with the Contract Documents or where such information is available to the Contractor from a careful study and comparison of the Contract Documents, field conditions, other Owner-provided information, Contractor-prepared coordination drawings, or prior Project correspondence or documentation;

.3   Preparing Change Orders and Construction Change Directives that require evaluation of Contractor's proposals and supporting data, or the preparation or revision of Instruments of Service;

.4   Evaluating an extensive number of Claims as the Initial Decision Maker;

.5   Evaluating substitutions proposed by the Owner or Contractor and making subsequent revisions to Instruments of Service resulting therefrom; or

.6   To the extent the Architect's Basic Services are affected, providing Construction Phase Services 60 days after (1) the date of Substantial Completion of the Work or (2) the anticipated date of Substantial Completion, identified in Initial Information, whichever is earlier.

§ 3.3.3 The Architect shall provide Construction Phase Services exceeding the limits set forth below as Additional Services. When the limits below are reached, the Architect shall notify the Owner:

.1   (   ) reviews of each Shop Drawing, Product Data item, sample and similar submittal of the Contractor

.2   (   ) visits to the site by the Architect over the duration of the Project during construction

.3   (   ) inspections for any portion of the Work to determine whether such portion of the Work is substantially complete in accordance with the requirements of the Contract Documents

.4   (   ) inspections for any portion of the Work to determine final completion

§ 3.3.4 If the services covered by this Agreement have not been completed within   (   ) months of the date of this Agreement, through no fault of the Architect, extension of the Architect's services beyond that time shall be compensated as Additional Services.

ARTICLE 4 OWNER'S RESPONSIBILITIES

§ 4.1 The Owner shall establish and periodically update the Owner's budget for the Project, including (1) the budget for the Cost of the Work as defined in Section 5.1; (2) the Owner's other costs; and, (3) reasonable contingencies related to all of these costs. If the Owner significantly increases or decreases the Owner's budget for the Cost of the Work, the Owner shall notify the Architect. The Owner and the Architect shall thereafter agree to a corresponding change in the Project's scope and quality.

§ 4.2 The Owner shall furnish surveys to describe physical characteristics, legal limitations and utility locations for the site of the Project, and a written legal description of the site. The surveys and legal information shall include, as applicable, grades and lines of streets, alleys, pavements and adjoining property and structures; designated wetlands; adjacent drainage; rights-of-way, restrictions, easements, encroachments, zoning, deed restrictions, boundaries and contours of the site; locations, dimensions and necessary data with respect to existing buildings, other improvements and trees; and information concerning available utility services and lines, both public and private, above and below grade, including inverts and depths. All the information on the survey shall be referenced to a Project benchmark.

§ 4.3 The Owner shall furnish services of geotechnical engineers, which may include but are not limited to test borings, test pits, determinations of soil bearing values, percolation tests, evaluations of hazardous materials, seismic evaluation, ground corrosion tests and resistivity tests, including necessary operations for anticipating subsoil conditions, with written reports and appropriate recommendations.

§ 4.4 The Owner shall furnish tests, inspections and reports required by law or the Contract Documents, such as structural, mechanical, and chemical tests, tests for air and water pollution, and tests for hazardous materials.

§ 4.5 Except as otherwise provided in this Agreement, or when direct communications have been specially authorized, the Owner shall endeavor to communicate with the Contractor and the Architect's consultants through the Architect about matters arising out of or relating to the Contract Documents. The Owner shall promptly notify the Architect of any direct communications that may affect the Architect's services.

§ 4.6 Before executing the Contract for Construction, the Owner shall coordinate the Architect's duties and responsibilities set forth in the Contract for Construction with the Architect's services set forth in this Agreement. The Owner shall provide the Architect a copy of the executed agreement between the Owner and Contractor, including the General Conditions of the Contract for Construction.

§ 4.7 The Owner shall provide the Architect access to the Project site prior to commencement of the Work and shall obligate the Contractor to provide the Architect access to the Work wherever it is in preparation or progress.

ARTICLE 5 COST OF THE WORK

§ 5.1 For purposes of this Agreement, the Cost of the Work shall be the total cost to the Owner to construct all elements of the Project designed or specified by the Architect and shall include contractors' general conditions costs, overhead and profit. The Cost of the Work does not include the compensation of the Architect, the costs of the land, rights-of-way, financing, contingencies for changes in the Work or other costs that are the responsibility of the Owner.

§ 5.2 The Owner's budget for the Cost of the Work is provided in Initial Information, and may be adjusted throughout the Project as required under Sections 4.1, 5.4 and 5.5. Evaluations of the Owner's budget for the Cost of the Work, the preliminary estimate of the Cost of the Work and updated estimates of the Cost of the Work prepared by the Architect, represent the Architect's judgment as a design professional. It is recognized, however, that neither the Architect nor the Owner has control over the cost of labor, materials or equipment; the Contractor's methods of determining bid prices; or competitive bidding, market or negotiating conditions. Accordingly, the Architect cannot and does not warrant or represent that bids or negotiated prices will not vary from the Owner's budget for the Cost of the Work or from any estimate of the Cost of the Work or evaluation prepared or agreed to by the Architect.

§ 5.3 In preparing estimates of the Cost of Work, the Architect shall be permitted to include contingencies for design, bidding and price escalation; to determine what materials, equipment, component systems and types of construction are to be included in the Contract Documents; to make reasonable adjustments in the program and scope of the Project; and to include in the Contract Documents alternate bids as may be necessary to adjust the estimated Cost of the Work to meet the Owner's budget for the Cost of the Work. The Architect's estimate of the Cost of the Work shall be based on current area, volume or similar conceptual estimating techniques. If the Owner requests detailed cost estimating services, the Architect shall provide such services as an Additional Service under Article 3.

§ 5.4 If the Bidding or Negotiation Phase has not commenced within 90 days after the Architect submits the Construction Documents to the Owner, through no fault of the Architect, the Owner's budget for the Cost of the Work shall be adjusted to reflect changes in the general level of prices in the applicable construction market.

§ 5.5 If at any time the Architect's estimate of the Cost of the Work exceeds the Owner's budget for the Cost of the Work, the Architect shall make appropriate recommendations to the Owner to adjust the Project's size, quality or budget for the Cost of the Work, and the Owner shall cooperate with the Architect in making such adjustments.

§ 5.6 If the Owner's budget for the Cost of the Work at the conclusion of the Construction Documents Phase Services is exceeded by the lowest bona fide bid or negotiated proposal, the Owner shall
.1 give written approval of an increase in the budget for the Cost of the Work;
.2 authorize rebidding or renegotiating of the Project within a reasonable time;
.3 terminate in accordance with Section 5.5 of AIA Document B102–2007;
.4 in consultation with the Architect, revise the Project program, scope, or quality as required to reduce the Cost of the Work; or
.5 implement any other mutually acceptable alternative.

§ 5.7 If the Owner chooses to proceed under Section 5.6.4, the Architect, without additional compensation, shall modify the Construction Documents as necessary to comply with the Owner's budget for the Cost of the Work at the conclusion of the Construction Documents Phase Services, or the budget as adjusted under Section 6.6.1. The Architect's modification of the Construction Documents shall be the limit of the Architect's responsibility under this Article 5.

ARTICLE 6 COMPENSATION

§ 6.1 For the Architect's Basic Services described under Article 2, the Owner shall compensate the Architect as follows:
(Insert amount of, or basis for, compensation.)

§ 6.2 For Additional Services designated in Section 3.1, the Owner shall compensate the Architect as follows:
(Insert amount of, or basis for, compensation. If necessary, list specific services to which particular methods of compensation apply.)

§ 6.3 For Additional Services that may arise during the course of the Project, including those under Section 3.3, during the course of the Project, the Owner shall compensate the Architect as follows:
(Insert amount of, or basis for, compensation.)

§ 6.4 Compensation for Additional Services of the Architect's consultants when not included in Section 6.2 or 6.3, shall be the amount invoiced to the Architect plus a fee of _____ percent (_____ %), or as otherwise stated below:

§ 6.5 Where compensation for Basic Services is based on a stipulated sum or percentage of the Cost of the Work, the compensation for each phase of services shall be as follows:

Schematic Design Phase:	percent (	%)
Design Development Phase:	percent (	%)
Construction Documents Phase:	percent (	%)
Bidding or Negotiation Phase:	percent (	%)
Construction Phase:	percent (	%)
Total Basic Compensation	one hundred percent (	100.00%)

§ 6.6 When compensation is based on a percentage of the Cost of the Work and any portions of the Project are deleted or otherwise not constructed, compensation for those portions of the Project shall be payable to the extent services are performed on those portions, in accordance with the schedule set forth in Section 6.5 based on (1) the lowest bona fide bid or negotiated proposal, or (2) if no such bid or proposal is received, the most recent estimate of the Cost of the Work for such portions of the Project. The Architect shall be entitled to compensation in accordance with this Agreement for all services performed whether or not the Construction Phase is commenced.

§ 6.7 The hourly billing rates for services of the Architect and the Architect's consultants, if any, are set forth below. The rates shall be adjusted in accordance with the Architect's and Architect's consultants' normal review practices.
(If applicable, attach an exhibit of hourly billing rates or insert them below.)

ARTICLE 7 ATTACHMENTS AND EXHIBITS
The following attachments and exhibits, if any, are incorporated herein by reference:
(List other documents, if any, including Exhibit A, Initial Information, and any exhibits relied on in Section 3.1.)

Document B201™ – 2007 Exhibit A

Initial Information

for the following PROJECT:
(Name and location or address)

THE OWNER:
(Name and address)

THE ARCHITECT:
(Name and address)

This Agreement is based on the following information.
(Note the disposition for the following items by inserting the requested information or a statement such as "not applicable," "unknown at time of execution" or "to be determined later by mutual agreement.")

> This document has important legal consequences. Consultation with an attorney is encouraged with respect to its completion or modification.

ARTICLE A.1 PROJECT INFORMATION

§ A.1.1 The Owner's program for the Project:
(Identify documentation or state the manner in which the program will be developed.)

§ A.1.2 The Project's physical characteristics:
(Identify or describe, if appropriate, size, location, dimensions, or other pertinent information, such as geotechnical reports; site, boundary and topographic surveys; traffic and utility studies; availability of public and private utilities and services; legal description of the site; etc.)

§ A.1.3 The Owner's budget for the Cost of the Work, as defined in Section 6.1:
(Provide total and, if known, a line item break down.)

§ A.1.4 The Owner's other anticipated scheduling information, if any, not provided in Section 1.2:

§ A.1.5 The Owner intends the following procurement or delivery method for the Project:
(Identify method such as competitive bid, negotiated contract, or construction management.)

§ A.1.6 Other Project information:
(Identify special characteristics or needs of the Project not provided elsewhere, such as environmentally responsible design or historic preservation requirements.)

ARTICLE A.2 PROJECT TEAM
§ A.2.1 The Owner identifies the following representative in accordance with Section 5.3:
(List name, address and other information.)

§ A.2.2 The persons or entities, in addition to the Owner's representative, who are required to review the Architect's submittals to the Owner are as follows:
(List name, address and other information.)

§ A.2.3 The Owner will retain the following consultants and contractors:
(List discipline and, if known, identify them by name and address.)

§ A.2.4 The Architect identifies the following representative in accordance with Section 2.3:
(List name, address and other information.)

§ A.2.5 The Architect will retain the consultants identified in Sections A.2.5.1 and A.2.5.2.
(List discipline and, if known, identify them by name and address.)

§ A.2.5.1 Consultants retained under Basic Services:
 .1 Structural Engineer

 .2 Mechanical Engineer

 .3 Electrical Engineer

§ A.2.5.2 Consultants retained under Additional Services:

§ A.2.6 Other Initial Information on which the Agreement is based:
(Provide other Initial Information.)

APPENDIX E

AIA Document
A201-2007

AIA® Document A201™ – 2007

General Conditions of the Contract for Construction

for the following PROJECT:
(Name and location or address)

THE OWNER:
(Name and address)

THE ARCHITECT:
(Name and address)

This document has important legal consequences. Consultation with an attorney is encouraged with respect to its completion or modification.

TABLE OF ARTICLES

INDEX
(Numbers and Topics in Bold are Section Headings)

Init.

/

Init.

/

3

Init.

/

5

6

621

Init.

/

ARTICLE 1 GENERAL PROVISIONS
§ 1.1 BASIC DEFINITIONS
§ 1.1.1 THE CONTRACT DOCUMENTS

The Contract Documents are enumerated in the Agreement between the Owner and Contractor (hereinafter the Agreement) and consist of the Agreement, Conditions of the Contract (General, Supplementary and other Conditions), Drawings, Specifications, Addenda issued prior to execution of the Contract, other documents listed in the Agreement and Modifications issued after execution of the Contract. A Modification is (1) a written amendment to the Contract signed by both parties, (2) a Change Order, (3) a Construction Change Directive or (4) a written order for a minor change in the Work issued by the Architect. Unless specifically enumerated in the Agreement, the Contract Documents do not include the advertisement or invitation to bid, Instructions to Bidders, sample forms, other information furnished by the Owner in anticipation of receiving bids or proposals, the Contractor's bid or proposal, or portions of Addenda relating to bidding requirements.

§ 1.1.2 THE CONTRACT

The Contract Documents form the Contract for Construction. The Contract represents the entire and integrated agreement between the parties hereto and supersedes prior negotiations, representations or agreements, either written or oral. The Contract may be amended or modified only by a Modification. The Contract Documents shall not be construed to create a contractual relationship of any kind (1) between the Contractor and the Architect or the Architect's consultants, (2) between the Owner and a Subcontractor or a Sub-subcontractor, (3) between the Owner and the Architect or the Architect's consultants or (4) between any persons or entities other than the Owner and the Contractor. The Architect shall, however, be entitled to performance and enforcement of obligations under the Contract intended to facilitate performance of the Architect's duties.

§ 1.1.3 THE WORK

The term "Work" means the construction and services required by the Contract Documents, whether completed or partially completed, and includes all other labor, materials, equipment and services provided or to be provided by the Contractor to fulfill the Contractor's obligations. The Work may constitute the whole or a part of the Project.

§ 1.1.4 THE PROJECT

The Project is the total construction of which the Work performed under the Contract Documents may be the whole or a part and which may include construction by the Owner and by separate contractors.

§ 1.1.5 THE DRAWINGS

The Drawings are the graphic and pictorial portions of the Contract Documents showing the design, location and dimensions of the Work, generally including plans, elevations, sections, details, schedules and diagrams.

§ 1.1.6 THE SPECIFICATIONS

The Specifications are that portion of the Contract Documents consisting of the written requirements for materials, equipment, systems, standards and workmanship for the Work, and performance of related services.

§ 1.1.7 INSTRUMENTS OF SERVICE

Instruments of Service are representations, in any medium of expression now known or later developed, of the tangible and intangible creative work performed by the Architect and the Architect's consultants under their respective professional services agreements. Instruments of Service may include, without limitation, studies, surveys, models, sketches, drawings, specifications, and other similar materials.

§ 1.1.8 INITIAL DECISION MAKER

The Initial Decision Maker is the person identified in the Agreement to render initial decisions on Claims in accordance with Section 15.2 and certify termination of the Agreement under Section 14.2.2.

§ 1.2 CORRELATION AND INTENT OF THE CONTRACT DOCUMENTS

§ 1.2.1 The intent of the Contract Documents is to include all items necessary for the proper execution and completion of the Work by the Contractor. The Contract Documents are complementary, and what is required by one shall be as binding as if required by all; performance by the Contractor shall be required only to the extent consistent with the Contract Documents and reasonably inferable from them as being necessary to produce the indicated results.

§ 1.2.2 Organization of the Specifications into divisions, sections and articles, and arrangement of Drawings shall not control the Contractor in dividing the Work among Subcontractors or in establishing the extent of Work to be performed by any trade.

§ 1.2.3 Unless otherwise stated in the Contract Documents, words that have well-known technical or construction industry meanings are used in the Contract Documents in accordance with such recognized meanings.

§ 1.3 CAPITALIZATION
Terms capitalized in these General Conditions include those that are (1) specifically defined, (2) the titles of numbered articles or (3) the titles of other documents published by the American Institute of Architects.

§ 1.4 INTERPRETATION
In the interest of brevity the Contract Documents frequently omit modifying words such as "all" and "any" and articles such as "the" and "an," but the fact that a modifier or an article is absent from one statement and appears in another is not intended to affect the interpretation of either statement.

§ 1.5 OWNERSHIP AND USE OF DRAWINGS, SPECIFICATIONS AND OTHER INSTRUMENTS OF SERVICE
§ 1.5.1 The Architect and the Architect's consultants shall be deemed the authors and owners of their respective Instruments of Service, including the Drawings and Specifications, and will retain all common law, statutory and other reserved rights, including copyrights. The Contractor, Subcontractors, Sub-subcontractors, and material or equipment suppliers shall not own or claim a copyright in the Instruments of Service. Submittal or distribution to meet official regulatory requirements or for other purposes in connection with this Project is not to be construed as publication in derogation of the Architect's or Architect's consultants' reserved rights.

§ 1.5.2 The Contractor, Subcontractors, Sub-subcontractors and material or equipment suppliers are authorized to use and reproduce the Instruments of Service provided to them solely and exclusively for execution of the Work. All copies made under this authorization shall bear the copyright notice, if any, shown on the Instruments of Service. The Contractor, Subcontractors, Sub-subcontractors, and material or equipment suppliers may not use the Instruments of Service on other projects or for additions to this Project outside the scope of the Work without the specific written consent of the Owner, Architect and the Architect's consultants.

§ 1.6 TRANSMISSION OF DATA IN DIGITAL FORM
If the parties intend to transmit Instruments of Service or any other information or documentation in digital form, they shall endeavor to establish necessary protocols governing such transmissions, unless otherwise already provided in the Agreement or the Contract Documents.

ARTICLE 2 OWNER
§ 2.1 GENERAL
§ 2.1.1 The Owner is the person or entity identified as such in the Agreement and is referred to throughout the Contract Documents as if singular in number. The Owner shall designate in writing a representative who shall have express authority to bind the Owner with respect to all matters requiring the Owner's approval or authorization. Except as otherwise provided in Section 4.2.1, the Architect does not have such authority. The term "Owner" means the Owner or the Owner's authorized representative.

§ 2.1.2 The Owner shall furnish to the Contractor within fifteen days after receipt of a written request, information necessary and relevant for the Contractor to evaluate, give notice of or enforce mechanic's lien rights. Such information shall include a correct statement of the record legal title to the property on which the Project is located, usually referred to as the site, and the Owner's interest therein.

§ 2.2 INFORMATION AND SERVICES REQUIRED OF THE OWNER
§ 2.2.1 Prior to commencement of the Work, the Contractor may request in writing that the Owner provide reasonable evidence that the Owner has made financial arrangements to fulfill the Owner's obligations under the Contract. Thereafter, the Contractor may only request such evidence if (1) the Owner fails to make payments to the Contractor as the Contract Documents require; (2) a change in the Work materially changes the Contract Sum; or (3) the Contractor identifies in writing a reasonable concern regarding the Owner's ability to make payment when due. The Owner shall furnish such evidence as a condition precedent to commencement or continuation of the Work or the portion of the Work affected by a material change. After the Owner furnishes the evidence, the Owner shall not materially vary such financial arrangements without prior notice to the Contractor.

§ 2.2.2 Except for permits and fees that are the responsibility of the Contractor under the Contract Documents, including those required under Section 3.7.1, the Owner shall secure and pay for necessary approvals, easements, assessments and charges required for construction, use or occupancy of permanent structures or for permanent changes in existing facilities.

§ 2.2.3 The Owner shall furnish surveys describing physical characteristics, legal limitations and utility locations for the site of the Project, and a legal description of the site. The Contractor shall be entitled to rely on the accuracy of information furnished by the Owner but shall exercise proper precautions relating to the safe performance of the Work.

§ 2.2.4 The Owner shall furnish information or services required of the Owner by the Contract Documents with reasonable promptness. The Owner shall also furnish any other information or services under the Owner's control and relevant to the Contractor's performance of the Work with reasonable promptness after receiving the Contractor's written request for such information or services.

§ 2.2.5 Unless otherwise provided in the Contract Documents, the Owner shall furnish to the Contractor one copy of the Contract Documents for purposes of making reproductions pursuant to Section 1.5.2.

§ 2.3 OWNER'S RIGHT TO STOP THE WORK
If the Contractor fails to correct Work that is not in accordance with the requirements of the Contract Documents as required by Section 12.2 or repeatedly fails to carry out Work in accordance with the Contract Documents, the Owner may issue a written order to the Contractor to stop the Work, or any portion thereof, until the cause for such order has been eliminated; however, the right of the Owner to stop the Work shall not give rise to a duty on the part of the Owner to exercise this right for the benefit of the Contractor or any other person or entity, except to the extent required by Section 6.1.3.

§ 2.4 OWNER'S RIGHT TO CARRY OUT THE WORK
If the Contractor defaults or neglects to carry out the Work in accordance with the Contract Documents and fails within a ten-day period after receipt of written notice from the Owner to commence and continue correction of such default or neglect with diligence and promptness, the Owner may, without prejudice to other remedies the Owner may have, correct such deficiencies. In such case an appropriate Change Order shall be issued deducting from payments then or thereafter due the Contractor the reasonable cost of correcting such deficiencies, including Owner's expenses and compensation for the Architect's additional services made necessary by such default, neglect or failure. Such action by the Owner and amounts charged to the Contractor are both subject to prior approval of the Architect. If payments then or thereafter due the Contractor are not sufficient to cover such amounts, the Contractor shall pay the difference to the Owner.

ARTICLE 3   CONTRACTOR
§ 3.1 GENERAL
§ 3.1.1 The Contractor is the person or entity identified as such in the Agreement and is referred to throughout the Contract Documents as if singular in number. The Contractor shall be lawfully licensed, if required in the jurisdiction where the Project is located. The Contractor shall designate in writing a representative who shall have express authority to bind the Contractor with respect to all matters under this Contract. The term "Contractor" means the Contractor or the Contractor's authorized representative.

§ 3.1.2 The Contractor shall perform the Work in accordance with the Contract Documents.

§ 3.1.3 The Contractor shall not be relieved of obligations to perform the Work in accordance with the Contract Documents either by activities or duties of the Architect in the Architect's administration of the Contract, or by tests, inspections or approvals required or performed by persons or entities other than the Contractor.

§ 3.2 REVIEW OF CONTRACT DOCUMENTS AND FIELD CONDITIONS BY CONTRACTOR
§ 3.2.1 Execution of the Contract by the Contractor is a representation that the Contractor has visited the site, become generally familiar with local conditions under which the Work is to be performed and correlated personal observations with requirements of the Contract Documents.

§ 3.2.2 Because the Contract Documents are complementary, the Contractor shall, before starting each portion of the Work, carefully study and compare the various Contract Documents relative to that portion of the Work, as well as the information furnished by the Owner pursuant to Section 2.2.3, shall take field measurements of any existing conditions related to that portion of the Work, and shall observe any conditions at the site affecting it. These obligations are for the purpose of facilitating coordination and construction by the Contractor and are not for the purpose of discovering errors, omissions, or inconsistencies in the Contract Documents; however, the Contractor shall promptly report to the Architect any errors, inconsistencies or omissions discovered by or made known to the Contractor as a request for information in such form as the Architect may require. It is recognized that the Contractor's review is made in the Contractor's capacity as a contractor and not as a licensed design professional, unless otherwise specifically provided in the Contract Documents.

§ 3.2.3 The Contractor is not required to ascertain that the Contract Documents are in accordance with applicable laws, statutes, ordinances, codes, rules and regulations, or lawful orders of public authorities, but the Contractor shall promptly report to the Architect any nonconformity discovered by or made known to the Contractor as a request for information in such form as the Architect may require.

§ 3.2.4 If the Contractor believes that additional cost or time is involved because of clarifications or instructions the Architect issues in response to the Contractor's notices or requests for information pursuant to Sections 3.2.2 or 3.2.3, the Contractor shall make Claims as provided in Article 15. If the Contractor fails to perform the obligations of Sections 3.2.2 or 3.2.3, the Contractor shall pay such costs and damages to the Owner as would have been avoided if the Contractor had performed such obligations. If the Contractor performs those obligations, the Contractor shall not be liable to the Owner or Architect for damages resulting from errors, inconsistencies or omissions in the Contract Documents, for differences between field measurements or conditions and the Contract Documents, or for nonconformities of the Contract Documents to applicable laws, statutes, ordinances, codes, rules and regulations, and lawful orders of public authorities.

§ 3.3 SUPERVISION AND CONSTRUCTION PROCEDURES

§ 3.3.1 The Contractor shall supervise and direct the Work, using the Contractor's best skill and attention. The Contractor shall be solely responsible for, and have control over, construction means, methods, techniques, sequences and procedures and for coordinating all portions of the Work under the Contract, unless the Contract Documents give other specific instructions concerning these matters. If the Contract Documents give specific instructions concerning construction means, methods, techniques, sequences or procedures, the Contractor shall evaluate the jobsite safety thereof and, except as stated below, shall be fully and solely responsible for the jobsite safety of such means, methods, techniques, sequences or procedures. If the Contractor determines that such means, methods, techniques, sequences or procedures may not be safe, the Contractor shall give timely written notice to the Owner and Architect and shall not proceed with that portion of the Work without further written instructions from the Architect. If the Contractor is then instructed to proceed with the required means, methods, techniques, sequences or procedures without acceptance of changes proposed by the Contractor, the Owner shall be solely responsible for any loss or damage arising solely from those Owner-required means, methods, techniques, sequences or procedures.

§ 3.3.2 The Contractor shall be responsible to the Owner for acts and omissions of the Contractor's employees, Subcontractors and their agents and employees, and other persons or entities performing portions of the Work for, or on behalf of, the Contractor or any of its Subcontractors.

§ 3.3.3 The Contractor shall be responsible for inspection of portions of Work already performed to determine that such portions are in proper condition to receive subsequent Work.

§ 3.4 LABOR AND MATERIALS

§ 3.4.1 Unless otherwise provided in the Contract Documents, the Contractor shall provide and pay for labor, materials, equipment, tools, construction equipment and machinery, water, heat, utilities, transportation, and other facilities and services necessary for proper execution and completion of the Work, whether temporary or permanent and whether or not incorporated or to be incorporated in the Work.

§ 3.4.2 Except in the case of minor changes in the Work authorized by the Architect in accordance with Sections 3.12.8 or 7.4, the Contractor may make substitutions only with the consent of the Owner, after evaluation by the Architect and in accordance with a Change Order or Construction Change Directive.

§ 3.4.3 The Contractor shall enforce strict discipline and good order among the Contractor's employees and other persons carrying out the Work. The Contractor shall not permit employment of unfit persons or persons not properly skilled in tasks assigned to them.

§ 3.5 WARRANTY
The Contractor warrants to the Owner and Architect that materials and equipment furnished under the Contract will be of good quality and new unless the Contract Documents require or permit otherwise. The Contractor further warrants that the Work will conform to the requirements of the Contract Documents and will be free from defects, except for those inherent in the quality of the Work the Contract Documents require or permit. Work, materials, or equipment not conforming to these requirements may be considered defective. The Contractor's warranty excludes remedy for damage or defect caused by abuse, alterations to the Work not executed by the Contractor, improper or insufficient maintenance, improper operation, or normal wear and tear and normal usage. If required by the Architect, the Contractor shall furnish satisfactory evidence as to the kind and quality of materials and equipment.

§ 3.6 TAXES
The Contractor shall pay sales, consumer, use and similar taxes for the Work provided by the Contractor that are legally enacted when bids are received or negotiations concluded, whether or not yet effective or merely scheduled to go into effect.

§ 3.7 PERMITS, FEES, NOTICES, AND COMPLIANCE WITH LAWS
§ 3.7.1 Unless otherwise provided in the Contract Documents, the Contractor shall secure and pay for the building permit as well as for other permits, fees, licenses, and inspections by government agencies necessary for proper execution and completion of the Work that are customarily secured after execution of the Contract and legally required at the time bids are received or negotiations concluded.

§ 3.7.2 The Contractor shall comply with and give notices required by applicable laws, statutes, ordinances, codes, rules and regulations, and lawful orders of public authorities applicable to performance of the Work.

§ 3.7.3 If the Contractor performs Work knowing it to be contrary to applicable laws, statutes, ordinances, codes, rules and regulations, or lawful orders of public authorities, the Contractor shall assume appropriate responsibility for such Work and shall bear the costs attributable to correction.

§ 3.7.4 Concealed or Unknown Conditions. If the Contractor encounters conditions at the site that are (1) subsurface or otherwise concealed physical conditions that differ materially from those indicated in the Contract Documents or (2) unknown physical conditions of an unusual nature that differ materially from those ordinarily found to exist and generally recognized as inherent in construction activities of the character provided for in the Contract Documents, the Contractor shall promptly provide notice to the Owner and the Architect before conditions are disturbed and in no event later than 21 days after first observance of the conditions. The Architect will promptly investigate such conditions and, if the Architect determines that they differ materially and cause an increase or decrease in the Contractor's cost of, or time required for, performance of any part of the Work, will recommend an equitable adjustment in the Contract Sum or Contract Time, or both. If the Architect determines that the conditions at the site are not materially different from those indicated in the Contract Documents and that no change in the terms of the Contract is justified, the Architect shall promptly notify the Owner and Contractor in writing, stating the reasons. If either party disputes the Architect's determination or recommendation, that party may proceed as provided in Article 15.

§ 3.7.5 If, in the course of the Work, the Contractor encounters human remains or recognizes the existence of burial markers, archaeological sites or wetlands not indicated in the Contract Documents, the Contractor shall immediately suspend any operations that would affect them and shall notify the Owner and Architect. Upon receipt of such notice, the Owner promptly take any action necessary to obtain governmental authorization required to resume the operations. The Contractor shall continue to suspend such operations until otherwise instructed by the Owner but shall continue with all other operations that do not affect those remains or features. Requests for adjustments in the Contract Sum and Contract Time arising from the existence of such remains or features may be made as provided in Article 15.

§ 3.8 ALLOWANCES
§ 3.8.1 The Contractor shall include in the Contract Sum all allowances stated in the Contract Documents. Items covered by allowances shall be supplied for such amounts and by such persons or entities as the Owner may direct,

but the Contractor shall not be required to employ persons or entities to whom the Contractor has reasonable objection.

§ 3.8.2 Unless otherwise provided in the Contract Documents,

    .1   allowances shall cover the cost to the Contractor of materials and equipment delivered at the site and all required taxes, less applicable trade discounts;

    .2   Contractor's costs for unloading and handling at the site, labor, installation costs, overhead, profit and other expenses contemplated for stated allowance amounts shall be included in the Contract Sum but not in the allowances; and

    .3   whenever costs are more than or less than allowances, the Contract Sum shall be adjusted accordingly by Change Order. The amount of the Change Order shall reflect (1) the difference between actual costs and the allowances under Section 3.8.2.1 and (2) changes in Contractor's costs under Section 3.8.2.2.

§ 3.8.3 Materials and equipment under an allowance shall be selected by the Owner with reasonable promptness.

§ 3.9 SUPERINTENDENT

§ 3.9.1 The Contractor shall employ a competent superintendent and necessary assistants who shall be in attendance at the Project site during performance of the Work. The superintendent shall represent the Contractor, and communications given to the superintendent shall be as binding as if given to the Contractor.

§ 3.9.2 The Contractor, as soon as practicable after award of the Contract, shall furnish in writing to the Owner through the Architect the name and qualifications of a proposed superintendent. The Architect may reply within 14 days to the Contractor in writing stating (1) whether the Owner or the Architect has reasonable objection to the proposed superintendent or (2) that the Architect requires additional time to review. Failure of the Architect to reply within the 14 day period shall constitute notice of no reasonable objection.

§ 3.9.3 The Contractor shall not employ a proposed superintendent to whom the Owner or Architect has made reasonable and timely objection. The Contractor shall not change the superintendent without the Owner's consent, which shall not unreasonably be withheld or delayed.

§ 3.10 CONTRACTOR'S CONSTRUCTION SCHEDULES

§ 3.10.1 The Contractor, promptly after being awarded the Contract, shall prepare and submit for the Owner's and Architect's information a Contractor's construction schedule for the Work. The schedule shall not exceed time limits current under the Contract Documents, shall be revised at appropriate intervals as required by the conditions of the Work and Project, shall be related to the entire Project to the extent required by the Contract Documents, and shall provide for expeditious and practicable execution of the Work.

§ 3.10.2 The Contractor shall prepare a submittal schedule, promptly after being awarded the Contract and thereafter as necessary to maintain a current submittal schedule, and shall submit the schedule(s) for the Architect's approval. The Architect's approval shall not unreasonably be delayed or withheld. The submittal schedule shall (1) be coordinated with the Contractor's construction schedule, and (2) allow the Architect reasonable time to review submittals. If the Contractor fails to submit a submittal schedule, the Contractor shall not be entitled to any increase in Contract Sum or extension of Contract Time based on the time required for review of submittals.

§ 3.10.3 The Contractor shall perform the Work in general accordance with the most recent schedules submitted to the Owner and Architect.

§ 3.11 DOCUMENTS AND SAMPLES AT THE SITE

The Contractor shall maintain at the site for the Owner one copy of the Drawings, Specifications, Addenda, Change Orders and other Modifications, in good order and marked currently to indicate field changes and selections made during construction, and one copy of approved Shop Drawings, Product Data, Samples and similar required submittals. These shall be available to the Architect and shall be delivered to the Architect for submittal to the Owner upon completion of the Work as a record of the Work as constructed.

§ 3.12 SHOP DRAWINGS, PRODUCT DATA AND SAMPLES

§ 3.12.1 Shop Drawings are drawings, diagrams, schedules and other data specially prepared for the Work by the Contractor or a Subcontractor, Sub-subcontractor, manufacturer, supplier or distributor to illustrate some portion of the Work.

§ 3.12.2 Product Data are illustrations, standard schedules, performance charts, instructions, brochures, diagrams and other information furnished by the Contractor to illustrate materials or equipment for some portion of the Work.

§ 3.12.3 Samples are physical examples that illustrate materials, equipment or workmanship and establish standards by which the Work will be judged.

§ 3.12.4 Shop Drawings, Product Data, Samples and similar submittals are not Contract Documents. Their purpose is to demonstrate the way by which the Contractor proposes to conform to the information given and the design concept expressed in the Contract Documents for those portions of the Work for which the Contract Documents require submittals. Review by the Architect is subject to the limitations of Section 4.2.7. Informational submittals upon which the Architect is not expected to take responsive action may be so identified in the Contract Documents. Submittals that are not required by the Contract Documents may be returned by the Architect without action.

§ 3.12.5 The Contractor shall review for compliance with the Contract Documents, approve and submit to the Architect Shop Drawings, Product Data, Samples and similar submittals required by the Contract Documents in accordance with the submittal schedule approved by the Architect or, in the absence of an approved submittal schedule, with reasonable promptness and in such sequence as to cause no delay in the Work or in the activities of the Owner or of separate contractors.

§ 3.12.6 By submitting Shop Drawings, Product Data, Samples and similar submittals, the Contractor represents to the Owner and Architect that the Contractor has (1) reviewed and approved them, (2) determined and verified materials, field measurements and field construction criteria related thereto, or will do so and (3) checked and coordinated the information contained within such submittals with the requirements of the Work and of the Contract Documents.

§ 3.12.7 The Contractor shall perform no portion of the Work for which the Contract Documents require submittal and review of Shop Drawings, Product Data, Samples or similar submittals until the respective submittal has been approved by the Architect.

§ 3.12.8 The Work shall be in accordance with approved submittals except that the Contractor shall not be relieved of responsibility for deviations from requirements of the Contract Documents by the Architect's approval of Shop Drawings, Product Data, Samples or similar submittals unless the Contractor has specifically informed the Architect in writing of such deviation at the time of submittal and (1) the Architect has given written approval to the specific deviation as a minor change in the Work, or (2) a Change Order or Construction Change Directive has been issued authorizing the deviation. The Contractor shall not be relieved of responsibility for errors or omissions in Shop Drawings, Product Data, Samples or similar submittals by the Architect's approval thereof.

§ 3.12.9 The Contractor shall direct specific attention, in writing or on resubmitted Shop Drawings, Product Data, Samples or similar submittals, to revisions other than those requested by the Architect on previous submittals. In the absence of such written notice, the Architect's approval of a resubmission shall not apply to such revisions.

§ 3.12.10 The Contractor shall not be required to provide professional services that constitute the practice of architecture or engineering unless such services are specifically required by the Contract Documents for a portion of the Work or unless the Contractor needs to provide such services in order to carry out the Contractor's responsibilities for construction means, methods, techniques, sequences and procedures. The Contractor shall not be required to provide professional services in violation of applicable law. If professional design services or certifications by a design professional related to systems, materials or equipment are specifically required of the Contractor by the Contract Documents, the Owner and the Architect will specify all performance and design criteria that such services must satisfy. The Contractor shall cause such services or certifications to be provided by a properly licensed design professional, whose signature and seal shall appear on all drawings, calculations, specifications, certifications, Shop Drawings and other submittals prepared by such professional. Shop Drawings and other submittals related to the Work designed or certified by such professional, if prepared by others, shall bear such professional's written approval when submitted to the Architect. The Owner and the Architect shall be entitled to rely upon the adequacy, accuracy and completeness of the services, certifications and approvals performed or provided by such design professionals, provided the Owner and Architect have specified to the Contractor all performance and design criteria that such services must satisfy. Pursuant to this Section 3.12.10, the Architect will review, approve or take other appropriate action on submittals only for the limited purpose of checking for conformance with information given and the design concept expressed in the Contract Documents. The Contractor shall not be responsible for the adequacy of the performance and design criteria specified in the Contract Documents.

§ 3.13 USE OF SITE
The Contractor shall confine operations at the site to areas permitted by applicable laws, statutes, ordinances, codes, rules and regulations, and lawful orders of public authorities and the Contract Documents and shall not unreasonably encumber the site with materials or equipment.

§ 3.14 CUTTING AND PATCHING
§ 3.14.1 The Contractor shall be responsible for cutting, fitting or patching required to complete the Work or to make its parts fit together properly. All areas requiring cutting, fitting and patching shall be restored to the condition existing prior to the cutting, fitting and patching, unless otherwise required by the Contract Documents.

§ 3.14.2 The Contractor shall not damage or endanger a portion of the Work or fully or partially completed construction of the Owner or separate contractors by cutting, patching or otherwise altering such construction, or by excavation. The Contractor shall not cut or otherwise alter such construction by the Owner or a separate contractor except with written consent of the Owner and of such separate contractor; such consent shall not be unreasonably withheld. The Contractor shall not unreasonably withhold from the Owner or a separate contractor the Contractor's consent to cutting or otherwise altering the Work.

§ 3.15 CLEANING UP
§ 3.15.1 The Contractor shall keep the premises and surrounding area free from accumulation of waste materials or rubbish caused by operations under the Contract. At completion of the Work, the Contractor shall remove waste materials, rubbish, the Contractor's tools, construction equipment, machinery and surplus materials from and about the Project.

§ 3.15.2 If the Contractor fails to clean up as provided in the Contract Documents, the Owner may do so and Owner shall be entitled to reimbursement from the Contractor.

§ 3.16 ACCESS TO WORK
The Contractor shall provide the Owner and Architect access to the Work in preparation and progress wherever located.

§ 3.17 ROYALTIES, PATENTS AND COPYRIGHTS
The Contractor shall pay all royalties and license fees. The Contractor shall defend suits or claims for infringement of copyrights and patent rights and shall hold the Owner and Architect harmless from loss on account thereof, but shall not be responsible for such defense or loss when a particular design, process or product of a particular manufacturer or manufacturers is required by the Contract Documents, or where the copyright violations are contained in Drawings, Specifications or other documents prepared by the Owner or Architect. However, if the Contractor has reason to believe that the required design, process or product is an infringement of a copyright or a patent, the Contractor shall be responsible for such loss unless such information is promptly furnished to the Architect.

§ 3.18 INDEMNIFICATION
§ 3.18.1 To the fullest extent permitted by law the Contractor shall indemnify and hold harmless the Owner, Architect, Architect's consultants, and agents and employees of any of them from and against claims, damages, losses and expenses, including but not limited to attorneys' fees, arising out of or resulting from performance of the Work, provided that such claim, damage, loss or expense is attributable to bodily injury, sickness, disease or death, or to injury to or destruction of tangible property (other than the Work itself), but only to the extent caused by the negligent acts or omissions of the Contractor, a Subcontractor, anyone directly or indirectly employed by them or anyone for whose acts they may be liable, regardless of whether or not such claim, damage, loss or expense is caused in part by a party indemnified hereunder. Such obligation shall not be construed to negate, abridge, or reduce

other rights or obligations of indemnity that would otherwise exist as to a party or person described in this Section 3.18.

§ 3.18.2 In claims against any person or entity indemnified under this Section 3.18 by an employee of the Contractor, a Subcontractor, anyone directly or indirectly employed by them or anyone for whose acts they may be liable, the indemnification obligation under Section 3.18.1 shall not be limited by a limitation on amount or type of damages, compensation or benefits payable by or for the Contractor or a Subcontractor under workers' compensation acts, disability benefit acts or other employee benefit acts.

ARTICLE 4 ARCHITECT
§ 4.1 GENERAL
§ 4.1.1 The Owner shall retain an architect lawfully licensed to practice architecture or an entity lawfully practicing architecture in the jurisdiction where the Project is located. That person or entity is identified as the Architect in the Agreement and is referred to throughout the Contract Documents as if singular in number.

§ 4.1.2 Duties, responsibilities and limitations of authority of the Architect as set forth in the Contract Documents shall not be restricted, modified or extended without written consent of the Owner, Contractor and Architect. Consent shall not be unreasonably withheld.

§ 4.1.3 If the employment of the Architect is terminated, the Owner shall employ a successor architect as to whom the Contractor has no reasonable objection and whose status under the Contract Documents shall be that of the Architect.

§ 4.2 ADMINISTRATION OF THE CONTRACT
§ 4.2.1 The Architect will provide administration of the Contract as described in the Contract Documents and will be an Owner's representative during construction until the date the Architect issues the final Certificate For Payment. The Architect will have authority to act on behalf of the Owner only to the extent provided in the Contract Documents.

§ 4.2.2 The Architect will visit the site at intervals appropriate to the stage of construction, or as otherwise agreed with the Owner, to become generally familiar with the progress and quality of the portion of the Work completed, and to determine in general if the Work observed is being performed in a manner indicating that the Work, when fully completed, will be in accordance with the Contract Documents. However, the Architect will not be required to make exhaustive or continuous on-site inspections to check the quality or quantity of the Work. The Architect will not have control over, charge of, or responsibility for, the construction means, methods, techniques, sequences or procedures, or for the safety precautions and programs in connection with the Work, since these are solely the Contractor's rights and responsibilities under the Contract Documents, except as provided in Section 3.3.1.

§ 4.2.3 On the basis of the site visits, the Architect will keep the Owner reasonably informed about the progress and quality of the portion of the Work completed, and report to the Owner (1) known deviations from the Contract Documents and from the most recent construction schedule submitted by the Contractor, and (2) defects and deficiencies observed in the Work. The Architect will not be responsible for the Contractor's failure to perform the Work in accordance with the requirements of the Contract Documents. The Architect will not have control over or charge of and will not be responsible for acts or omissions of the Contractor, Subcontractors, or their agents or employees, or any other persons or entities performing portions of the Work.

§ 4.2.4 COMMUNICATIONS FACILITATING CONTRACT ADMINISTRATION
Except as otherwise provided in the Contract Documents or when direct communications have been specially authorized, the Owner and Contractor shall endeavor to communicate with each other through the Architect about matters arising out of or relating to the Contract. Communications by and with the Architect's consultants shall be through the Architect. Communications by and with Subcontractors and material suppliers shall be through the Contractor. Communications by and with separate contractors shall be through the Owner.

§ 4.2.5 Based on the Architect's evaluations of the Contractor's Applications for Payment, the Architect will review and certify the amounts due the Contractor and will issue Certificates for Payment in such amounts.

§ 4.2.6 The Architect has authority to reject Work that does not conform to the Contract Documents. Whenever the Architect considers it necessary or advisable, the Architect will have authority to require inspection or testing of the

Work in accordance with Sections 13.5.2 and 13.5.3, whether or not such Work is fabricated, installed or completed. However, neither this authority of the Architect nor a decision made in good faith either to exercise or not to exercise such authority shall give rise to a duty or responsibility of the Architect to the Contractor, Subcontractors, material and equipment suppliers, their agents or employees, or other persons or entities performing portions of the Work.

§ 4.2.7 The Architect will review and approve, or take other appropriate action upon, the Contractor's submittals such as Shop Drawings, Product Data and Samples, but only for the limited purpose of checking for conformance with information given and the design concept expressed in the Contract Documents. The Architect's action will be taken in accordance with the submittal schedule approved by the Architect or, in the absence of an approved submittal schedule, with reasonable promptness while allowing sufficient time in the Architect's professional judgment to permit adequate review. Review of such submittals is not conducted for the purpose of determining the accuracy and completeness of other details such as dimensions and quantities, or for substantiating instructions for installation or performance of equipment or systems, all of which remain the responsibility of the Contractor as required by the Contract Documents. The Architect's review of the Contractor's submittals shall not relieve the Contractor of the obligations under Sections 3.3, 3.5 and 3.12. The Architect's review shall not constitute approval of safety precautions or, unless otherwise specifically stated by the Architect, of any construction means, methods, techniques, sequences or procedures. The Architect's approval of a specific item shall not indicate approval of an assembly of which the item is a component.

§ 4.2.8 The Architect will prepare Change Orders and Construction Change Directives, and may authorize minor changes in the Work as provided in Section 7.4. The Architect will investigate and make determinations and recommendations regarding concealed and unknown conditions as provided in Section 3.7.4.

§ 4.2.9 The Architect will conduct inspections to determine the date or dates of Substantial Completion and the date of final completion; issue Certificates of Substantial Completion pursuant to Section 9.8; receive and forward to the Owner, for the Owner's review and records, written warranties and related documents required by the Contract and assembled by the Contractor pursuant to Section 9.10; and issue a final Certificate for Payment pursuant to Section 9.10.

§ 4.2.10 If the Owner and Architect agree, the Architect will provide one or more project representatives to assist in carrying out the Architect's responsibilities at the site. The duties, responsibilities and limitations of authority of such project representatives shall be as set forth in an exhibit to be incorporated in the Contract Documents.

§ 4.2.11 The Architect will interpret and decide matters concerning performance under, and requirements of, the Contract Documents on written request of either the Owner or Contractor. The Architect's response to such requests will be made in writing within any time limits agreed upon or otherwise with reasonable promptness.

§ 4.2.12 Interpretations and decisions of the Architect will be consistent with the intent of, and reasonably inferable from, the Contract Documents and will be in writing or in the form of drawings. When making such interpretations and decisions, the Architect will endeavor to secure faithful performance by both Owner and Contractor, will not show partiality to either and will not be liable for results of interpretations or decisions rendered in good faith.

§ 4.2.13 The Architect's decisions on matters relating to aesthetic effect will be final if consistent with the intent expressed in the Contract Documents.

§ 4.2.14 The Architect will review and respond to requests for information about the Contract Documents. The Architect's response to such requests will be made in writing within any time limits agreed upon or otherwise with reasonable promptness. If appropriate, the Architect will prepare and issue supplemental Drawings and Specifications in response to the requests for information.

ARTICLE 5 SUBCONTRACTORS
§ 5.1 DEFINITIONS
§ 5.1.1 A Subcontractor is a person or entity who has a direct contract with the Contractor to perform a portion of the Work at the site. The term "Subcontractor" is referred to throughout the Contract Documents as if singular in number and means a Subcontractor or an authorized representative of the Subcontractor. The term "Subcontractor" does not include a separate contractor or subcontractors of a separate contractor.

§ 5.1.2 A Sub-subcontractor is a person or entity who has a direct or indirect contract with a Subcontractor to perform a portion of the Work at the site. The term "Sub-subcontractor" is referred to throughout the Contract Documents as if singular in number and means a Sub-subcontractor or an authorized representative of the Sub-subcontractor.

§ 5.2 AWARD OF SUBCONTRACTS AND OTHER CONTRACTS FOR PORTIONS OF THE WORK

§ 5.2.1 Unless otherwise stated in the Contract Documents or the bidding requirements, the Contractor, as soon as practicable after award of the Contract, shall furnish in writing to the Owner through the Architect the names of persons or entities (including those who are to furnish materials or equipment fabricated to a special design) proposed for each principal portion of the Work. The Architect may reply within 14 days to the Contractor in writing stating (1) whether the Owner or the Architect has reasonable objection to any such proposed person or entity or (2) that the Architect requires additional time for review. Failure of the Owner or Architect to reply within the 14-day period shall constitute notice of no reasonable objection.

§ 5.2.2 The Contractor shall not contract with a proposed person or entity to whom the Owner or Architect has made reasonable and timely objection. The Contractor shall not be required to contract with anyone to whom the Contractor has made reasonable objection.

§ 5.2.3 If the Owner or Architect has reasonable objection to a person or entity proposed by the Contractor, the Contractor shall propose another to whom the Owner or Architect has no reasonable objection. If the proposed but rejected Subcontractor was reasonably capable of performing the Work, the Contract Sum and Contract Time shall be increased or decreased by the difference, if any, occasioned by such change, and an appropriate Change Order shall be issued before commencement of the substitute Subcontractor's Work. However, no increase in the Contract Sum or Contract Time shall be allowed for such change unless the Contractor has acted promptly and responsively in submitting names as required.

§ 5.2.4 The Contractor shall not substitute a Subcontractor, person or entity previously selected if the Owner or Architect makes reasonable objection to such substitution.

§ 5.3 SUBCONTRACTUAL RELATIONS

By appropriate agreement, written where legally required for validity, the Contractor shall require each Subcontractor, to the extent of the Work to be performed by the Subcontractor, to be bound to the Contractor by terms of the Contract Documents, and to assume toward the Contractor all the obligations and responsibilities, including the responsibility for safety of the Subcontractor's Work, which the Contractor, by these Documents, assumes toward the Owner and Architect. Each subcontract agreement shall preserve and protect the rights of the Owner and Architect under the Contract Documents with respect to the Work to be performed by the Subcontractor so that subcontracting thereof will not prejudice such rights, and shall allow to the Subcontractor, unless specifically provided otherwise in the subcontract agreement, the benefit of all rights, remedies and redress against the Contractor that the Contractor, by the Contract Documents, has against the Owner. Where appropriate, the Contractor shall require each Subcontractor to enter into similar agreements with Sub-subcontractors. The Contractor shall make available to each proposed Subcontractor, prior to the execution of the subcontract agreement, copies of the Contract Documents to which the Subcontractor will be bound, and, upon written request of the Subcontractor, identify to the Subcontractor terms and conditions of the proposed subcontract agreement that may be at variance with the Contract Documents. Subcontractors will similarly make copies of applicable portions of such documents available to their respective proposed Sub-subcontractors.

§ 5.4 CONTINGENT ASSIGNMENT OF SUBCONTRACTS

§ 5.4.1 Each subcontract agreement for a portion of the Work is assigned by the Contractor to the Owner, provided that

.1   assignment is effective only after termination of the Contract by the Owner for cause pursuant to Section 14.2 and only for those subcontract agreements that the Owner accepts by notifying the Subcontractor and Contractor in writing; and

.2   assignment is subject to the prior rights of the surety, if any, obligated under bond relating to the Contract.

When the Owner accepts the assignment of a subcontract agreement, the Owner assumes the Contractor's rights and obligations under the subcontract.

§ **5.4.2** Upon such assignment, if the Work has been suspended for more than 30 days, the Subcontractor's compensation shall be equitably adjusted for increases in cost resulting from the suspension.

§ **5.4.3** Upon such assignment to the Owner under this Section 5.4, the Owner may further assign the subcontract to a successor contractor or other entity. If the Owner assigns the subcontract to a successor contractor or other entity, the Owner shall nevertheless remain legally responsible for all of the successor contractor's obligations under the subcontract.

ARTICLE 6 CONSTRUCTION BY OWNER OR BY SEPARATE CONTRACTORS
§ 6.1 OWNER'S RIGHT TO PERFORM CONSTRUCTION AND TO AWARD SEPARATE CONTRACTS

§ **6.1.1** The Owner reserves the right to perform construction or operations related to the Project with the Owner's own forces, and to award separate contracts in connection with other portions of the Project or other construction or operations on the site under Conditions of the Contract identical or substantially similar to these including those portions related to insurance and waiver of subrogation. If the Contractor claims that delay or additional cost is involved because of such action by the Owner, the Contractor shall make such Claim as provided in Article 15.

§ **6.1.2** When separate contracts are awarded for different portions of the Project or other construction or operations on the site, the term "Contractor" in the Contract Documents in each case shall mean the Contractor who executes each separate Owner-Contractor Agreement.

§ **6.1.3** The Owner shall provide for coordination of the activities of the Owner's own forces and of each separate contractor with the Work of the Contractor, who shall cooperate with them. The Contractor shall participate with other separate contractors and the Owner in reviewing their construction schedules. The Contractor shall make any revisions to the construction schedule deemed necessary after a joint review and mutual agreement. The construction schedules shall then constitute the schedules to be used by the Contractor, separate contractors and the Owner until subsequently revised.

§ **6.1.4** Unless otherwise provided in the Contract Documents, when the Owner performs construction or operations related to the Project with the Owner's own forces, the Owner shall be deemed to be subject to the same obligations and to have the same rights that apply to the Contractor under the Conditions of the Contract, including, without excluding others, those stated in Article 3, this Article 6 and Articles 10, 11 and 12.

§ 6.2 MUTUAL RESPONSIBILITY

§ **6.2.1** The Contractor shall afford the Owner and separate contractors reasonable opportunity for introduction and storage of their materials and equipment and performance of their activities, and shall connect and coordinate the Contractor's construction and operations with theirs as required by the Contract Documents.

§ **6.2.2** If part of the Contractor's Work depends for proper execution or results upon construction or operations by the Owner or a separate contractor, the Contractor shall, prior to proceeding with that portion of the Work, promptly report to the Architect apparent discrepancies or defects in such other construction that would render it unsuitable for such proper execution and results. Failure of the Contractor so to report shall constitute an acknowledgment that the Owner's or separate contractor's completed or partially completed construction is fit and proper to receive the Contractor's Work, except as to defects not then reasonably discoverable.

§ **6.2.3** The Contractor shall reimburse the Owner for costs the Owner incurs that are payable to a separate contractor because of the Contractor's delays, improperly timed activities or defective construction. The Owner shall be responsible to the Contractor for costs the Contractor incurs because of a separate contractor's delays, improperly timed activities, damage to the Work or defective construction.

§ **6.2.4** The Contractor shall promptly remedy damage the Contractor wrongfully causes to completed or partially completed construction or to property of the Owner, separate contractors as provided in Section 10.2.5.

§ **6.2.5** The Owner and each separate contractor shall have the same responsibilities for cutting and patching as are described for the Contractor in Section 3.14.

§ 6.3 OWNER'S RIGHT TO CLEAN UP

If a dispute arises among the Contractor, separate contractors and the Owner as to the responsibility under their respective contracts for maintaining the premises and surrounding area free from waste materials and rubbish, the Owner may clean up and the Architect will allocate the cost among those responsible.

ARTICLE 7   CHANGES IN THE WORK
§ 7.1 GENERAL

§ 7.1.1 Changes in the Work may be accomplished after execution of the Contract, and without invalidating the Contract, by Change Order, Construction Change Directive or order for a minor change in the Work, subject to the limitations stated in this Article 7 and elsewhere in the Contract Documents.

§ 7.1.2 A Change Order shall be based upon agreement among the Owner, Contractor and Architect; a Construction Change Directive requires agreement by the Owner and Architect and may or may not be agreed to by the Contractor; an order for a minor change in the Work may be issued by the Architect alone.

§ 7.1.3 Changes in the Work shall be performed under applicable provisions of the Contract Documents, and the Contractor shall proceed promptly, unless otherwise provided in the Change Order, Construction Change Directive or order for a minor change in the Work.

§ 7.2 CHANGE ORDERS

§ 7.2.1 A Change Order is a written instrument prepared by the Architect and signed by the Owner, Contractor and Architect stating their agreement upon all of the following:

- .1 The change in the Work;
- .2 The amount of the adjustment, if any, in the Contract Sum; and
- .3 The extent of the adjustment, if any, in the Contract Time.

§ 7.3 CONSTRUCTION CHANGE DIRECTIVES

§ 7.3.1 A Construction Change Directive is a written order prepared by the Architect and signed by the Owner and Architect, directing a change in the Work prior to agreement on adjustment, if any, in the Contract Sum or Contract Time, or both. The Owner may by Construction Change Directive, without invalidating the Contract, order changes in the Work within the general scope of the Contract consisting of additions, deletions or other revisions, the Contract Sum and Contract Time being adjusted accordingly.

§ 7.3.2 A Construction Change Directive shall be used in the absence of total agreement on the terms of a Change Order.

§ 7.3.3 If the Construction Change Directive provides for an adjustment to the Contract Sum, the adjustment shall be based on one of the following methods:

- .1 Mutual acceptance of a lump sum properly itemized and supported by sufficient substantiating data to permit evaluation;
- .2 Unit prices stated in the Contract Documents or subsequently agreed upon;
- .3 Cost to be determined in a manner agreed upon by the parties and a mutually acceptable fixed or percentage fee; or
- .4 As provided in Section 7.3.7.

§ 7.3.4 If unit prices are stated in the Contract Documents or subsequently agreed upon, and if quantities originally contemplated are materially changed in a proposed Change Order or Construction Change Directive so that application of such unit prices to quantities of Work proposed will cause substantial inequity to the Owner or Contractor, the applicable unit prices shall be equitably adjusted.

§ 7.3.5 Upon receipt of a Construction Change Directive, the Contractor shall promptly proceed with the change in the Work involved and advise the Architect of the Contractor's agreement or disagreement with the method, if any, provided in the Construction Change Directive for determining the proposed adjustment in the Contract Sum or Contract Time.

§ 7.3.6 A Construction Change Directive signed by the Contractor indicates the Contractor's agreement therewith, including adjustment in Contract Sum and Contract Time or the method for determining them. Such agreement shall be effective immediately and shall be recorded as a Change Order.

21

§ 7.3.7 If the Contractor does not respond promptly or disagrees with the method for adjustment in the Contract Sum, the Architect shall determine the method and the adjustment on the basis of reasonable expenditures and savings of those performing the Work attributable to the change, including, in case of an increase in the Contract Sum, an amount for overhead and profit as set forth in the Agreement, or if no such amount is set forth in the Agreement, a reasonable amount. In such case, and also under Section 7.3.3.3, the Contractor shall keep and present, in such form as the Architect may prescribe, an itemized accounting together with appropriate supporting data. Unless otherwise provided in the Contract Documents, costs for the purposes of this Section 7.3.7 shall be limited to the following:

.1 Costs of labor, including social security, old age and unemployment insurance, fringe benefits required by agreement or custom, and workers' compensation insurance;

.2 Costs of materials, supplies and equipment, including cost of transportation, whether incorporated or consumed;

.3 Rental costs of machinery and equipment, exclusive of hand tools, whether rented from the Contractor or others;

.4 Costs of premiums for all bonds and insurance, permit fees, and sales, use or similar taxes related to the Work; and

.5 Additional costs of supervision and field office personnel directly attributable to the change.

§ 7.3.8 The amount of credit to be allowed by the Contractor to the Owner for a deletion or change that results in a net decrease in the Contract Sum shall be actual net cost as confirmed by the Architect. When both additions and credits covering related Work or substitutions are involved in a change, the allowance for overhead and profit shall be figured on the basis of net increase, if any, with respect to that change.

§ 7.3.9 Pending final determination of the total cost of a Construction Change Directive to the Owner, the Contractor may request payment for Work completed under the Construction Change Directive in Applications for Payment. The Architect will make an interim determination for purposes of monthly certification for payment for those costs and certify for payment the amount that the Architect determines, in the Architect's professional judgment, to be reasonably justified. The Architect's interim determination of cost shall adjust the Contract Sum on the same basis as a Change Order, subject to the right of either party to disagree and assert a Claim in accordance with Article 15.

§ 7.3.10 When the Owner and Contractor agree with a determination made by the Architect concerning the adjustments in the Contract Sum and Contract Time, or otherwise reach agreement upon the adjustments, such agreement shall be effective immediately and the Architect will prepare a Change Order. Change Orders may be issued for all or any part of a Construction Change Directive.

§ 7.4 MINOR CHANGES IN THE WORK
The Architect has authority to order minor changes in the Work not involving adjustment in the Contract Sum or extension of the Contract Time and not inconsistent with the intent of the Contract Documents. Such changes will be effected by written order signed by the Architect and shall be binding on the Owner and Contractor.

ARTICLE 8 TIME
§ 8.1 DEFINITIONS
§ 8.1.1 Unless otherwise provided, Contract Time is the period of time, including authorized adjustments, allotted in the Contract Documents for Substantial Completion of the Work.

§ 8.1.2 The date of commencement of the Work is the date established in the Agreement.

§ 8.1.3 The date of Substantial Completion is the date certified by the Architect in accordance with Section 9.8.

§ 8.1.4 The term "day" as used in the Contract Documents shall mean calendar day unless otherwise specifically defined.

§ 8.2 PROGRESS AND COMPLETION
§ 8.2.1 Time limits stated in the Contract Documents are of the essence of the Contract. By executing the Agreement the Contractor confirms that the Contract Time is a reasonable period for performing the Work.

§ 8.2.2 The Contractor shall not knowingly, except by agreement or instruction of the Owner in writing, prematurely commence operations on the site or elsewhere prior to the effective date of insurance required by Article 11 to be furnished by the Contractor and Owner. The date of commencement of the Work shall not be changed by the effective date of such insurance.

§ 8.2.3 The Contractor shall proceed expeditiously with adequate forces and shall achieve Substantial Completion within the Contract Time.

§ 8.3 DELAYS AND EXTENSIONS OF TIME
§ 8.3.1 If the Contractor is delayed at any time in the commencement or progress of the Work by an act or neglect of the Owner or Architect, or of an employee of either, or of a separate contractor employed by the Owner; or by changes ordered in the Work; or by labor disputes, fire, unusual delay in deliveries, unavoidable casualties or other causes beyond the Contractor's control; or by delay authorized by the Owner pending mediation and arbitration; or by other causes that the Architect determines may justify delay, then the Contract Time shall be extended by Change Order for such reasonable time as the Architect may determine.

§ 8.3.2 Claims relating to time shall be made in accordance with applicable provisions of Article 15.

§ 8.3.3 This Section 8.3 does not preclude recovery of damages for delay by either party under other provisions of the Contract Documents.

ARTICLE 9 PAYMENTS AND COMPLETION
§ 9.1 CONTRACT SUM
The Contract Sum is stated in the Agreement and, including authorized adjustments, is the total amount payable by the Owner to the Contractor for performance of the Work under the Contract Documents.

§ 9.2 SCHEDULE OF VALUES
Where the Contract is based on a stipulated sum or Guaranteed Maximum Price, the Contractor shall submit to the Architect, before the first Application for Payment, a schedule of values allocating the entire Contract Sum to the various portions of the Work and prepared in such form and supported by such data to substantiate its accuracy as the Architect may require. This schedule, unless objected to by the Architect, shall be used as a basis for reviewing the Contractor's Applications for Payment.

§ 9.3 APPLICATIONS FOR PAYMENT
§ 9.3.1 At least ten days before the date established for each progress payment, the Contractor shall submit to the Architect an itemized Application for Payment prepared in accordance with the schedule of values, if required under Section 9.2., for completed portions of the Work. Such application shall be notarized, if required, and supported by such data substantiating the Contractor's right to payment as the Owner or Architect may require, such as copies of requisitions from Subcontractors and material suppliers, and shall reflect retainage if provided for in the Contract Documents.

§ 9.3.1.1 As provided in Section 7.3.9, such applications may include requests for payment on account of changes in the Work that have been properly authorized by Construction Change Directives, or by interim determinations of the Architect, but not yet included in Change Orders.

§ 9.3.1.2 Applications for Payment shall not include requests for payment for portions of the Work for which the Contractor does not intend to pay a Subcontractor or material supplier, unless such Work has been performed by others whom the Contractor intends to pay.

§ 9.3.2 Unless otherwise provided in the Contract Documents, payments shall be made on account of materials and equipment delivered and suitably stored at the site for subsequent incorporation in the Work. If approved in advance by the Owner, payment may similarly be made for materials and equipment suitably stored off the site at a location agreed upon in writing. Payment for materials and equipment stored on or off the site shall be conditioned upon compliance by the Contractor with procedures satisfactory to the Owner to establish the Owner's title to such materials and equipment or otherwise protect the Owner's interest, and shall include the costs of applicable insurance, storage and transportation to the site for such materials and equipment stored off the site.

§ 9.3.3 The Contractor warrants that title to all Work covered by an Application for Payment will pass to the Owner no later than the time of payment. The Contractor further warrants that upon submittal of an Application for Payment all Work for which Certificates for Payment have been previously issued and payments received from the

Owner shall, to the best of the Contractor's knowledge, information and belief, be free and clear of liens, claims, security interests or encumbrances in favor of the Contractor, Subcontractors, material suppliers, or other persons or entities making a claim by reason of having provided labor, materials and equipment relating to the Work.

§ 9.4 CERTIFICATES FOR PAYMENT

§ 9.4.1 The Architect will, within seven days after receipt of the Contractor's Application for Payment, either issue to the Owner a Certificate for Payment, with a copy to the Contractor, for such amount as the Architect determines is properly due, or notify the Contractor and Owner in writing of the Architect's reasons for withholding certification in whole or in part as provided in Section 9.5.1.

§ 9.4.2 The issuance of a Certificate for Payment will constitute a representation by the Architect to the Owner, based on the Architect's evaluation of the Work and the data comprising the Application for Payment, that, to the best of the Architect's knowledge, information and belief, the Work has progressed to the point indicated and that the quality of the Work is in accordance with the Contract Documents. The foregoing representations are subject to an evaluation of the Work for conformance with the Contract Documents upon Substantial Completion, to results of subsequent tests and inspections, to correction of minor deviations from the Contract Documents prior to completion and to specific qualifications expressed by the Architect. The issuance of a Certificate for Payment will further constitute a representation that the Contractor is entitled to payment in the amount certified. However, the issuance of a Certificate for Payment will not be a representation that the Architect has (1) made exhaustive or continuous on-site inspections to check the quality or quantity of the Work, (2) reviewed construction means, methods, techniques, sequences or procedures, (3) reviewed copies of requisitions received from Subcontractors and material suppliers and other data requested by the Owner to substantiate the Contractor's right to payment, or (4) made examination to ascertain how or for what purpose the Contractor has used money previously paid on account of the Contract Sum.

§ 9.5 DECISIONS TO WITHHOLD CERTIFICATION

§ 9.5.1 The Architect may withhold a Certificate for Payment in whole or in part, to the extent reasonably necessary to protect the Owner, if in the Architect's opinion the representations to the Owner required by Section 9.4.2 cannot be made. If the Architect is unable to certify payment in the amount of the Application, the Architect will notify the Contractor and Owner as provided in Section 9.4.1. If the Contractor and Architect cannot agree on a revised amount, the Architect will promptly issue a Certificate for Payment for the amount for which the Architect is able to make such representations to the Owner. The Architect may also withhold a Certificate for Payment or, because of subsequently discovered evidence, may nullify the whole or a part of a Certificate for Payment previously issued, to such extent as may be necessary in the Architect's opinion to protect the Owner from loss for which the Contractor is responsible, including loss resulting from acts and omissions described in Section 3.3.2, because of

.1 defective Work not remedied;

.2 third party claims filed or reasonable evidence indicating probable filing of such claims unless security acceptable to the Owner is provided by the Contractor;

.3 failure of the Contractor to make payments properly to Subcontractors or for labor, materials or equipment;

.4 reasonable evidence that the Work cannot be completed for the unpaid balance of the Contract Sum;

.5 damage to the Owner or a separate contractor;

.6 reasonable evidence that the Work will not be completed within the Contract Time, and that the unpaid balance would not be adequate to cover actual or liquidated damages for the anticipated delay; or

.7 repeated failure to carry out the Work in accordance with the Contract Documents.

§ 9.5.2 When the above reasons for withholding certification are removed, certification will be made for amounts previously withheld.

§ 9.5.3 If the Architect withholds certification for payment under Section 9.5.1.3, the Owner may, at its sole option, issue joint checks to the Contractor and to any Subcontractor or material or equipment suppliers to whom the Contractor failed to make payment for Work properly performed or material or equipment suitably delivered. If the Owner makes payments by joint check, the Owner shall notify the Architect and the Architect will reflect such payment on the next Certificate for Payment.

§ 9.6 PROGRESS PAYMENTS

§ 9.6.1 After the Architect has issued a Certificate for Payment, the Owner shall make payment in the manner and within the time provided in the Contract Documents, and shall so notify the Architect.

§ 9.6.2 The Contractor shall pay each Subcontractor no later than seven days after receipt of payment from the Owner the amount to which the Subcontractor is entitled, reflecting percentages actually retained from payments to the Contractor on account of the Subcontractor's portion of the Work. The Contractor shall, by appropriate agreement with each Subcontractor, require each Subcontractor to make payments to Sub-subcontractors in a similar manner.

§ 9.6.3 The Architect will, on request, furnish to a Subcontractor, if practicable, information regarding percentages of completion or amounts applied for by the Contractor and action taken thereon by the Architect and Owner on account of portions of the Work done by such Subcontractor.

§ 9.6.4 The Owner has the right to request written evidence from the Contractor that the Contractor has properly paid Subcontractors and material and equipment suppliers amounts paid by the Owner to the Contractor for subcontracted Work. If the Contractor fails to furnish such evidence within seven days, the Owner shall have the right to contact Subcontractors to ascertain whether they have been properly paid. Neither the Owner nor Architect shall have an obligation to pay or to see to the payment of money to a Subcontractor, except as may otherwise be required by law.

§ 9.6.5 Contractor payments to material and equipment suppliers shall be treated in a manner similar to that provided in Sections 9.6.2, 9.6.3 and 9.6.4.

§ 9.6.6 A Certificate for Payment, a progress payment, or partial or entire use or occupancy of the Project by the Owner shall not constitute acceptance of Work not in accordance with the Contract Documents.

§ 9.6.7 Unless the Contractor provides the Owner with a payment bond in the full penal sum of the Contract Sum, payments received by the Contractor for Work properly performed by Subcontractors and suppliers shall be held by the Contractor for those Subcontractors or suppliers who performed Work or furnished materials, or both, under contract with the Contractor for which payment was made by the Owner. Nothing contained herein shall require money to be placed in a separate account and not commingled with money of the Contractor, shall create any fiduciary liability or tort liability on the part of the Contractor for breach of trust or shall entitle any person or entity to an award of punitive damages against the Contractor for breach of the requirements of this provision.

§ 9.7 FAILURE OF PAYMENT

If the Architect does not issue a Certificate for Payment, through no fault of the Contractor, within seven days after receipt of the Contractor's Application for Payment, or if the Owner does not pay the Contractor within seven days after the date established in the Contract Documents the amount certified by the Architect or awarded by binding dispute resolution, then the Contractor may, upon seven additional days' written notice to the Owner and Architect, stop the Work until payment of the amount owing has been received. The Contract Time shall be extended appropriately and the Contract Sum shall be increased by the amount of the Contractor's reasonable costs of shut-down, delay and start-up, plus interest as provided for in the Contract Documents.

§ 9.8 SUBSTANTIAL COMPLETION

§ 9.8.1 Substantial Completion is the stage in the progress of the Work when the Work or designated portion thereof is sufficiently complete in accordance with the Contract Documents so that the Owner can occupy or utilize the Work for its intended use.

§ 9.8.2 When the Contractor considers that the Work, or a portion thereof which the Owner agrees to accept separately, is substantially complete, the Contractor shall prepare and submit to the Architect a comprehensive list of items to be completed or corrected prior to final payment. Failure to include an item on such list does not alter the responsibility of the Contractor to complete all Work in accordance with the Contract Documents.

§ 9.8.3 Upon receipt of the Contractor's list, the Architect will make an inspection to determine whether the Work or designated portion thereof is substantially complete. If the Architect's inspection discloses any item, whether or not included on the Contractor's list, which is not sufficiently complete in accordance with the Contract Documents so that the Owner can occupy or utilize the Work or designated portion thereof for its intended use, the Contractor shall, before issuance of the Certificate of Substantial Completion, complete or correct such item upon notification by the Architect. In such case, the Contractor shall then submit a request for another inspection by the Architect to determine Substantial Completion.

§ 9.8.4 When the Work or designated portion thereof is substantially complete, the Architect will prepare a Certificate of Substantial Completion that shall establish the date of Substantial Completion, shall establish responsibilities of the Owner and Contractor for security, maintenance, heat, utilities, damage to the Work and insurance, and shall fix the time within which the Contractor shall finish all items on the list accompanying the Certificate. Warranties required by the Contract Documents shall commence on the date of Substantial Completion of the Work or designated portion thereof unless otherwise provided in the Certificate of Substantial Completion.

§ 9.8.5 The Certificate of Substantial Completion shall be submitted to the Owner and Contractor for their written acceptance of responsibilities assigned to them in such Certificate. Upon such acceptance and consent of surety, if any, the Owner shall make payment of retainage applying to such Work or designated portion thereof. Such payment shall be adjusted for Work that is incomplete or not in accordance with the requirements of the Contract Documents.

§ 9.9 PARTIAL OCCUPANCY OR USE

§ 9.9.1 The Owner may occupy or use any completed or partially completed portion of the Work at any stage when such portion is designated by separate agreement with the Contractor, provided such occupancy or use is consented to by the insurer as required under Section 11.3.1.5 and authorized by public authorities having jurisdiction over the Project. Such partial occupancy or use may commence whether or not the portion is substantially complete, provided the Owner and Contractor have accepted in writing the responsibilities assigned to each of them for payments, retainage, if any, security, maintenance, heat, utilities, damage to the Work and insurance, and have agreed in writing concerning the period for correction of the Work and commencement of warranties required by the Contract Documents. When the Contractor considers a portion substantially complete, the Contractor shall prepare and submit a list to the Architect as provided under Section 9.8.2. Consent of the Contractor to partial occupancy or use shall not be unreasonably withheld. The stage of the progress of the Work shall be determined by written agreement between the Owner and Contractor or, if no agreement is reached, by decision of the Architect.

§ 9.9.2 Immediately prior to such partial occupancy or use, the Owner, Contractor and Architect shall jointly inspect the area to be occupied or portion of the Work to be used in order to determine and record the condition of the Work.

§ 9.9.3 Unless otherwise agreed upon, partial occupancy or use of a portion or portions of the Work shall not constitute acceptance of Work not complying with the requirements of the Contract Documents.

§ 9.10 FINAL COMPLETION AND FINAL PAYMENT

§ 9.10.1 Upon receipt of the Contractor's written notice that the Work is ready for final inspection and acceptance and upon receipt of a final Application for Payment, the Architect will promptly make such inspection and, when the Architect finds the Work acceptable under the Contract Documents and the Contract fully performed, the Architect will promptly issue a final Certificate for Payment stating that to the best of the Architect's knowledge, information and belief, and on the basis of the Architect's on-site visits and inspections, the Work has been completed in accordance with terms and conditions of the Contract Documents and that the entire balance found to be due the Contractor and noted in the final Certificate is due and payable. The Architect's final Certificate for Payment will constitute a further representation that conditions listed in Section 9.10.2 as precedent to the Contractor's being entitled to final payment have been fulfilled.

§ 9.10.2 Neither final payment nor any remaining retained percentage shall become due until the Contractor submits to the Architect (1) an affidavit that payrolls, bills for materials and equipment, and other indebtedness connected with the Work for which the Owner or the Owner's property might be responsible or encumbered (less amounts withheld by Owner) have been paid or otherwise satisfied, (2) a certificate evidencing that insurance required by the Contract Documents to remain in force after final payment is currently in effect and will not be canceled or allowed to expire until at least 30 days' prior written notice has been given to the Owner, (3) a written statement that the Contractor knows of no substantial reason that the insurance will not be renewable to cover the period required by the Contract Documents, (4) consent of surety, if any, to final payment and (5), if required by the Owner, other data establishing payment or satisfaction of obligations, such as receipts, releases and waivers of liens, claims, security interests or encumbrances arising out of the Contract, to the extent and in such form as may be designated by the Owner. If a Subcontractor refuses to furnish a release or waiver required by the Owner, the Contractor may furnish a bond satisfactory to the Owner to indemnify the Owner against such lien. If such lien remains unsatisfied after payments are made, the Contractor shall refund to the Owner all money that the Owner may be compelled to pay in discharging such lien, including all costs and reasonable attorneys' fees.

§ 9.10.3 If, after Substantial Completion of the Work, final completion thereof is materially delayed through no fault of the Contractor or by issuance of Change Orders affecting final completion, and the Architect so confirms, the Owner shall, upon application by the Contractor and certification by the Architect, and without terminating the Contract, make payment of the balance due for that portion of the Work fully completed and accepted. If the remaining balance for Work not fully completed or corrected is less than retainage stipulated in the Contract Documents, and if bonds have been furnished, the written consent of surety to payment of the balance due for that portion of the Work fully completed and accepted shall be submitted by the Contractor to the Architect prior to certification of such payment. Such payment shall be made under terms and conditions governing final payment, except that it shall not constitute a waiver of claims.

§ 9.10.4 The making of final payment shall constitute a waiver of Claims by the Owner except those arising from

- .1 liens, Claims, security interests or encumbrances arising out of the Contract and unsettled;
- .2 failure of the Work to comply with the requirements of the Contract Documents; or
- .3 terms of special warranties required by the Contract Documents.

§ 9.10.5 Acceptance of final payment by the Contractor, a Subcontractor or material supplier shall constitute a waiver of claims by that payee except those previously made in writing and identified by that payee as unsettled at the time of final Application for Payment.

ARTICLE 10 PROTECTION OF PERSONS AND PROPERTY
§ 10.1 SAFETY PRECAUTIONS AND PROGRAMS
The Contractor shall be responsible for initiating, maintaining and supervising all safety precautions and programs in connection with the performance of the Contract.

§ 10.2 SAFETY OF PERSONS AND PROPERTY
§ 10.2.1 The Contractor shall take reasonable precautions for safety of, and shall provide reasonable protection to prevent damage, injury or loss to

- .1 employees on the Work and other persons who may be affected thereby;
- .2 the Work and materials and equipment to be incorporated therein, whether in storage on or off the site, under care, custody or control of the Contractor or the Contractor's Subcontractors or Sub-subcontractors; and
- .3 other property at the site or adjacent thereto, such as trees, shrubs, lawns, walks, pavements, roadways, structures and utilities not designated for removal, relocation or replacement in the course of construction.

§ 10.2.2 The Contractor shall comply with and give notices required by applicable laws, statutes, ordinances, codes, rules and regulations, and lawful orders of public authorities bearing on safety of persons or property or their protection from damage, injury or loss.

§ 10.2.3 The Contractor shall erect and maintain, as required by existing conditions and performance of the Contract, reasonable safeguards for safety and protection, including posting danger signs and other warnings against hazards, promulgating safety regulations and notifying owners and users of adjacent sites and utilities.

§ 10.2.4 When use or storage of explosives or other hazardous materials or equipment or unusual methods are necessary for execution of the Work, the Contractor shall exercise utmost care and carry on such activities under supervision of properly qualified personnel.

§ 10.2.5 The Contractor shall promptly remedy damage and loss (other than damage or loss insured under property insurance required by the Contract Documents) to property referred to in Sections 10.2.1.2 and 10.2.1.3 caused in whole or in part by the Contractor, a Subcontractor, a Sub-subcontractor, or anyone directly or indirectly employed by any of them, or by anyone for whose acts they may be liable and for which the Contractor is responsible under Sections 10.2.1.2 and 10.2.1.3, except damage or loss attributable to acts or omissions of the Owner or Architect or anyone directly or indirectly employed by either of them, or by anyone for whose acts either of them may be liable, and not attributable to the fault or negligence of the Contractor. The foregoing obligations of the Contractor are in addition to the Contractor's obligations under Section 3.18.

§ 10.2.6 The Contractor shall designate a responsible member of the Contractor's organization at the site whose duty shall be the prevention of accidents. This person shall be the Contractor's superintendent unless otherwise designated by the Contractor in writing to the Owner and Architect.

§ 10.2.7 The Contractor shall not permit any part of the construction or site to be loaded so as to cause damage or create an unsafe condition.

§ 10.2.8 INJURY OR DAMAGE TO PERSON OR PROPERTY
If either party suffers injury or damage to person or property because of an act or omission of the other party, or of others for whose acts such party is legally responsible, written notice of such injury or damage, whether or not insured, shall be given to the other party within a reasonable time not exceeding 21 days after discovery. The notice shall provide sufficient detail to enable the other party to investigate the matter.

§ 10.3 HAZARDOUS MATERIALS
§ 10.3.1 The Contractor is responsible for compliance with any requirements included in the Contract Documents regarding hazardous materials. If the Contractor encounters a hazardous material or substance not addressed in the Contract Documents and if reasonable precautions will be inadequate to prevent foreseeable bodily injury or death to persons resulting from a material or substance, including but not limited to asbestos or polychlorinated biphenyl (PCB), encountered on the site by the Contractor, the Contractor shall, upon recognizing the condition, immediately stop Work in the affected area and report the condition to the Owner and Architect in writing.

§ 10.3.2 Upon receipt of the Contractor's written notice, the Owner shall obtain the services of a licensed laboratory to verify the presence or absence of the material or substance reported by the Contractor and, in the event such material or substance is found to be present, to cause it to be rendered harmless. Unless otherwise required by the Contract Documents, the Owner shall furnish in writing to the Contractor and Architect the names and qualifications of persons or entities who are to perform tests verifying the presence or absence of such material or substance or who are to perform the task of removal or safe containment of such material or substance. The Contractor and the Architect will promptly reply to the Owner in writing stating whether or not either has reasonable objection to the persons or entities proposed by the Owner. If either the Contractor or Architect has an objection to a person or entity proposed by the Owner, the Owner shall propose another to whom the Contractor and the Architect have no reasonable objection. When the material or substance has been rendered harmless, Work in the affected area shall resume upon written agreement of the Owner and Contractor. By Change Order, the Contract Time shall be extended appropriately and the Contract Sum shall be increased in the amount of the Contractor's reasonable additional costs of shut-down, delay and start-up.

§ 10.3.3 To the fullest extent permitted by law, the Owner shall indemnify and hold harmless the Contractor, Subcontractors, Architect, Architect's consultants and agents and employees of any of them from and against claims, damages, losses and expenses, including but not limited to attorneys' fees, arising out of or resulting from performance of the Work in the affected area if in fact the material or substance presents the risk of bodily injury or death as described in Section 10.3.1 and has not been rendered harmless, provided that such claim, damage, loss or expense is attributable to bodily injury, sickness, disease or death, or to injury to or destruction of tangible property (other than the Work itself), except to the extent that such damage, loss or expense is due to the fault or negligence of the party seeking indemnity.

§ 10.3.4 The Owner shall not be responsible under this Section 10.3 for materials or substances the Contractor brings to the site unless such materials or substances are required by the Contract Documents. The Owner shall be responsible for materials or substances required by the Contract Documents, except to the extent of the Contractor's fault or negligence in the use and handling of such materials or substances.

§ 10.3.5 The Contractor shall indemnify the Owner for the cost and expense the Owner incurs (1) for remediation of a material or substance the Contractor brings to the site and negligently handles, or (2) where the Contractor fails to perform its obligations under Section 10.3.1, except to the extent that the cost and expense are due to the Owner's fault or negligence.

§ 10.3.6 If, without negligence on the part of the Contractor, the Contractor is held liable by a government agency for the cost of remediation of a hazardous material or substance solely by reason of performing Work as required by the Contract Documents, the Owner shall indemnify the Contractor for all cost and expense thereby incurred.

§ 10.4 EMERGENCIES

In an emergency affecting safety of persons or property, the Contractor shall act, at the Contractor's discretion, to prevent threatened damage, injury or loss. Additional compensation or extension of time claimed by the Contractor on account of an emergency shall be determined as provided in Article 15 and Article 7.

ARTICLE 11 INSURANCE AND BONDS
§ 11.1 CONTRACTOR'S LIABILITY INSURANCE

§ 11.1.1 The Contractor shall purchase from and maintain in a company or companies lawfully authorized to do business in the jurisdiction in which the Project is located such insurance as will protect the Contractor from claims set forth below which may arise out of or result from the Contractor's operations and completed operations under the Contract and for which the Contractor may be legally liable, whether such operations be by the Contractor or by a Subcontractor or by anyone directly or indirectly employed by any of them, or by anyone for whose acts any of them may be liable:

  .1   Claims under workers' compensation, disability benefit and other similar employee benefit acts that are applicable to the Work to be performed;

  .2   Claims for damages because of bodily injury, occupational sickness or disease, or death of the Contractor's employees;

  .3   Claims for damages because of bodily injury, sickness or disease, or death of any person other than the Contractor's employees;

  .4   Claims for damages insured by usual personal injury liability coverage;

  .5   Claims for damages, other than to the Work itself, because of injury to or destruction of tangible property, including loss of use resulting therefrom;

  .6   Claims for damages because of bodily injury, death of a person or property damage arising out of ownership, maintenance or use of a motor vehicle;

  .7   Claims for bodily injury or property damage arising out of completed operations; and

  .8   Claims involving contractual liability insurance applicable to the Contractor's obligations under Section 3.18.

§ 11.1.2 The insurance required by Section 11.1.1 shall be written for not less than limits of liability specified in the Contract Documents or required by law, whichever coverage is greater. Coverages, whether written on an occurrence or claims-made basis, shall be maintained without interruption from the date of commencement of the Work until the date of final payment and termination of any coverage required to be maintained after final payment, and, with respect to the Contractor's completed operations coverage, until the expiration of the period for correction of Work or for such other period for maintenance of completed operations coverage as specified in the Contract Documents.

§ 11.1.3 Certificates of insurance acceptable to the Owner shall be filed with the Owner prior to commencement of the Work and thereafter upon renewal or replacement of each required policy of insurance. These certificates and the insurance policies required by this Section 11.1 shall contain a provision that coverages afforded under the policies will not be canceled or allowed to expire until at least 30 days' prior written notice has been given to the Owner. An additional certificate evidencing continuation of liability coverage, including coverage for completed operations, shall be submitted with the final Application for Payment as required by Section 9.10.2 and thereafter upon renewal or replacement of such coverage until the expiration of the time required by Section 11.1.2. Information concerning reduction of coverage on account of revised limits or claims paid under the General Aggregate, or both, shall be furnished by the Contractor with reasonable promptness.

§ 11.1.4 The Contractor shall cause the commercial liability coverage required by the Contract Documents to include (1) the Owner, the Architect and the Architect's Consultants as additional insureds for claims caused in whole or in part by the Contractor's negligent acts or omissions during the Contractor's operations; and (2) the Owner as an additional insured for claims caused in whole or in part by the Contractor's negligent acts or omissions during the Contractor's completed operations.

§ 11.2 OWNER'S LIABILITY INSURANCE

The Owner shall be responsible for purchasing and maintaining the Owner's usual liability insurance.

§ 11.3 PROPERTY INSURANCE

§ 11.3.1 Unless otherwise provided, the Owner shall purchase and maintain, in a company or companies lawfully authorized to do business in the jurisdiction in which the Project is located, property insurance written on a builder's risk "all-risk" or equivalent policy form in the amount of the initial Contract Sum, plus value of subsequent Contract Modifications and cost of materials supplied or installed by others, comprising total value for the entire Project at the site on a replacement cost basis without optional deductibles. Such property insurance shall be maintained, unless otherwise provided in the Contract Documents or otherwise agreed in writing by all persons and entities who are beneficiaries of such insurance, until final payment has been made as provided in Section 9.10 or until no person or entity other than the Owner has an insurable interest in the property required by this Section 11.3 to be covered, whichever is later. This insurance shall include interests of the Owner, the Contractor, Subcontractors and Sub-subcontractors in the Project.

§ 11.3.1.1 Property insurance shall be on an "all-risk" or equivalent policy form and shall include, without limitation, insurance against the perils of fire (with extended coverage) and physical loss or damage including, without duplication of coverage, theft, vandalism, malicious mischief, collapse, earthquake, flood, windstorm, falsework, testing and startup, temporary buildings and debris removal including demolition occasioned by enforcement of any applicable legal requirements, and shall cover reasonable compensation for Architect's and Contractor's services and expenses required as a result of such insured loss.

§ 11.3.1.2 If the Owner does not intend to purchase such property insurance required by the Contract and with all of the coverages in the amount described above, the Owner shall so inform the Contractor in writing prior to commencement of the Work. The Contractor may then effect insurance that will protect the interests of the Contractor, Subcontractors and Sub-subcontractors in the Work, and by appropriate Change Order the cost thereof shall be charged to the Owner. If the Contractor is damaged by the failure or neglect of the Owner to purchase or maintain insurance as described above, without so notifying the Contractor in writing, then the Owner shall bear all reasonable costs properly attributable thereto.

§ 11.3.1.3 If the property insurance requires deductibles, the Owner shall pay costs not covered because of such deductibles.

§ 11.3.1.4 This property insurance shall cover portions of the Work stored off the site, and also portions of the Work in transit.

§ 11.3.1.5 Partial occupancy or use in accordance with Section 9.9 shall not commence until the insurance company or companies providing property insurance have consented to such partial occupancy or use by endorsement or otherwise. The Owner and the Contractor shall take reasonable steps to obtain consent of the insurance company or companies and shall, without mutual written consent, take no action with respect to partial occupancy or use that would cause cancellation, lapse or reduction of insurance.

§ 11.3.2 BOILER AND MACHINERY INSURANCE
The Owner shall purchase and maintain boiler and machinery insurance required by the Contract Documents or by law, which shall specifically cover such insured objects during installation and until final acceptance by the Owner; this insurance shall include interests of the Owner, Contractor, Subcontractors and Sub-subcontractors in the Work, and the Owner and Contractor shall be named insureds.

§ 11.3.3 LOSS OF USE INSURANCE
The Owner, at the Owner's option, may purchase and maintain such insurance as will insure the Owner against loss of use of the Owner's property due to fire or other hazards, however caused. The Owner waives all rights of action against the Contractor for loss of use of the Owner's property, including consequential losses due to fire or other hazards however caused.

§ 11.3.4 If the Contractor requests in writing that insurance for risks other than those described herein or other special causes of loss be included in the property insurance policy, the Owner shall, if possible, include such insurance, and the cost thereof shall be charged to the Contractor by appropriate Change Order.

§ 11.3.5 If during the Project construction period the Owner insures properties, real or personal or both, at or adjacent to the site by property insurance under policies separate from those insuring the Project, or if after final payment property insurance is to be provided on the completed Project through a policy or policies other than those insuring the Project during the construction period, the Owner shall waive all rights in accordance with the terms of Section 11.3.7 for damages caused by fire or other causes of loss covered by this separate property insurance. All separate policies shall provide this waiver of subrogation by endorsement or otherwise.

Init.

/

30

§ 11.3.6 Before an exposure to loss may occur, the Owner shall file with the Contractor a copy of each policy that includes insurance coverages required by this Section 11.3. Each policy shall contain all generally applicable conditions, definitions, exclusions and endorsements related to this Project. Each policy shall contain a provision that the policy will not be canceled or allowed to expire, and that its limits will not be reduced, until at least 30 days' prior written notice has been given to the Contractor.

§ 11.3.7 WAIVERS OF SUBROGATION

The Owner and Contractor waive all rights against (1) each other and any of their subcontractors, sub-subcontractors, agents and employees, each of the other, and (2) the Architect, Architect's consultants, separate contractors described in Article 6, if any, and any of their subcontractors, sub-subcontractors, agents and employees, for damages caused by fire or other causes of loss to the extent covered by property insurance obtained pursuant to this Section 11.3 or other property insurance applicable to the Work, except such rights as they have to proceeds of such insurance held by the Owner as fiduciary. The Owner or Contractor, as appropriate, shall require of the Architect, Architect's consultants, separate contractors described in Article 6, if any, and the subcontractors, sub-subcontractors, agents and employees of any of them, by appropriate agreements, written where legally required for validity, similar waivers each in favor of other parties enumerated herein. The policies shall provide such waivers of subrogation by endorsement or otherwise. A waiver of subrogation shall be effective as to a person or entity even though that person or entity would otherwise have a duty of indemnification, contractual or otherwise, did not pay the insurance premium directly or indirectly, and whether or not the person or entity had an insurable interest in the property damaged.

§ 11.3.8 A loss insured under the Owner's property insurance shall be adjusted by the Owner as fiduciary and made payable to the Owner as fiduciary for the insureds, as their interests may appear, subject to requirements of any applicable mortgagee clause and of Section 11.3.10. The Contractor shall pay Subcontractors their just shares of insurance proceeds received by the Contractor, and by appropriate agreements, written where legally required for validity, shall require Subcontractors to make payments to their Sub-subcontractors in similar manner.

§ 11.3.9 If required in writing by a party in interest, the Owner as fiduciary shall, upon occurrence of an insured loss, give bond for proper performance of the Owner's duties. The cost of required bonds shall be charged against proceeds received as fiduciary. The Owner shall deposit in a separate account proceeds so received, which the Owner shall distribute in accordance with such agreement as the parties in interest may reach, or as determined in accordance with the method of binding dispute resolution selected in the Agreement between the Owner and Contractor. If after such loss no other special agreement is made and unless the Owner terminates the Contract for convenience, replacement of damaged property shall be performed by the Contractor after notification of a Change in the Work in accordance with Article 7.

§ 11.3.10 The Owner as fiduciary shall have power to adjust and settle a loss with insurers unless one of the parties in interest shall object in writing within five days after occurrence of loss to the Owner's exercise of this power; if such objection is made, the dispute shall be resolved in the manner selected by the Owner and Contractor as the method of binding dispute resolution in the Agreement. If the Owner and Contractor have selected arbitration as the method of binding dispute resolution, the Owner as fiduciary shall make settlement with insurers or, in the case of a dispute over distribution of insurance proceeds, in accordance with the directions of the arbitrators.

§ 11.4 PERFORMANCE BOND AND PAYMENT BOND

§ 11.4.1 The Owner shall have the right to require the Contractor to furnish bonds covering faithful performance of the Contract and payment of obligations arising thereunder as stipulated in bidding requirements or specifically required in the Contract Documents on the date of execution of the Contract.

§ 11.4.2 Upon the request of any person or entity appearing to be a potential beneficiary of bonds covering payment of obligations arising under the Contract, the Contractor shall promptly furnish a copy of the bonds or shall authorize a copy to be furnished.

ARTICLE 12 UNCOVERING AND CORRECTION OF WORK
§ 12.1 UNCOVERING OF WORK

§ 12.1.1 If a portion of the Work is covered contrary to the Architect's request or to requirements specifically expressed in the Contract Documents, it must, if requested in writing by the Architect, be uncovered for the Architect's examination and be replaced at the Contractor's expense without change in the Contract Time.

§ 12.1.2 If a portion of the Work has been covered that the Architect has not specifically requested to examine prior to its being covered, the Architect may request to see such Work and it shall be uncovered by the Contractor. If such Work is in accordance with the Contract Documents, costs of uncovering and replacement shall, by appropriate Change Order, be at the Owner's expense. If such Work is not in accordance with the Contract Documents, such costs and the cost of correction shall be at the Contractor's expense unless the condition was caused by the Owner or a separate contractor in which event the Owner shall be responsible for payment of such costs.

§ 12.2 CORRECTION OF WORK
§ 12.2.1 BEFORE OR AFTER SUBSTANTIAL COMPLETION
The Contractor shall promptly correct Work rejected by the Architect or failing to conform to the requirements of the Contract Documents, whether discovered before or after Substantial Completion and whether or not fabricated, installed or completed. Costs of correcting such rejected Work, including additional testing and inspections, the cost of uncovering and replacement, and compensation for the Architect's services and expenses made necessary thereby, shall be at the Contractor's expense.

§ 12.2.2 AFTER SUBSTANTIAL COMPLETION
§ 12.2.2.1 In addition to the Contractor's obligations under Section 3.5, if, within one year after the date of Substantial Completion of the Work or designated portion thereof or after the date for commencement of warranties established under Section 9.9.1, or by terms of an applicable special warranty required by the Contract Documents, any of the Work is found to be not in accordance with the requirements of the Contract Documents, the Contractor shall correct it promptly after receipt of written notice from the Owner to do so unless the Owner has previously given the Contractor a written acceptance of such condition. The Owner shall give such notice promptly after discovery of the condition. During the one-year period for correction of Work, if the Owner fails to notify the Contractor and give the Contractor an opportunity to make the correction, the Owner waives the rights to require correction by the Contractor and to make a claim for breach of warranty. If the Contractor fails to correct nonconforming Work within a reasonable time during that period after receipt of notice from the Owner or Architect, the Owner may correct it in accordance with Section 2.4.

§ 12.2.2.2 The one-year period for correction of Work shall be extended with respect to portions of Work first performed after Substantial Completion by the period of time between Substantial Completion and the actual completion of that portion of the Work.

§ 12.2.2.3 The one-year period for correction of Work shall not be extended by corrective Work performed by the Contractor pursuant to this Section 12.2.

§ 12.2.3 The Contractor shall remove from the site portions of the Work that are not in accordance with the requirements of the Contract Documents and are neither corrected by the Contractor nor accepted by the Owner.

§ 12.2.4 The Contractor shall bear the cost of correcting destroyed or damaged construction, whether completed or partially completed, of the Owner or separate contractors caused by the Contractor's correction or removal of Work that is not in accordance with the requirements of the Contract Documents.

§ 12.2.5 Nothing contained in this Section 12.2 shall be construed to establish a period of limitation with respect to other obligations the Contractor has under the Contract Documents. Establishment of the one-year period for correction of Work as described in Section 12.2.2 relates only to the specific obligation of the Contractor to correct the Work, and has no relationship to the time within which the obligation to comply with the Contract Documents may be sought to be enforced, nor to the time within which proceedings may be commenced to establish the Contractor's liability with respect to the Contractor's obligations other than specifically to correct the Work.

§ 12.3 ACCEPTANCE OF NONCONFORMING WORK
If the Owner prefers to accept Work that is not in accordance with the requirements of the Contract Documents, the Owner may do so instead of requiring its removal and correction, in which case the Contract Sum will be reduced as appropriate and equitable. Such adjustment shall be effected whether or not final payment has been made.

ARTICLE 13   MISCELLANEOUS PROVISIONS
§ 13.1 GOVERNING LAW
The Contract shall be governed by the law of the place where the Project is located except that, if the parties have selected arbitration as the method of binding dispute resolution, the Federal Arbitration Act shall govern Section 15.4.

§ 13.2 SUCCESSORS AND ASSIGNS
§ **13.2.1** The Owner and Contractor respectively bind themselves, their partners, successors, assigns and legal representatives to covenants, agreements and obligations contained in the Contract Documents. Except as provided in Section 13.2.2, neither party to the Contract shall assign the Contract as a whole without written consent of the other. If either party attempts to make such an assignment without such consent, that party shall nevertheless remain legally responsible for all obligations under the Contract.

§ **13.2.2** The Owner may, without consent of the Contractor, assign the Contract to a lender providing construction financing for the Project, if the lender assumes the Owner's rights and obligations under the Contract Documents. The Contractor shall execute all consents reasonably required to facilitate such assignment.

§ 13.3 WRITTEN NOTICE
Written notice shall be deemed to have been duly served if delivered in person to the individual, to a member of the firm or entity, or to an officer of the corporation for which it was intended; or if delivered at, or sent by registered or certified mail or by courier service providing proof of delivery to, the last business address known to the party giving notice.

§ 13.4 RIGHTS AND REMEDIES
§ **13.4.1** Duties and obligations imposed by the Contract Documents and rights and remedies available thereunder shall be in addition to and not a limitation of duties, obligations, rights and remedies otherwise imposed or available by law.

§ **13.4.2** No action or failure to act by the Owner, Architect or Contractor shall constitute a waiver of a right or duty afforded them under the Contract, nor shall such action or failure to act constitute approval of or acquiescence in a breach there under, except as may be specifically agreed in writing.

§ 13.5 TESTS AND INSPECTIONS
§ **13.5.1** Tests, inspections and approvals of portions of the Work shall be made as required by the Contract Documents and by applicable laws, statutes, ordinances, codes, rules and regulations or lawful orders of public authorities. Unless otherwise provided, the Contractor shall make arrangements for such tests, inspections and approvals with an independent testing laboratory or entity acceptable to the Owner, or with the appropriate public authority, and shall bear all related costs of tests, inspections and approvals. The Contractor shall give the Architect timely notice of when and where tests and inspections are to be made so that the Architect may be present for such procedures. The Owner shall bear costs of (1) tests, inspections or approvals that do not become requirements until after bids are received or negotiations concluded, and (2) tests, inspections or approvals where building codes or applicable laws or regulations prohibit the Owner from delegating their cost to the Contractor.

§ **13.5.2** If the Architect, Owner or public authorities having jurisdiction determine that portions of the Work require additional testing, inspection or approval not included under Section 13.5.1, the Architect will, upon written authorization from the Owner, instruct the Contractor to make arrangements for such additional testing, inspection or approval by an entity acceptable to the Owner, and the Contractor shall give timely notice to the Architect of when and where tests and inspections are to be made so that the Architect may be present for such procedures. Such costs, except as provided in Section 13.5.3, shall be at the Owner's expense.

§ **13.5.3** If such procedures for testing, inspection or approval under Sections 13.5.1 and 13.5.2 reveal failure of the portions of the Work to comply with requirements established by the Contract Documents, all costs made necessary by such failure including those of repeated procedures and compensation for the Architect's services and expenses shall be at the Contractor's expense.

§ **13.5.4** Required certificates of testing, inspection or approval shall, unless otherwise required by the Contract Documents, be secured by the Contractor and promptly delivered to the Architect.

§ 13.5.5 If the Architect is to observe tests, inspections or approvals required by the Contract Documents, the Architect will do so promptly and, where practicable, at the normal place of testing.

§ 13.5.6 Tests or inspections conducted pursuant to the Contract Documents shall be made promptly to avoid unreasonable delay in the Work.

§ 13.6 INTEREST
Payments due and unpaid under the Contract Documents shall bear interest from the date payment is due at such rate as the parties may agree upon in writing or, in the absence thereof, at the legal rate prevailing from time to time at the place where the Project is located.

§ 13.7 TIME LIMITS ON CLAIMS
The Owner and Contractor shall commence all claims and causes of action, whether in contract, tort, breach of warranty or otherwise, against the other arising out of or related to the Contract in accordance with the requirements of the final dispute resolution method selected in the Agreement within the time period specified by applicable law, but in any case not more than 10 years after the date of Substantial Completion of the Work. The Owner and Contractor waive all claims and causes of action not commenced in accordance with this Section 13.7.

ARTICLE 14   TERMINATION OR SUSPENSION OF THE CONTRACT
§ 14.1 TERMINATION BY THE CONTRACTOR
§ 14.1.1 The Contractor may terminate the Contract if the Work is stopped for a period of 30 consecutive days through no act or fault of the Contractor or a Subcontractor, Sub-subcontractor or their agents or employees or any other persons or entities performing portions of the Work under direct or indirect contract with the Contractor, for any of the following reasons:

    .1   Issuance of an order of a court or other public authority having jurisdiction that requires all Work to be stopped;

    .2   An act of government, such as a declaration of national emergency that requires all Work to be stopped;

    .3   Because the Architect has not issued a Certificate for Payment and has not notified the Contractor of the reason for withholding certification as provided in Section 9.4.1, or because the Owner has not made payment on a Certificate for Payment within the time stated in the Contract Documents; or

    .4   The Owner has failed to furnish to the Contractor promptly, upon the Contractor's request, reasonable evidence as required by Section 2.2.1.

§ 14.1.2 The Contractor may terminate the Contract if, through no act or fault of the Contractor or a Subcontractor, Sub-subcontractor or their agents or employees or any other persons or entities performing portions of the Work under direct or indirect contract with the Contractor, repeated suspensions, delays or interruptions of the entire Work by the Owner as described in Section 14.3 constitute in the aggregate more than 100 percent of the total number of days scheduled for completion, or 120 days in any 365-day period, whichever is less.

§ 14.1.3 If one of the reasons described in Section 14.1.1 or 14.1.2 exists, the Contractor may, upon seven days' written notice to the Owner and Architect, terminate the Contract and recover from the Owner payment for Work executed, including reasonable overhead and profit, costs incurred by reason of such termination, and damages.

§ 14.1.4 If the Work is stopped for a period of 60 consecutive days through no act or fault of the Contractor or a Subcontractor or their agents or employees or any other persons performing portions of the Work under contract with the Contractor because the Owner has repeatedly failed to fulfill the Owner's obligations under the Contract Documents with respect to matters important to the progress of the Work, the Contractor may, upon seven additional days' written notice to the Owner and the Architect, terminate the Contract and recover from the Owner as provided in Section 14.1.3.

§ 14.2 TERMINATION BY THE OWNER FOR CAUSE
§ 14.2.1 The Owner may terminate the Contract if the Contractor

    .1   repeatedly refuses or fails to supply enough properly skilled workers or proper materials;

    .2   fails to make payment to Subcontractors for materials or labor in accordance with the respective agreements between the Contractor and the Subcontractors;

    .3   repeatedly disregards applicable laws, statutes, ordinances, codes, rules and regulations, or lawful orders of a public authority; or

    .4   otherwise is guilty of substantial breach of a provision of the Contract Documents.

§ 14.2.2 When any of the above reasons exist, the Owner, upon certification by the Initial Decision Maker that sufficient cause exists to justify such action, may without prejudice to any other rights or remedies of the Owner and after giving the Contractor and the Contractor's surety, if any, seven days' written notice, terminate employment of the Contractor and may, subject to any prior rights of the surety:

.1 Exclude the Contractor from the site and take possession of all materials, equipment, tools, and construction equipment and machinery thereon owned by the Contractor;

.2 Accept assignment of subcontracts pursuant to Section 5.4; and

.3 Finish the Work by whatever reasonable method the Owner may deem expedient. Upon written request of the Contractor, the Owner shall furnish to the Contractor a detailed accounting of the costs incurred by the Owner in finishing the Work.

§ 14.2.3 When the Owner terminates the Contract for one of the reasons stated in Section 14.2.1, the Contractor shall not be entitled to receive further payment until the Work is finished.

§ 14.2.4 If the unpaid balance of the Contract Sum exceeds costs of finishing the Work, including compensation for the Architect's services and expenses made necessary thereby, and other damages incurred by the Owner and not expressly waived, such excess shall be paid to the Contractor. If such costs and damages exceed the unpaid balance, the Contractor shall pay the difference to the Owner. The amount to be paid to the Contractor or Owner, as the case may be, shall be certified by the Initial Decision Maker, upon application, and this obligation for payment shall survive termination of the Contract.

§ 14.3 SUSPENSION BY THE OWNER FOR CONVENIENCE

§ 14.3.1 The Owner may, without cause, order the Contractor in writing to suspend, delay or interrupt the Work in whole or in part for such period of time as the Owner may determine.

§ 14.3.2 The Contract Sum and Contract Time shall be adjusted for increases in the cost and time caused by suspension, delay or interruption as described in Section 14.3.1. Adjustment of the Contract Sum shall include profit. No adjustment shall be made to the extent

.1 that performance is, was or would have been so suspended, delayed or interrupted by another cause for which the Contractor is responsible; or

.2 that an equitable adjustment is made or denied under another provision of the Contract.

§ 14.4 TERMINATION BY THE OWNER FOR CONVENIENCE

§ 14.4.1 The Owner may, at any time, terminate the Contract for the Owner's convenience and without cause.

§ 14.4.2 Upon receipt of written notice from the Owner of such termination for the Owner's convenience, the Contractor shall

.1 cease operations as directed by the Owner in the notice;

.2 take actions necessary, or that the Owner may direct, for the protection and preservation of the Work; and

.3 except for Work directed to be performed prior to the effective date of termination stated in the notice, terminate all existing subcontracts and purchase orders and enter into no further subcontracts and purchase orders.

§ 14.4.3 In case of such termination for the Owner's convenience, the Contractor shall be entitled to receive payment for Work executed, and costs incurred by reason of such termination, along with reasonable overhead and profit on the Work not executed.

ARTICLE 15 CLAIMS AND DISPUTES

§ 15.1 CLAIMS

§ 15.1.1 DEFINITION

A Claim is a demand or assertion by one of the parties seeking, as a matter of right, payment of money, or other relief with respect to the terms of the Contract. The term "Claim" also includes other disputes and matters in question between the Owner and Contractor arising out of or relating to the Contract. The responsibility to substantiate Claims shall rest with the party making the Claim.

§ 15.1.2 NOTICE OF CLAIMS

Claims by either the Owner or Contractor must be initiated by written notice to the other party and to the Initial Decision Maker with a copy sent to the Architect, if the Architect is not serving as the Initial Decision Maker.

Init.

/

Claims by either party must be initiated within 21 days after occurrence of the event giving rise to such Claim or within 21 days after the claimant first recognizes the condition giving rise to the Claim, whichever is later.

§ 15.1.3 CONTINUING CONTRACT PERFORMANCE
Pending final resolution of a Claim, except as otherwise agreed in writing or as provided in Section 9.7 and Article 14, the Contractor shall proceed diligently with performance of the Contract and the Owner shall continue to make payments in accordance with the Contract Documents. The Architect will prepare Change Orders and issue Certificates for Payment in accordance with the decisions of the Initial Decision Maker.

§ 15.1.4 CLAIMS FOR ADDITIONAL COST
If the Contractor wishes to make a Claim for an increase in the Contract Sum, written notice as provided herein shall be given before proceeding to execute the Work. Prior notice is not required for Claims relating to an emergency endangering life or property arising under Section 10.4.

§ 15.1.5 CLAIMS FOR ADDITIONAL TIME
§ 15.1.5.1 If the Contractor wishes to make a Claim for an increase in the Contract Time, written notice as provided herein shall be given. The Contractor's Claim shall include an estimate of cost and of probable effect of delay on progress of the Work. In the case of a continuing delay, only one Claim is necessary.

§ 15.1.5.2 If adverse weather conditions are the basis for a Claim for additional time, such Claim shall be documented by data substantiating that weather conditions were abnormal for the period of time, could not have been reasonably anticipated and had an adverse effect on the scheduled construction.

§ 15.1.6 CLAIMS FOR CONSEQUENTIAL DAMAGES
The Contractor and Owner waive Claims against each other for consequential damages arising out of or relating to this Contract. This mutual waiver includes
.1   damages incurred by the Owner for rental expenses, for losses of use, income, profit, financing, business and reputation, and for loss of management or employee productivity or of the services of such persons; and
.2   damages incurred by the Contractor for principal office expenses including the compensation of personnel stationed there, for losses of financing, business and reputation, and for loss of profit except anticipated profit arising directly from the Work.

This mutual waiver is applicable, without limitation, to all consequential damages due to either party's termination in accordance with Article 14. Nothing contained in this Section 15.1.6 shall be deemed to preclude an award of liquidated damages, when applicable, in accordance with the requirements of the Contract Documents.

§ 15.2 INITIAL DECISION
§ 15.2.1 Claims, excluding those arising under Sections 10.3, 10.4, 11.3.9, and 11.3.10, shall be referred to the Initial Decision Maker for initial decision. The Architect will serve as the Initial Decision Maker, unless otherwise indicated in the Agreement. Except for those Claims excluded by this Section 15.2.1, an initial decision shall be required as a condition precedent to mediation of any Claim arising prior to the date final payment is due, unless 30 days have passed after the Claim has been referred to the Initial Decision Maker with no decision having been rendered. Unless the Initial Decision Maker and all affected parties agree, the Initial Decision Maker will not decide disputes between the Contractor and persons or entities other than the Owner.

§ 15.2.2 The Initial Decision Maker will review Claims and within ten days of the receipt of a Claim take one or more of the following actions: (1) request additional supporting data from the claimant or a response with supporting data from the other party, (2) reject the Claim in whole or in part, (3) approve the Claim, (4) suggest a compromise, or (5) advise the parties that the Initial Decision Maker is unable to resolve the Claim if the Initial Decision Maker lacks sufficient information to evaluate the merits of the Claim or if the Initial Decision Maker concludes that, in the Initial Decision Maker's sole discretion, it would be inappropriate for the Initial Decision Maker to resolve the Claim.

§ 15.2.3 In evaluating Claims, the Initial Decision Maker may, but shall not be obligated to, consult with or seek information from either party or from persons with special knowledge or expertise who may assist the Initial Decision Maker in rendering a decision. The Initial Decision Maker may request the Owner to authorize retention of such persons at the Owner's expense.

§ 15.2.4 If the Initial Decision Maker requests a party to provide a response to a Claim or to furnish additional supporting data, such party shall respond, within ten days after receipt of such request, and shall either (1) provide a response on the requested supporting data, (2) advise the Initial Decision Maker when the response or supporting data will be furnished or (3) advise the Initial Decision Maker that no supporting data will be furnished. Upon receipt of the response or supporting data, if any, the Initial Decision Maker will either reject or approve the Claim in whole or in part.

§ 15.2.5 The Initial Decision Maker will render an initial decision approving or rejecting the Claim, or indicating that the Initial Decision Maker is unable to resolve the Claim. This initial decision shall (1) be in writing; (2) state the reasons therefor; and (3) notify the parties and the Architect, if the Architect is not serving as the Initial Decision Maker, of any change in the Contract Sum or Contract Time or both. The initial decision shall be final and binding on the parties but subject to mediation and, if the parties fail to resolve their dispute through mediation, to binding dispute resolution.

§ 15.2.6 Either party may file for mediation of an initial decision at any time, subject to the terms of Section 15.2.6.1.

§ 15.2.6.1 Either party may, within 30 days from the date of an initial decision, demand in writing that the other party file for mediation within 60 days of the initial decision. If such a demand is made and the party receiving the demand fails to file for mediation within the time required, then both parties waive their rights to mediate or pursue binding dispute resolution proceedings with respect to the initial decision.

§ 15.2.7 In the event of a Claim against the Contractor, the Owner may, but is not obligated to, notify the surety, if any, of the nature and amount of the Claim. If the Claim relates to a possibility of a Contractor's default, the Owner may, but is not obligated to, notify the surety and request the surety's assistance in resolving the controversy.

§ 15.2.8 If a Claim relates to or is the subject of a mechanic's lien, the party asserting such Claim may proceed in accordance with applicable law to comply with the lien notice or filing deadlines.

§ 15.3 MEDIATION

§ 15.3.1 Claims, disputes, or other matters in controversy arising out of or related to the Contract except those waived as provided for in Sections 9.10.4, 9.10.5, and 15.1.6 shall be subject to mediation as a condition precedent to binding dispute resolution.

§ 15.3.2 The parties shall endeavor to resolve their Claims by mediation which, unless the parties mutually agree otherwise, shall be administered by the American Arbitration Association in accordance with its Construction Industry Mediation Procedures in effect on the date of the Agreement. A request for mediation shall be made in writing, delivered to the other party to the Contract, and filed with the person or entity administering the mediation. The request may be made concurrently with the filing of binding dispute resolution proceedings but, in such event, mediation shall proceed in advance of binding dispute resolution proceedings, which shall be stayed pending mediation for a period of 60 days from the date of filing, unless stayed for a longer period by agreement of the parties or court order. If an arbitration is stayed pursuant to this Section 15.3.2, the parties may nonetheless proceed to the selection of the arbitrator(s) and agree upon a schedule for later proceedings.

§ 15.3.3 The parties shall share the mediator's fee and any filing fees equally. The mediation shall be held in the place where the Project is located, unless another location is mutually agreed upon. Agreements reached in mediation shall be enforceable as settlement agreements in any court having jurisdiction thereof.

§ 15.4 ARBITRATION

§ 15.4.1 If the parties have selected arbitration as the method for binding dispute resolution in the Agreement, any Claim subject to, but not resolved by, mediation shall be subject to arbitration which, unless the parties mutually agree otherwise, shall be administered by the American Arbitration Association in accordance with its Construction Industry Arbitration Rules in effect on the date of the Agreement. A demand for arbitration shall be made in writing, delivered to the other party to the Contract, and filed with the person or entity administering the arbitration. The party filing a notice of demand for arbitration must assert in the demand all Claims then known to that party on which arbitration is permitted to be demanded.

§ 15.4.1.1 A demand for arbitration shall be made no earlier than concurrently with the filing of a request for mediation, but in no event shall it be made after the date when the institution of legal or equitable proceedings based on the Claim would be barred by the applicable statute of limitations. For statute of limitations purposes, receipt of a written demand for arbitration by the person or entity administering the arbitration shall constitute the institution of legal or equitable proceedings based on the Claim.

§ 15.4.2 The award rendered by the arbitrator or arbitrators shall be final, and judgment may be entered upon it in accordance with applicable law in any court having jurisdiction thereof.

§ 15.4.3 The foregoing agreement to arbitrate and other agreements to arbitrate with an additional person or entity duly consented to by parties to the Agreement shall be specifically enforceable under applicable law in any court having jurisdiction thereof.

§ 15.4.4 CONSOLIDATION OR JOINDER

§ 15.4.4.1 Either party, at its sole discretion, may consolidate an arbitration conducted under this Agreement with any other arbitration to which it is a party provided that (1) the arbitration agreement governing the other arbitration permits consolidation, (2) the arbitrations to be consolidated substantially involve common questions of law or fact, and (3) the arbitrations employ materially similar procedural rules and methods for selecting arbitrator(s).

§ 15.4.4.2 Either party, at its sole discretion, may include by joinder persons or entities substantially involved in a common question of law or fact whose presence is required if complete relief is to be accorded in arbitration, provided that the party sought to be joined consents in writing to such joinder. Consent to arbitration involving an additional person or entity shall not constitute consent to arbitration of any claim, dispute or other matter in question not described in the written consent.

§ 15.4.4.3 The Owner and Contractor grant to any person or entity made a party to an arbitration conducted under this Section 15.4, whether by joinder or consolidation, the same rights of joinder and consolidation as the Owner and Contractor under this Agreement.

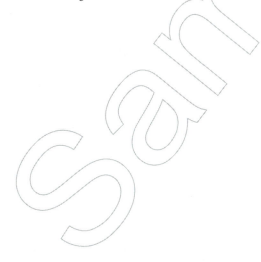

AIA Document
A310-1970

Document A310™ – 1970

Bid Bond

KNOW ALL MEN BY THESE PRESENTS, that we
(Here insert full name and address or legal title of Contractor)

as Principal, hereinafter called the Principal, and
(Here insert full name and address or legal title of Surety)

a corporation duly organized under the laws of the State of
as Surety, hereinafter called the Surety, are held and firmly bound unto
(Here insert full name and address or legal title of Owner)

as Obligee, hereinafter called the Obligee, in the sum of

Dollars ($), for the payment
of which sum well and truly to be made, the said Principal and the said Surety, bind ourselves, our heirs, executors, administrators, successors and assigns, jointly and severally, firmly by these presents.
WHEREAS, the Principal has submitted a bid for
(Here insert full name, address and description of project)

NOW, THEREFORE, if the Obligee shall accept the bid of the Principal and the Principal shall enter into a Contract with the Obligee in accordance with the terms of such bid, and give such bond or bonds as may be specified in the bidding or Contract Documents with good and sufficient surety for the faithful performance of such Contract and for the prompt payment of labor and material furnished in the prosecution thereof, or in the event of the failure of the Principal to enter such Contract and give such bond or bonds, if the Principal shall pay to the Obligee the difference not to exceed the penalty hereof between the amount specified in said bid and such larger amount for which the Obligee may in good faith contract with another party to perform the Work covered by said bid, then this obligation shall be null and void, otherwise to remain in full force and effect.

Signed and sealed this   day of   20

_____   _____
(Witness)   (Principal)   (Seal)

                                    _____
                                    (Title)

                                    _____
_____   (Surety)   (Seal)
(Witness)
                                    _____
                                    (Title)

AIA Document
A312-1984

Document A312™ – 1984

Performance Bond

CONTRACTOR *(Name and Address)*: **SURETY** *(Name and Principal Place of Business)*:

OWNER *(Name and Address)*:

Any singular reference to Contract, Surety, Owner or other party shall be considered plural where applicable.

CONSTRUCTION CONTRACT
Date:

Amount:

Description *(Name and Location)*:

BOND
Date *(Not earlier than Construction Contract Date)*:

Amount:

Modifications to this Bond: ☐ None ☐ See page 4

CONTRACTOR AS PRINCIPAL		SURETY	
Company:	*(Corporate Seal)*	Company:	*(Corporate Seal)*
Signature:		Signature:	
Name and Title:		Name and Title:	

(Any additional signatures appear on page 4)

(FOR INFORMATION ONLY - Name, Address and Telephone)
AGENT or **BROKER**: **OWNER'S REPRESENTATIVE** *(Architect, Engineer or other party)*:

§ 1 The Contractor and the Surety, jointly and severally, bind themselves, their heirs, executors, administrators, successors and assigns to the Owner for the performance of the Construction Contract, which is incorporated herein by reference.

§ 2 If the Contractor performs the Construction Contract, the Surety and the Contractor shall have no obligation under this Bond, except to participate in conferences as provided in Section 3.1.

§ 3 If there is no Owner Default, the Surety's obligation under this Bond shall arise after:
§ 3.1 The Owner has notified the Contractor and the Surety at its address described in Section 10 below that the Owner is considering declaring a Contractor Default and has requested and attempted to arrange a conference with the Contractor and the Surety to be held not later than fifteen days after receipt of such notice to discuss methods of performing the Construction Contract. If the Owner, the Contractor and the Surety agree, the Contractor shall be allowed a reasonable time to perform the Construction Contract, but such an agreement shall not waive the Owner's right, if any, subsequently to declare a Contractor Default; and

§ 3.2 The Owner has declared a Contractor Default and formally terminated the Contractor's right to complete the contract. Such Contractor Default shall not be declared earlier than twenty days after the Contractor and the Surety have received notice as provided in Section 3.1; and

§ 3.3 The Owner has agreed to pay the Balance of the Contract Price to the Surety in accordance with the terms of the Construction Contract or to a contractor selected to perform the Construction Contract in accordance with the terms of the contract with the Owner.

§ 4 When the Owner has satisfied the conditions of Section 3, the Surety shall promptly and at the Surety's expense take one of the following actions:
§ 4.1 Arrange for the Contractor, with consent of the Owner, to perform and complete the Construction Contract; or

§ 4.2 Undertake to perform and complete the Construction Contract itself, through its agents or through independent contractors; or

§ 4.3 Obtain bids or negotiated proposals from qualified contractors acceptable to the Owner for a contract for performance and completion of the Construction Contract, arrange for a contract to be prepared for execution by the Owner and the contractor selected with the Owner's concurrence, to be secured with performance and payment bonds executed by a qualified surety equivalent to the bonds issued on the Construction Contract, and pay to the Owner the amount of damages as described in Section 6 in excess of the Balance of the Contract Price incurred by the Owner resulting from the Contractor's default; or

§ 4.4 Waive its right to perform and complete, arrange for completion, or obtain a new contractor and with reasonable promptness under the circumstances:
    .1   After investigation, determine the amount for which it may be liable to the Owner and, as soon as practicable after the amount is determined, tender payment therefor to the Owner; or
    .2   Deny liability in whole or in part and notify the Owner citing reasons therefor.

§ 5 If the Surety does not proceed as provided in Section 4 with reasonable promptness, the Surety shall be deemed to be in default on this Bond fifteen days after receipt of an additional written notice from the Owner to the Surety demanding that the Surety perform its obligations under this Bond, and the Owner shall be entitled to enforce any remedy available to the Owner. If the Surety proceeds as provided in Section 4.4, and the Owner refuses the payment tendered or the Surety has denied liability, in whole or in part, without further notice the Owner shall be entitled to enforce any remedy available to the Owner.

§ 6 After the Owner has terminated the Contractor's right to complete the Construction Contract, and if the Surety elects to act under Section 4.1, 4.2, or 4.3 above, then the responsibilities of the Surety to the Owner shall not be greater than those of the Contractor under the Construction Contract, and the responsibilities of the Owner to the Surety shall not be greater than those of the Owner under the Construction Contract. To the limit of the amount of this Bond, but subject to commitment by the Owner of the Balance of the Contract Price to mitigation of costs and damages on the Construction Contract, the Surety is obligated without duplication for:
§ 6.1 The responsibilities of the Contractor for correction of defective work and completion of the Construction Contract;

§ 6.2 Additional legal, design professional and delay costs resulting from the Contractor's Default, and resulting from the actions or failure to act of the Surety under Section 4; and

§ 6.3 Liquidated damages, or if no liquidated damages are specified in the Construction Contract, actual damages caused by delayed performance or non-performance of the Contractor.

§ 7 The Surety shall not be liable to the Owner or others for obligations of the Contractor that are unrelated to the Construction Contract, and the Balance of the Contract Price shall not be reduced or set off on account of any such unrelated obligations. No right of action shall accrue on this Bond to any person or entity other than the Owner or its heirs, executors, administrators or successors.

§ 8 The Surety hereby waives notice of any change, including changes of time, to the Construction Contract or to related subcontracts, purchase orders and other obligations.

§ 9 Any proceeding, legal or equitable, under this Bond may be instituted in any court of competent jurisdiction in the location in which the work or part of the work is located and shall be instituted within two years after Contractor Default or within two years after the Contractor ceased working or within two years after the Surety refuses or fails to perform its obligations under this Bond, whichever occurs first. If the provisions of this Paragraph are void or prohibited by law, the minimum period of limitation available to sureties as a defense in the jurisdiction of the suit shall be applicable.

§ 10 Notice to the Surety, the Owner or the Contractor shall be mailed or delivered to the address shown on the signature page.

§ 11 When this Bond has been furnished to comply with a statutory or other legal requirement in the location where the construction was to be performed, any provision in this Bond conflicting with said statutory or legal requirement shall be deemed deleted here from and provisions conforming to such statutory or other legal requirement shall be deemed incorporated herein. The intent is that this Bond shall be construed as a statutory bond and not as a common law bond.

§ 12 DEFINITIONS

§ 12.1 Balance of the Contract Price: The total amount payable by the Owner to the Contractor under the Construction Contract after all proper adjustments have been made, including allowance to the Contractor of any amounts received or to be received by the Owner in settlement of insurance or other claims for damages to which the Contractor is entitled, reduced by all valid and proper payments made to or on behalf of the Contractor under the Construction Contract.

§ 12.2 Construction Contract: The agreement between the Owner and the Contractor identified on the signature page, including all Contract Documents and changes thereto.

§ 12.3 Contractor Default: Failure of the Contractor, which has neither been remedied nor waived, to perform or otherwise to comply with the terms of the Construction Contract.

§ 12.4 Owner Default: Failure of the Owner, which has neither been remedied nor waived, to pay the Contractor as required by the Construction Contract or to perform and complete or comply with the other terms thereof.

§ 13 MODIFICATIONS TO THIS BOND ARE AS FOLLOWS:

(Space is provided below for additional signatures of added parties, other than those appearing on the cover page.)

CONTRACTOR AS PRINCIPAL		SURETY	
Company:	*(Corporate Seal)*	Company:	*(Corporate Seal)*
Signature: _____		Signature: _____	
Name and Title:		Name and Title:	
Address:		Address:	

Payment Bond

CONTRACTOR *(Name and Address):*   SURETY *(Name and Principal Place of Business):*

OWNER *(Name and Address):*

Any singular reference to Contract, Surety, Owner or other party shall be considered plural where applicable.

CONSTRUCTION CONTRACT
Date:

Amount:

Description *(Name and Location):*

BOND
Date *(Not earlier than Construction Contract Date):*

Amount:

Modifications to this Bond: ☐ None   ☐ See page 4

CONTRACTOR AS PRINCIPAL	SURETY
Company: *(Corporate Seal)*	Company: *(Corporate Seal)*
Signature:	Signature:
Name and Title: _____	Name and Title: _____

(Any additional signatures appear on page 4)

(FOR INFORMATION ONLY - Name, Address and Telephone)
AGENT or BROKER: OWNER'S REPRESENTATIVE *(Architect, Engineer or other party):*

§ 1 The Contractor and the Surety, jointly and severally bind themselves, their heirs, executors, administrators, successors and assigns to the Owner to pay for labor, materials and equipment furnished for use in the performance of the Construction Contract, which is incorporated herein by reference.

§ 2 With respect to the Owner, this obligation shall be null and void if the Contractor:
§ 2.1 Promptly makes payment, directly or indirectly, for all sums due Claimants, and

§ 2.2 Defends, indemnifies and holds harmless the Owner from claims, demands, liens or suits by any person or entity whose claim, demand, lien or suit is for the payment for labor, materials or equipment furnished for use in the performance of the Construction Contract, provided the Owner has promptly notified the Contractor and the Surety (at the address described in Section 12) of any claims, demands, liens or suits and tendered defense of such claims, demands, liens or suits to the Contractor and the Surety, and provided there is no Owner Default.

§ 3 With respect to Claimants, this obligation shall be null and void if the Contractor promptly makes payment, directly or indirectly, for all sums due.

§ 4 The Surety shall have no obligation to Claimants under this Bond until:
§ 4.1 Claimants who are employed by or have a direct contract with the Contractor have given notice to the Surety (at the address described in Section 12) and sent a copy, or notice thereof, to the Owner, stating that a claim is being made under this Bond and, with substantial accuracy, the amount of the claim.

§ 4.2 Claimants who do not have a direct contract with the Contractor:
- .1 Have furnished written notice to the Contractor and sent a copy, or notice thereof, to the Owner, within 90 days after having last performed labor or last furnished materials or equipment included in the claim stating, with substantial accuracy, the amount of the claim and the name of the party to whom the materials were furnished or supplied or for whom the labor was done or performed; and
- .2 Have either received a rejection in whole or in part from the Contractor, or not received within 30 days of furnishing the above notice any communication from the Contractor by which the Contractor has indicated the claim will be paid directly or indirectly; and
- .3 Not having been paid within the above 30 days, have sent a written notice to the Surety (at the address described in Section 12) and sent a copy, or notice thereof, to the Owner, stating that a claim is being made under this Bond and enclosing a copy of the previous written notice furnished to the Contractor.

§ 5 If a notice required by Section 4 is given by the Owner to the Contractor or to the Surety, that is sufficient compliance.

§ 6 When the Claimant has satisfied the conditions of Section 4, the Surety shall promptly and at the Surety's expense take the following actions:
§ 6.1 Send an answer to the Claimant, with a copy to the Owner, within 45 days after receipt of the claim, stating the amounts that are undisputed and the basis for challenging any amounts that are disputed.

§ 6.2 Pay or arrange for payment of any undisputed amounts.

§ 7 The Surety's total obligation shall not exceed the amount of this Bond, and the amount of this Bond shall be credited for any payments made in good faith by the Surety.

§ 8 Amounts owed by the Owner to the Contractor under the Construction Contract shall be used for the performance of the Construction Contract and to satisfy claims, if any, under any Construction Performance Bond. By the Contractor furnishing and the Owner accepting this Bond, they agree that all funds earned by the Contractor in the performance of the Construction Contract are dedicated to satisfy obligations of the Contractor and the Surety under this Bond, subject to the Owner's priority to use the funds for the completion of the work.

§ 9 The Surety shall not be liable to the Owner, Claimants or others for obligations of the Contractor that are unrelated to the Construction Contract. The Owner shall not be liable for payment of any costs or expenses of any Claimant under this Bond, and shall have under this Bond no obligations to make payments to, give notices on behalf of, or otherwise have obligations to Claimants under this Bond.

§ 10 The Surety hereby waives notice of any change, including changes of time, to the Construction Contract or to related subcontracts, purchase orders and other obligations.

§ 11 No suit or action shall be commenced by a Claimant under this Bond other than in a court of competent jurisdiction in the location in which the work or part of the work is located or after the expiration of one year from the date (1) on which the Claimant gave the notice required by Section 4.1 or Section 4.2.3, or (2) on which the last labor or service was performed by anyone or the last materials or equipment were furnished by anyone under the Construction Contract, whichever of (1) or (2) first occurs. If the provisions of this Paragraph are void or prohibited by law, the minimum period of limitation available to sureties as a defense in the jurisdiction of the suit shall be applicable.

§ 12 Notice to the Surety, the Owner or the Contractor shall be mailed or delivered to the address shown on the signature page. Actual receipt of notice by Surety, the Owner or the Contractor, however accomplished, shall be sufficient compliance as of the date received at the address shown on the signature page.

§ 13 When this Bond has been furnished to comply with a statutory or other legal requirement in the location where the construction was to be performed, any provision in this Bond conflicting with said statutory or legal requirement shall be deemed deleted herefrom and provisions conforming to such statutory or other legal requirement shall be deemed incorporated herein. The intent is that this Bond shall be construed as a statutory bond and not as a common law bond.

§ 14 Upon request by any person or entity appearing to be a potential beneficiary of this Bond, the Contractor shall promptly furnish a copy of this Bond or shall permit a copy to be made.

§ 15 DEFINITIONS

§ 15.1 Claimant: An individual or entity having a direct contract with the Contractor or with a subcontractor of the Contractor to furnish labor, materials or equipment for use in the performance of the Contract. The intent of this Bond shall be to include without limitation in the terms "labor, materials or equipment" that part of water, gas, power, light, heat, oil, gasoline, telephone service or rental equipment used in the Construction Contract, architectural and engineering services required for performance of the work of the Contractor and the Contractor's subcontractors, and all other items for which a mechanic's lien may be asserted in the jurisdiction where the labor, materials or equipment were furnished.

§ 15.2 Construction Contract: The agreement between the Owner and the Contractor identified on the signature page, including all Contract Documents and changes thereto.

§ 15.3 Owner Default: Failure of the Owner, which has neither been remedied nor waived, to pay the Contractor as required by the Construction Contract or to perform and complete or comply with the other terms thereof.

§ 16 MODIFICATIONS TO THIS BOND ARE AS FOLLOWS:

(Space is provided below for additional signatures of added parties, other than those appearing on the cover page.)

CONTRACTOR AS PRINCIPAL

Company: *(Corporate Seal)*

Signature: _____

Name and Title:

Address:

SURETY

Company: *(Corporate Seal)*

Signature: _____

Name and Title:

Address:

APPENDIX

AIA Document
A401-2007

AIA® Document A401™ – 2007

Standard Form of Agreement Between Contractor and Subcontractor

AGREEMENT made as of the   day of
in the year
(In words, indicate day, month and year)

BETWEEN the Contractor:
(Name, address and other information)

and the Subcontractor:
(Name, address and other information)

The Contractor has made a contract for construction (hereinafter, the Prime Contract) dated:

with the Owner:
(Name, address and other information)

for the following Project:
(Name, location and detailed description)

The Prime Contract provides for the furnishing of labor, materials, equipment and services in connection with the construction of the Project. A copy of the Prime Contract, consisting of the Agreement Between Owner and Contractor (from which compensation amounts may be deleted) and the other Contract Documents enumerated therein, has been made available to the Subcontractor.

The Architect for the Project:
(Name, address and other information)

The Contractor and the Subcontractor agree as follows.

TABLE OF ARTICLES

ARTICLE 1   THE SUBCONTRACT DOCUMENTS

§ 1.1 The Subcontract Documents consist of (1) this Agreement; (2) the Prime Contract, consisting of the Agreement between the Owner and Contractor and the other Contract Documents enumerated therein; (3) Modifications issued subsequent to the execution of the Agreement between the Owner and Contractor, whether before or after the execution of this Agreement; (4) other documents listed in Article 16 of this Agreement; and (5) Modifications to this Subcontract issued after execution of this Agreement. These form the Subcontract, and are as fully a part of the Subcontract as if attached to this Agreement or repeated herein. The Subcontract represents the entire and integrated agreement between the parties hereto and supersedes prior negotiations, representations or agreements, either written or oral. An enumeration of the Subcontract Documents, other than Modifications issued subsequent to the execution of this Agreement, appears in Article 16.

§ 1.2 Except to the extent of a conflict with a specific term or condition contained in the Subcontract Documents, the General Conditions governing this Subcontract shall be the AIA Document A201™–2007, General Conditions of the Contract for Construction.

§ 1.3 The Subcontract may be amended or modified only by a Modification. The Subcontract Documents shall not be construed to create a contractual relationship of any kind (1) between the Architect and the Subcontractor, (2) between the Owner and the Subcontractor, or (3) between any persons or entities other than the Contractor and Subcontractor.

§ 1.4 The Contractor shall make available the Subcontract Documents to the Subcontractor prior to execution of this Agreement, and thereafter, upon request, but the Contractor may charge the Subcontractor for the reasonable cost of reproduction.

Init.

/

2

ARTICLE 2 MUTUAL RIGHTS AND RESPONSIBILITIES

The Contractor and Subcontractor shall be mutually bound by the terms of this Agreement and, to the extent that the provisions of AIA Document A201–2007 apply to this Agreement pursuant to Section 1.2 and provisions of the Prime Contract apply to the Work of the Subcontractor, the Contractor shall assume toward the Subcontractor all obligations and responsibilities that the Owner, under such documents, assumes toward the Contractor, and the Subcontractor shall assume toward the Contractor all obligations and responsibilities which the Contractor, under such documents, assumes toward the Owner and the Architect. The Contractor shall have the benefit of all rights, remedies and redress against the Subcontractor that the Owner, under such documents, has against the Contractor, and the Subcontractor shall have the benefit of all rights, remedies and redress against the Contractor that the Contractor, under such documents, has against the Owner, insofar as applicable to this Subcontract. Where a provision of such documents is inconsistent with a provision of this Agreement, this Agreement shall govern.

ARTICLE 3 CONTRACTOR
§ 3.1 SERVICES PROVIDED BY THE CONTRACTOR

§ 3.1.1 The Contractor shall cooperate with the Subcontractor in scheduling and performing the Contractor's Work to avoid conflicts or interference in the Subcontractor's Work and shall expedite written responses to submittals made by the Subcontractor in accordance with Section 4.1 and Article 5. Promptly after execution of this Agreement, the Contractor shall provide the Subcontractor copies of the Contractor's construction schedule and schedule of submittals, together with such additional scheduling details as will enable the Subcontractor to plan and perform the Subcontractor's Work properly. The Contractor shall promptly notify the Subcontractor of subsequent changes in the construction and submittal schedules and additional scheduling details.

§ 3.1.2 The Contractor shall provide suitable areas for storage of the Subcontractor's materials and equipment during the course of the Work. Additional costs to the Subcontractor resulting from relocation of such storage areas at the direction of the Contractor, except as previously agreed upon, shall be reimbursed by the Contractor.

§ 3.1.3 Except as provided in Article 14, the Contractor's equipment will be available to the Subcontractor only at the Contractor's discretion and on mutually satisfactory terms.

§ 3.2 COMMUNICATIONS

§ 3.2.1 The Contractor shall promptly make available to the Subcontractor information, including information received from the Owner, that affects this Subcontract and that becomes available to the Contractor subsequent to execution of this Subcontract.

§ 3.2.2 The Contractor shall not give instructions or orders directly to the Subcontractor's employees or to the Subcontractor's Sub-subcontractors or material suppliers unless such persons are designated as authorized representatives of the Subcontractor.

§ 3.2.3 The Contractor shall permit the Subcontractor to request directly from the Architect information regarding the percentages of completion and the amount certified on account of Work done by the Subcontractor.

§ 3.2.4 If hazardous substances of a type of which an employer is required by law to notify its employees are being used on the site by the Contractor, a subcontractor or anyone directly or indirectly employed by them (other than the Subcontractor), the Contractor shall, prior to harmful exposure of the Subcontractor's employees to such substance, give written notice of the chemical composition thereof to the Subcontractor in sufficient detail and time to permit the Subcontractor's compliance with such laws.

§ 3.2.5 The Contractor shall furnish to the Subcontractor within 30 days after receipt of a written request, or earlier if so required by law, information necessary and relevant for the Subcontractor to evaluate, give notice of or enforce mechanic's lien rights. Such information shall include a correct statement of the record legal title to the property, usually referred to as the site, on which the Project is located and the Owner's interest therein.

§ 3.2.6 If the Contractor asserts or defends a claim against the Owner that relates to the Work of the Subcontractor, the Contractor shall promptly make available to the Subcontractor all information relating to the portion of the claim that relates to the Work of the Subcontractor.

3

§ 3.3 CLAIMS BY THE CONTRACTOR

§ 3.3.1 Liquidated damages for delay, if provided for in Section 9.3 of this Agreement, shall be assessed against the Subcontractor only to the extent caused by the Subcontractor or any person or entity for whose acts the Subcontractor may be liable, and in no case for delays or causes arising outside the scope of this Subcontract.

§ 3.3.2 The Contractor's claims for the costs of services or materials provided due to the Subcontractor's failure to execute the Work shall require

.1   seven days' written notice prior to the Contractor's providing services or materials, except in an emergency; and

.2   written compilations to the Subcontractor of services and materials provided by the Contractor and charges for such services and materials no later than the fifteenth day of the month following the Contractor's providing such services or materials.

§ 3.4 CONTRACTOR'S REMEDIES

If the Subcontractor defaults or neglects to carry out the Work in accordance with this Agreement and fails within five working days after receipt of written notice from the Contractor to commence and continue correction of such default or neglect with diligence and promptness, the Contractor may, by appropriate Modification, and without prejudice to any other remedy the Contractor may have, make good such deficiencies and may deduct the reasonable cost thereof from the payments then or thereafter due the Subcontractor.

ARTICLE 4   SUBCONTRACTOR
§ 4.1 EXECUTION AND PROGRESS OF THE WORK

§ 4.1.1 For all Work the Subcontractor intends to subcontract, the Subcontractor shall enter into written agreements with Sub-subcontractors performing portions of the Work of this Subcontract by which the Subcontractor and the Sub-subcontractor are mutually bound, to the extent of the Work to be performed by the Sub-subcontractor, assuming toward each other all obligations and responsibilities that the Contractor and Subcontractor assume toward each other and having the benefit of all rights, remedies and redress each against the other that the Contractor and Subcontractor have by virtue of the provisions of this Agreement.

§ 4.1.2 The Subcontractor shall supervise and direct the Subcontractor's Work, and shall cooperate with the Contractor in scheduling and performing the Subcontractor's Work to avoid conflict, delay in or interference with the Work of the Contractor, other subcontractors, the Owner, or separate contractors.

§ 4.1.3 The Subcontractor shall promptly submit Shop Drawings, Product Data, Samples and similar submittals required by the Subcontract Documents with reasonable promptness and in such sequence as to cause no delay in the Work or in the activities of the Contractor or other subcontractors.

§ 4.1.4 The Subcontractor shall furnish to the Contractor periodic progress reports on the Work of this Subcontract as mutually agreed, including information on the status of materials and equipment that may be in the course of preparation, manufacture, or transit.

§ 4.1.5 The Subcontractor agrees that the Contractor and the Architect each have the authority to reject Work of the Subcontractor that does not conform to the Prime Contract. The Architect's decisions on matters relating to aesthetic effect shall be final and binding on the Subcontractor if consistent with the intent expressed in the Prime Contract.

§ 4.1.6 The Subcontractor shall pay for all materials, equipment and labor used in connection with the performance of this Subcontract through the period covered by previous payments received from the Contractor, and shall furnish satisfactory evidence, when requested by the Contractor, to verify compliance with the above requirements.

§ 4.1.7 The Subcontractor shall take necessary precautions to protect properly the work of other subcontractors from damage caused by operations under this Subcontract.

§ 4.1.8 The Subcontractor shall cooperate with the Contractor, other subcontractors, the Owner, and separate contractors whose work might interfere with the Subcontractor's Work. The Subcontractor shall participate in the preparation of coordinated drawings in areas of congestion, if required by the Prime Contract, specifically noting and advising the Contractor of potential conflicts between the Work of the Subcontractor and that of the Contractor, other subcontractors, the Owner, or separate contractors.

4

§ 4.2 PERMITS, FEES, NOTICES, AND COMPLIANCE WITH LAWS

§ 4.2.1 The Subcontractor shall give notices and comply with applicable laws, statutes, ordinances, codes, rules and regulations, and lawful orders of public authorities bearing on performance of the Work of this Subcontract. The Subcontractor shall secure and pay for permits, fees, licenses and inspections by government agencies necessary for proper execution and completion of the Subcontractor's Work, the furnishing of which is required of the Contractor by the Prime Contract.

§ 4.2.2 The Subcontractor shall comply with Federal, state and local tax laws, social security acts, unemployment compensation acts and workers' compensation acts insofar as applicable to the performance of this Subcontract.

§ 4.3 SAFETY PRECAUTIONS AND PROCEDURES

§ 4.3.1 The Subcontractor shall take reasonable safety precautions with respect to performance of this Subcontract, shall comply with safety measures initiated by the Contractor and with applicable laws, statutes, ordinances, codes, rules and regulations, and lawful orders of public authorities for the safety of persons and property in accordance with the requirements of the Prime Contract. The Subcontractor shall report to the Contractor within three days an injury to an employee or agent of the Subcontractor which occurred at the site.

§ 4.3.2 If hazardous substances of a type of which an employer is required by law to notify its employees are being used on the site by the Subcontractor, the Subcontractor's Sub-subcontractors or anyone directly or indirectly employed by them, the Subcontractor shall, prior to harmful exposure of any employees on the site to such substance, give written notice of the chemical composition thereof to the Contractor in sufficient detail and time to permit compliance with such laws by the Contractor, other subcontractors and other employers on the site.

§ 4.3.3 If reasonable precautions will be inadequate to prevent foreseeable bodily injury or death to persons resulting from a hazardous material or substance, including but not limited to asbestos or polychlorinated biphenyl (PCB), encountered on the site by the Subcontractor, the Subcontractor shall, upon recognizing the condition, immediately stop Work in the affected area and promptly report the condition to the Contractor in writing. When the material or substance has been rendered harmless, the Subcontractor's Work in the affected area shall resume upon written agreement of the Contractor and Subcontractor. The Subcontract Time shall be extended appropriately and the Subcontract Sum shall be increased in the amount of the Subcontractor's reasonable additional costs of demobilization, delay and remobilization, which adjustments shall be accomplished as provided in Article 5 of this Agreement.

§ 4.3.4 To the fullest extent permitted by law, the Contractor shall indemnify and hold harmless the Subcontractor, the Subcontractor's Sub-subcontractors, and agents and employees of any of them from and against claims, damages, losses and expenses, including but not limited to attorneys' fees, arising out of or resulting from performance of the Work in the affected area if in fact the material or substance presents the risk of bodily injury or death as described in Section 4.3.3 and has not been rendered harmless, provided that such claim, damage, loss or expense is attributable to bodily injury, sickness, disease or death, or to injury to or destruction of tangible property (other than the Work itself) except to the extent that such damage, loss or expense is due to the fault or negligence of the party seeking indemnity.

§ 4.3.5 The Subcontractor shall indemnify the Contractor for the cost and expense the Contractor incurs 1) for remediation of a material or substance brought to the site and negligently handled by the Subcontractor or 2) where the Subcontractor fails to perform its obligations under Section 4.3.3, except to the extent that the cost and expense are due to the Contractor's fault or negligence.

§ 4.4 CLEANING UP

§ 4.4.1 The Subcontractor shall keep the premises and surrounding area free from accumulation of waste materials or rubbish caused by operations performed under this Subcontract. The Subcontractor shall not be held responsible for conditions caused by other contractors or subcontractors.

§ 4.4.2 As provided under Section 3.3.2, if the Subcontractor fails to clean up as provided in the Subcontract Documents, the Contractor may charge the Subcontractor for the Subcontractor's appropriate share of cleanup costs.

§ 4.5 WARRANTY

The Subcontractor warrants to the Owner, Architect, and Contractor that materials and equipment furnished under this Subcontract will be of good quality and new unless the Subcontract Documents require or permit otherwise. The Subcontractor further warrants that the Work will conform to the requirements of the Subcontract Documents and will be free from defects, except for those inherent in the quality of the Work the Subcontract Documents require or permit. Work, materials, or equipment not conforming to these requirements may be considered defective. The Subcontractor's warranty excludes remedy for damage or defect caused by abuse, alterations to the Work not executed by the Subcontractor, improper or insufficient maintenance, improper operation, or normal wear and tear under normal usage. If required by the Architect and Contractor, the Subcontractor shall furnish satisfactory evidence as to the kind and quality of materials and equipment.

§ 4.6 INDEMNIFICATION

§ 4.6.1 To the fullest extent permitted by law, the Subcontractor shall indemnify and hold harmless the Owner, Contractor, Architect, Architect's consultants, and agents and employees of any of them from and against claims, damages, losses and expenses, including but not limited to attorney's fees, arising out of or resulting from performance of the Subcontractor's Work under this Subcontract, provided that any such claim, damage, loss or expense is attributable to bodily injury, sickness, disease or death, or to injury to or destruction of tangible property (other than the Work itself), but only to the extent caused by the negligent acts or omissions of the Subcontractor, the Subcontractor's Sub-subcontractors, anyone directly or indirectly employed by them or anyone for whose acts they may be liable, regardless of whether or not such claim, damage, loss or expense is caused in part by a party indemnified hereunder. Such obligation shall not be construed to negate, abridge, or otherwise reduce other rights or obligations of indemnity which would otherwise exist as to a party or person described in this Section 4.6.

§ 4.6.2 In claims against any person or entity indemnified under this Section 4.6 by an employee of the Subcontractor, the Subcontractor's Sub-subcontractors, anyone directly or indirectly employed by them or anyone for whose acts they may be liable, the indemnification obligation under Section 4.6.1 shall not be limited by a limitation on the amount or type of damages, compensation or benefits payable by or for the Subcontractor or the Subcontractor's Sub-subcontractors under workers' compensation acts, disability benefit acts or other employee benefit acts.

§ 4.7 REMEDIES FOR NONPAYMENT

If the Contractor does not pay the Subcontractor through no fault of the Subcontractor, within seven days from the time payment should be made as provided in this Agreement, the Subcontractor may, without prejudice to any other available remedies, upon seven additional days' written notice to the Contractor, stop the Work of this Subcontract until payment of the amount owing has been received. The Subcontract Sum shall, by appropriate Modification, be increased by the amount of the Subcontractor's reasonable costs of demobilization, delay and remobilization.

ARTICLE 5   CHANGES IN THE WORK

§ 5.1 The Owner may make changes in the Work by issuing Modifications to the Prime Contract. Upon receipt of such a Modification issued subsequent to the execution of the Subcontract Agreement, the Contractor shall promptly notify the Subcontractor of the Modification. Unless otherwise directed by the Contractor, the Subcontractor shall not thereafter order materials or perform Work that would be inconsistent with the changes made by the Modification to the Prime Contract.

§ 5.2 The Subcontractor may be ordered in writing by the Contractor, without invalidating this Subcontract, to make changes in the Work within the general scope of this Subcontract consisting of additions, deletions or other revisions, including those required by Modifications to the Prime Contract issued subsequent to the execution of this Agreement, the Subcontract Sum and the Subcontract Time being adjusted accordingly. The Subcontractor, prior to the commencement of such changed or revised Work, shall submit promptly to the Contractor written copies of a claim for adjustment to the Subcontract Sum and Subcontract Time for such revised Work in a manner consistent with requirements of the Subcontract Documents.

§ 5.3 The Subcontractor shall make all claims promptly to the Contractor for additional cost, extensions of time and damages for delays or other causes in accordance with the Subcontract Documents. A claim which will affect or become part of a claim which the Contractor is required to make under the Prime Contract within a specified time period or in a specified manner shall be made in sufficient time to permit the Contractor to satisfy the requirements of the Prime Contract. Such claims shall be received by the Contractor not less than two working days preceding the time by which the Contractor's claim must be made. Failure of the Subcontractor to make such a timely claim shall bind the Subcontractor to the same consequences as those to which the Contractor is bound.

ARTICLE 6   MEDIATION AND BINDING DISPUTE RESOLUTION
§ 6.1 MEDIATION

§ 6.1.1 Any claim arising out of or related to this Subcontract, except those waived in this Subcontract, shall be subject to mediation as a condition precedent to binding dispute resolution.

§ 6.1.2 The parties shall endeavor to resolve their claims by mediation which, unless the parties mutually agree otherwise, shall be administered by the American Arbitration Association in accordance with its Construction Industry Mediation Procedures in effect on the date of the Agreement. A request for mediation shall be made in writing, delivered to the other party to this Subcontract and filed with the person or entity administering the mediation. The request may be made concurrently with the filing of binding dispute resolution proceedings but, in such event, mediation shall proceed in advance of binding dispute resolution proceedings, which shall be stayed pending mediation for a period of 60 days from the date of filing, unless stayed for a longer period by agreement of the parties or court order. If an arbitration is stayed pursuant to this Section, the parties may nonetheless proceed to the selection of the arbitrators(s) and agree upon a schedule for later proceedings.

§ 6.1.3 The parties shall share the mediator's fee and any filing fees equally. The mediation shall be held in the place where the Project is located, unless another location is mutually agreed upon. Agreements reached in mediation shall be enforceable as settlement agreements in any court having jurisdiction thereof.

§ 6.2 BINDING DISPUTE RESOLUTION

For any claim subject to, but not resolved by mediation pursuant to Section 6.1, the method of binding dispute resolution shall be as follows:
(Check the appropriate box. If the Contractor and Subcontractor do not select a method of binding dispute resolution below, or do not subsequently agree in writing to a binding dispute resolution method other than litigation, claims will be resolved by litigation in a court of competent jurisdiction.)

☐ Arbitration pursuant to Section 6.3 of this Agreement

☐ Litigation in a court of competent jurisdiction

☐ Other *(Specify)*

§ 6.3 ARBITRATION

§ 6.3.1 If the Contractor and Subcontractor have selected arbitration as the method of binding dispute resolution in Section 6.2, any claim subject to, but not resolved by, mediation shall be subject to arbitration which, unless the parties mutually agree otherwise, shall be administered by the American Arbitration Association in accordance with its Construction Industry Arbitration Rules in effect on the date of the Agreement. A demand for arbitration shall be made in writing, delivered to the other party to the Subcontract, and filed with the person or entity administering the arbitration. The party filing a notice of demand for arbitration must assert in the demand all claims then known to that party on which arbitration is permitted to be demanded.

§ 6.3.2 A demand for arbitration shall be made no earlier than concurrently with the filing of a request for meditation but in no event shall it be made after the date when the institution of legal or equitable proceedings based on the claim would be barred by the applicable statute of limitations. For statute of limitations purposes, receipt of a written demand for arbitration by the person or entity administering the arbitration shall constitute the institution of legal or equitable proceedings based on the claim.

§ 6.3.3 Either party, at its sole discretion, may consolidate an arbitration conducted under this Agreement with any other arbitration to which it is a party provided that (1) the arbitration agreement governing the other arbitration permits consolidation; (2) the arbitrations to be consolidated substantially involve common questions of law or fact; and (3) the arbitrations employ materially similar procedural rules and methods for selecting arbitrator(s).

§ 6.3.4 Either party, at its sole discretion, may include by joinder persons or entities substantially involved in a common question of law or fact whose presence is required if complete relief is to be accorded in arbitration, provided that the party sought to be joined consents in writing to such joinder. Consent to arbitration involving an additional person or entity shall not constitute consent to arbitration of a claim not described in the written consent.

§ 6.3.5 The Contractor and Subcontractor grant to any person or entity made a party to an arbitration conducted under this Section 6.3, whether by joinder or consolidation, the same rights of joinder and consolidation as the Contractor and Subcontractor under this Agreement.

7

§ 6.3.6 This agreement to arbitrate and any other written agreement to arbitrate with an additional person or persons referred to herein shall be specifically enforceable under applicable law in any court having jurisdiction thereof. The award rendered by the arbitrator or arbitrators shall be final, and judgment may be entered upon it in accordance with applicable law in any court having jurisdiction thereof.

ARTICLE 7   TERMINATION, SUSPENSION OR ASSIGNMENT OF THE SUBCONTRACT
§ 7.1 TERMINATION BY THE SUBCONTRACTOR

The Subcontractor may terminate the Subcontract for the same reasons and under the same circumstances and procedures with respect to the Contractor as the Contractor may terminate with respect to the Owner under the Prime Contract, or for nonpayment of amounts due under this Subcontract for 60 days or longer. In the event of such termination by the Subcontractor for any reason which is not the fault of the Subcontractor, Sub-subcontractors or their agents or employees or other persons performing portions of the Work under contract with the Subcontractor, the Subcontractor shall be entitled to recover from the Contractor payment for Work executed and for proven loss with respect to materials, equipment, tools, and construction equipment and machinery, including reasonable overhead, profit and damages.

§ 7.2 TERMINATION BY THE CONTRACTOR

§ 7.2.1 If the Subcontractor repeatedly fails or neglects to carry out the Work in accordance with the Subcontract Documents or otherwise to perform in accordance with this Subcontract and fails within a ten-day period after receipt of written notice to commence and continue correction of such default or neglect with diligence and promptness, the Contractor may, by written notice to the Subcontractor and without prejudice to any other remedy the Contractor may have, terminate the Subcontract and finish the Subcontractor's Work by whatever method the Contractor may deem expedient. If the unpaid balance of the Subcontract Sum exceeds the expense of finishing the Subcontractor's Work and other damages incurred by the Contractor and not expressly waived, such excess shall be paid to the Subcontractor. If such expense and damages exceed such unpaid balance, the Subcontractor shall pay the difference to the Contractor.

§ 7.2.2 If the Owner terminates the Contract for the Owner's convenience, the Contractor shall promptly deliver written notice to the Subcontractor.

§ 7.2.3 Upon receipt of written notice of termination, the Subcontractor shall

    .1   cease operations as directed by the Contractor in the notice;

    .2   take actions necessary, or that the Contractor may direct, for the protection and preservation of the Work; and

    .3   except for Work directed to be performed prior to the effective date of termination stated in the notice, terminate all existing Sub-subcontracts and purchase orders and enter into no further Sub-subcontracts and purchase orders.

§ 7.2.4 In case of such termination for the Owner's convenience, the Subcontractor shall be entitled to receive payment for Work executed, and costs incurred by reason of such termination, along with reasonable overhead and profit on the Work not executed.

§ 7.3 SUSPENSION BY THE CONTRACTOR FOR CONVENIENCE

§ 7.3.1 The Contractor may, without cause, order the Subcontractor in writing to suspend, delay or interrupt the Work of this Subcontract in whole or in part for such period of time as the Contractor may determine. In the event of suspension ordered by the Contractor, the Subcontractor shall be entitled to an equitable adjustment of the Subcontract Time and Subcontract Sum.

§ 7.3.2 An adjustment shall be made for increases in the Subcontract Time and Subcontract Sum, including profit on the increased cost of performance, caused by suspension, delay or interruption. No adjustment shall be made to the extent that

    .1   performance is, was or would have been so suspended, delayed or interrupted by another cause for which the Subcontractor is responsible; or

    .2   an equitable adjustment is made or denied under another provision of this Subcontract.

§ 7.4 ASSIGNMENT OF THE SUBCONTRACT

§ 7.4.1 In the event the Owner terminates the Prime Contract for cause, this Subcontract is assigned to the Owner pursuant to Section 5.4 of A201–2007 provided the Owner accepts the assignment.

§ 7.4.2 Without the Contractor's written consent, the Subcontractor shall not assign the Work of this Subcontract, subcontract the whole of this Subcontract, or subcontract portions of this Subcontract.

ARTICLE 8   THE WORK OF THIS SUBCONTRACT

The Subcontractor shall execute the following portion of the Work described in the Subcontract Documents, including all labor, materials, equipment, services and other items required to complete such portion of the Work, except to the extent specifically indicated in the Subcontract Documents to be the responsibility of others.
(Insert a precise description of the Work of this Subcontract, referring where appropriate to numbers of Drawings, sections of Specifications and pages of Addenda, Modifications and accepted alternates.)

ARTICLE 9   DATE OF COMMENCEMENT AND SUBSTANTIAL COMPLETION

§ 9.1 Subcontract Time is the period of time, including authorized adjustments, allotted in the Subcontract Documents for Substantial Completion of the Work described in the Subcontract Documents. The Subcontractor's date of commencement is the date from which the Subcontract Time of Section 9.3 is measured; it shall be the date of this Agreement, as first written above, unless a different date is stated below or provision is made for the date to be fixed in a notice to proceed issued by the Contractor.
(Insert the date of commencement, if it differs from the date of this Agreement or, if applicable, state that the date will be fixed in a notice to proceed.)

§ 9.2 Unless the date of commencement is established by a notice to proceed issued by the Contractor, or the Contractor has commenced visible Work at the site under the Prime Contract, the Subcontractor shall notify the Contractor in writing not less than five days before commencing the Subcontractor's Work to permit the timely filing of mortgages, mechanic's liens and other security interests.

§ 9.3 The Work of this Subcontract shall be substantially completed not later than
(Insert the calendar date or number of calendar days after the Subcontractor's date of commencement. Also insert any requirements for earlier substantial completion of certain portions of the Subcontractor's Work, if not stated elsewhere in the Subcontract Documents.)

, subject to adjustments of this Subcontract Time as provided in the Subcontract Documents.
(Insert provisions, if any, for liquidated damages relating to failure to complete on time.)

§ 9.4 With respect to the obligations of both the Contractor and the Subcontractor, time is of the essence of this Subcontract.

§ 9.5 No extension of time will be valid without the Contractor's written consent after claim made by the Subcontractor in accordance with Section 5.3.

ARTICLE 10   SUBCONTRACT SUM

§ 10.1 The Contractor shall pay the Subcontractor in current funds for performance of the Subcontract the Subcontract Sum of   Dollars
($), subject to additions and deductions as provided in the Subcontract Documents.

§ 10.2 The Subcontract Sum is based upon the following alternates, if any, which are described in the Subcontract Documents and have been accepted by the Owner and the Contractor:
(Insert the numbers or other identification of accepted alternates.)

§ 10.3 Unit prices, if any:
(Identify and state the unit price, and state the quantity limitations, if any, to which the unit price will be applicable.)

Item	Units and Limitations	Price per Unit

§ 10.4 Allowances included in the Subcontract Sum, if any:
(Identify allowance and state exclusions, if any, from the allowance price.)

Item	Price

ARTICLE 11 PROGRESS PAYMENTS

§ 11.1 Based upon applications for payment submitted to the Contractor by the Subcontractor, corresponding to applications for payment submitted by the Contractor to the Architect, and certificates for payment issued by the Architect, the Contractor shall make progress payments on account of the Subcontract Sum to the Subcontractor as provided below and elsewhere in the Subcontract Documents. Unless the Contractor provides the Owner with a payment bond in the full penal sum of the Contract Sum, payments received by the Contractor and Subcontractor for Work properly performed by their contractors and suppliers shall be held by the Contractor and Subcontractor for those contractors or suppliers who performed Work or furnished materials, or both, under contract with the Contractor or Subcontractor for which payment was made to the Contractor by the Owner or to the Subcontractor by the Contractor, as applicable. Nothing contained herein shall require money to be placed in a separate account and not commingled with money of the Contractor or Subcontractor, shall create any fiduciary liability or tort liability on the part of the Contractor or Subcontractor for breach of trust or shall entitle any person or entity to an award of punitive damages against the Contractor or Subcontractor for breach of the requirements of this provision.

§ 11.2 The period covered by each application for payment shall be one calendar month ending on the last day of the month, or as follows:

§ 11.3 Provided an application for payment is received by the Contractor not later than the   day of a month, the Contractor shall include the Subcontractor's Work covered by that application in the next application for payment which the Contractor is entitled to submit to the Architect. The Contractor shall pay the Subcontractor each progress payment no later than seven working days after the Contractor receives payment from the Owner. If the Architect does not issue a certificate for payment or the Contractor does not receive payment for any cause which is not the fault of the Subcontractor, the Contractor shall pay the Subcontractor, on demand, a progress payment computed as provided in Sections 11.7, 11.8 and 11.9.

§ 11.4 If the Subcontractor's application for payment is received by the Contractor after the application date fixed above, the Subcontractor's Work covered by it shall be included by the Contractor in the next application for payment submitted to the Architect.

§ 11.5 The Subcontractor shall submit to the Contractor a schedule of values prior to submitting the Subcontractor's first Application for Payment. Each subsequent application for payment shall be based upon the most recent schedule of values submitted by the Subcontractor in accordance with the Subcontract Documents. The schedule of values shall allocate the entire Subcontract Sum among the various portions of the Subcontractor's Work and be prepared in such form and supported by such data to substantiate its accuracy as the Contractor may require. This schedule, unless objected to by the Contractor, shall be used as a basis for reviewing the Subcontractor's applications for payment.

§ **11.6** Applications for payment submitted by the Subcontractor shall indicate the percentage of completion of each portion of the Subcontractor's Work as of the end of the period covered by the application for payment.

§ **11.7** Subject to the provisions of the Subcontract Documents, the amount of each progress payment shall be computed as set forth in the sections below.

§ **11.7.1** Take that portion of the Subcontract Sum properly allocable to completed Work as determined by multiplying the percentage completion of each portion of the Subcontractor's Work by the share of the total Subcontract Sum allocated to that portion of the Subcontractor's Work in the schedule of values, less that percentage actually retained, if any, from payments to the Contractor on account of the Work of the Subcontractor. Pending final determination of cost to the Contractor of changes in the Work that have been properly authorized by the Contractor, amounts not in dispute shall be included to the same extent provided in the Prime Contract, even though the Subcontract Sum has not yet been adjusted;

§ **11.7.2** Add that portion of the Subcontract Sum properly allocable to materials and equipment delivered and suitably stored at the site by the Subcontractor for subsequent incorporation in the Subcontractor's Work or, if approved by the Contractor, suitably stored off the site at a location agreed upon in writing, less the same percentage retainage required by the Prime Contract to be applied to such materials and equipment in the Contractor's application for payment;

§ **11.7.3** Subtract the aggregate of previous payments made by the Contractor; and

§ **11.7.4** Subtract amounts, if any, calculated under Section 11.7.1 or 11.7.2 that are related to Work of the Subcontractor for which the Architect has withheld or nullified, in whole or in part, a certificate of payment for a cause that is the fault of the Subcontractor.

§ **11.8** Upon the partial or entire disapproval by the Contractor of the Subcontractor's application for payment, the Contractor shall provide written notice to the Subcontractor. When the basis for the disapproval has been remedied, the Subcontractor shall be paid the amounts withheld.

§ **11.9 SUBSTANTIAL COMPLETION**
When the Subcontractor's Work or a designated portion thereof is substantially complete and in accordance with the requirements of the Prime Contract, the Contractor shall, upon application by the Subcontractor, make prompt application for payment for such Work. Within 30 days following issuance by the Architect of the certificate for payment covering such substantially completed Work, the Contractor shall, to the full extent allowed in the Prime Contract, make payment to the Subcontractor, deducting any portion of the funds for the Subcontractor's Work withheld in accordance with the certificate to cover costs of items to be completed or corrected by the Subcontractor. Such payment to the Subcontractor shall be the entire unpaid balance of the Subcontract Sum if a full release of retainage is allowed under the Prime Contract for the Subcontractor's Work prior to the completion of the entire Project. If the Prime Contract does not allow for a full release of retainage, then such payment shall be an amount which, when added to previous payments to the Subcontractor, will reduce the retainage on the Subcontractor's substantially completed Work to the same percentage of retainage as that on the Contractor's Work covered by the certificate.

ARTICLE 12 FINAL PAYMENT
§ **12.1** Final payment, constituting the entire unpaid balance of the Subcontract Sum, shall be made by the Contractor to the Subcontractor when the Subcontractor's Work is fully performed in accordance with the requirements of the Subcontract Documents, the Architect has issued a certificate for payment covering the Subcontractor's completed Work and the Contractor has received payment from the Owner. If, for any cause which is not the fault of the Subcontractor, a certificate for payment is not issued or the Contractor does not receive timely payment or does not pay the Subcontractor within seven days after receipt of payment from the Owner, final payment to the Subcontractor shall be made upon demand.
(Insert provisions for earlier final payment to the Subcontractor, if applicable.)

§ 12.2 Before issuance of the final payment, the Subcontractor, if required, shall submit evidence satisfactory to the Contractor that all payrolls, bills for materials and equipment, and all known indebtedness connected with the Subcontractor's Work have been satisfied. Acceptance of final payment by the Subcontractor shall constitute a waiver of claims by the Subcontractor, except those previously made in writing and identified by the Subcontractor as unsettled at the time of final application for payment.

ARTICLE 13 INSURANCE AND BONDS

§ 13.1 The Subcontractor shall purchase and maintain insurance of the following types of coverage and limits of liability as will protect the Subcontractor from claims that may arise out of, or result from, the Subcontractor's operations and completed operations under the Subcontract:

§ 13.2 Coverages, whether written on an occurrence or claims-made basis, shall be maintained without interruption from the date of commencement of the Subcontractor's Work until the date of final payment and termination of any coverage required to be maintained after final payment to the Subcontractor, and, with respect to the Subcontractor's completed operations coverage, until the expiration of the period for correction of Work or for such other period for maintenance of completed operations coverage as specified in the Prime Contract.

§ 13.3 Certificates of insurance acceptable to the Contractor shall be filed with the Contractor prior to commencement of the Subcontractor's Work. These certificates and the insurance policies required by this Article 13 shall contain a provision that coverages afforded under the policies will not be canceled or allowed to expire until at least 30 days' prior written notice has been given to the Contractor. If any of the foregoing insurance coverages are required to remain in force after final payment and are reasonably available, an additional certificate evidencing continuation of such coverage shall be submitted with the final application for payment as required in Article 12. If any information concerning reduction of coverage is not furnished by the insurer, it shall be furnished by the Subcontractor with reasonable promptness according to the Subcontractor's information and belief.

§ 13.4 The Subcontractor shall cause the commercial liability coverage required by the Subcontract Documents to include: (1) the Contractor, the Owner, the Architect and the Architect's consultants as additional insureds for claims caused in whole or in part by the Subcontractor's negligent acts or omissions during the Subcontractor's operations; and (2) the Contractor as an additional insured for claims caused in whole or in part by the Subcontractor's negligent acts or omissions during the Subcontractor's completed operations.

§ 13.5 The Contractor shall furnish to the Subcontractor satisfactory evidence of insurance required of the Contractor under the Prime Contract.

§ 13.6 The Contractor shall promptly, upon request of the Subcontractor, furnish a copy or permit a copy to be made of any bond covering payment of obligations arising under the Subcontract.

§ 13.7 Performance Bond and Payment Bond:
(If the Subcontractor is to furnish bonds, insert the specific requirements here.)

§ 13.8 PROPERTY INSURANCE

§ 13.8.1 When requested in writing, the Contractor shall provide the Subcontractor with copies of the property and equipment policies in effect for the Project. The Contractor shall notify the Subcontractor if the required property insurance policies are not in effect.

12

§ 13.8.2 If the required property insurance is not in effect for the full value of the Subcontractor's Work, then the Subcontractor shall purchase insurance for the value of the Subcontractor's Work, and the Subcontractor shall be reimbursed for the cost of the insurance by an adjustment in the Subcontract Sum.

§ 13.8.3 Property insurance for the Subcontractor's materials and equipment required for the Subcontractor's Work, stored off site or in transit and not covered by the Project property insurance, shall be paid for through the application for payment process.

§ 13.9 WAIVERS OF SUBROGATION

The Contractor and Subcontractor waive all rights against (1) each other and any of their subcontractors, sub-subcontractors, agents and employees, each of the other, and (2) the Owner, the Architect, the Architect's consultants, separate contractors, and any of their subcontractors, sub-subcontractors, agents and employees for damages caused by fire or other causes of loss to the extent covered by property insurance provided under the Prime Contract or other property insurance applicable to the Work, except such rights as they may have to proceeds of such insurance held by the Owner as a fiduciary. The Subcontractor shall require of the Subcontractor's Sub-subcontractors, agents and employees, by appropriate agreements, written where legally required for validity, similar waivers in favor of the parties enumerated herein. The policies shall provide such waivers of subrogation by endorsement or otherwise. A waiver of subrogation shall be effective as to a person or entity even though that person or entity would otherwise have a duty of indemnification, contractual or otherwise, did not pay the insurance premium directly or indirectly, and whether or not the person or entity had an insurable interest in the property damaged.

ARTICLE 14   TEMPORARY FACILITIES AND WORKING CONDITIONS

§ 14.1 The Contractor shall furnish and make available at no cost to the Subcontractor the Contractor's temporary facilities, equipment and services, except as noted below:

§ 14.2 Specific working conditions:
(Insert any applicable arrangements concerning working conditions and labor matters for the Project.)

ARTICLE 15   MISCELLANEOUS PROVISIONS

§ 15.1 Where reference is made in this Subcontract to a provision of another Subcontract Document, the reference refers to that provision as amended or supplemented by other provisions of the Subcontract Documents.

§ 15.2 Payments due and unpaid under this Subcontract shall bear interest from the date payment is due at such rate as the parties may agree upon in writing or, in the absence thereof, at the legal rate prevailing from time to time at the place where the Project is located.
(Insert rate of interest agreed upon, if any.)

§ 15.3 Retainage and any reduction thereto is as follows:

§ 15.4 The Contractor and Subcontractor waive claims against each other for consequential damages arising out of or relating to this Subcontract, including without limitation, any consequential damages due to either party's termination in accordance with Article 7.

13

ARTICLE 16 ENUMERATION OF SUBCONTRACT DOCUMENTS
§ 16.1 The Subcontract Documents, except for Modifications issued after execution of this Subcontract, are enumerated in the sections below.

§ 16.1.1 This executed AIA Document A401–2007, Standard Form of Agreement Between Contractor and Subcontractor.

§ 16.1.2 The Prime Contract, consisting of the Agreement between the Owner and Contractor dated as first entered above and the other Contract Documents enumerated in the Owner-Contractor Agreement.

§ 16.1.3 The following Modifications to the Prime Contract, if any, issued subsequent to the execution of the Owner-Contractor Agreement but prior to the execution of this Agreement:

Modification   Date

§ 16.1.4 Additional Documents, if any, forming part of the Subcontract Documents:
  .1   AIA Document E201™–2007, Digital Data Protocol Exhibit, if completed by the parties, or the following:

  .2   Other documents:
            *(List here any additional documents that are intended to form part of the Subcontract Documents.
            Requests for proposal and the Subcontractor's bid or proposal should be listed here only if intended to be
            made part of the Subcontract Documents.)*

This Agreement entered into as of the day and year first written above.

_____   _____
CONTRACTOR *(Signature)*   **SUBCONTRACTOR** *(Signature)*

_____   _____
(Printed name and title)   *(Printed name and title)*

CAUTION: You should sign an original AIA Contract Document, on which this text appears in RED. An original assures that changes will not be obscured.

14

I

AIA Document
G701-2007

▓AIA® Document G701™ – 2001

Change Order

PROJECT: *(Name and address)*	**CHANGE ORDER NUMBER:**	OWNER ☐
	DATE:	ARCHITECT ☐
	ARCHITECT'S PROJECT NUMBER:	CONTRACTOR ☐
		FIELD ☐
TO CONTRACTOR: *(Name and address)*	**CONTRACT DATE:**	OTHER ☐
	CONTRACT FOR:	

The Contract is changed as follows:
(Include, where applicable, any undisputed amount attributable to previously executed Construction Change Directives)

The original (Contract Sum) (Guaranteed Maximum Price) was $ _____

The net change by previously authorized Change Orders $ _____

The (Contract Sum) (Guaranteed Maximum Price) prior to this Change Order was $ _____

The (Contract Sum) (Guaranteed Maximum Price) will be (increased) (decreased) (unchanged)

by this Change Order in the amount of $ _____

The new (Contract Sum) (Guaranteed Maximum Price) including this Change Order will be $ _____

The Contract Time will be (increased) (decreased) (unchanged) by   (   ) days

The date of Substantial Completion as of the date of this Change Order therefore is

(Note: This Change Order does not include changes in the Contract Sum, Contract Time or Guaranteed Maximum Price which have been authorized by Construction Change Directive until the cost and time have been agreed upon by both the Owner and Contractor, in which case a Change Order is executed to supersede the Construction Change Directive.)

NOT VALID UNTIL SIGNED BY THE ARCHITECT, CONTRACTOR AND OWNER.

ARCHITECT *(Firm name)*	CONTRACTOR *(Firm name)*	OWNER *(Firm name)*
ADDRESS	ADDRESS	ADDRESS
BY *(Signature)*	BY *(Signature)*	BY *(Signature)*
(Typed name)	*(Typed name)*	*(Typed name)*
DATE	DATE	DATE

APPENDIX J

AIA Document
G702-1992

AIA® Document G702™ – 1992

Application and Certificate for Payment

TO OWNER:	PROJECT:	APPLICATION NO:	Distribution to:
		PERIOD TO:	OWNER ☐
		CONTRACT FOR:	ARCHITECT ☐
FROM CONTRACTOR:	VIA ARCHITECT:	CONTRACT DATE:	CONTRACTOR ☐
		PROJECT NOS:	FIELD ☐
			OTHER ☐

CONTRACTOR'S APPLICATION FOR PAYMENT

Application is made for payment, as shown below, in connection with the Contract.
Continuation Sheet, AIA Document G703, is attached.

1. ORIGINAL CONTRACT SUM .... $ _____

2. Net change by Change Orders .... $ _____

3. CONTRACT SUM TO DATE (Line 1 ± 2) .... $ _____

4. TOTAL COMPLETED & STORED TO DATE (Column G on G703) ...... $ _____

5. RETAINAGE:
  a. _____ % of Completed Work
    (Column D + E on G703) $ _____
  b. _____ % of Stored Material
    (Column F on G703) $ _____

  Total Retainage (Lines 5a + 5b or Total in Column I of G703) ..... $ _____

6. TOTAL EARNED LESS RETAINAGE .... $ _____
  (Line 4 Less Line 5 Total)

7. LESS PREVIOUS CERTIFICATES FOR PAYMENT .... $ _____
  (Line 6 from prior Certificate)

8. CURRENT PAYMENT DUE .... $ _____

9. BALANCE TO FINISH, INCLUDING RETAINAGE .... $ _____
  (Line 3 less Line 6)

CHANGE ORDER SUMMARY	ADDITIONS	DEDUCTIONS
Total changes approved in previous months by Owner	$	$
Total approved this Month	$	$
TOTALS	$	$
NET CHANGES by Change Order		$

The undersigned Contractor certifies that to the best of the Contractor's knowledge, information and belief the Work covered by this Application for Payment has been completed in accordance with the Contract Documents, that all amounts have been paid by the Contractor for Work for which previous Certificates for Payment were issued and payments received from the Owner, and that current payment shown herein is now due.

CONTRACTOR:

By: _____ Date: _____

State of:
County of:
Subscribed and sworn to before
me this _____ day of _____
Notary Public: _____
My Commission expires:

ARCHITECT'S CERTIFICATE FOR PAYMENT

In accordance with the Contract Documents, based on on-site observations and the data comprising this application, the Architect certifies to the Owner that to the best of the Architect's knowledge, information and belief the Work has progressed as indicated, the quality of the Work is in accordance with the Contract Documents, and the Contractor is entitled to payment of the AMOUNT CERTIFIED.

AMOUNT CERTIFIED .... $ _____

(Attach explanation if amount certified differs from the amount applied. Initial all figures on this Application and on the Continuation Sheet that are changed to conform with the amount certified.)

ARCHITECT:

By: _____ Date: _____

This Certificate is not negotiable. The AMOUNT CERTIFIED is payable only to the Contractor named herein. Issuance, payment and acceptance of payment are without prejudice to any rights of the Owner or Contractor under this Contract

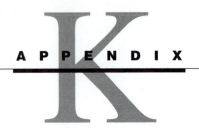

APPENDIX

AIA Document
G810-2001

 **Contract Administration**
G810 Transmittal Letter
(Instructions on the reverse side)

PROJECT *(Name and address):*

TO *(Name and address):*

FROM *(Name and address):*

WE TRANSMIT: ☐ Attached ☐ Under separate cover

VIA: ☐ Overnight delivery ☐ Mail ☐ E-mail ☐ Courier ☐ Fax ☐ Other

FOR: ☐ Approval / Action ☐ Information ☐ Use as requested

☐ Comment ☐ Distribution ☐ Other

THE FOLLOWING: ☐ Drawings ☐ Specifications ☐ Digital files ☐ Submittals ☐ Other

NO. OF COPIES	DATE	FORMAT	DESCRIPTION

REMARKS:

BY:

COPIES TO:

INDEX